FAST TRACK TO A 5

Preparing for the AP® Chemistry Examination

CENGAGE Learning®

Australia · Brazil · Mexico · Singapore · United Kingdom · United States

CENGAGE Learning

National Geographic Learning/Cengage Learning is pleased to offer our college-level materials to high schools for Advanced Placement®, honors, and electives courses.
To contact your National Geographic Learning representative, please call us toll-free at **1-888-915-3276** or visit us at **http://ngl.cengage.com**.

For permission to use material from this text or product, submit all requests online at **www.cengage.com/permissions**
Further permissions questions can be emailed to **permissionrequest@cengage.com.**

ISBN: 978-1-337-62911-9

Cengage Learning
20 Channel Center Street
Boston, MA 02210
USA

Cengage Learning is a leading provider of customized learning solutions with office locations around the globe, including Singapore, the United Kingdom, Australia, Mexico, Brazil, and Japan. Locate your local office at: **www.cengage.com/global**.

Cengage Learning products are represented in Canada by Nelson Education, Ltd.

To learn more about Cengage Learning Solutions, visit **www.cengage.com**.

To find online supplements and other instructional support, please visit **www.cengagebrain.com**.

Printed in the United States of America
Print Number: 01 Print Year: 2017

CONTENTS

Part I

Strategies for the AP® Exam

PREPARING FOR THE AP®* CHEMISTRY EXAMINATION

Whether you are taking an AP® course at your school or you are working on AP® independently, the stage is set for a great intellectual experience.

But sometime in January, when the examination begins to loom on a very real horizon, Advanced Placement can seem downright intimidating. If you are nervous about taking the exam, that is only normal.

The best way to deal with an AP® examination is to master it, not let it master you. If you can think of these examinations as a way to show off how much chemistry you know, you have a leg up. Attitude and confidence do help. If you are not one of those students, there is still a great deal you can do to sideline your anxiety. This book is designed to put you on a fast track. Focused review and practice time will help you master the examination so that you can walk in with confidence and score a 5.

WHAT'S IN THIS BOOK

This book can be used with all textbooks. It is divided into three sections. Part I offers suggestions for getting yourself ready, from signing up to take the test and sharpening your pencils to organizing a free-response essay. This is followed by the list of AP® Chemistry Learning Objectives from the College Board Curriculum Framework. At the end of Part I, you will find a Diagnostic Test that has all of the elements of the AP® Chemistry Examination.

Part II is made up of 14 chapters—again following the College Board Curriculum Framework. These chapters are not a substitute for your textbook and class discussion; they simply review the AP® Chemistry course. At the end of each chapter, you will find 15 multiple-choice questions and 2 free-response questions based on the material in that chapter. At the end of each chapter, you will find answers to these questions keyed to the AP® Chemistry Learning Objective number covering that point.

Part III has two complete AP® Chemistry examinations. At the end of each test, you will find the answers, explanations, and the AP® Chemistry Learning Objectives for the multiple-choice and free-response questions.

---

*AP® is a trademark registered by the College Board, which is not affiliated with, and does not endorse, this product.

Being successful in chemistry usually involves careful reading of your textbook and working as many different types of problems as possible. The successful student is not one who memorizes a bunch of facts but one who is able to synthesize and analyze the problem. Chemistry is like a lot of subjects in that an understanding of new material often depends on an understanding of previous concepts.

SETTING UP A REVIEW SCHEDULE

If you have been doing your homework steadily and keeping up with the course work, you are in good shape. Organize your notes, homework, and handouts from class by topic. Reference these materials as well as your textbook and this study guide when you have difficulty in a specific section. But even if you have done all that—or if it is too late to do all that—there are more ways to get it all together.

To begin, read Part I of this book. You will be much more comfortable going into the exam if you understand the format of the exam questions and how best to approach them. Then take the Diagnostic Test and see where you are right now.

Take out a calendar and set up a schedule for yourself. If you begin studying early, you can chip away at the review chapters in Part II. You will be surprised—and pleased—by how much material you can cover in half an hour a day of study for a month or so before the test. Look carefully at the sections of the Diagnostic Test; if you missed a number of questions in one particular area, allow more time for the chapters that cover that area of the course. The practice tests in Part III will give you more experience with different kinds of multiple-choice questions and the wide range of free-response questions.

If time is short, skip reading the review chapters (although you might read through the chapter subheadings) and work on the multiple-choice and free-response questions at the end of each review. This will give you a good idea of your understanding of that particular topic. Then take the tests in Part III.

If time is *really* short, go straight from Part I to Part III. Taking practice tests over and over again is the fastest, most practical way to prepare. You cannot study chemistry by reading it like a novel. You must actively solve problems to gain understanding and excel in your performance. Athletes do not perform well just by reading books about their sport or by watching others. They must get up and practice. Just like athletes, you must practice, practice, practice if you want to do your best!

BEFORE THE EXAMINATION

By February, long before the exam, you need to make sure that you are registered to take the test. Many schools take care of the paperwork and handle the fees for their AP® students, but check with your teacher or the AP® coordinator to make sure that you are on the list. This is especially important if you have a documented disability and need test accommodations. If you are studying AP® independently, call AP® Services at the College Board for the name of the local AP® coordinator, who will help you through the registration process.

The evening before the exam is not a great time for cramming for the exam. If you like, look over class notes or drift through your textbook, concentrating on the broad outlines, not the small details, of the course. You might also want to skim through this book and read the AP® tips.

The evening before the exam *is* a great time to get your things organized for the next day. Sharpen a fistful of no. 2 pencils with good erasers; bring a scientific calculator with fresh batteries. Certain types of calculators are not allowed, so be sure to verify with your teacher or the College Board that your model is acceptable. For example, you cannot use a calculator with a typewriter-style keyboard or a cell phone. Cell phones are not even allowed in the testing room, so you will need a watch (without Internet connections—no smartwatches) and be certain to turn off the alarm if it has one. Bring a piece of fruit or a power bar and a bottle of water for the break. Make sure you have your Social Security number and whatever photo identification and admission ticket are required. Then relax. And get a good night's sleep.

On the day of the examination, plan to arrive early. It is wise not to skip breakfast—studies show that students who eat a hot breakfast before testing earn higher grades. Be careful not to drink a lot of liquids, necessitating a trip to the bathroom during the exam. Breakfast will give you the energy to power through the exam—and more. You will spend some time waiting while everyone is seated in the right room for the right exam before the test has even begun. With a short break between Sections I and II, the AP® Chemistry exam lasts for more than three and a half hours. So be prepared for a long morning. You do not want to be distracted by a growling stomach or hunger pangs.

Be sure to wear comfortable clothes, taking along a sweater in case the heating or air conditioning is erratic. Be sure, too, to wear clothes you like—everyone performs better when they think they look better—and by all means wear your lucky socks.

You have been on the fast track. Now go get a 5!

TAKING THE AP®* CHEMISTRY EXAMINATION

The AP® Chemistry curriculum emphasizes the importance of inquiry and reasoning in science, and requires depth of understanding more than memorization of facts. On the exam, you may be asked to perform a range of tasks, such as interpreting lab data or making predictions based on your knowledge of chemical principles.

Some content in your text will not be tested in the examination, although your teacher may choose to include that material in your course.

The AP® Chemistry exam will consist of two sections. The sections are described below.

Section I: The main features of the multiple-choice sections are given in the following table.

	Format
Number of questions	60
Time to complete	90 minutes
Number of choices	4
Calculator	No
Periodic table and formula chart	Yes
Weighting	50%

The questions will often refer to "real world" situations and will contain several sets of questions. Some sample questions are given in the "Types of Multiple-Choice Questions" section that follows.

Section II: After Section I is collected, you will have a short break. The main features of the free-response sections are given in the following table.

*AP® is a trademark registered by the College Board, which is not affiliated with, and does not endorse, this product.

	Format
Number of questions	7 The questions are made up of three long questions [with between four and eight main parts, designated with letters, (a), (b), etc., and often with subparts designated (i), (ii), etc.] and four short questions (with up to four parts/subparts).
Point value	This section is worth up to 46 points. Each long question is worth 10 points. Each short question is worth 4 points.
Time to complete	105 minutes
Calculator	Yes
Periodic table and formula chart	Yes
Weighting	50%

In this section, you may be asked to:

- Design an experiment to test a hypothesis, given a specific set of laboratory equipment.
- Interpret and draw conclusions from data.
- Use models to describe physical phenomena.
- Follow a logical or analytical pathway to solve a problem.

When answering these questions, you will need to be able to explain what is happening at the atomic or molecular level and you may be asked to draw diagrams of the particle level to demonstrate that you can make models of systems and understand concepts.

The focus on laboratory work as a tool of scientific inquiry means that you will need to practice writing laboratory procedures, understand the correct use of laboratory equipment to gather data, and incorporate safe practices.

PLAN OF ATTACK FOR THE MULTIPLE-CHOICE SECTION

Each question in the multiple-choice section is worth one point—whether it takes you 5 minutes or 15 seconds, it counts the same. Your goal is to first answer 45 questions and feel confident that you have those 45 questions correct. That will generally give you a 3 on the exam just from the multiple-choice questions. Here are some ideas for attacking this section of the exam.

- Go through the test completely once, answering the questions that are easy for you to answer.
- When you run into a problem that you know you can do, but that will take you several minutes, put a star (★) next to the problem number and move on.

- If you are a slow test taker, especially when mathematical calculations are involved, mark these questions with a star in the margin and go back after finishing the rest of the questions.
- When you run into a problem that you have no idea how to solve, put a circle around the number and go on.
- When you have read through the entire test and have done all of the simple questions for you, go back and start working on the questions that you starred.
- Look at the number of questions you have answered. Remember, your goal is 45 questions and you should be there at this point. Take a deep breath, relax, and look at the circled questions again … maybe a second time through will jog your memory. If it doesn't, eliminate as many wrong choices as possible and then guess.
- Make sure that you keep a watch with you (or check the room clock). When five minutes of test time are left, bubble in any blanks with your favorite letter (since there is no penalty for guessing), except for the problem that you are working on. After filling in the blanks, continue with the problem that you are working on.
- **Bubble the answer grid as you go and be sure to allow sufficient time to make any last-minute guesses for questions that you did not complete. Don't leave anything blank!**
- Make sure that as you turn the page, you are "bubbling" the correct question. That way, if you get off in bubbling, you only have a couple of questions to look through. You do not want to get to question 60 and find out that you are bubbling in question 59!

STRATEGIES FOR ANSWERING THE MULTIPLE-CHOICE QUESTIONS

- **Read the question carefully.** Pressured for time, many students make the mistake of reading the questions too quickly or merely skimming them. By reading a question carefully, you may already have some idea about the correct answer. You can then look for it in the responses. Look for key terms in the stem of the question.
- **Eliminate any answer you know is wrong.** You can write on the multiple-choice questions in the test book. As you read through the responses, draw a line through any answer you know is wrong.
- **With a little bit of content knowledge, the majority of multiple choice answers can be narrowed down to two choices.** Many questions have the answers set up in the format of two sets [(A) and (B) as a set followed by (C) and (D)], and they follow the pattern: answer – reason. Choices A and B have the same answer and different reasons, and Choices C and D have the opposite answer and the same reasons. So if you

know the answer or the reason, you can narrow your choice to two possible answers.

- **Read all of the possible answers, and then choose the most accurate response.** AP® examinations are written to test your precise knowledge of a subject. Some of the responses may be partially correct, but there will only be one response that is completely true.
- **Mark tough questions.** If you are hung up on a question, make an educated guess, but mark it in the margin of the question book and come back to review it later if you have time.
- **Check the question number with the scantron number every 10 questions.**

TYPES OF MULTIPLE-CHOICE QUESTIONS

The exam will include independent (stand-alone) and set questions. The latter will ask several questions about the same set of data or situation. Here are some suggestions for approaching each of the various kinds of multiple-choice questions.

CLASSIC/BEST ANSWER QUESTIONS This is the most common type of multiple-choice question. It simply requires you to read the question and select the most correct answer.

1. Given the following reduction potentials, which metal would be best (considering reactivity, not cost or strength) for a pipeline carrying dilute hydrochloric acid?

Reaction	E°, V
$Pt^{2+} + 2\,e^- \rightarrow Pt$	+1.20
$Zn^{2+} + 2\,e^- \rightarrow Zn$	–0.76
$Cu^{2+} + 2\,e^- \rightarrow Cu$	+0.34
$Fe^{2+} + 2\,e^- \rightarrow Fe$	–0.44

(A) Ti
(B) Zn
(C) Cu
(D) Fe

ANSWER: A. You are looking for the metal that is least likely to be oxidized by the protons in a solution of the strong acid. The higher (more positive) the reduction potential, the less likely the metal will be oxidized. Eliminate (B) and (D) because their reduction potentials are negative, indicating that they are more likely to be oxidized. Of the two remaining, (A) has the largest positive value. (AP® Chemistry Learning Objective 3.12—henceforth in this book, AP® Chemistry Learning Objectives will be designated as LO followed by the number.)

NONCALCULATOR COMPUTATIONS These questions require computation without the use of a calculator—the numbers provided in the problem will allow the mathematics to be relatively simple. Estimation and rounding are often the keys to this type of problem.

2. A weak acid, HA, has a K_a value of 1.0×10^{-6}. A student adds 10.0 mL of 0.200 *M* NaOH to 40.0 mL of 0.100 *M* HA and the two react completely. Calculate the final pH of the solution.
   (A) 2.00
   (B) 6.00
   (C) 7.00
   (D) 8.00

ANSWER: B. To answer this question, you must consider what point in the titration the question references. Because the base, NaOH, is twice as concentrated as the acid, HA, it will take half as much NaOH to reach the equivalence point, or 20.0 mL. We are not at the equivalence point. Half the amount of the NaOH, 10.0 mL, needed to reach the equivalence point has been added. Halfway to the equivalence point, the pH equals the pK_a, which equals $-\log(K_a)$ or 6.00. (LO 6.13)

SET QUESTIONS

Questions 3 and 4 refer to the following data.
A student performing an experiment to determine the relative intermolecular forces of various liquids immerses an electronic temperature probe covered with filter paper in a specific liquid, removes the probe, and suspends it from a support. The student monitors the resulting temperature change over a three-minute interval while the liquid on the filter paper evaporates. She then uses the initial temperature and the lowest temperature that each liquid reaches to calculate the change in temperature. The data she obtains are:

Liquid	ΔT (°C)
Ethanol	8.0
1-Propanol	6.2
1-Butanol	3.3
Water	2.1

3. Based on these data, which substance has the strongest intermolecular forces?
   (A) Ethanol (C_2H_5OH)
   (B) 1-Propanol (C_3H_7OH)
   (C) 1-Butanol (C_4H_9OH)
   (D) Water

ANSWER: D. Substances with strong intermolecular forces do not readily change phase from liquid to gas. Temperature change, specifically cooling, is related to the amount of substance that evaporates. A low temperature change indicates less evaporation and higher intermolecular forces. Water exhibits the smallest temperature change so water must have the strongest intermolecular forces. (LO 2.13)

4. Why does 1-propanol have a greater temperature change than 1-butanol?
   (A) 1-Propanol has stronger London dispersion forces because it has a longer carbon chain.
   (B) 1-Butanol has stronger London dispersion forces because it has a longer carbon chain.
   (C) The hydrogen bonds are stronger in 1-propanol than in 1-butanol.
   (D) The hydrogen bonds are stronger in 1-butanol than in 1-propanol.

ANSWER: A. Both substances are able to form the same number of hydrogen bonds per molecule, so this does not explain the difference in temperature change and eliminates choices (C) and (D). 1-Butanol has more electrons than 1-propanol, leading to stronger dispersion forces due to the greater polarizability of the molecule. Coupled with the hydrogen bonds, the IMFs are stronger in 1-butanol than in 1-propanol, causing a slower rate of evaporation. (LO 2.16)

PLAN OF ATTACK FOR THE FREE-RESPONSE SECTION

Section II of the AP® exam comes with a periodic table and a table of equations and constants. The needed portions of the standard reduction potentials, $E°_{red}$, will be provided within the questions.

- Free-response questions have lots of subparts and can generally be done in any order. Start with a question that you feel confident in answering.
- Answer *all* parts of a question. When the question ends, there will be a short space followed by the line space for your answer. Sometimes a question is long enough to cover two pages. Check just above the lined section to see the last subpart letter, and double check that you have answered all parts of the question.
- Show all of your work and use units in your answers. Partial credit may be awarded for problems if the correct work is shown but the answer is not present or is incorrect. For problems involving calculations, circle or box your final answer.
- Cross out incorrect answers with an "X" rather than spending time erasing.
- Be clear, neat, and organized in your work. If a grader cannot clearly understand your work, you may not receive full credit.
- All of the questions will have multiple parts, with the first three questions being longer. Attempt to solve each part of each question. Even if your answer to the first part is incorrect, you still may be awarded points for the remaining parts of the question if the work is correct for those parts. The question parts do not necessarily become more difficult as you progress through them.
- Units are important in your answer. Keeping track of units throughout calculations, and performing unit cancellation where possible, will help guide you to your answer. Points may be deducted for missing or incorrect units in the answer.

- You do not need to work the questions in order. In addition, questions are broken into parts, such as (a) and (b). Clearly label your work with (a), (b), and (i), (ii), etc.

STRATEGIES FOR ANSWERING FREE-RESPONSE QUESTIONS

Free-response questions will ask you to explain, justify, compare, or predict. Performing calculations, showing mathematical relationships, or drawing graphs or structures may also be involved. Free-response questions do not require an introduction or conclusion. Many of these answers may be written in a bulleted or short-answer format. You must know the material very well because these are targeted questions. Examination readers want specifics. They are looking for accurate information presented in clear, concise prose. The use of appropriate vocabulary terms is strongly suggested as this may make the difference between the awarding of a point or not.

To be successful in writing free-response answers for the AP® Chemistry exam, be sure to get straight to the answer and use key terms in your explanations. If you ramble on and on, you might accidentally state an incorrect fact and lose a point (and valuable time). Points will not be awarded for incorrect or extraneous information. For mathematical questions, clearly show your work (with units) … for clarity, use two lines for conversions and box your answer. For short answer questions, be **CLEVER**—make your **CL**aim, state your **EV**idenc**E**, and provide your **R**easoning.

Be sure to work all of the free-response questions found in this book, along with any free-response questions found in your textbook.

TYPES OF FREE-RESPONSE QUESTIONS

The seven free-response questions are comprised of 3 long and 4 short. Each long question includes four to eight parts (with five or six being the most common) and are worth 10 points. They are followed by four short questions, which include up to four parts and are worth 4 points each.

Skills assessed by the free-response section are as follows (The long questions, especially, will assess more than one of these skills.):

- Experimental design
- Quantitative/qualitative translation
- Analysis of authentic lab data and observations to identify patterns or explain phenomena
- Creating or analyzing atomic and molecular views to explain observations
- Following a logical/analytical pathway to solve a problem (mathematical calculations in most cases)

EXAMPLE OF A LONG FREE-RESPONSE QUESTION

Beryllium is a unique element. It is a metal, but often bonds covalently. It is not common, yet it is not quite rare, being the 32nd most abundant element in the earth's crust. Elemental beryllium is not found by itself. It must be combined with other elements in a compound.

(a) What properties of beryllium would cause it to not be found free in nature? Justify your answer using its atomic structure.

(b) Beryllium is often combined with copper to produce an alloy called "beryllium bronze" that has very unique properties. When struck, this very hard alloy does not produce a spark. This property makes it valuable for use in oil field applications and other explosive environments. This alloy contains only a small percentage of Be.

(i) Circle the correct representation of this alloy. Explain your answer.

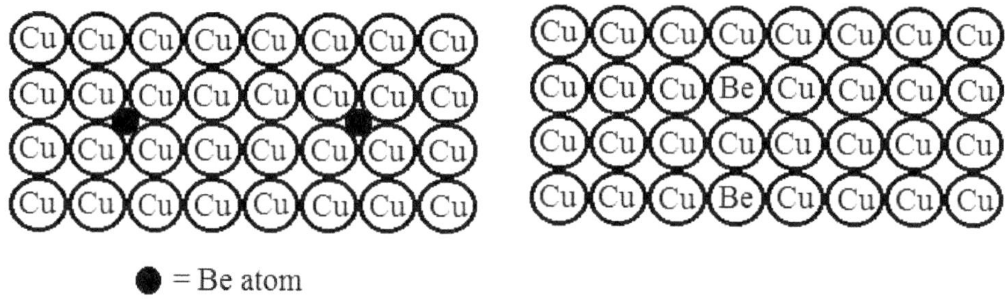

(ii) Is the formation of an alloy a chemical change or a physical change? Explain.

(c) Beryllium can be obtained in commercial quantities by electrolysis of molten $BeCl_2$, the equation of which is shown below.

$$BeCl_2(l) \rightarrow Be(s) + Cl_2(g)$$

(i) What is being reduced in this reaction?

(ii) If a current of 10.0 amps is run through a solution of $BeCl_2(l)$ for 10.0 minutes, how many grams of solid beryllium can be produced?

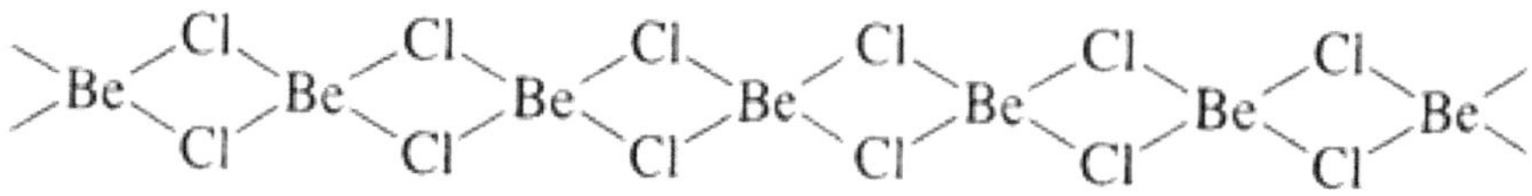

(iii) The structure of solid beryllium chloride is actually a polymer as represented by the diagram shown above. Use formal charges to explain how each beryllium atom can form four covalent bonds.

(d) Beryllium and boron's first three ionization energies are shown below.

	Ionization energy (kJ/mol)		
	First	Second	Third
Beryllium	899	1757	14,849
Boron	800	2427	3660

Explain each of the following in terms of atomic structure. In your responses, make reference to both elements.

(i) Beryllium has a lower second ionization energy than boron.

(ii) Beryllium has a higher third ionization energy than boron.

(e) Beryllium was discovered in 1798, but the charge of its ion was not known for almost a hundred years. There was debate in the chemistry community between those that believed that beryllium's ion had a 2+ charge (Be^{2+}) while others claimed it had a 3+ charge (Be^{3+}). The debate was settled when beryllium was reacted with the anion $C_5H_7O_2^-$ and the density of the resulting gaseous product was measured. Using the data below, verify that beryllium's charge is 2+.

Mass	0.3527 g
Volume	41.2 mL
Temperature	23°C
Pressure	1.00 atm Hg

ANSWER

(a) Beryllium is in Group 2 of the periodic table and has two valence electrons. Metals with one or two valence electrons can lose those electrons very easily (low ionization energies) and react quickly to form compounds. They are rarely, if ever, found free in nature (LO 1.10).

(b) (i) The atomic radius of the beryllium atom is much less than that of the copper atom. The beryllium atom will fit into holes in the copper crystal lattice, forming an interstitial alloy (the alloy shown on the left) (LO 2.25).

(ii) An alloy is a mixture of two or more elements (usually metals). The mixture can form with any ratio of the elements and does not have a chemical formula. This is a physical change (LO 3.10).

(c) (i) Beryllium is being reduced as it is gaining two electrons. The oxidation number of beryllium is decreasing from +2 to 0 (LO 3.8).

(ii)

$$10.0\text{ min} \times \frac{60\text{ s}}{1\text{ min}} \times \frac{10.0\text{ C}}{1\text{ s}} \times \frac{1\text{ mol }e^-}{96{,}485\text{ C}} \times \frac{1\text{ mol Be}}{2\text{ mol }e^-} \times \frac{9.01\text{ g Be}}{1\text{ mol Be}} = 0.280\text{ g Be}$$

(LO 3.12)

(iii) The formal charge of an atom in a molecule gives an indication as to where the electrons in a bond come from. The formal charge of each beryllium atom is –2, because the beryllium atom has 2 valence electrons and the atom has 4 bonds in the structure. To calculate formal charge subtract ½ the number of bonded electrons plus any unshared electrons from the number of valence electrons: 2 – (8/2) = –2 Chlorine has a formal charge of +1. Chlorine has 7 valence electrons, and each chlorine atom also has two unshared pair of electrons (not drawn out) in the structure plus two bonds: $7 - (4 + 4/2) = +1$. The negative formal charge on beryllium indicates that it is using more than its fair share of electrons in the bonds with chlorine. Chlorine is providing both electrons in half of the bonds (i.e., coordinate covalent bond is formed where the Cl is providing the two electrons) (LO 2.32).

(d) (i) The electron removed with the second ionization energy is in the second energy level of both atoms. Beryllium has a lower second ionization energy than boron because boron has one more proton than does beryllium. Because of the extra proton, the nuclear charge pulling on the second energy level of electrons in boron is greater than that of beryllium, making it easier to remove the second electron from beryllium than from boron (LO 1.6).

(ii) Beryllium has a higher third ionization energy than boron because the third electron removed in beryllium is located in 1*s*, while the third electron removed in boron is located in 2*s*. It is very difficult to remove an electron from an inner energy level that is closer to the nucleus and more tightly held (LO 1.6).

(e) $PV = nRT$ 1.00 atm (0.0412 L) = n (0.08206 × 296 K)
$n = 1.70 \times 10^{-3}$ mol $Be(C_5H_7O_2)_x$
0.3527 g $Be(C_5H_7O_2)_x$ /1.70×10^{-3} mol = 207 g/mol $Be(C_5H_7O_2)_x$
Molar mass of $Be(C_5H_7O_2)_2$ = 207 g/mol
Molar mass of $Be(C_5H_7O_2)_3$ = 306 g/mol
The molar mass matches that of $Be(C_5H_7O_2)_2$.
(LO 1.3 and LO 2.6)

AP® Chemistry Learning Objectives

The following is the list of learning objectives from the College Board curriculum. You may want to check off those that you feel you have mastered.

- ❑ LEARNING OBJECTIVE 1.1 The student can justify the observation that the ratio of the masses of the constituent elements in any pure sample of that compound is always identical on the basis of the atomic molecular theory.
- ❑ LEARNING OBJECTIVE 1.2 The student is able to select and apply mathematical routines to mass data to identify or infer the composition of pure substances and/or mixtures.
- ❑ LEARNING OBJECTIVE 1.3 The student is able to select and apply mathematical relationships to mass data in order to justify a claim regarding the identity and/or estimated purity of a substance.
- ❑ LEARNING OBJECTIVE 1.4 The student is able to connect the number of particles, moles, mass, and volume of substances to one another, both qualitatively and quantitatively.
- ❑ LEARNING OBJECTIVE 1.5 The student is able to explain the distribution of electrons in an atom or ion based upon data.
- ❑ LEARNING OBJECTIVE 1.6 The student is able to analyze data relating to electron energies for patterns and relationships.
- ❑ LEARNING OBJECTIVE 1.7 The student is able to describe the electronic structure of the atom, using PES data, ionization energy data, and/or Coulomb's law to construct explanations of how the energies of electrons within shells in atoms vary.
- ❑ LEARNING OBJECTIVE 1.8 The student is able to explain the distribution of electrons using Coulomb's law to analyze measured energies.
- ❑ LEARNING OBJECTIVE 1.9 The student is able to predict and/or justify trends in atomic properties based on location on the periodic table and/or the shell model.
- ❑ LEARNING OBJECTIVE 1.10 Students can justify with evidence the arrangement of the periodic table and can apply periodic properties to chemical reactivity.
- ❑ LEARNING OBJECTIVE 1.11 The student can analyze data, based on periodicity and the properties of binary compounds, to identify patterns and generate hypotheses related to the molecular design of compounds for which data are not supplied.

- ☐ LEARNING OBJECTIVE 1.12 The student is able to explain why a given set of data suggests, or does not suggest, the need to refine the atomic model from a classical shell model with the quantum mechanical model.
- ☐ LEARNING OBJECTIVE 1.13 Given information about a particular model of the atom, the student is able to determine if the model is consistent with specified evidence.
- ☐ LEARNING OBJECTIVE 1.14 The student is able to use data from mass spectrometry to identify the elements and the masses of individual atoms of a specific element.
- ☐ LEARNING OBJECTIVE 1.15 The student can justify the selection of a particular type of spectroscopy to measure properties associated with vibrational or electronic motions of molecules.
- ☐ LEARNING OBJECTIVE 1.16 The student can design and/or interpret the results of an experiment regarding the absorption of light to determine the concentration of an absorbing species in a solution.
- ☐ LEARNING OBJECTIVE 1.17 The student is able to express the law of conservation of mass quantitatively and qualitatively using symbolic representations and particulate drawings.
- ☐ LEARNING OBJECTIVE 1.18 The student is able to apply conservation of atoms to the rearrangement of atoms in various processes.
- ☐ LEARNING OBJECTIVE 1.19 The student can design, and/or interpret data from, an experiment that uses gravimetric analysis to determine the concentration of an analyte in a solution.
- ☐ LEARNING OBJECTIVE 1.20 The student can design, and/or interpret data from, an experiment that uses titration to determine the concentration of an analyte in a solution.
- ☐ LEARNING OBJECTIVE 2.1 Students can predict properties of substances based on their chemical formulas, and provide explanations of their properties based on particle views.
- ☐ LEARNING OBJECTIVE 2.2 The student is able to explain the relative strengths of acids and bases based on molecular structure, interparticle forces, and solution equilibrium.
- ☐ LEARNING OBJECTIVE 2.3 The student is able to use aspects of particulate models (i.e., particle spacing, motion, and forces of attraction) to reason about observed differences between solid and liquid phases and among solid and liquid materials.
- ☐ LEARNING OBJECTIVE 2.4 The student is able to use KMT and concepts of intermolecular forces to make predictions about the macroscopic properties of gases, including both ideal and nonideal behaviors.
- ☐ LEARNING OBJECTIVE 2.5 The student is able to refine multiple representations of a sample of matter in the gas phase to accurately represent the effect of changes in macroscopic properties on the sample.

- ❑ LEARNING OBJECTIVE 2.6 The student can apply mathematical relationships or estimation to determine macroscopic variables for ideal gases.
- ❑ LEARNING OBJECTIVE 2.7 The student is able to explain how solutes can be separated by chromatography based on intermolecular interactions.
- ❑ LEARNING OBJECTIVE 2.8 The student can draw and/or interpret representations of solutions that show the interactions between the solute and solvent.
- ❑ LEARNING OBJECTIVE 2.9 The student is able to create or interpret representations that link the concept of molarity with particle views of solutions.
- ❑ LEARNING OBJECTIVE 2.10 The student can design and/or interpret the results of a separation experiment (filtration, paper chromatography, column chromatography, or distillation) in terms of the relative strength of interactions among and between the components.
- ❑ LEARNING OBJECTIVE 2.11 The student is able to explain the trends in properties and/or predict properties of samples consisting of particles with no permanent dipole on the basis of London dispersion forces.
- ❑ LEARNING OBJECTIVE 2.12 The student can qualitatively analyze data regarding real gases to identify deviations from ideal behavior and relate these to molecular interactions.
- ❑ LEARNING OBJECTIVE 2.13 The student is able to describe the relationships between the structural features of polar molecules and the forces of attraction between the particles.
- ❑ LEARNING OBJECTIVE 2.14 The student is able to apply Coulomb's law qualitatively (including using representations) to describe the interactions of ions, and the attractions between ions and solvents to explain the factors that contribute to the solubility of ionic compounds.
- ❑ LEARNING OBJECTIVE 2.15 The student is able to explain observations regarding the solubility of ionic solids and molecules in water and other solvents on the basis of particle views that include intermolecular interactions and entropic effects.
- ❑ LEARNING OBJECTIVE 2.16 The student is able to explain the properties (phase, vapor pressure, viscosity, etc.) of small and large molecular compounds in terms of the strengths and types of intermolecular forces.
- ❑ LEARNING OBJECTIVE 2.17 The student can predict the type of bonding present between two atoms in a binary compound based on position in the periodic table and the electronegativity of the elements.
- ❑ LEARNING OBJECTIVE 2.18 The student is able to rank and justify the ranking of bond polarity on the basis of the locations of the bonded atoms in the periodic table.
- ❑ LEARNING OBJECTIVE 2.19 The student can create visual representations of ionic substances that connect the microscopic structure to macroscopic properties, and/or use representations to connect the

microscopic structure to macroscopic properties (e.g., boiling point, solubility, hardness, brittleness, low volatility, lack of malleability, ductility, or conductivity).

- ❑ LEARNING OBJECTIVE 2.20 The student is able to explain how a bonding model involving delocalized electrons is consistent with macroscopic properties of metals (e.g., conductivity, malleability, ductility, and low volatility) and the shell model of the atom.
- ❑ LEARNING OBJECTIVE 2.21 The student is able to use Lewis diagrams and VSEPR to predict the geometry of molecules, identify hybridization, and make predictions about polarity.
- ❑ LEARNING OBJECTIVE 2.22 The student is able to design or evaluate a plan to collect and/or interpret data needed to deduce the type of bonding in a sample of a solid.
- ❑ LEARNING OBJECTIVE 2.23 The student can create a representation of an ionic solid that shows essential characteristics of the structure and interactions present in the substance.
- ❑ LEARNING OBJECTIVE 2.24 The student is able to explain a representation that connects properties of an ionic solid to its structural attributes and to the interactions present at the atomic level.
- ❑ LEARNING OBJECTIVE 2.25 The student is able to compare the properties of metal alloys with their constituent elements to determine if an alloy has formed, identify the type of alloy formed, and explain the differences in properties using particulate level reasoning.
- ❑ LEARNING OBJECTIVE 2.26 Students can use the electron sea model of metallic bonding to predict or make claims about the macroscopic properties of metals or alloys.
- ❑ LEARNING OBJECTIVE 2.27 The student can create a representation of a metallic solid that shows essential characteristics of the structure and interactions present in the substance.
- ❑ LEARNING OBJECTIVE 2.28 The student is able to explain a representation that connects properties of a metallic solid to its structural attributes and to the interactions present at the atomic level.
- ❑ LEARNING OBJECTIVE 2.29 The student can create a representation of a covalent solid that shows essential characteristics of the structure and interactions present in the substance.
- ❑ LEARNING OBJECTIVE 2.30 The student is able to explain a representation that connects properties of a covalent solid to its structural attributes and to the interactions present at the atomic level.
- ❑ LEARNING OBJECTIVE 2.31 The student can create a representation of a molecular solid that shows essential characteristics of the structure and interactions present in the substance.
- ❑ LEARNING OBJECTIVE 2.32 The student is able to explain a representation that connects properties of a molecular solid to its structural attributes and to the interactions present at the atomic level.
- ❑ LEARNING OBJECTIVE 3.1 Students can translate among macroscopic observations of change, chemical equations, and particle views.

- ❑ LEARNING OBJECTIVE 3.2 The student can translate an observed chemical change into a balanced chemical equation and justify the choice of equation type (molecular, ionic, or net ionic) in terms of utility for the given circumstances.
- ❑ LEARNING OBJECTIVE 3.3 The student is able to use stoichiometric calculations to predict the results of performing a reaction in the laboratory and/or to analyze deviations from the expected results.
- ❑ LEARNING OBJECTIVE 3.4 The student is able to relate quantities (measured mass of substances, volumes of solutions, or volumes and pressures of gases) to identify stoichiometric relationships for a reaction, including situations involving limiting reactants and situations in which the reaction has not gone to completion.
- ❑ LEARNING OBJECTIVE 3.5 The student is able to design a plan in order to collect data on the synthesis or decomposition of a compound to confirm the conservation of matter and the law of definite proportions.
- ❑ LEARNING OBJECTIVE 3.6 The student is able to use data from synthesis or decomposition of a compound to confirm the conservation of matter and the law of definite proportions.
- ❑ LEARNING OBJECTIVE 3.7 The student is able to identify compounds as Brønsted-Lowry acids, bases, and/or conjugate acid-base pairs, using proton-transfer reactions to justify the identification.
- ❑ LEARNING OBJECTIVE 3.8 The student is able to identify redox reactions and justify the identification in terms of electron transfer.
- ❑ LEARNING OBJECTIVE 3.9 The student is able to design and/or interpret the results of an experiment involving a redox titration.
- ❑ LEARNING OBJECTIVE 3.10 The student is able to evaluate the classification of a process as a physical change, chemical change, or ambiguous change based on both macroscopic observations and the distinction between rearrangement of covalent interactions and noncovalent interactions.
- ❑ LEARNING OBJECTIVE 3.11 The student is able to interpret observations regarding macroscopic energy changes associated with a reaction or process to generate a relevant symbolic and/or graphical representation of the energy changes.
- ❑ LEARNING OBJECTIVE 3.12 The student can make qualitative or quantitative predictions about galvanic or electrolytic reactions based on half-cell reactions and potentials and/or Faraday's laws.
- ❑ LEARNING OBJECTIVE 3.13 The student can analyze data regarding galvanic or electrolytic cells to identify properties of the underlying redox reactions.
- ❑ LEARNING OBJECTIVE 4.1 The student is able to design and/or interpret the results of an experiment regarding the factors (i.e., temperature, concentration, surface area) that may influence the rate of a reaction.
- ❑ LEARNING OBJECTIVE 4.2 The student is able to analyze concentration vs. time data to determine the rate law for a zeroth-, first-, or second-order reaction.

- ❑ LEARNING OBJECTIVE 4.3 The student is able to connect the half-life of a reaction to the rate constant of a first-order reaction and justify the use of this relation in terms of the reaction being a first-order reaction.
- ❑ LEARNING OBJECTIVE 4.4 The student is able to connect the rate law for an elementary reaction to the frequency and success of molecular collisions, including connecting the frequency and success to the order and rate constant, respectively.
- ❑ LEARNING OBJECTIVE 4.5 The student is able to explain the difference between collisions that convert reactants to products and those that do not in terms of energy distributions and molecular orientation.
- ❑ LEARNING OBJECTIVE 4.6 The student is able to use representations of the energy profile for an elementary reaction (from the reactants, through the transition state, to the products) to make qualitative predictions regarding the relative temperature dependence of the reaction rate.
- ❑ LEARNING OBJECTIVE 4.7 The student is able to evaluate alternative explanations, as expressed by reaction mechanisms, to determine which are consistent with data regarding the overall rate of a reaction, and data that can be used to infer the presence of a reaction intermediate.
- ❑ LEARNING OBJECTIVE 4.8 The student can translate among reaction energy profile representations, particulate representations, and symbolic representations (chemical equations) of a chemical reaction occurring in the presence and absence of a catalyst.
- ❑ LEARNING OBJECTIVE 4.9 The student is able to explain changes in reaction rates arising from the use of acid-base catalysts, surface catalysts, or enzyme catalysts, including selecting appropriate mechanisms with or without the catalyst present.
- ❑ LEARNING OBJECTIVE 5.1 The student is able to create or use graphical representations in order to connect the dependence of potential energy to the distance between atoms and factors, such as bond order (for covalent interactions) and polarity (for intermolecular interactions), which influence the interaction strength.
- ❑ LEARNING OBJECTIVE 5.2 The student is able to relate temperature to the motions of particles, either via particulate representations, such as drawings of particles with arrows indicating velocities, and/or via representations of average kinetic energy and distribution of kinetic energies of the particles, such as plots of the Maxwell-Boltzmann distribution.
- ❑ LEARNING OBJECTIVE 5.3 The student can generate explanations or make predictions about the transfer of thermal energy between systems based on this transfer being due to a kinetic energy transfer between systems arising from molecular collisions.
- ❑ LEARNING OBJECTIVE 5.4 The student is able to use conservation of energy to relate the magnitudes of the energy changes occurring in two or more interacting systems, including identification of the systems, the type (heat versus work), or the direction of energy flow.

- ❑ LEARNING OBJECTIVE 5.5 The student is able to use conservation of energy to relate the magnitudes of the energy changes when two nonreacting substances are mixed or brought into contact with one another.
- ❑ LEARNING OBJECTIVE 5.6 The student is able to use calculations or estimations to relate energy changes associated with heating/cooling a substance to the heat capacity, relate energy changes associated with a phase transition to the enthalpy of fusion/vaporization, relate energy changes associated with a chemical reaction to the enthalpy of the reaction, and relate energy changes to $P\Delta V$ work.
- ❑ LEARNING OBJECTIVE 5.7 The student is able to design and/or interpret the results of an experiment in which calorimetry is used to determine the change in enthalpy of a chemical process (heating/cooling, phase transition, or chemical reaction) at constant pressure.
- ❑ LEARNING OBJECTIVE 5.8 The student is able to draw qualitative and quantitative connections between the reaction enthalpy and the energies involved in the breaking and formation of chemical bonds.
- ❑ LEARNING OBJECTIVE 5.9 The student is able to make claims and/or predictions regarding relative magnitudes of the forces acting within collections of interacting molecules based on the distribution of electrons within the molecules and the types of intermolecular forces through which the molecules interact.
- ❑ LEARNING OBJECTIVE 5.10 The student can support the claim about whether a process is a chemical or physical change (or may be classified as both) based on whether the process involves changes in intramolecular versus intermolecular interactions.
- ❑ LEARNING OBJECTIVE 5.11 The student is able to identify the noncovalent interactions within and between large molecules, and/or connect the shape and function of the large molecule to the presence and magnitude of these interactions.
- ❑ LEARNING OBJECTIVE 5.12 The student is able to use representations and models to predict the sign and relative magnitude of the entropy change associated with chemical or physical processes.
- ❑ LEARNING OBJECTIVE 5.13 The student is able to predict whether or not a physical or chemical process is thermodynamically favored by determination of (either quantitatively or qualitatively) the signs of both $\Delta H°$ and $\Delta S°$, and calculation or estimation of $\Delta G°$ when needed.
- ❑ LEARNING OBJECTIVE 5.14 The student is able to determine whether a chemical or physical process is thermodynamically favorable by calculating the change in standard Gibbs free energy.
- ❑ LEARNING OBJECTIVE 5.15 The student is able to explain how the application of external energy sources or the coupling of favorable with unfavorable reactions can be used to cause processes that are not thermodynamically favorable to become favorable.
- ❑ LEARNING OBJECTIVE 5.16 The student can use Le Chatelier's principle to make qualitative predictions for systems in which coupled reactions that share a common intermediate drive formation of a product.

- ☐ LEARNING OBJECTIVE 5.17 The student can make quantitative predictions for systems involving coupled reactions that share a common intermediate, based on the equilibrium constant for the combined reaction.
- ☐ LEARNING OBJECTIVE 5.18 The student can explain why a thermodynamically favored chemical reaction may not produce large amounts of product (based on consideration of both initial conditions and kinetic effects), or why a thermodynamically unfavored chemical reaction can produce large amounts of product for certain sets of initial conditions.
- ☐ LEARNING OBJECTIVE 6.1 The student is able to, given a set of experimental observations regarding physical, chemical, biological, or environmental processes that are reversible, construct an explanation that connects the observations to the reversibility of the underlying chemical reactions or processes.
- ☐ LEARNING OBJECTIVE 6.2 The student can, given a manipulation of a chemical reaction or set of reactions (e.g., reversal of reaction or addition of two reactions), determine the effects of that manipulation on Q or K.
- ☐ LEARNING OBJECTIVE 6.3 The student can connect kinetics to equilibrium by using reasoning about equilibrium, such as Le Chatelier's principle, to infer the relative rates of the forward and reverse reactions.
- ☐ LEARNING OBJECTIVE 6.4 The student can, given a set of initial conditions (concentrations or partial pressures) and the equilibrium constant, K, use the tendency of Q to approach K to predict and justify the prediction as to whether the reaction will proceed toward products or reactants as equilibrium is approached.
- ☐ LEARNING OBJECTIVE 6.5 The student can, given data (tabular, graphical, etc.) from which the state of a system at equilibrium can be obtained, calculate the equilibrium constant, K.
- ☐ LEARNING OBJECTIVE 6.6 The student can, given a set of initial conditions (concentrations or partial pressures) and the equilibrium constant, K, use stoichiometric relationships and the law of mass action (Q equals K at equilibrium) to determine qualitatively and/or quantitatively the conditions at equilibrium for a system involving a single reversible reaction.
- ☐ LEARNING OBJECTIVE 6.7 The student is able, for a reversible reaction that has a large or small K, to determine which chemical species will have very large versus very small concentrations at equilibrium.
- ☐ LEARNING OBJECTIVE 6.8 The student is able to use Le Chatelier's principle to predict the direction of the shift resulting from various possible stresses on a system at chemical equilibrium.
- ☐ LEARNING OBJECTIVE 6.9 The student is able to use Le Chatelier's principle to design a set of conditions that will optimize a desired outcome, such as product yield.

- ❑ LEARNING OBJECTIVE 6.10 The student is able to connect Le Chatelier's principle to the comparison of Q to K by explaining the effects of the stress on Q and K.
- ❑ LEARNING OBJECTIVE 6.11 The student can generate or use a particulate representation of an acid (strong or weak or polyprotic) and a strong base to explain the species that will have large versus small concentrations at equilibrium.
- ❑ LEARNING OBJECTIVE 6.12 The student can reason about the distinction between strong and weak acid solutions with similar values of pH, including the percent ionization of the acids, the concentrations needed to achieve the same pH, and the amount of base needed to reach the equivalence point in a titration.
- ❑ LEARNING OBJECTIVE 6.13 The student can interpret titration data for monoprotic or polyprotic acids involving titration of a weak or strong acid by a strong base (or a weak or strong base by a strong acid) to determine the concentration of the titrant and the pK_a for a weak acid, or the pK_b for a weak base.
- ❑ LEARNING OBJECTIVE 6.14 The student can, based on the dependence of K_w on temperature, reason that neutrality requires $[H^+] = [OH^-]$ as opposed to requiring pH = 7, including especially the applications to biological systems.
- ❑ LEARNING OBJECTIVE 6.15 The student can identify a given solution as containing a mixture of strong acids and/or bases and calculate or estimate the pH (and concentrations of all chemical species) in the resulting solution.
- ❑ LEARNING OBJECTIVE 6.16 The student can identify a given solution as being the solution of a monoprotic weak acid or base (including salts in which one ion is a weak acid or base), calculate the pH and concentration of all species in the solution, and/or infer the relative strengths of the weak acids or bases from given equilibrium concentrations.
- ❑ LEARNING OBJECTIVE 6.17 The student can, given an arbitrary mixture of weak and strong acids and bases (including polyprotic systems), determine which species will react strongly with one another (i.e., with $K > 1$) and what species will be present in large concentrations at equilibrium.
- ❑ LEARNING OBJECTIVE 6.18 The student can design a buffer solution with a target pH and buffer capacity by selecting an appropriate conjugate acid-base pair and estimating the concentrations needed to achieve the desired capacity.
- ❑ LEARNING OBJECTIVE 6.19 The student can relate the predominant form of a chemical species involving a labile proton (i.e., protonated/deprotonated form of a weak acid) to the pH of a solution and the pK_a associated with the labile proton.
- ❑ LEARNING OBJECTIVE 6.20 The student can identify a solution as being a buffer solution and explain the buffer mechanism in terms of the reactions that would occur on addition of acid or base.

- ❑ LEARNING OBJECTIVE 6.21 The student can predict the solubility of a salt, or rank the solubility of salts, given the relevant K_{sp} values.
- ❑ LEARNING OBJECTIVE 6.22 The student can interpret data regarding solubility of salts to determine, or rank, the relevant K_{sp} values.
- ❑ LEARNING OBJECTIVE 6.23 The student can interpret data regarding the relative solubility of salts in terms of factors (common ions, pH) that influence the solubility.
- ❑ LEARNING OBJECTIVE 6.24 The student can analyze the enthalpic and entropic changes associated with the dissolution of a salt, using particulate level interactions and representations.
- ❑ LEARNING OBJECTIVE 6.25 The student is able to express the equilibrium constant in terms of $\Delta G°$ and RT and use this relationship to estimate the magnitude of K and, consequently, the thermodynamic favorability of the process.

A Diagnostic Test

The purpose of this test is to give you an indication of how well you will perform on the AP® Chemistry exam. These questions are representative of the AP® Chemistry examination, but bear in mind it is impossible to predict exactly how well you will do on the actual exam. Calculators may not be used for answering questions in the first section of this test. The first section is 50% of your total test grade. Time yourself to finish this part in 90 minutes. Select the best answer in each case.

AP® CHEMISTRY EXAMINATION
Section I: Multiple-Choice Questions
Time: 90 minutes
Number of Questions: 60

DIRECTIONS: Each of the questions or incomplete statements below is followed by four suggested answers or completions. Select the one that is best in each case. **No calculators can be used in this section.** A periodic table and a formula chart with constants are provided at the back of the book. Assume that the conditions are a temperature of 298 K and pressure of 1.0 atm with all solutions aqueous unless otherwise specified.

Questions 1–3 refer to the following information.

Various masses of calcium chloride were dissolved in water and added to 6.00 mL of a 1.7 *M* sodium carbonate solution. The resulting precipitate was separated, dried, and weighed. The graph of the resulting data is shown below.

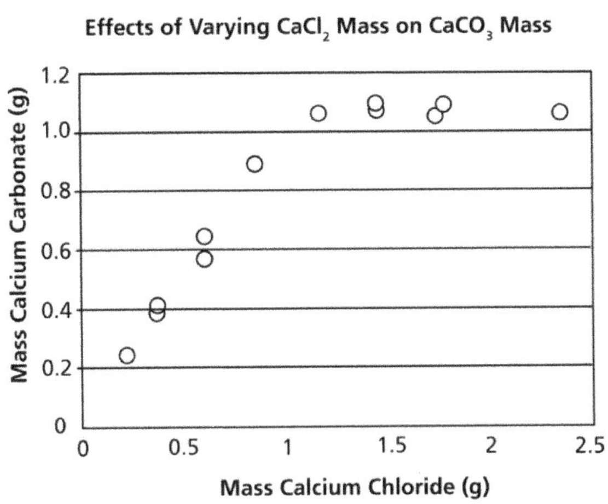

1. Which technique would best separate calcium carbonate precipitate from the rest of the substances in the reaction?
(A) column chromatography
(B) distillation
(C) evaporation
(D) filtration

2. Which substance was present in the lowest concentration in the reaction vessel after 0.5 g of calcium chloride reacted?
(A) Ca^{2+}
(B) Na^{+}
(C) calcium carbonate
(D) sodium chloride

GO ON TO NEXT PAGE

3. Which statement best explains what occurs when approximately 1.2 g of $CaCl_2$ are added?
   (A) That is where the indicator changed color.
   (B) The excess reactant became the limiting reactant.
   (C) The additional calcium carbonate that was being formed was dissolving to form a supersaturated solution.
   (D) Calcium chloride and sodium carbonate reached equilibrium.

4. The complete photoelectron spectra (PES) for an element shows three peaks of identical size. What does this indicate about the element?
   (A) The element is in the third period of the periodic table.
   (B) The element is lithium (Li).
   (C) The element is in Group 13 (3A) of the periodic table.
   (D) The element has electrons in three sublevels, 1*s*, 2*s*, and 2*p*.

$$Q + 2\ R \rightarrow R_2Q$$

5. The reaction shown above is thought to proceed via a single step elementary reaction. Which statement is most probable for this reaction?
   (A) Single step reactions occur very quickly. The rate law is Rate = $k[Q][R]^2$.
   (B) Ternary collisions are very rare so the reaction rate would be low. The rate law is Rate = $k[Q][R]^2$.
   (C) All particles that collide will react so the reaction rate would be high. The rate law is Rate = $k[R]^2$.
   (D) The reaction involves a catalyst and would proceed quickly. The rate law is Rate = $k[R]^2$.

6. Iodine-131 is used in medicine as a treatment for thyroid cancer. It is injected into the body in the form of an aqueous solution of NaI, which decays by first-order kinetics and has a half-life of 8 days. If 20.0 μg of I-131 ion is injected into the body, approximately how many days have passed if the amount of I-131 has fallen to 2.00 μg?
   (A) 24 days
   (B) 27 days
   (C) 32 days
   (D) 36 days

Questions 7–9 use the following information.

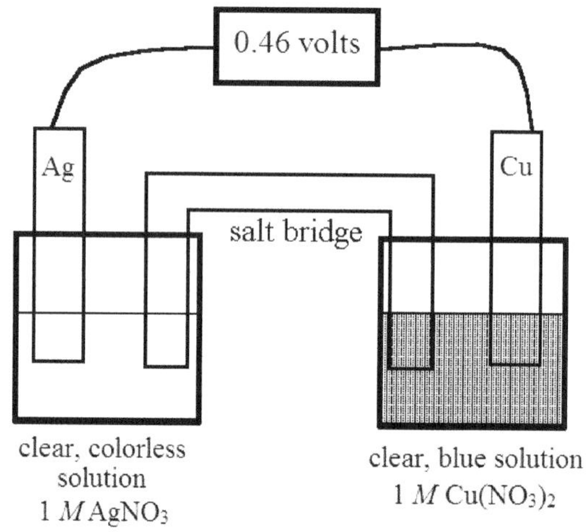

The electrochemical cell above involves the following reaction:

$$2\ Ag^+(aq) + Cu(s) \rightarrow Cu^{2+}(aq) + 2\ Ag(s)$$

The measured voltage is +0.46 V at 25°C.

7. Which will result if an additional 25 mL of 1.0 *M* silver nitrate solution is added to the silver half-cell?
   (A) There will be a decrease in blue color in the copper cell.
   (B) The measured voltage will increase.
   (C) The concentration of copper(II) ions will decrease.
   (D) There will be no change in the cell voltage.

8. What reaction is occurring at the anode?
   (A) $Ag^+(aq) + e^- \rightarrow Ag(s)$
   (B) $Ag(s) \rightarrow Ag^+(aq) + e^-$
   (C) $Cu(s) \rightarrow Cu^{2+}(aq) + 2\ e^-$
   (D) $Cu^{2+}(aq) + 2\ e^- \rightarrow Cu(s)$

9. The concentration of the copper(II) nitrate solution remaining after the electrochemical cell has been running for 60 minutes is to be determined by visual light spectroscopy. The student set the spectrophotometer to a wavelength of 470 nm (blue light). The absorbance of the sample was found to be too low to measure. How can the student achieve a measurable absorbance?
   (A) Set the spectrophotometer at 630 nm (orange light).
   (B) Measure concentration of the copper(II) nitrate solution after 24 hours (rather than 60 minutes) when it is more concentrated.
   (C) Dilute the copper(II) nitrate solution with water.
   (D) Make sure that the test tube or cuvette used for the sample is clear of scratches and fingerprints.

Questions 10 and 11 use the following information.

A sample of clean copper (Cu) wire (molar mass = 63.6 g) is placed in a crucible and completely covered with powdered sulfur, S_8 (molar mass = 256.5 g). The crucible is then heated under a fume hood with the lid ajar, causing excess sulfur to burn with a blue flame. Below is the data table from the student's laboratory notebook.

Mass of crucible and cover	17.004 g
Mass of crucible, cover, and sample of Cu	17.640 g
Mass of crucible, cover, and product after heating	17.960 g

10. What is the empirical formula for the copper sulfide compound?
    (A) CuS
    (B) Cu_2S
    (C) CuS_2
    (D) CuS_8

11. Another group forgets to use a crucible cover and has considerable spattering of the sulfur coming from the crucible. However, their results were the same as those of the rest of the class. What is the reason for their results?
    (A) The amount of copper that didn't react with the sulfur formed an alloy with the product.
    (B) Equal masses of sulfur and copper were lost from the crucible, so the mole ratio remained the same.
    (C) The amount of sulfur that reacted would be too low, so the amount of copper that reacted was too low.
    (D) The sulfur that was lost was an excess reagent, so it did not affect the results.

Questions 12–14 use the following information.

$2\,NO(g) + O_2(g) \rightleftarrows 2\,NO_2(g)$
At a certain temperature, $K_c = 100$

12. 0.20 mol of NO(*g*) and 0.10 mol of $O_2(g)$ are placed in an evacuated rigid container and allowed to reach equilibrium at constant temperature. How does the original pressure, before any reaction occurs, compare with the equilibrium pressure?
    (A) The original pressure is higher than the equilibrium pressure because the reaction has 3 moles of gaseous reactants for every 2 moles of gaseous products.
    (B) The original pressure is higher than the equilibrium pressure because the reaction produces very little product at equilibrium.
    (C) The equilibrium pressure is higher than the original pressure because the reaction produces mostly products at equilibrium.
    (D) The equilibrium pressure is higher than the original pressure because the reaction will have 5 total moles of gas.

GO ON TO NEXT PAGE

13. At the same temperature, what is the K_c for the reaction
$NO_2(g) \rightleftarrows NO(g) + \frac{1}{2} O_2(g)$?
(A) 0.01
(B) 0.1
(C) 10
(D) 100

14. $\Delta H_{rxn} = -112$ kJ for the reaction producing nitrogen dioxide gas. What must be true about this reaction?
(A) More energy is required to break the bonds of the reactants than is released when new bonds are formed.
(B) Less energy is required to break the bonds of the reactants than is released when new bonds are formed.
(C) More energy is released when the bonds of the reactants are broken than is required to form the products.
(D) Less energy is released when the bonds of the reactants are broken than is required to form the products.

Ion	Ionic Radius
Mg^{2+}	65 pm
Ca^{2+}	99 pm
O^{2-}	140 pm

15. Calcium oxide has a melting point of 2572°C, while magnesium oxide has a melting point of 2852°C. Given the table of ionic radii above, which statement gives the best reason for the difference in melting points of calcium oxide and magnesium oxide?
(A) higher charge of Mg^{2+} than of Ca^{2+}
(B) greater ionic size of O^{2-} than of Mg^{2+}
(C) greater ionic size of Ca^{2+} than of Mg^{2+}
(D) greater ionic size of Mg^{2+} than of Ca^{2+}

16. Which system would maintain equal rates of its forward and reverse reactions in the time immediately following an increase in volume of the system at constant temperature?
(A) $2\ O_3(g) \rightleftarrows 3\ O_2(g)$
(B) $PCl_5(g) \rightleftarrows PCl_3(g) + Cl_2(g)$
(C) $H_2(g) + I_2(g) \rightleftarrows 2\ HI(g)$
(D) $2\ NO_2(g) \rightleftarrows N_2O_4(g)$

17. A solution of 50.0 mL of 0.0010 *M* $Ba(OH)_2$ is slowly titrated with 50.0 mL of 0.0030 *M* H_2SO_4. A precipitate is observed to form during the titration. What will happen to the conductivity of this solution as the titration progresses?
(A) The conductivity decreases to near zero and then increases.
(B) The conductivity decreases to near zero and remains very low.
(C) The conductivity increases as the acid is added and then becomes constant at a high value.
(D) The conductivity increases as the acid is added and then slowly becomes very low.

$$N_2(g) + 2\ O_2(g) \rightarrow 2\ NO_2(g)$$

18. Calculate $\Delta G°$ for the reaction shown above and predict whether the reaction is thermodynamically favored given the following information.
$N_2(g) + O_2(g) \rightarrow 2\ NO(g)$
$\Delta G° = 174$ kJ/mol
$2\ NO(g) + O_2(g) \rightarrow 2\ NO_2(g)$
$\Delta G° = -70$ kJ/mol
(A) +104 kJ/mol; thermodynamically favored
(B) −104 kJ/mol; thermodynamically favored
(C) +104 kJ/mol; thermodynamically unfavored
(D) −104 kJ/mol; thermodynamically unfavored

$HX + H_2O \rightleftarrows H_3O^+ + X^-$ $K_a = 1.0 \times 10^{-9}$

19. If 0.050 mol NaX is placed in a 500-mL volumetric flask and dissolved in water, what is the pH of the resulting solution?
    (A) 3.00
    (B) 6.00
    (C) 8.00
    (D) 11.00

Questions 20–22 refer to the following information.

Strontium-90 is a common radioactive isotope formed in nuclear fission reactions and has a half-life of 28.79 years. Sr-90 is often found after nuclear weapons testing and nuclear accidents.

20. How does a sample of nonradioactive strontium chloride compare to a sample of radioactive (strontium-90) strontium chloride?
    (A) The nonradioactive sample has a higher molar mass.
    (B) The nonradioactive sample has a lower molar mass.
    (C) The nonradioactive sample has the same molar mass, but the number of neutrons in the nonradioactive sample is greater.
    (D) The nonradioactive sample has the same molar mass, but the number of neutrons in the nonradioactive sample is less.

21. Strontium-90 exposure is dangerous in particular because it tends to replace certain elements in the human body. What is it most likely to replace?
    (A) Calcium in the bones
    (B) Iron in blood
    (C) Phosphorus in bones
    (D) Carbon in DNA

22. Strontium (both radioactive and nonradioactive isotopes) can be tentatively identified with a flame test. When valence electrons become excited by added energy, they absorb the energy and reach a higher energy level. Unable to sustain this high energy state, they release energy (sometimes in the form of visible light) as they fall back to their ground state. The production of red-colored light by strontium ions when placed in a flame disproves which historical model of the atom?
    (A) Dalton's model—atoms are solid, indivisible spheres
    (B) Rutherford's model—atoms consist of electrons in a positively charged material
    (C) Bohr's model—atoms consist of a nucleus that is orbited by electrons
    (D) Quantum mechanical model—atoms consist of a small positively charged nucleus with electrons in orbitals outside the nucleus

Questions 23 and 24 refer to the following information.

This pH curve represents the titration of 50.0 mL of 0.100 *M* acid with 0.100 *M* base.

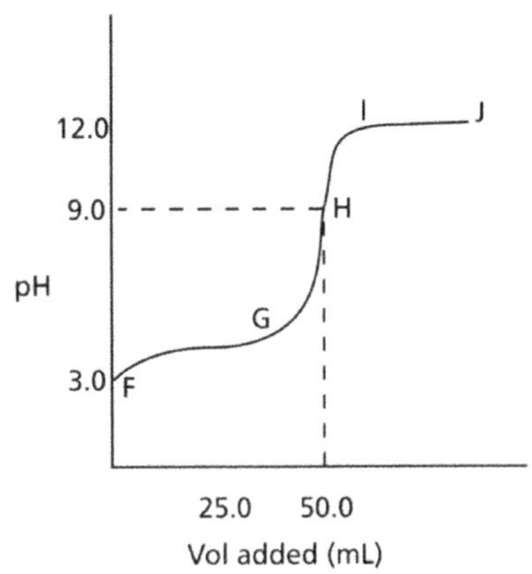

23. What type of titration does this curve represent?
    (A) strong acid titrated by a strong base
    (B) strong acid titrated by a weak base
    (C) weak acid titrated by a strong base
    (D) weak acid titrated by its conjugate base

GO ON TO NEXT PAGE

24. Buffering is most effective
   (A) between points F and G.
   (B) between points G and H.
   (C) at point H between points I and J.
   (D) between both F and G and between I and J.

25. 1 2 3

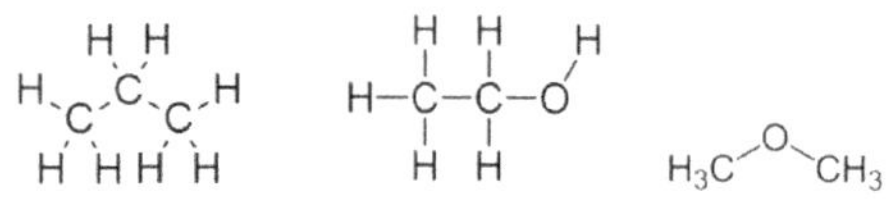

Arrange the above liquids in order of increasing vapor pressure:
   (A) 2 < 3 < 1
   (B) 2 < 1 < 3
   (C) 1 < 3 < 2
   (D) 1 < 2 < 3

26. The bond angle in OCl_2 has been found to be 110.9°. The following statements are all true. Which best explains this observation?
   (A) Oxygen is sp^3 hybridized in the molecule.
   (B) Chlorine atoms have a large radius with many electrons.
   (C) The molecule has a bent molecular geometry.
   (D) Unshared electron pairs generally repel more than shared pairs.

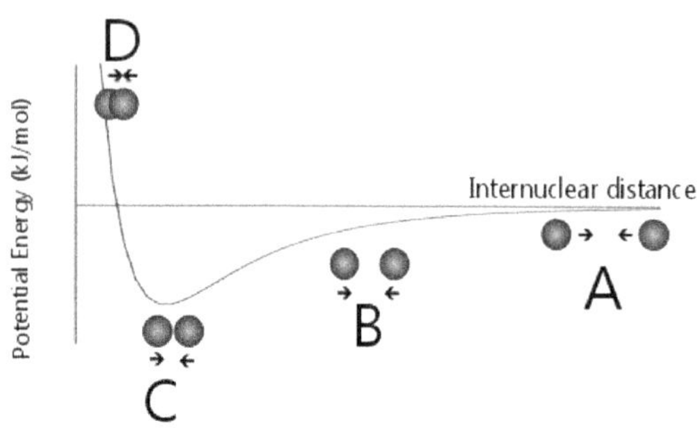

27. The potential energy versus internuclear distance of a 2-atom system is shown in the diagram above. At what point is the interaction between the 2 nuclei the predominate interaction between the 2 atoms?
   (A) A
   (B) B
   (C) C
   (D) D

$$C_2H_4 + H_2 \rightarrow C_2H_6$$

<u>Mechanism 1</u>

$H_2 \rightleftarrows 2H$
$C_2H_4 + H \rightarrow C_2H_5$
$C_2H_5 + H \rightarrow C_2H_6$

<u>Mechanism 2</u>

$H_2 + 2\ Pd \rightleftarrows 2\ Pd\text{-}H$
$C_2H_4 + Pd\text{-}H \rightarrow C_2H_5\text{-}Pd$
$C_2H_5\text{-}Pd + Pd\text{-}H \rightarrow C_2H_6 + 2\ Pd$

28. The hydrogenation reaction that converts ethene (C_2H_4) to ethane (C_2H_6) is shown above. Following it are two possible mechanisms for the reaction. Which statement is most likely correct?
   (A) The reaction will have a higher rate with mechanism 1 as it involves the collision of fewer particles than does mechanism 2.
   (B) The reaction will have a higher rate with mechanism 1 as mechanism 2 involves an intermediate.
   (C) The reaction will have a higher rate with mechanism 2 than with mechanism 1 because mechanism 2 involves a solid-state catalyst.
   (D) The rates of reaction will be the same with both mechanisms. However, mechanism 2 will have a lower activation energy because it contains a catalyst.

29. The electronic and molecular geometries of a PH_3 molecule are correctly described by the VSEPR model as

	Electronic Geometry	**Molecular Geometry**
(A)	Trigonal planar	Trigonal planar
(B)	Trigonal pyramidal	Trigonal pyramidal
(C)	Tetrahedral	Trigonal pyramidal
(D)	Tetrahedral	Tetrahedral

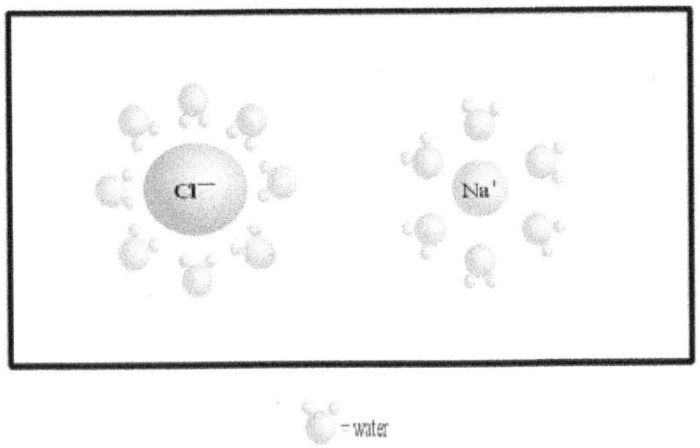

30. The dissolution of sodium chloride shown in the diagram above can be considered to be both a chemical and a physical change. Justify this statement.

	Chemical Change	**Physical Change**
(A)	Na^+ and Cl^- bonds are broken.	H–O–H bonds are broken.
(B)	Na^+ and Cl^- form ion–dipole attractions to water.	Hydrogen bonds between water molecules are broken.
(C)	H–O–H bonds are broken.	Na^+ and Cl^- bonds are broken.
(D)	Hydrogen bonds between water molecules are broken.	Na^+ and Cl^- form ion–dipole attractions to water.

31. A student determines that a reaction is second order with respect to A. Which standard kinetics graph would show a straight line in this reaction?
(A) [A] versus time
(B) ln[A] versus time
(C) $[A]^{-1}$ versus time
(D) ln k versus temperature^{-1}

$$A + B \rightleftarrows 2\,C$$

32. The equilibrium concentrations of reactants and products for the reaction above are represented in the box below. Each particle represents 1.0 mol.

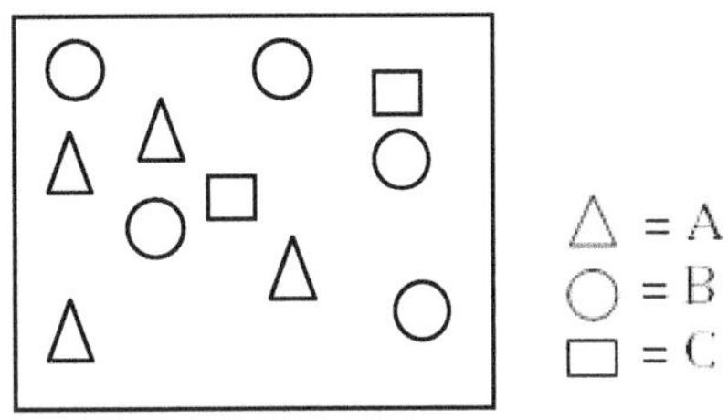

What is the value of the equilibrium constant, K?
(A) 0.10
(B) 0.20
(C) 5.0
(D) 10.

Trial	[A] (mol/L)	[B] (mol/L)	[C] (mol/L)	Initial Rate of Appearance of D (mol L^{-1} s^{-1})
1	0.125	0.125	0.5	12
2	0.250	0.125	1	48
3	0.125	0.125	1	24
4	0.250	0.500	1	192

33. A student is conducting an investigation to determine the rate law for the reaction $A + 2\,B + C \rightarrow D$ and collects the data shown in the table above. What is the overall order of the rate law for this reaction?
(A) 2
(B) 3
(C) 4
(D) 5

34. Which of these sets of two elements would be predicted to have the greatest difference in electronegativity?
(A) F–F
(B) Sc–Ti
(C) Cr–Br
(D) P–Cl

GO ON TO NEXT PAGE

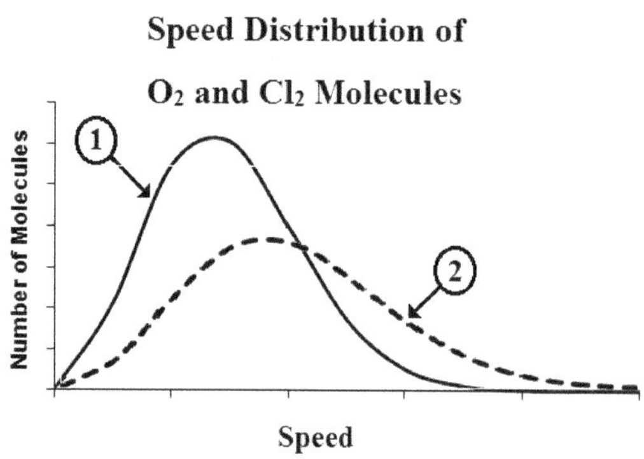

35. The graph above shows the speed distribution of two samples of gas (oxygen and chlorine) in two containers under identical conditions. Which statement correctly identifies the oxygen gas and provides a correct explanation?
(A) Gas 1 is oxygen because the molar mass of oxygen is less. If the particles are at the same temperature and have the same average kinetic energy, then the average speed of oxygen will have to be greater as shown by the higher peak.
(B) Gas 1 is oxygen because the double bond in oxygen would absorb more energy in order to vibrate, so more molecules will be at each speed and the shape will be squeezed together.
(C) Gas 2 is oxygen. The molar mass of oxygen is less than that of chlorine. Therefore, although the particles have the same average kinetic energy, the average speed of oxygen will be greater resulting in a graph that is shorter and more spread out.
(D) Gas 2 is oxygen because the double bond in oxygen would absorb more energy in order to vibrate, so it will have more kinetic energy and a greater speed. The graph will be shorter and more spread out.

Isotope	Atomic Mass (amu)	Percent Abundance
^{48}X	48.0	10.0
^{50}X	50.0	90.0

36. What is the average atomic mass of element X based on the data above?
(A) 49.0 amu
(B) 49.5 amu
(C) 49.8 amu
(D) 50.0 amu

37. We know that no matter where a sample of pure water is found, it always has the formula H_2O; however, the average molar mass of the water can vary slightly, depending upon where it is found. Since oxygen has three stable isotopes and hydrogen has two stable isotopes, there are nine different possible combinations of isotopes in a water molecule with masses varying from approximately 18 to 22 g/mol. While deep ocean water has a very uniform makeup, fresh water has a widely varying isotopic makeup, depending on the location and sources of the water. What type of spectroscopy can analyze a sample of water to determine its isotopic composition?
(A) infrared (IR) spectroscopy
(B) mass spectrometry
(C) photoelectron spectroscopy (PES)
(D) visual light spectroscopy

38. The pH of a 0.001 *M* HBr solution is
(A) 1.0.
(B) 3.0.
(C) 7.0.
(D) 11.0.

X010012884

39. The solubility of $Zn(OH)_2$ is 2.0×10^{-6} M at a certain temperature. Determine the value of the K_{sp} at this same temperature.
(A) 4.0×10^{-6}
(B) 8.0×10^{-6}
(C) 1.6×10^{-17}
(D) 3.2×10^{-17}

40. A rigid cylinder contains CO_2 gas at a constant temperature. Some of the gas is allowed to escape. Which of the following applies to the remaining CO_2 in the cylinder?
(A) The pressure of the gas increases because the gas molecules are less crowded and can experience more collisions.
(B) The volume of the gas decreases because there are fewer moles of gas in the cylinder.
(C) The volume of the gas increases because the molecules have more room to expand.
(D) The distance between CO_2 molecules is increased because there are fewer molecules in the same volume.

41. The first-order rate constant for nuclear decay of ^{60}Co is 0.13 yr^{-1}, and for ^{90}Sr it is 0.24 yr^{-1}. Which statement is correct?
(A) The half-life of Sr is shorter than that of Co.
(B) The half-lives of both isotopes depend on the starting amount and therefore cannot be determined.
(C) Since Co has the neutron : proton ratio closest to 1.0, the stability of the Co nucleus is greater.
(D) Since Sr has more neutrons, the stability of the Sr nucleus is greater.

$$2\ NH_3(g) \rightleftarrows N_2(g) + 3\ H_2(g)$$

42. When ammonia is placed into a closed system of constant volume and temperature, the above reaction occurs. The original pressure of ammonia is 0.80 atm. When equilibrium is established, the total pressure is 1.20 atm. What is the pressure change of the hydrogen gas?
(A) increases by 0.20 atm
(B) increases by 0.40 atm
(C) increases by 0.60 atm
(D) increases by 1.20 atm

43. A student is asked to determine the empirical formula for a hydrate of cobalt(II) chloride ($CoCl_2 \bullet x\ H_2O$). He set up the following experimental setup. He recorded the mass of the hydrate before the reaction and will again take the mass of the anhydrous product after heating gently for 10 minutes. What additional step could the student do to **directly** verify the law of conservation of mass?

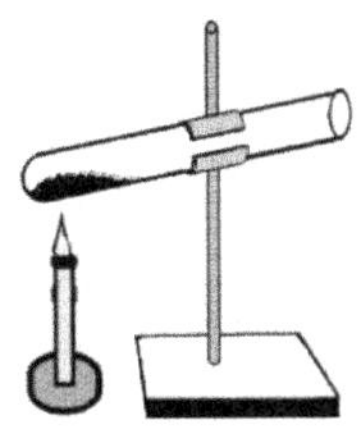

(A) Take the mass of the test tube with the hydrate before the experiment begins.
(B) Measure the amount of heat energy it takes to heat the sample up and convert that heat energy into mass using the equation $q = mc\Delta T$.
(C) Mass the amount of oxygen gas that reacts.
(D) Condense the water vapor produced by the reaction and mass it.

GO ON TO NEXT PAGE

Questions 44–46 use the following information.

$$SO_2Cl_2(g) \rightleftarrows SO_2(g) + Cl_2(g)$$

Sulfuryl chloride gas, $SO_2Cl_2(g)$, decomposes according to the equation above. A sample of sulfuryl chloride is placed into a container with a movable piston, as shown below. A table of thermodynamic values is also provided.

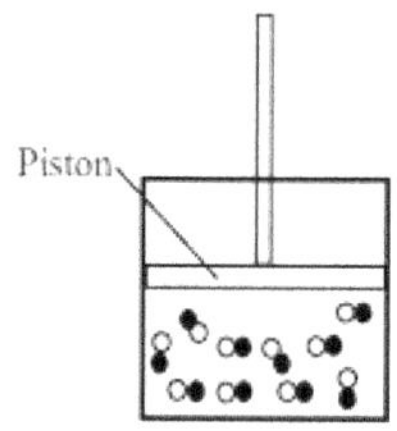

Substance	$\Delta H_f°$ (kJ/mol$_{rxn}$)
$SO_2Cl_2(g)$	–213
$SO_2(g)$	–297

44. The sample is allowed to come to equilibrium. Which is the correct diagram of the piston, along with the correct statement concerning the work done as the reaction comes to equilibrium?

(A) 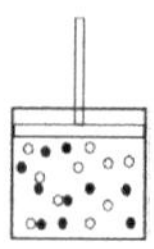 As work was done on the surroundings by the system, the sign for work will be positive.

(B) 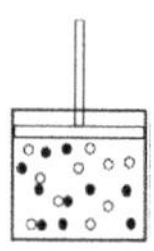 As work was done on the surroundings by the system, the sign for work will be negative.

(C) 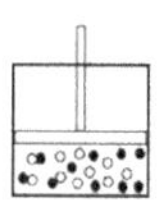 As work was done on the system by the surroundings, the sign for work will be positive.

(D) 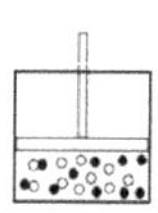 As work was done on the system by the surroundings, the sign for work will be negative.

45. What is the $\Delta H°_{rxn}$ for the decomposition of the sulfuryl chloride?
   (A) –510 kJ/mol$_{rxn}$
   (B) –84 kJ/mol$_{rxn}$
   (C) 84 kJ/mol$_{rxn}$
   (D) 510 kJ/mol$_{rxn}$

46. Which statement is correct concerning entropy change in the sulfuryl chloride decomposition?
   (A) The sign of the change of entropy is positive because 1 mole of gas decomposed into 2 moles of gas.
   (B) The sign of the change of entropy is negative because 1 mole of gas decomposed into 2 moles of gas.
   (C) The sign of the change of entropy is positive because the two single bonds between the S and O atoms in $SO_2Cl_2(g)$ became two double bonds between the S and O in $SO_2(g)$.
   (D) The sign of the change of entropy is negative because the two single bonds between the S and O atoms in $SO_2Cl_2(g)$ became two double bonds between the S and O in $SO_2(g)$.

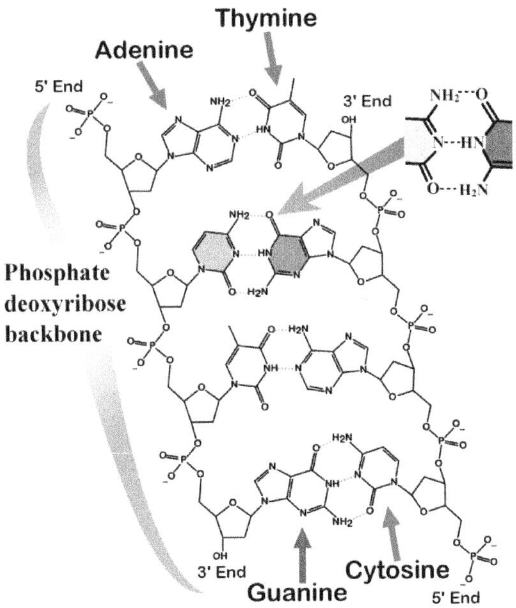

47. The double helix structure of deoxyribonucleic acid (DNA) is shown above, with an enlarged section for clarity. When DNA replicates, the two strands must "unzip." Which statement describes the energy requirements and forces involved?
   (A) Energy must be added to break the covalent bonds.
   (B) Energy must be added to break the hydrogen bonds.
   (C) Energy is released when the covalent bonds break.
   (D) Energy is released when the hydrogen bonds break.

48. A student tests the pH of an aqueous solution of a cleaning product known as "washing soda." This product is actually sodium carbonate (Na_2CO_3). The pH of the solution is found to be 8.79. Which net ionic equation correctly shows why this salt solution does not have a neutral pH?
   (A) $Na_2CO_3(s) \rightarrow$ $2\ Na^+(aq) + CO_3^{2-}(aq)$
   (B) $Na(s) + 2\ H_2O(l) \rightarrow$ $Na^+(aq) + OH^-(aq) + H_2(g)$
   (C) $CO_3^{2-}(aq) + H_2O(l) \rightarrow$ $HCO_3^-(aq) + OH^-(aq)$
   (D) $CO_3^{2-}(aq) + 2\ H^+(aq) \rightarrow$ $H_2O(l) + CO_2(g)$

49. Which statement best describes any changes in atomic radii as the atomic number increases from lithium to fluorine across the periodic table?
   (A) The radii do not change because the electrons are being added to the same energy level.
   (B) The radii increase due to greater electron–electron repulsion.
   (C) The radii decrease because the greater nuclear charge exerts a larger attraction on the electrons.
   (D) The radii decrease because as additional electrons are added, the orbitals become more stable due to electrons having opposite spins.

50. Assume that a reaction has an overall equation of $2\ A + B \rightarrow D$. The rate determining step in the mechanism is found to be: $2\ A \rightarrow C$. If 3.0 moles of A and 2.0 moles of B are placed in an evacuated 1.0-L flask, after five minutes the concentration of D reaches 1.0 *M* and the rate will have
   (A) decreased by a factor of 2.
   (B) decreased by a factor of 4.
   (C) decreased by a factor of 8.
   (D) decreased by a factor of 9.

GO ON TO NEXT PAGE

51. A student was asked to investigate different types of chemical reactions. One of the experiments was to look at the decomposition of copper(II) sulfate pentahydrate. The equation and the experimental setup are shown below. An excerpt of the student's laboratory notebook is shown below.

- When heat was applied, the blue crystal made a popping noise and a gas was observed coming out of the top of the test tube. Droplets of liquid were seen on the inside of the test tube.
- After the test tube was cooled to room temperature, 5 drops of water were poured into the test tube. The off-white anhydrous crystals made a hissing sound, the crystals changed color to a deep blue color, and the bottom of the test tube became extremely hot.

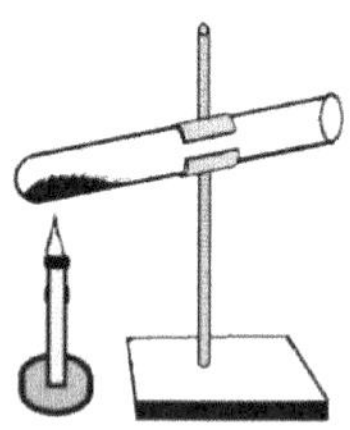

$CuSO_4 \cdot 5\,H_2O(s) \rightarrow CuSO_4(s) + 5\,H_2O(g)$

Which of the following diagrams best represents the energy change occurring in the reaction described in the equation above?

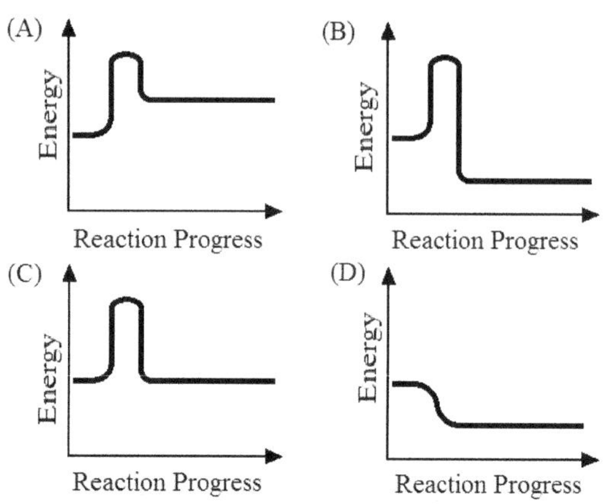

52. Which statement correctly gives the major reason that an increase in temperature causes an increase in reaction rate?
(A) The activation energy changes with temperature.
(B) The fraction of high energy molecules increases.
(C) The higher temperature changes the pathway that the reaction undergoes.
(D) Catalysts become more effective.

53. Which statement is true for zero-order reactions?
(A) The reaction rate is independent of time.
(B) The rate constant equals zero.
(C) The concentration of reactants does not change over time.
(D) The activation energy is very low.

54. Laboratory investigations find that a sample of an unknown substance is brittle with no conductivity as a solid and has a high melting point. Which of these substances is the unknown sample likely to be?
(A) RbF
(B) CCl_4
(C) Pd
(D) SiO_2

$BrO_3^-(aq) + 5\,Br^-(aq) + 6\,H^+(aq) \rightarrow 3\,H_2O(l) + 3\,Br_2(l)$

55. To determine the order with respect to Br^- in the reaction shown above, solutions should be prepared which differ in
(A) temperature.
(B) concentration of all reactants.
(C) concentration of Br^-.
(D) presence of a catalyst.

56. Identify the process that produces the greatest change in entropy per mole of substance.
(A) $H_2O(l) \rightarrow H_2O(g)$
(B) $I_2(s) \rightarrow I_2(aq)$
(C) $C_{10}H_8(s) \rightarrow C_{10}H_8(l)$
(D) C (*graphite*) $\rightarrow$ C (*diamond*)

57. Which of the following is in order of increasing boiling points?
(A) $RbCl < CH_3Cl < CH_3OH < CH_4$
(B) $CH_4 < CH_3Cl < CH_3OH < RbCl$
(C) $CH_4 < CH_3OH < CH_3Cl < RbCl$
(D) $RbCl < CH_3OH < CH_3Cl < CH_4$

58. A 100. g cube of steel is heated to 100.°C in a boiling water bath and is immediately dropped into a metal cup containing 100. g of water (specific heat capacity 4.18 J/g°C) at a temperature of 20.°C. The water plus steel is stirred and the temperature of the water increases to 25°C. The specific heat capacity of the steel sample is then calculated and compared with the accepted value of 0.500 J/g°C. Which statement is most correct?
(A) The calculated value is lower than the accepted value because the metal cup absorbed some of the heat, allowing energy to flow from the system to the surroundings.
(B) The calculated value is lower than the accepted value because the cube of steel was incorrectly massed and was actually heavier than 100. g.
(C) The calculated value is higher than the accepted value because the metal cup allowed heat energy to flow from the surroundings to the system.
(D) The calculated value is higher than the accepted value because heat from the cube is lost to the surrounding air.

Questions 59 and 60 use the following information.

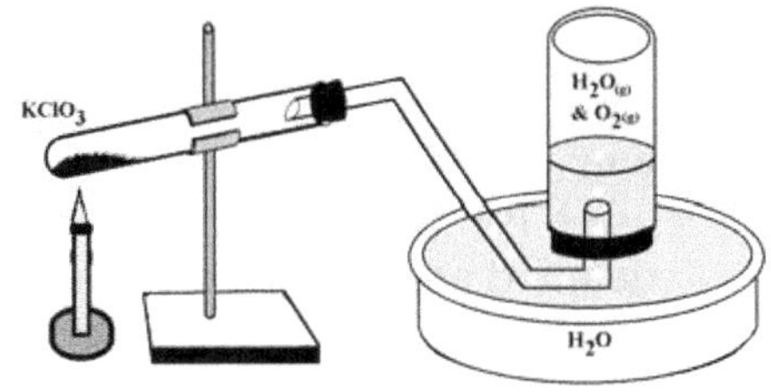

A solid mixture of potassium chlorate ($KClO_3$) and an inert substance is heated to produce potassium chloride and oxygen gas. The oxygen gas is collected by water displacement, using the experimental setup shown above.

59. Which equation best represents this reaction?
(A) $2\ KClO_3 \rightarrow 2\ KCl + 3\ O_2$
(B) $2\ K^+ + 2\ ClO_3^- \rightarrow 2\ KCl + 3\ O_2$
(C) $2\ ClO_3^- \rightarrow 2\ Cl^- + 3\ O_2$
(D) $2\ KClO_3 \rightarrow 2\ K^+ + 2\ Cl^- + 3\ O_2$

60. A student is asked to determine the percent of potassium chlorate in the original sample. Since the oxygen gas was collected by water displacement, the water vapor pressure is needed. Which measurement below is needed to determine the vapor pressure of water?
(A) mass of the original sample and test tube
(B) mass of the sample and test tube after heating
(C) volume of oxygen gas produced
(D) temperature of the water

GO ON TO NEXT PAGE

INTRODUCTION TO SECTION II: FREE-RESPONSE QUESTIONS

Section II of the AP® Chemistry examination counts for 50% of the total test grade and involves several parts. Answering these questions gives you an opportunity to demonstrate your ability to present your material in clear, orderly, and convincing language. Your answers will be graded on the basis of accuracy, the kinds of information you include to support your responses, and the importance of the descriptive material used. Be specific; general, all-encompassing answers will not be graded as well as detailed answers with examples and equations. CLEARLY SHOW THE METHOD USED AND THE STEPS INVOLVED IN ARRIVING AT YOUR ANSWERS. It is to your advantage to do this, since you may obtain partial credit if you do and you will receive little or no credit if you do not. Attention should be paid to significant figures. On the AP® exam, be sure to write all your answers to the questions on the lined pages following each question in the test booklet. Do not write your answers in the white space between questions.

Section II: Free-Response Questions
Time: 105 minutes
Number of Questions: 7

DIRECTIONS: You may use a calculator, the formula chart, and the periodic table located at the back of this book throughout this section.

1. $2\ SO_3(g) \rightleftarrows O_2(g) + 2\ SO_2(g)$ $\quad \Delta H° = +791.44\ kJ/\ mol_{rxn}$

   A 3.21 g sample of sulfur trioxide (SO_3, molar mass = 80.06 g) is placed in a 2.25-L cylinder and allowed to reach equilibrium at a constant temperature of 500. K, as shown in the above equation. Analysis shows 1.23×10^{-2} mol of sulfur dioxide at equilibrium.

   (a) Sketch the reaction profile (or progress) on the axes given.

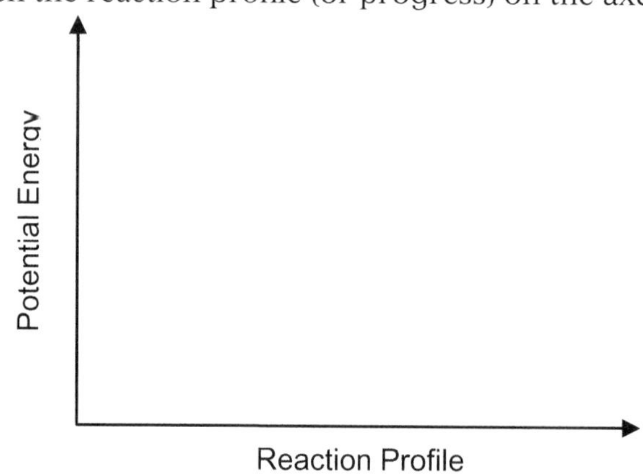

   (b) Write the equilibrium constant expression (K_c) for this system.
   (c) Calculate the initial molarity of the $SO_3(g)$.
   (d) Calculate the concentration of all three gases at equilibrium.
   (e) Calculate the K_c value for this system.

(f) In a separate experiment, all of the variables remained the same except the temperature was raised to 750. K.
  (i) Predict how the equilibrium concentrations of each of the reactants and products will change. Justify your answer.
  (ii) After reaching equilibrium at the new temperature of 750. K, the student decreases the volume of the container by compressing it. Which direction does the reaction shift? Justify your answer.
  (iii) Which of the two gases produced would behave most ideally at 750. K? Justify your answer in terms of the deviations from ideal gas behavior.

2. Pure liquid perbromic acid, $HBrO_4$, can react with pure liquid sulfuric acid, H_2SO_4, according to the equation shown below.

$$HBrO_4 + H_2SO_4 \rightleftarrows H_3SO_4^+ + BrO_4^-$$

(a) The K value for this reaction is <1.
  (i) Indicate one of the conjugate acid–base pairs in this reaction by drawing lines to connect the pair and label the acid and the conjugate base in the pair.
  (ii) Which is the stronger acid, $HBrO_4$ or $H_3SO_4^+$? Identify the stronger acid by circling it. Justify your choice.
  (iii) Draw a Lewis structure for the $H_3SO_4^+$ ion.
  (iv) Identify the hybridization of the sulfur atom in the structure of the $H_3SO_4^+$ ion.

(b) 100.0 mL of a 0.100 M solution of H_2SO_4 needs to be prepared for use in a laboratory investigation. The solution is to be prepared under a fume hood, by a person wearing goggles and gloves, from 3.00 M H_2SO_4. The available equipment is listed below (it is not necessary to use all of it). Write out a step-by-step procedure for preparation of the 0.100 M solution. Show any calculations needed.

  100-mL graduated cylinder
  10-mL graduated cylinder
  10-mL graduated pipet
  100-mL volumetric flask and stopper
  250-mL beaker
  Stirring rod

(c) 50.0 mL of 3.00 M H_2SO_4 at a temperature of 20.0°C is added to an insulated cup (calorimeter). 50.0 mL of 3.00 M NaOH, also at 20.0°C, is added to the same cup. The solution is stirred. The temperature of the solution is monitored for 5 minutes, and the highest temperature reached by the reaction is found to be 38.0 °C. Assume that the density of each solution is 1.0 g/mL, the specific heat of each solution is 4.18 J/g°C, and that the volume of the solutions is additive.
  (i) How many joules of heat were released when the two solutions were mixed?
  (ii) What was the heat of neutralization, in kJ/mol of water produced?
  (iii) Theoretically, the ΔH_{neut} for $H^+ + OH^- \rightarrow H_2O$ should be –55.8 kJ/mol. Give two reasons why your value differs from this.

GO ON TO NEXT PAGE

3. Analysis of an iron(II) ion solution by titration with a standardized potassium permanganate solution can be used to determine the percent iron in a sample. The purple color of the permanganate ion acts as an indicator, changing to a faint pink (or colorless) solution when manganese(II) ions are present. When performing this laboratory work, the sample of known mass containing iron(II) is dissolved in 25.0 mL of water.
   (a) Explain how and why each of the following affects the reported percentage of iron in the unknown solid:
      (i) The student fills the buret with the standardized $KMnO_4$ solution after rinsing with only deionized water.
      (ii) An air bubble appears in the buret tip before titration begins.
      (iii) More than 25.0 mL of water is used to dissolve the solid.
      (iv) Each time the volume of fluid is measured in the buret, measurement is made to the top of the meniscus, rather than the bottom of the meniscus.
   (b) Assuming that the accepted value for the percentage of iron in the original sample was 7.77% and that your experiment gave a result of 6.896%, determine the percent error in your work.
   (c) Write a balanced net ionic equation for the reaction that occurs between iron(II) ions and the permanganate ion in acidic solution.
   (d) Identify the oxidation number change of the manganese atom in the reaction.
   (e) In a separate experiment with a potassium permanganate solution of unknown concentration, the objective was to use a spectrophotometer to determine the concentration of the unknown sample. A standard curve was to be prepared with solutions of known concentration ranging from 1.00×10^{-4} *M* to 3.00×10^{-4} *M*. Given a standardized 3.00×10^{-4} *M* solution of potassium permanganate, describe how you would prepare 10 mL each of three other concentrations to obtain four total samples in the range of approximately 1.00×10^{-4} *M* to 3.00×10^{-4} *M*.

4. The melting points of carboxylic acids found in milk are shown in the graph below. In this case, "saturated" means that all carbon–carbon bonds in the molecule are single bonds.

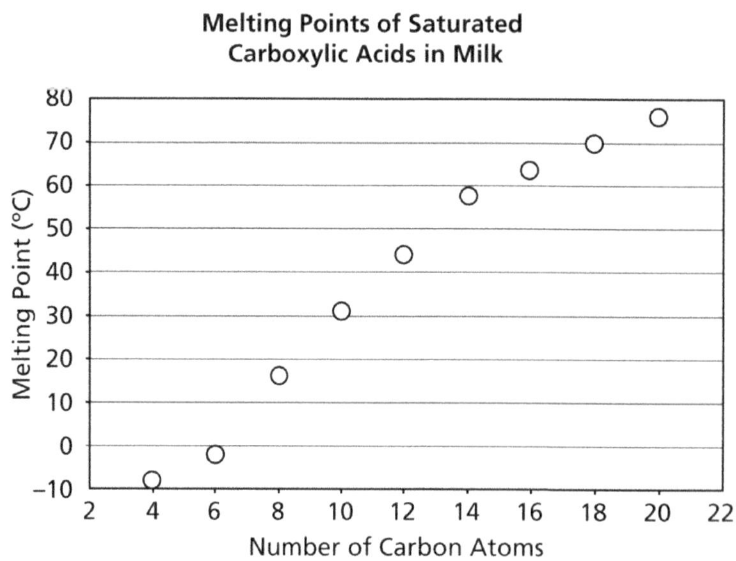

(a) Identify and explain the relationship between the number of carbon atoms and the melting point of the acid in terms of the intermolecular forces present.
(b) Explain why these carboxylic acids are soluble in the water which makes up 87% of milk.

There are several unsaturated carboxylic acids present in milk. "Unsaturated" means that at least one carbon–carbon multiple bond is present. Data for three C_{18} carboxylic acids are listed below.

Acid Name	Number of Double Bonds	Melting Point (°C)	
Oleic	1	13	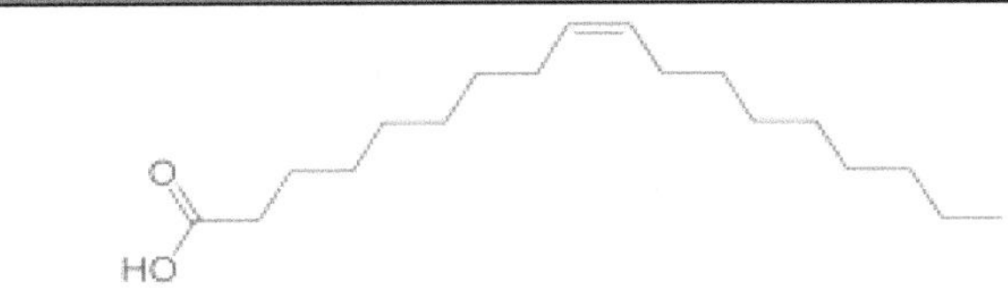
Linoleic	2	–5	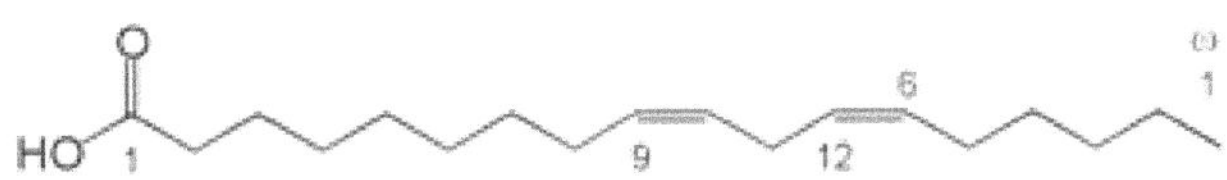
Linolenic	3	–11	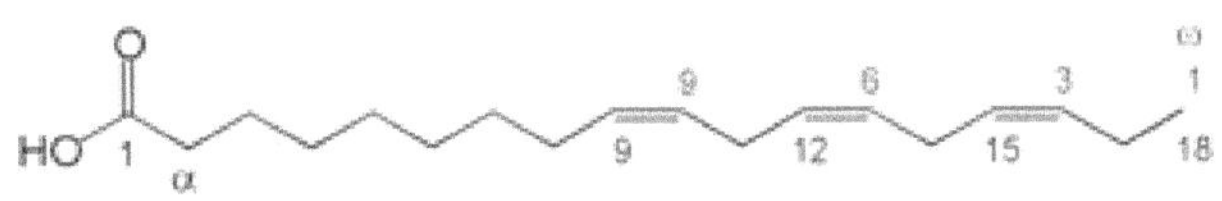

(c) Identify and explain the relationship between the number of double bonds and the melting point of the acid in terms of the intermolecular forces and molecular shape.

(d) Determine the bond angle of the O-C-O bond in the carboxyl functional group.

5. A laboratory investigation of the kinetics of the following reaction was conducted.

$$2\,A + B \rightarrow C + 2\,D$$

In the process of creating the laboratory report, the following graphs were created.

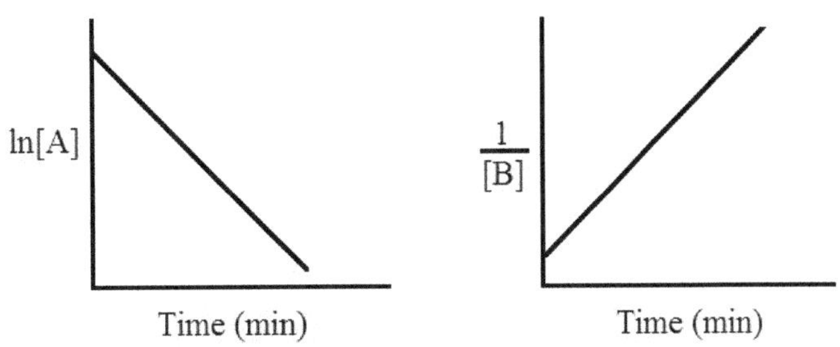

(a) What is the order for A? Justify your answer.
(b) What is the order for B? Justify your answer.

When [A] = 0.24 *M* and [B] = 0.16 *M*, the initial rate of the reaction was measured as 0.709 *M* min^{-1}.
(c) What is the rate constant? Specify its units.

GO ON TO NEXT PAGE

(d) The following mechanism was proposed for this reaction.

$A + B \rightleftarrows E$	Fast Equilibrium
$E + B \rightarrow C + F$	Slow
$F \rightarrow D$	Fast

Is this mechanism consistent with the results of this laboratory investigation? Justify your answer.

6. Diborane, B_2H_6, is a gas used in the doping of semiconductors and in the production of pure boron which is used in the manufacturing of semiconductors.

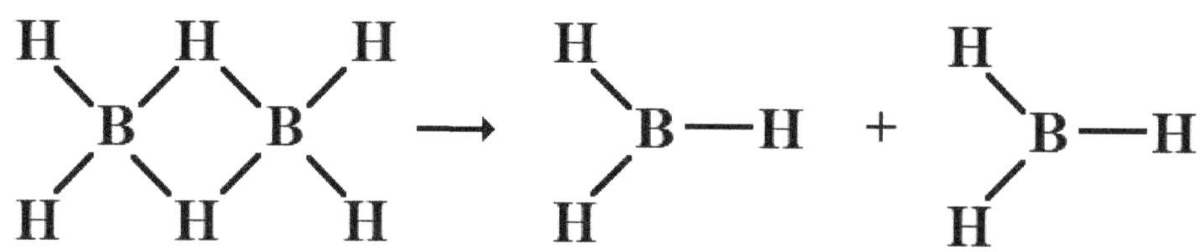

Diborane can decompose to borane gas, as shown above.

(a) Predict the sign of ΔH_{rxn}. Justify your answer.

(b) Predict the sign of ΔS_{rxn}. Justify your answer.

(c) Is this reaction expected to be thermodynamically favored at low temperature, high temperature, or at all temperatures? Justify your answer.

The valence shell electron pair repulsion (VSEPR) theory and the valence bond (VB) theory, like all models, have limitations.

(d) Based on the reaction above, state one area where the VSEPR and/or VB model does not adequately describe the bonding in diborane.

7. Effervescent cold relief tablets are composed of 1.000 g of citric acid ($H_3C_6H_5O_7$) and 1.916 g of sodium hydrogen carbonate ($NaHCO_3$). When placed into water, the two solids dissolve and react completely with each other.

(a) Write the balanced net ionic equation for this reaction assuming that all acidic hydrogens react.

(b) Determine the theoretical volume of gas produced at STP.

In a calorimetry experiment, two cold relief tablets placed into 75.0 g of water at 21.2°C lower the temperature of the water to 13.6°C. Assume no heat exchange with the surroundings.

(c) Calculate the energy absorbed by the water (specific heat capacity of $H_2O = 4.18\ J\ g^{-1}\ °C^{-1}$).

(d) Calculate $\Delta H°$ for the balanced chemical equation.

END OF EXAMINATION

SECTION I: MULTIPLE-CHOICE QUESTIONS

Score your test using the table below.

Determine how many questions you answered correctly. You will find explanations of the answers on the following pages.

1. D	13. B	25. A	37. B	49. C
2. A	14. B	26. B	38. B	50. D
3. B	15. C	27. D	39. D	51. A
4. D	16. C	28. C	40. D	52. B
5. B	17. A	29. C	41. A	53. A
6. B	18. C	30. B	42. C	54. A
7. D	19. D	31. C	43. D	55. C
8. C	20. B	32. B	44. B	56. A
9. A	21. A	33. B	45. B	57. B
10. A	22. A	34. C	46. A	58. A
11. D	23. C	35. C	47. B	59. A
12. A	24. A	36. C	48. C	60. D

CALCULATE YOUR SCORE:

Number answered correctly: ______

WHAT YOUR SCORE MEANS:

Each year, since the test is different, the scoring is a little different. But generally, if you scored 23 or more on the multiple-choice questions, you'll most likely get a 3 or better on the test. If you scored 33 or more, you'll probably score a 4 or better. And if you scored 42 or more, you'll most likely get a 5. Keep in mind that the multiple-choice section is worth 50% of your final grade, and the free-response section is worth 50% of your final grade. To learn more about the scoring for the free-response questions, turn to the last page of this section.

To help you narrow your review, you can use the following grid to see how well you understand each of the big ideas of chemistry. *Note:* On this Diagnostic Test, the Big Ideas were evenly distributed (10 questions each) to better aid you in assessing your weak areas. On the Practice Tests at the end of the book, the Big Idea distribution will mirror that of the most recent released AP® exams.

Big Idea	Put an "X" through the number if you got the question *correct*	Number correct for this big idea
1	4, 9, 17, 20, 21, 22, 36, 37, 49, 60	
2	1, 15, 25, 26, 29, 34, 40, 42, 54, 57	
3	2, 3, 7, 8, 10, 11, 30, 43, 51, 59	
4	5, 6, 28, 31, 33, 41, 50, 52, 53, 55	
5	14, 18, 27, 35, 44, 45, 46, 47, 56, 58	
6	12, 13, 16, 19, 23, 24, 32, 38, 39, 48	

ANSWERS AND EXPLANATIONS

SECTION I: MULTIPLE-CHOICE QUESTIONS

1. ANSWER: D Filtration can be used to separate a heterogeneous mixture, while the other techniques can be used to separate homogeneous mixtures (LO 2.10).

2. ANSWER: A The positive slope at the left part of the graph indicates a direct relationship between the mass of calcium chloride and the mass of calcium carbonate. This means that $CaCl_2$ is the limiting reagent and is determining the mass of $CaCO_3$ produced. At 0.5 g of $CaCl_2$, all of the calcium ions are forming calcium carbonate so the $[Ca^{2+}]$ will be lowest in concentration (LO 3.4).

3. ANSWER: B In the horizontal portion of the graph, the mass of $CaCl_2$ is not related to the mass of $CaCO_3$ produced. Therefore, the constant amount of Na_2CO_3 (6.00 mL of 1.7 *M*) solution must be responsible for the horizontal portion of the graph and must be the limiting reagent (LO 3.4).

4. ANSWER: D Since the spectrum has three peaks of equal sizes, the element must have 2 electrons in its first three sublevels and will have a configuration of $1s^2 2s^2 2p^2$. The element is carbon (LO 1.7).

5. ANSWER: B Elementary reactions occur in a single step, and their rate law agrees with their molecularity (coefficients). Since there are 3 molecules as reactants in this reaction, a ternary (3-way) collision must be occurring. The chance of 3 particles colliding all with sufficient energy and proper molecular orientation is extremely low; therefore, the rate of reaction would be extremely low. Writing the rate law from the coefficients results in Rate = $k[Q][R]^2$ (LO 4.4).

6. ANSWER: B The initial concentration is 20.0 μg, and the final concentration is 2.0 μg. To reach 2 μg, the original 20 μg must be cut in half more than 3 times but less than 4 times. The time must be between 24 and 32 days. (20→10→5→2.5→1.25) To obtain a

more exact answer, you would need a calculator and would use $t_{1/2} = 0.693/k$ along with the first-order integrated rate law formula (LO 4.3).

7. ANSWER: D Adding additional $AgNO_3$ solution of the same concentration will not affect the molarity in the cell so the cell potential will not change. If the additional Ag^+ had been in the solid form, the concentration would have increased, causing an increase in cell potential (LO 3.12).

8. ANSWER: C By definition, the anode is the site of oxidation. Oxidation is the loss of electrons. Copper must lose 2 electrons to become copper(II) ions. Half-reactions involving a decrease in positive charge, like responses (A) and (D), require a gain of electrons (LO 3.13).

9. ANSWER: A Light of the same color as the solution will have very little, if any, absorbance. Light of a complementary color will result in the highest absorbance. Orange is a complementary color to blue (LO 1.16).

10. ANSWER: A An empirical formula is the ratio by moles of each element in a formula. 0.010 mole of copper reacted. The final mass of 17.960 – 17.640 = 0.320 mole of S (not moles of S_8, but when working with empirical formulas, we use the elemental mass). This is approximately 0.01 mole of sulfur. Cu : S = 0.010 : 0.10 or 1 : 1. The empirical formula is CuS (LO 3.6).

11. ANSWER: D Since excess sulfur was used, any extra that spattered out must not have affected results. There must have been sufficient sulfur to react completely with the copper. Excess sulfur is meant to vaporize during the experiment (LO 3.6).

12. ANSWER: A The equilibrium constant of 100 indicates that the reaction goes far to the right and results in mostly products. Since there are 2 moles of gaseous products and 3 moles of gaseous reactants, the equilibrium pressure will be lower than the initial pressure. The quantities of reactants are stoichiometric. Initial moles of gas before the reaction will be 0.30 mol. If the reaction went all the way to completion (with a K of 100 it will be close to completion), the number of moles of gas will be 0.20 (LO 6.7).

13. ANSWER: B Note that the second equation is the reverse of the first, and the coefficients have been halved. Therefore, the K_c for the second equation will be the reciprocal of the square root of the first: $1/\sqrt{100} = 0.1$ (LO 6.2).

14. ANSWER: B Energy is always required to break bonds, and energy is always released when bonds are formed. Since the ΔH value is negative, this is an exothermic reaction. More energy is released when new bonds are formed than was added to break apart the reactants (LO 5.8).

15. ANSWER: C Both CaO and MgO are ionically bonded. Both involve oxygen; therefore, the difference in melting points must be due to the difference in coulombic attraction of Mg for O, compared to that of Ca for O. Both are 2+ ions, but magnesium ions are smaller. With equal charge, smaller ions will be closer together and have a greater attraction and thus a higher melting point (LO 2.24).

16. ANSWER: C Equilibrium tends to shift to the side with the least number of moles of gas to relieve pressure. When the number of gaseous moles is equal, there will be no change to the system. The system remains at equilibrium with equal rates of the forward and reverse reactions (LO 6.3).

17. ANSWER: A The reaction forms a $BaSO_4(s)$ precipitate which means that ions are removed from solution, thereby making the solution less conductive. However, continued addition of the acid causes the conductivity to increase again with the excess of ions (LO 1.20).

18. ANSWER: C To calculate $\Delta G°$ for
$N_2(g) + 2\ O_2(g) \rightarrow 2\ NO_2(g)$, you must add the two equations together.

$N_2(g) + O_2(g) \rightarrow \cancel{2\ NO}(g)$	$\Delta G° = 174$ kJ/mol
$\cancel{2\ NO}(g) + O_2(g) \rightarrow 2\ NO_2(g)$	$\Delta G° = -70$ kJ/mol
$N_2(g) + 2\ O_2(g) \rightarrow 2\ NO_2(g)$	$\Delta G° = 104$ kJ/mol

Positive values for $\Delta G°$ indicate thermodynamically unfavored reactions (LO 5.17).

19. ANSWER: D
Since the K_a given is for HX, the K_b for X^- would be
$1 \times 10^{-14}/1 \times 10^{-9} = 1 \times 10^{-5}$

Using the general equation $B + H_2O \rightleftarrows HB^+ + OH^-$,
$K_b = ([HB^+][OH^-]) / [B]$

For NaX, this would be $X^- + H_2O \rightleftarrows HX + OH^-$,
$K_b = ([HX][OH^-]) / [X^-]$

Since 0.050 mol of X^-/0.500 L = 0.10 M X^-
$1.0 \times 10^{-5} = x^2/0.10$
$x^2 = 1.0 \times 10^{-6}$
$[OH^-] = 1.0 \times 10^{-3}\ M$ pOH = 3.00
pH = 14.00 – 3.00 = 11.00
(LO 6.16)

20. ANSWER: B Using the periodic table, the molar mass of nonradioactive strontium is 87.62 g/mol. This is less than the mass of strontium-90, which has additional neutrons (LO 1.13).

21. ANSWER: A Chemical reactivity is very similar within a group in the periodic table because all group members have the same number of valence electrons. Strontium and calcium are in the same group and would form the same compounds. Radioactive strontium is known to replace calcium in the bones and in other

compounds. This results in long-term exposure to the radiation produced by the strontium (LO 1.10).

22. ANSWER: A Electrons being excited into higher energy levels contradicts the Dalton model of the atom which existed before electrons were discovered (LO 1.13).

23. ANSWER: C A good clue as to the type of titration curve is the position of the equivalence point, which is the center of the vertical section of the graph showing the very rapid change of pH. In this case, that is in the basic region (above 9). At this point, the amount of added OH^- equals the original amount of acid (H^+ ions) initially present (LO 6.13).

24. ANSWER: A The leveling-off shown between F and G is caused by buffering. Optimal buffering occurs when [HA] = [A^-], which would be at a volume of 25 mL of base in this case (LO 6.20).

25. ANSWER: A

	Name	Formula	IMF Present
1	propane	$CH_3CH_2CH_3$	London dispersion
2	ethanol	CH_3CH_2OH	London dispersion, dipole–dipole, hydrogen bonding
3	dimethyl ether	CH_3OCH_3	London dispersion, dipole–dipole

These molecules all have similar molar masses (similar number of electrons) and would thus have similar London dispersion forces. The more easily a substance can attain the vapor state, the more molecules will be available to exert vapor pressure. Propane molecules only have to overcome relatively weak London dispersion forces to go into the gas phase. The other molecules have stronger IMF to overcome (LO 2.16).

26. ANSWER: B Using standard bonding models, OCl_2 would be expected to have a bond angle of < 109.5° because there are four regions of electron density and the two unshared pair of electrons on the oxygen are expected to take up more space. This anomaly of a larger than expected bond angle must be due to the large size and number of electrons in chlorine. The chlorine atoms repel each other forcing the bond angle to be greater than 109.5°. This is not an exception that you would be expected to know, but when given data that shows an exception, you should be able to choose a justification for that exception (LO 2.21).

27. ANSWER: D At points A and B, the atoms are too far apart for significant attraction and are at higher energy. Point C represents a balance between attractions (nucleus for electrons) and repulsions (nucleus for nucleus and electrons for electrons), giving the lowest energy and stable distance. Point D shows the atoms

too close together, and the repulsions between the nuclei and the electrons would result in higher energy (LO 5.1).

28. ANSWER: C Mechanism 2 involves palladium (Pd), a solid-state catalyst. The presence of this type of catalyst involves the adsorption of one of the reactants on its surface, assuring the correct molecular orientation of the collisions, lowering the activation energy, and increasing the reaction rate (LO 4.9).

29. ANSWER: C The VSEPR model is very helpful in describing almost all molecular shapes. PH_3 has 8 total valence electrons and the Lewis structure should have 3 single bonds from phosphorus to hydrogen, along with a lone pair of electrons on P. The electron geometry is said to be tetrahedral as there are four regions of electron density. Because one region involves a lone pair, the actual molecular geometry is trigonal pyramidal (the lone pair repels even more than a shared pair), producing bond angles of less than 109.5° (LO 2.21).

30. ANSWER: B Ionic bonds between Na^+ and Cl^- are broken when NaCl dissolves. Hydrogen bonds between some water molecules are broken and the ions forms ion–dipole bonds with water. The ion–dipole bonds are comparable in strength to covalent bonds (LO 3.10).

31. ANSWER: C For a zero-order reaction, a graph of concentration vs. time is linear. For a first-order reaction, a graph of ln (or log) of concentration vs. time is linear. For a second-order reaction, a graph of the reciprocal of concentration vs. time is linear (LO 4.2).

32. ANSWER: B

$$K = \frac{[C]^2}{[A]\,[B]} \qquad K = \frac{(2.0)^2}{(4.0 \times 5.0)} = 0.20$$

(LO 6.5)

33. ANSWER: B The general form of the rate law will be Rate = $k[A]^x[B]^y[C]^z$. Looking at the data, comparing experiment 3 to experiment 2, [B] and [C] were held constant, while [A] doubled from experiment 3 to experiment 2. This caused a doubling of the rate from 24 to 48. Since both concentration and rate double, A is first order. Comparing experiment 2 to experiment 4, [B] is four times greater, while [A] and [C] are held constant. This caused the rate to go from 48 to 192 (four times greater). Since the concentration and rate were changed by the same factor, [B] is also first order. Comparing experiment 1 to experiment 3, [C] doubled while the rate went from 12 to 24 (doubling). This shows that C is also first order. The rate law is Rate = k[A][B][C]. The sum of the exponents is the overall order so that equals 3 (LO 4.2).

34. ANSWER: C While using actual electronegativities would be helpful for this question, a good generalization is that the further apart the elements are on the periodic table the greater the difference in electronegativity and the more polar the bond. A metal and a nonmetal (Cr and Br) will usually have a large difference in electronegativity. Note also that this is a question

dealing with the polarity of an individual bond and not with the polarity of an entire molecule (LO 2.18).

35. ANSWER: C The conditions are equal, which indicates that both samples have the same average kinetic energy. When molecules have a higher speed due either to being at a higher temperature or having a lower molar mass, their average velocity will be higher and there will be a greater variety of velocities (graph will be more spread out). In this situation, the molar mass of O_2 is much less than that of Cl_2. Having the same kinetic energy (because they are at the same temperature), O_2 will have a greater average velocity along with a wider variety of velocities (LO 5.2).

36. ANSWER: C If both percent abundances were 50%, the average mass would be 48.5. Since the ^{50}X is 90%, the average mass must be very close to but less than 50. Calculations aren't needed here, but this formula can be used. (0.10)(48.0) + (0.90)(50) = 49.8 amu (LO 1.14).

37. ANSWER: B Mass spectrometry is used to determine the isotopic composition of elements and compounds by ionizing them and separating them by their mass to charge ratio (LO 1.15).

38. ANSWER: B Since HBr is a strong acid (completely dissociated), the $[H^+]$ would be 0.001 *M* which gives a pH = –log 0.001 = 3.0 (LO 6.15).

39. ANSWER: D From the equation $Zn(OH)_2(s) \rightleftarrows Zn^{2+}(aq) + 2\ OH^-(aq)$, you can see that if 2.0×10^{-6} mol/L of the zinc hydroxide dissolves, that will result in the same concentration of zinc ions (2.0×10^{-6} mol/L) and twice that much of hydroxide ions (2.0×10^{-6} mol/L $\times$ 2 = 4.0×10^{-6} mol/L). Then: $K_{sp} = [Zn^{2+}][OH^-]^2 = (2.0 \times 10^{-6})(4.0 \times 10^{-6})^2 = 3.2 \times 10^{-17}$ (LO 6.22).

40. ANSWER: D If gas is allowed to escape, then there are fewer molecules left within the cylinder; these fewer molecules exert less pressure (gas pressure is due to the collision of gas particles with the sides of the container). These remaining gas molecules occupy the entire container, so there must be more room between the individual molecules (LO 2.6).

41. ANSWER: A This is an application of the half-life expression for first-order rate laws, $t_{1/2} = 0.693/k$. Note that half-life (time) and the rate constant are inversely related; hence, the element with the lower rate constant will have the longer half-life (LO 4.3).

42. ANSWER: C While this may appear to be an equilibrium question, it is actually just asking you to show an understanding of the stoichiometry between ammonia and hydrogen gases as this reaction proceeds. A reaction chart is helpful. Note that the change in total pressure is +0.40 atm and this is also $+2x$. This means that $x = 0.20$ atm. Hydrogen has gone from a pressure of 0 to $3x$ or 0.60 atm.

Reaction	$2\ NH_3(g)$	$N_2(g)$	$3\ H_2(g)$	P_{total}
Initial Pressure	0.80 atm	0	0	0.80 atm
Change	$-2x$	$+x$	$+3x$	+ 0.40 atm and $+2x$ so $x = 0.20$ atm
Equilibrium	$0.80 - 2x$ 0.40 atm	x 0.20 atm	$3x$ 0.60 atm	1.20 atm

(LO 2.6)

43. ANSWER: D To verify the law of conservation of mass, you will need to mass both the reactants and the products directly. The problem states that the student already has the mass of the hydrated compound and the anhydrous compound, so the only substance that is still needed is the mass of the water vapor (LO 3.5).

44. ANSWER: B The reaction has 1 mole of gaseous reactant producing 2 moles of gaseous products. As the reaction progresses, pressure will tend to increase, but since the piston is moveable, the volume of gas will expand to maintain a constant pressure. When the gas expands, it does work on its surroundings. Since the system is doing the work, the work has a negative sign (LO 5.6).

45. ANSWER: B $\Delta H^\circ_{rxn} = \Sigma\ \Delta H_f^\circ{}_{products} - \Sigma\ \Delta H_f^\circ{}_{reactants}$

$$\Delta H^\circ_{rxn} = [(-297) + 0] - [-213] = -84\ kJ/mol_{rxn}$$

(LO 5.13)

46. ANSWER: A Since 1 mole of gas reacts to form 2 moles of gas, entropy increases. An increase in entropy means that it will have a + sign. More particles = more freedom of movement (LO 5.12).

47. ANSWER: B Hydrogen bonds exist between the nitrogenous bases in DNA. This relatively strong IMF must absorb energy to overcome the attraction between the atoms of the different strands of DNA (LO 5.11).

48. ANSWER: C Sodium carbonate is a basic salt. When dissolved, the carbonate ion reacts with water to produce a hydroxide ion and the bicarbonate ion (LO 6.16).

49. ANSWER: C The atomic radii decreases due to an increase in nuclear charge; note that the "added" electrons are going into the same principal energy level as you go from left to right across the same period (LO 1.9).

50. ANSWER: D Since the rate determining step is 2A → C, the reaction is second order in A. Although this is not an equilibrium reaction, a similar table can be set up to show the stoichiometry.

Reaction	2 A	B	D
Before	3.0 mol	2.0 mol	0
Change	$-2x$	$-x$	$+x$
After	$3.0 - 2x$	$2 - x$	$x = 1.0$ mol

Stoichiometry should always be done in moles. Since the volume is 1 L, the moles and molarity are the same. This shows that $x = 1.0$ mol. From the chart, A and B will also be 1.0 mol/L or 1.0 *M*. The [A] has dropped from 3.0 mol to 1.0 mol in 5 minutes. Its concentration was cut to 1/3 of its original concentration. Since it is second order, we must square this change $(1/3)^2 = 1/9$. The rate decreases by a factor of 9 (LO 4.2).

51. ANSWER: A The forward reaction is endothermic (as you have to add heat to remove the water of hydration) and the reverse reaction is exothermic, since adding water to the anhydrous produced heat. Graph A is an endothermic graph (LO 3.11).

52. ANSWER: B At a higher temperature, particles have greater kinetic energy. They collide more frequently and have more force so they are more likely to have the activation energy (LO 4.6).

53. ANSWER: A For a zero-order rate law, Rate = $k\,[A]^0$ and Rate = k (the rate constant). This means that the rate does not speed up or slow down over time (like most reactions do); it either takes place at a constant rate or does not take place at all (LO 4.2).

54. ANSWER: A Compounds with these properties are ionically bonded. This suggests elements from the far left side of the periodic table (Group 1 or 2) with elements from the far right side (Group 16 or 17) (LO 2.22).

55. ANSWER: C If you wish to determine how order is affected by a given concentration, then a series of reactions should be run with changes in only the concentration of that substance, with all other concentrations held constant (as well as all other conditions like temperature held constant) (LO 4.1).

56. ANSWER: A Gases have the most random organization and have the most microstates available. Gas formation is a driving force in chemical and physical processes due to the increase in entropy (LO 5.12).

57. ANSWER: B

Formula	Interparticle Forces Present
CH_4	London dispersion
CH_3Cl	London dispersion, dipole–dipole
CH_3OH	London dispersion, dipole–dipole, hydrogen bonding
RbCl	Ion–ion

The temperature at which substances boil is a function of the forces between the particles of that substance (interparticle forces). Small, nonpolar molecules boil at the lowest temperatures since they have the weakest IMF to overcome (London dispersion). As polarity increases, so does the boiling temperature. Ionic substances have the highest attraction, which means the boiling point is the highest (LO 2.1).

58. ANSWER: A $q = mc\Delta T$ With specific heat calorimetry, the heat lost by the metal is theoretically equal to the heat gained by the water, but often heat may be lost to some outside source (the surroundings). In this problem, q for water = (100 g)(4.18 J/g°C)(5°C) = a little over 2000 J. The q for steel = (100 g)(0.500 J/g°C)(75°C) = 3750 J. Since the metal lost more heat than the water gained, the heat must have been lost to something. Usually, an insulated cup is used to hold the water (as the calorimeter), but in this case a metal cup was used. Metal conducts rather than insulates. The heat gained by the water would quickly be used to warm up the metal cup rather than warming up the water, resulting in a lower measurement and lower calculation of the specific heat of the steel (LO 5.7).

59. ANSWER: A The original compounds are solids and are not in water, so they cannot be ionized. The products produced are still both a solid and a gas, so neither is dissolved in water and therefore cannot be ionized (LO 3.2).

60. ANSWER: D The vapor pressure of water is determined by the temperature of the water: the higher the temperature, the greater the vapor pressure since more particles will have the required energy to leave the liquid phase (LO 1.3).

SECTION II: FREE-RESPONSE QUESTIONS

Question 1: Answers

(a)

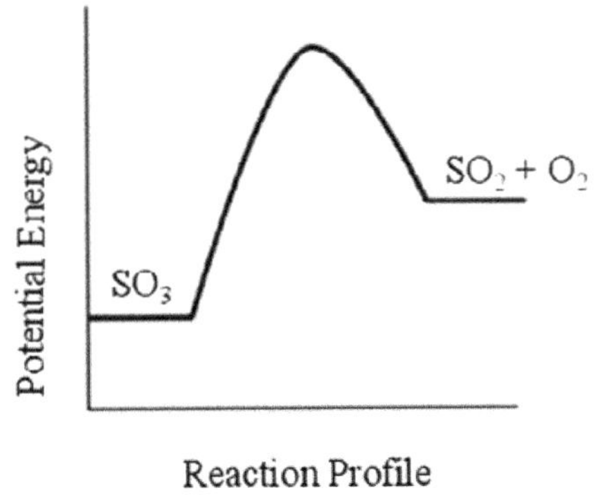

(LO 3.11)

(b) $K_C = \frac{[O_2][SO_2]^2}{[SO_3]^2}$

(LO 6.5)

(c) 3.21 g SO_3/80.06 = 0.0401 mol/2.25 L = 0.0178 *M*

	2 SO_3	O_2	2 SO_2
I	0.0178 *M*	0	0
C	$-2x$	$+x$	$+2x$
E	$0.0178 - 2x$ 0.0178 – 0.00547 = **0.0123 *M***	x x = (0.00547/2) = **0.00274 *M***	$2x$ 1.23×10^{-2} mol/2.25 L = **0.00547 *M***

(LO 1.4)

(d) $[SO_3]$ = 0.0124 *M* $[O_2]$ = 0.00274 *M* $[SO_2]$ = 0.00547 *M*
(LO 6.6)

(e) $K_C = (0.00274)(0.00547)^2/(0.0123)^2 = 5.42 \times 10^{-4}$
(LO 6.5)

(f) (i) $[SO_3]$ will decrease and $[SO_2]$ and $[O_2]$ will increase.
Since the reaction is endothermic, a higher temperature will force the reaction to products to relieve the stress of the increased heat. The higher temperature means that more particles will have sufficient energy to overcome the activation energy and produce products. Initially, the rate of the forward reaction will increase and the reverse reaction will decrease. As equilibrium is reestablished, the rates will again be equal (LO 6.10).

(ii) Decreasing the volume of the container (increases the pressure) causes the reaction to favor (shift towards) the side with fewer moles of gas to reestablish the equilibrium ratio. In this reaction, there are 2 moles of gaseous reactants and 3 moles of gaseous products; therefore, an increase in pressure favors the reactants. $[SO_3]$ will increase and $[SO_2]$ and $[O_2]$ will decrease (LO 6.8).

(iii) O_2 will behave more ideally than SO_2. O_2 has a smaller volume than SO_2, so the volume of O_2 is more negligible and won't substantially affect the volume of the container. O_2 is also a nonpolar molecule, having only London dispersion forces, while SO_2 has a bent shape and is polar with both London dispersion forces and dipole–dipole attractions. SO_2 molecules will attract each other to a greater extent, decreasing the number of collisions that they have (LO 2.4).

Question 2: Answers

(a) (i) The pairs are $HBrO_4$ (acid)/BrO_4^- (conjugate base) and H_2SO_4 (base)/$H_3SO_4^+$ (conjugate acid). Note that the acid form always has one more H^+ than its conjugate base (LO 3.7).

(ii) $H_3SO_4^+$ is the stronger acid. The equilibrium constant is less than 1 which means that the reverse reaction is favored, making $H_3SO_4^+$ stronger (greater proton donor) than $HBrO_4$ (LO 6.7).

(iii)

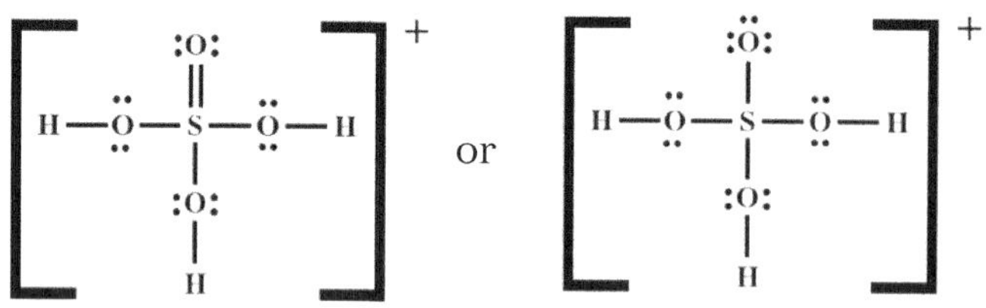

(LO 2.21)

(iv) sp^3 (LO 2.21)

(b) $M_1V_1 = M_2V_2$, (0.100 M)(100.0 mL) = (3.00 M)(x mL). Volume of 3.00 M sulfuric acid needed is 3.33 mL. Under the fume hood, add approximately 50 mL of distilled water to the 100-mL volumetric flask. Using the 10-mL graduated cylinder or graduated pipet, carefully measure 3.33 mL of 3.00 M sulfuric acid and add to the flask. Carefully swirl the solution to mix. Continue adding distilled water in several additions, stoppering and shaking the flask each time until the solution volume reaches the engraved line on the flask. Using the 10-mL graduated pipet provides more accuracy. In theory, the 10-mL graduated cylinder can be used to three significant figures, but it is very difficult to do accurately (LO 2.9).

(c) (i) $q = mc\Delta T$, q = 100.0 g(4.18 J/g°C)(18.0°C), q = 7520 J of heat released (LO 5.7).

(ii) $2\ NaOH + H_2SO_4 \rightarrow Na_2SO_4 + 2\ H_2O$ Since we have equal moles of NaOH and H_2SO_4, NaOH is limiting. Calculations should be done using the moles of NaOH, not H_2SO_4.

(0.0500 L)(3.00 M) = 0.150 mol of NaOH reacted = mol H_2O formed −7520 J/0.150 mol = -5.01×10^4 J/mol = −50.1 kJ/mol (LO 5.6).

(iii) −55.8 is a greater amount of heat released than −50.1. Heat may have been lost to the calorimeter or to the air or to the thermometer. Another source of error could be that the solutions were less concentrated than they were thought to be (LO 5.7).

Question 3: Answers

(a) (i) The $KMnO_4$ is actually less concentrated than you believe it to be since it has been diluted with the water left in the buret. That means that more of the standardized solution will be needed to react with the iron(II) solution leading you to report that the iron concentration is higher than it actually is (LO 3.9).

(ii) If the air bubble stays in the buret during the entire titration, then there is no effect. However, if it is replaced by the $KMnO_4$ solution during the titration, then less solution leaves the buret than you are reporting, leading to a higher than correct percentage of the iron(II) (LO 3.9).

(iii) If more water is added to the flask that contains the iron salt, it will not affect the number of moles of iron(II); therefore, it will not affect the reported result (LO 3.9).

(iv) If the same point on the meniscus is used every time as a reference, there will not be any effect on the percentage of iron reported (LO 3.9).

(b) $\% \text{ Error} = \frac{\text{(Your lab results – Accepted value)}}{\text{Accepted value}} \times 100\%$

$$\frac{(6.896 - 7.77)}{7.77} \times 100\% = -11\%$$

Please note: Sometimes the absolute value is taken and the answer would be reported as 11%. The sign simply tells you if your answer is below or above the accepted value. If you are only concerned with how far away your answer is from the accepted value, then the absolute value is taken. Note also that the answer is known only to two significant figures because the results of the subtraction in the numerator would have only two significant figures.

(c) $5\ Fe^{2+}(aq) + MnO_4^-(aq) + 8\ H^+(aq) \rightarrow 5\ Fe^{3+}(aq) + Mn^{2+}(aq) + 4\ H_2O(l)$ (LO 3.8)

(d) Manganese is reduced from +7 to +2 (LO 3.8).

(e) To prepare four additional solutions, the 3.00×10^{-4} *M* solution must be diluted. There are several dilutions possible. One way to do this is:
Sample 1: 10 mL of 3.00×10^{-4} *M* $KMnO_4$ = 3.00×10^{-4} *M* solution
Sample 2: 8 mL of 3.00×10^{-4} *M* $KMnO_4$ + 2 mL of water = 2.40×10^{-4}*M* solution
Sample 3: 6 mL of 3.00×10^{-4} *M* $KMnO_4$ + 4 mL of water = 1.80×10^{-4}*M* solution
Sample 4: 4 mL of 3.00×10^{-4} *M* $KMnO_4$ + 6 mL of water = 1.20×10^{-4}*M* solution
(LO 2.9)

Question 4: Answers

(a) Carboxylic acids have a nonpolar chain that has London dispersion forces as the main IMF and the functional group has a polar C=O (carbonyl group) that has dipole–dipole forces and an O–H (hydroxyl group) that can form hydrogen bonds. When the number of carbons is low, the polar end of the molecule with dipole–dipole and hydrogen bonding predominates. As the number of carbon atoms increases, the proportion of London dispersion forces also increases. As the strength of the London dispersion forces increases, the molecules will experience greater attractions. More energy is needed to overcome these attractions so the melting point is higher (LO 2.16).

(b) The solubility of the carboxylic acids in water decreases as the number of carbon atoms increases. The predominant IMF in water is hydrogen bonding. Water is able to interact with other molecules whose predominant IMFs are dipole–dipole or hydrogen bonding. As the number of carbon atoms increases, the London dispersion forces become predominant and water cannot interact with those molecules (LO 2.15).

(c) As the number of double bonds increases, the melting point of unsaturated carboxylic acid decreases. Even though the molecules are approximately equal in terms of number of electrons and therefore in terms of London dispersion forces, the shape of the molecules limits the locations of interactions between the molecules. The monounsaturated oleic acid has a greater chance of stacking and having many locations where electrons from each molecule can interact. The di- and tri-unsaturated acids have shapes that do not allow close packing and therefore they have fewer interactions and lower melting points. The more interactions molecules can form, the more energy must be added to overcome attractions, which leads to higher melting points (LO 2.16).

(d) 120° With three regions of electron density and no lone pairs, the bond angle is 120° (LO 2.21).

Question 5: Answers

(a) The reaction is first order in A. A linear graph for ln of concentration vs. time is linear for a first-order reaction (LO 4.2).

(b) The reaction is second order in B. A linear graph for the reciprocal of concentration vs. time is linear for a second-order reaction (LO 4.2).

(c)

$$\frac{\text{Rate}}{[\text{A}][\text{B}]} = k \Rightarrow k = \frac{0.709\ \text{mol}/_{\text{L}} \times \text{min}}{\left(0.24\ \text{mol}/_{\text{L}}\right)\left(0.16\ \text{mol}/_{\text{L}}\right)^2} = \frac{120\ \text{L}^2}{\text{mol}^2 \times \text{min}}$$

(LO 4.2)

(d) No. While the slow step does produce the rate law determined from the data in parts (a) and (b), the overall equation of the mechanism is A + 2 B → C + D, which does not match the overall equation given. You should always check two things to verity a mechanism: (1) the steps of the mechanism must add up to the overall equation, and (2) the slow step must be equivalent to the rate law (LO 4.7).

Question 6: Answers

(a) +, bonds are breaking, but no bonds are formed. Bond breaking is an endothermic process (LO 5.8).

(b) + , 1 mole of gas is producing 2 moles of gas (LO 5.12).

(c) high temperatures; Using $\Delta G = \Delta H - T\Delta S$, since ΔH_{rxn} is positive and ΔS_{rxn} is positive, for ΔG_{rxn} to be negative, the $T\Delta S$ term needs to be larger than ΔH_{rxn} (LO 5.13).

(d) Diborane has 2 H atoms that appear to have 4 valence electrons. Another possible answer is that when the bonds are broken in diborane, one would expect to see lone electrons on the products and there are none. The bonding in diborane is beyond the scope of the AP® Chemistry class, but it is known as an electron deficient bond, in which 2 electrons are shared between 3 atoms (LO 2.21).

Question 7: Answers

(a) $H_3C_6H_5O_7(aq) + 3\ HCO_3^-(aq) \rightleftarrows C_6H_5O_7^{3-}(aq) + 3\ CO_2(g) + 3\ H_2O(l)$ (LO 3.1)

(b) $\frac{1.000\ \text{g}\ H_3C_6H_5O_7}{192.12\ \text{g/mol}} = 0.005205\ \text{mol}$

$\frac{1.916\ \text{g}\ NaHCO_3}{84.01\ \text{g/mol}} = 0.02281\ \text{mol}$

	$H_3C_6H_5O_7$ +	$3\ HCO_3^-$ ⇄	$C_6H_5O_7^{3-}$ +	$3\ CO_2$ +	$3\ H_2O$
Initial mol	0.005205	0.02281	0	0	0
Change mol	$-x$	$-3x$	$+x$	$+3x$	$+3x$
Final mol	0	0.0072	0.005205	0.01561	0.01561

Since one of the reactants must be used up, $x = 0.005205$ or $x = 0.02281/3 = 0.007603$.

$x \neq 0.007603$ because neither reactant can have a negative number of moles at the end of the reaction. Therefore, $H_3C_6H_5O_7$ is the limiting reactant.

0.01561 mol CO_2 × 22.4 L mol^{-1} = 0.350 L (LO 2.6)

(c) Mass of a tablet: 1.000 g + 1.916 g = 2.916 g. Mass of two tablets = 2(2.916) = 5.832 g

$q = m \times c \times \Delta T = (5.832 + 75.0\ \text{g})(4.18\ \text{J g}^{-1}\ °\text{C}^{-1})(-7.6°\text{C}) = -2600\ \text{J}$ (LO 5.7)

(d) $\Delta H_{rxn} = \frac{q \text{ of the reaction}}{\text{Moles of limiting reactant}} = \frac{-2600 \text{ J}}{(2 \text{ tablets} \times \frac{0.005205 \text{ mol } H_3C_6H_5O_7}{1 \text{ tablet}})} = 250{,}000 \text{ J or } 250 \text{ kJ}$

Please note: The original reaction lowered the temperature, so the reaction is endothermic. The final answer needs to have a ΔH that is positive to reflect that it is endothermic (LO 5.7).

Note: This was real data collected by students; the assumption about no interaction with surroundings may not be valid.

SCORING THE FREE-RESPONSE QUESTIONS

It is difficult to come up with an exact score for this section of the test. However, if you compare your answers to the answers in this book (remembering that each part of the test you answer correctly is worth points even if the other parts of the answer are incorrect), you can get a general idea of the percentage of the questions for which you would get credit. If you believe that you got at least 40 percent of the possible credit, you would probably receive a 3 on this part of the test. If you believe that you would receive close to half or more of the available credit, your score would more likely be a 4 or a 5. Remember that questions 1, 2, and 3 (the long questions) are worth 10 points each and the short questions (questions 4, 5, 6, and 7) are worth 4 points each.

Part II

A Review of AP® Chemistry

PREREQUISITE CONCEPTS

While not covered in this test preparation book, you are expected to be competent in basic chemical foundations. These should have been mastered in your first year high school chemistry course.

You should be able to:

- Identify the number of significant figures in a given measurement.
- Use dimensional analysis to solve a variety of chemistry problems.
- Perform calculations, rounding answers to the correct number of significant figures.
- Differentiate between accuracy and precision as they apply to measurement.
- Determine the density of solids and liquids and calculate volumes or masses using the given density.
- Convert between units of temperature: degrees Celsius and Kelvin.
- Identify the characteristics of the states of matter: solids, liquids, and gases.
- Identify changes as being physical or chemical.
- Name and write formulas for binary compounds, ternary compounds (those with polyatomic ions), and acids.
- Differentiate between elements, compounds, and mixtures. Memorize the chemical formulas and charges of the common monatomic and polyatomic ions. (See Tables 2.3, 2.4, and 2.5 in Chapter 2 of the text.)

AP® Tip

Significant figures are important in calculations. A maximum of 1 point in the free-response sections will be deducted for errors in significant figures. Units are the key to problem solving. Include units in your calculated work to ensure that your answer has units that agree with the problem. Every measured number or number calculated from measurements has a unit except for equilibrium constants, which by convention are reported without units. When a calculation specifically asks for units to be included, you can be assured that the lack of units or incorrect units may result in a point deduction.

1

Big Idea 1: Atomic Structure and Periodicity

Big Idea 1

The chemical elements are the fundamental building materials of matter, and all matter can be understood in terms of arrangements of atoms. These atoms retain their identity in chemical reactions.

Atoms are the fundamental building blocks of the elements which comprise all matter. **Big Idea 1** relates to three main topics in this chapter: atomic structure, periodicity, and matter at the macroscale level. The structure of the atom is the basis for the organization of the elements on the periodic table. Throughout history, our "atomic model" has continually developed to match new scientific knowledge.

You should be able to:

- Use evidence to explain why atomic models have changed throughout history.
- Calculate the average atomic mass of an atom from mass spectrometry and isotopic data.
- Use photoelectron spectroscopy as evidence to show that electrons exist in quantized shells.
- Justify the arrangement of the periodic table based on the quantum mechanical model of the atom.
- Relate atomic structure to periodic trends and reactivities.
- Convert between mass, moles, and numbers of representative particles (atoms, molecules, formula units) of a substance.

- Calculate the molar mass, percent composition, and empirical formula of a compound.
- Express the law of conservation of mass using symbolic representations and particulate models.

PART 1: ATOMIC STRUCTURE

EVIDENCE FOR ATOMS

MODELS CHANGE OVER TIME

While an understanding of historical experiments and models that eventually led to our current model of the atom is very important, memorizing dates and names of scientists is not necessary for the exam.

Atomic models have been around since Democritus first thought about dividing matter by continually cutting it into smaller and smaller pieces until it could no longer be cut: *atomos*. The idea of atoms was opposed by Aristotle and it took about 2500 years until scientific evidence was used to revive the idea of atoms.

HISTORY OF THE ATOM

JOHN DALTON

Dalton's model of the atom was based on the work of many scientists. He used the law of definite proportion and the law of multiple proportions to provide evidence for his model.

Dalton's model stated that all matter is made up of atoms and that all atoms of a given element are identical, although atoms of different elements are different from each other. He also stated that atoms combine to form compounds and that a given compound always has the same relative number and types of atoms. Chemical reactions involve rearrangement of the ways that atoms are bonded together, but the atoms themselves are not changed.

THE LAW OF DEFINITE PROPORTION

The law of definite proportion states that a given compound always contains exactly the same proportions of elements by mass. For example, all samples of pure water are approximately 89% oxygen and 11% hydrogen by mass because water always contains 2 atoms of hydrogen and 1 atom of oxygen.

THE LAW OF MULTIPLE PROPORTIONS

The law of multiple proportions: When two elements form a series of compounds, the ratio of masses of the second element that combine with 1 gram of the first element can always be reduced to small whole numbers.

EXAMPLE: Sulfur and oxygen can react to form both sulfur dioxide and sulfur trioxide. In 1 mole of sulfur dioxide, there are 32.06 g sulfur and 32.00 g oxygen. In 1 mole of sulfur trioxide, there are 32.06 g sulfur combined with 48.00 g oxygen. What is the ratio of the masses of oxygen that combine with 32.06 g sulfur?

SOLUTION: 48.00 g/32.00 g = 1.5 or 3:2, demonstrating the law of multiple proportions

DISCOVERY OF THE ELECTRON

Dalton's characterization of atoms as solid, indivisible spheres that were identical for each element was quickly challenged. The model could not account for the discovery of subatomic particles, such as the electron. Mass spectrometry showed that there were atoms of the same element that had different masses. In fact, most elements were composed of isotopes that completely contradicted certain aspects of Dalton's model.

Historical Atomic Models	Summary	Experimental Evidence for Changing the Model
Dalton's Model	Atom is a hard, indivisible sphere	Existence of subatomic particles, isotopes
Thomson's Model	Atom has electrons randomly arranged in a positive material	Existence of the nucleus
Rutherford's Model	Atom has electrons surrounding a very small, very dense, positively charged nucleus	Electrons are not randomly arranged
Bohr's Model	Electrons are in definite paths (orbits) around the nucleus	Electrons are not in orbits (didn't work for elements beyond hydrogen)
Quantum Mechanical Model	Electrons are found in regions of electron density (orbitals) around the nucleus (mathematical model)	

ISOTOPES AND MASS SPECTROMETRY

Isotopes are atoms of the same element with different numbers of neutrons and therefore different atomic masses.

The average atomic mass of an element can be calculated from the percent abundance and mass of each isotope for that element.

Average atomic mass = Σ(Percent abundance of isotope × Atomic mass)

EXAMPLE: Element "E" is present with the following mass values and natural abundances.

Isotope	Atomic number (amu)	Percent Abundance (%)
^{10}E	10.01	19.78
^{11}E	11.01	80.22

What is the average atomic mass of element E? What is the actual name of the element?

SOLUTION: (0.1978)(10.01 amu) + (0.8022)(11.01 amu) = 10.812 amu

The element is boron.

Mass spectrometry is an analytical technique that is used to determine the mass spectrum of an element. Most elements exist as isotopes of different mass, and the mass spectrometer can determine the ratio of the masses.

An element sample is ionized, and the resulting fragments are passed through a magnetic field that deflects the ions based on mass and charge. The deflection separates the ions, and a unique spectrum is created.

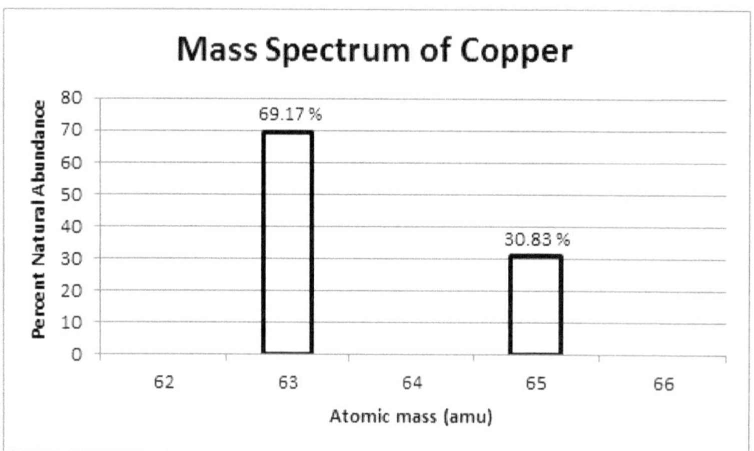

Isotope	Atomic Mass (amu)
^{63}Cu	62.930
^{65}Cu	64.928

The mass spectrum of copper is shown above. An average atomic mass could be calculated from these data.

$$[(0.6917 \times 62.930 \text{ amu}) + (0.3083 \times 64.928 \text{ amu})] = 63.546 \text{ amu}$$

Looking at the spectrum of copper, one can estimate the average atomic mass: we know that it would be between 63 and 65 and be much closer to 63 than to 65.

Another type of problem would be to estimate the percentages of the isotopes. The strategy shown below would be especially helpful when answering a multiple-choice question.

EXAMPLE: Chlorine has two stable isotopes, Cl-35 and Cl-37. What is the percent abundance of Cl-35 in nature?

SOLUTION: Create a chart like the one shown below. Start by putting in the ends (shown in bold) for the extreme percentages (100 and 0) and then add the middle—50% each (shown in italics). For each set of conditions, calculate the average atomic mass given those percentages. You can then fill in the 75–25% and 25–75% sections. Since the average atomic mass of Cl (found on the periodic table) is 35.45, the approximate percentage of Cl-35 is 75%.

	Assumed Percentage in Nature				
Cl-35	**100**	75	*50*	25	**0**
Cl-37	**0**	25	*50*	75	**100**
Average Atomic Mass	**35**	35.5	*36*	36.5	**37**

ELECTROMAGNETIC RADIATION

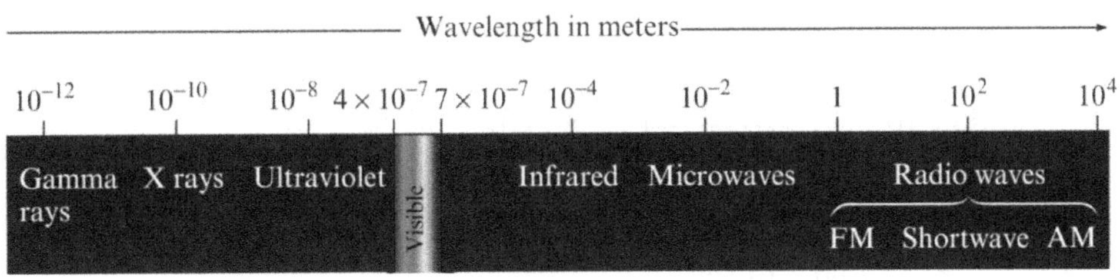

Electromagnetic radiation provides an important means of energy transfer. Light from the sun reaches the earth in the form of radiation that includes ultraviolet and visible radiation. Electromagnetic radiation, from the shortest to the longest wavelength, includes gamma rays, X-rays, ultraviolet, visible, infrared, microwaves, and radio waves. You should have a general idea of the order of energy of these types of electromagnetic radiation but do not need to memorize their wavelengths. All types of electromagnetic radiation exhibit both wave-like and particle-like behavior and travel at the speed of light in a vacuum.

Chemists use the interaction of light with matter to study the structure of the atom.

- Wavelength, λ, is the distance between two consecutive peaks or troughs in a wave.
- Frequency, v, is the number of waves (cycles) per second that passes a given point in space.

Wavelength and frequency are related by the equation:

$$v = \frac{c}{\lambda}$$

in which c is the speed of light in a vacuum, 2.998×10^8 m/s. Wavelength is measured in meters, and frequency is measured in 1/s, or hertz (Hz).

ENERGY OF ELECTROMAGNETIC RADIATION

Energy is quantized and can occur only in whole number multiples of hv. Each small packet of energy is called a quantum. A system can transfer energy only in discrete quanta. The change in energy for a system can be represented by

$$\Delta E = nhv$$

where n is an integer (1,2,3,...), h is Planck's constant which equals 6.626×10^{-34} J s, and v is the frequency.

Einstein viewed electromagnetic energy as a stream of particles called photons. And the energy of each photon can be given by

$$E_{photon} = \frac{hc}{\lambda} \quad \text{or} \quad E_{photon} = hv$$

EXAMPLE: The laser in an audio compact disc player uses light with a wavelength of 7.80×10^2 nm. Calculate the frequency of this radiation. What is the energy of this radiation per photon?

SOLUTION: 1 m = 10^9 nm

$$(7.80 \times 10^2 \text{nm}) \times \left(\frac{1 \text{ m}}{10^9 \text{ nm}}\right) = 7.80 \times 10^{-7} \text{ m}$$

$$v = \frac{c}{\lambda} = \frac{2.998 \times 10^8 \text{m/s}}{7.80 \times 10^{-7} \text{m}} = 3.84 \times 10^{14} \text{ s}^{-1}$$

$$E_{photon} = hv = (6.626 \times 10^{-34} \text{ J s}) \times (3.84 \times 10^{14} \text{ s}^{-1}) = 2.54 \times 10^{-19} \text{ J}$$

ATOMIC SPECTRUM

A continuous spectrum contains all wavelengths of visible light. A line spectrum contains only a few lines, each corresponding to a discrete wavelength. When atoms are excited by added energy, electrons move to higher energy levels as they absorb the energy. When the electrons return to their ground-state energies, the excess energy is given off by emitting light of various wavelengths to produce an emission spectrum or line spectrum.

The Emission lines for the element Hydrogen which correspond to the absorption lines for the same element

QUANTUM MECHANICAL MODEL

Our current model of the atom is the quantum mechanical model. In this model, electrons are treated as waves and the path of an electron cannot be described. An orbital is a three-dimensional electron density map in which there is a very high probability of finding the electron. The Heisenberg uncertainty principle is an important part of the model. It states that we cannot know precisely both the momentum and location of an electron. The quantum mechanical (QM) model is an improvement over the Bohr model because the QM model supports evidence of energy levels and sublevels within atoms. This model can be approximated using computers and is the basis for software that predicts molecular shape and reactivity.

ELECTROSTATIC FORCES HOLD THE ATOM TOGETHER

A neutral atom is composed of protons and neutrons in the nucleus and electrons in energy levels or "shells" outside of the nucleus. Coulomb's law states that the force between the two particles is proportional to the magnitude and inversely proportional to the square of the distance between them.

$$|F| = k_e \frac{|q_1 q_2|}{r^2}$$

where F = force, k_e = proportionality constant, q = charges on each particle, and r is the distance between the particles.

Sign of q_1	Sign of q_2	Type of Force	Where Found in Atom	Notes
+	–	Attraction	Between nucleus and electrons	Basis of all periodic trends
–	+	Attraction	Between nucleus and electrons	Basis of all periodic trends
–	–	Repulsion	Electron orbitals and subshells (electrons repel each other)	Basis of all periodic trends
+	+	Repulsion	Nucleus (protons repel each other)	Stability of nucleus—not covered on AP® exam

Note: Electron to electron repulsion between atoms is responsible for molecular geometry.

In the table, the first and second entries are the same—don't forget that electrons attract protons and protons attract electrons.

If the charges are of opposite signs, the force is attractive; if the charges are of the same sign, the force is repulsive. As the distance between the charges increases, the attractive force between the charges decreases. When comparing two atoms that have electrons approximately the same distance from the nucleus (such as nitrogen and oxygen), the nucleus with the higher nuclear charge (oxygen) will have a higher attractive force for the electrons.

PHOTOELECTRON SPECTROSCOPY

Photoelectron spectroscopy (PES) is a technique that can be used to provide evidence for electron shells and subshells in an atom and to measure ionization energies. UV or X-ray radiation is shined on an atom and the minimum frequency of energy needed to remove an electron is determined. Generally, UV radiation can be used to remove valence electrons, while X-ray radiation is necessary to remove core electrons. If the frequency of light is greater than the threshold needed to remove an electron, the extra energy is retained by the emitted electron as kinetic energy. Because energy is conserved, the energy of light that hits the atom (hv) must equal the energy required to remove the electron ($E_{electron}$) and the kinetic energy (KE) of the removed electron.

$$hv = E_{electron} + KE_{removed\ electron}$$

Since the frequency of the light hitting the atom is known and the kinetic energy of the removed electron can be measured, the binding energy of the electron can be calculated. Electrons that are farther away from the nucleus require less energy to remove because as energy levels get higher, they are larger and farther from the nucleus, experiencing less coulombic attraction. Inner electrons have a higher coulombic attraction to the nucleus (they are closer) and require greater energy to remove.

Both hydrogen and helium have a single electron binding energy that corresponds to the $n = 1$ energy level electrons ($1s$). Even though helium has 2 electrons, they are both approximately equally distant from the nucleus and therefore have the same energies. The energy required to remove an electron from helium is greater than that to remove an electron from hydrogen because helium has a higher nuclear charge (2 protons, rather than 1).

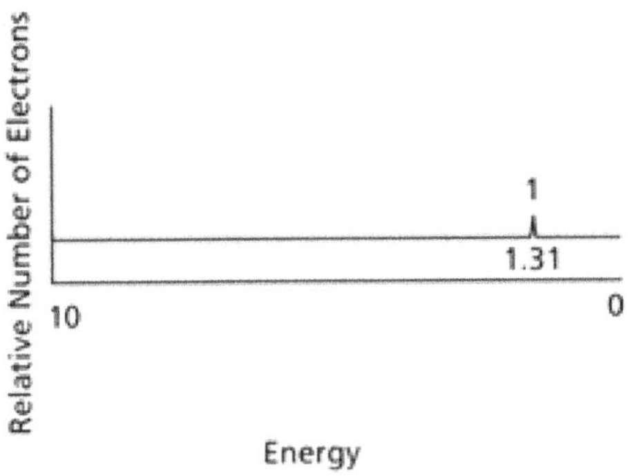

Hydrogen

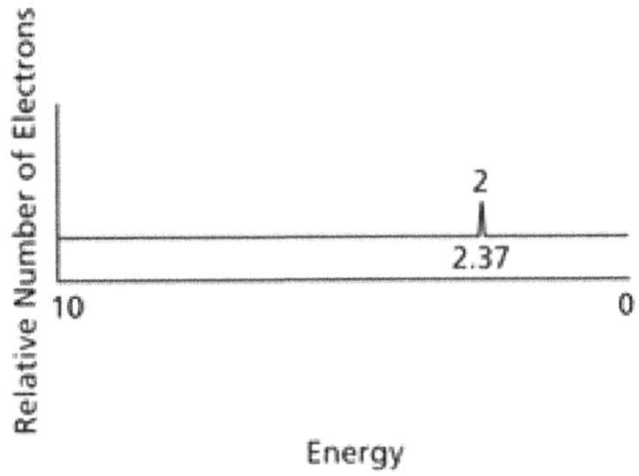

Helium

Lithium has 2 electron binding energies that correspond to the $n = 1$ and $n = 2$ energy levels (1*s* and 2*s*). The energy required to remove boron's 1*s* electrons is much higher than it is to remove its 2*s* electrons, while the energy required to remove the 2*p* electron is even less. This evidence supports the idea of 2 subshells in $n = 2$. The subshell closer to the nucleus is called 2*s*, and the one farther away is called 2*p*.

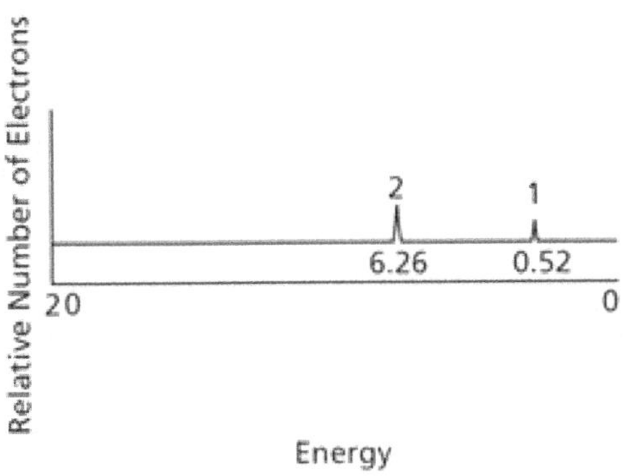

Lithium

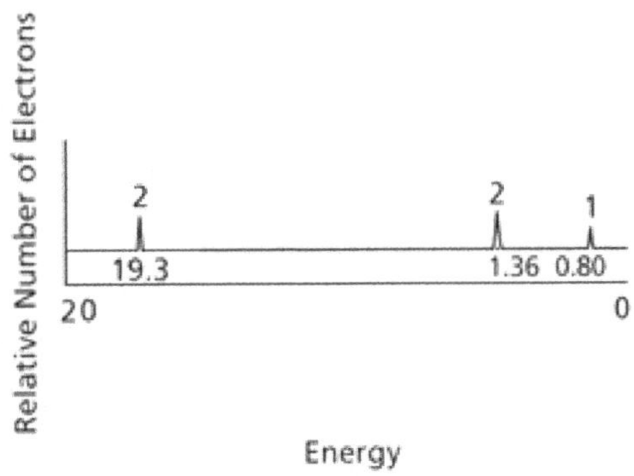

Boron

QUANTUM NUMBERS

Quantum numbers characterize various properties of the orbitals. While quantum numbers are not directly assessed on the exam, you must still understand the arrangement of electrons in shells (energy levels) and subshells (sublevels). Within each subshell, there exist orbitals (regions where electrons are likely to be). Each orbital can hold at most 2 electrons and these 2 electrons must be spinning in opposite directions.

The shell, *n*, is related to the size and energy of the orbital. As *n* increases, the orbital becomes larger, which also means it is higher in energy and the electrons are, on average, further from the nucleus and less tightly held.

The shell or energy level is often referred to as *n*, and is the number which appears in front of the types of orbital *s*, *p*, *d*, or *f*. In general, for periods (rows) 1, 2, and 3, the shell number, *n*, is the same as the row number. For periods 3, 4, 5, 6, and 7, the shell number written before *s* and *p* orbitals is the same as the row number for the main group of elements, those in columns 1–2 and 13–18 (1A through 8A). For the transition elements, the value of *n* which is written in front of the *d* orbital is 1 number less than the row number. For example, scandium, in period 4, has a value of *n* equal to 3 for its *d* electron. In the sixth row of the periodic table, the *f*-transition elements occur after lanthanum. For these elements, the value of *n* which is placed before the *f* is 2 less than the row in which they occur.

ORBITAL SHAPES AND ENERGIES

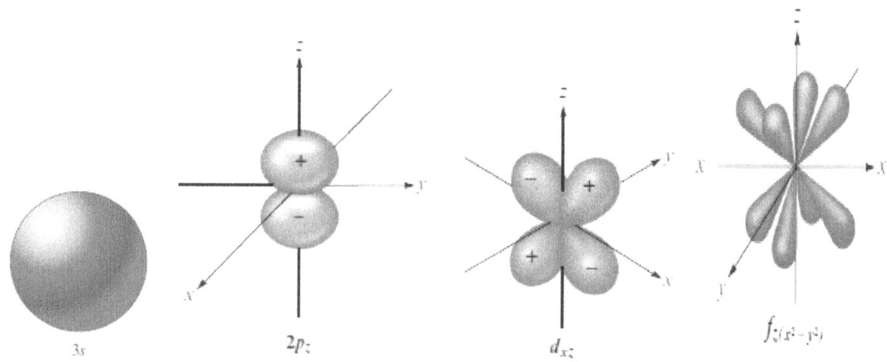

Shown is one example of each type of orbital.

s ORBITALS

The *s* orbitals have a spherical distribution of electron density. The *s* orbitals become larger as the value of *n* increases. Each subshell has only one *s* orbital.

p ORBITALS

There are three *p* orbitals which first appear in $n = 2$. The *p* orbitals have two lobes (peanut or dumbbell shape) and may be labeled n_{px}, n_{py}, and n_{pz}, according to the axis along which the lobes lie.

Degenerate orbitals are orbitals with the same value of *n* which have the same energy. The three *p* orbitals in the same energy level are degenerate or equal in energy.

d AND *f* ORBITALS

The *d* orbitals first occur when $n = 3$ but are located in the fourth period of the periodic table. The *f* orbitals first occur when $n = 4$ and are located in the sixth period.

EXAMPLE: Which of the following orbital designations is incorrect: 1*s*, 1*p*, 2*d*, 4*f* ?

SOLUTION: 1*p* is incorrect: *p* orbitals do not exist in energy level 1; $n = 2$ is the first energy level that has *p* orbitals.

2*d* is incorrect: *d* orbitals do not exist in energy level 2; the first energy level with *d* orbitals is energy level 3.

EXAMPLE: Give the maximum number of electrons in an atom that can exist in 2*p* orbitals.

SOLUTION: Each 2*p* orbital can hold 2 electrons with spins in opposite directions. Since three orientations of *p* orbitals exist, a total of 6 electrons can be in 2*p*.

PART 2: PERIODICITY

ARRANGING THE ELEMENTS

ELECTRONIC STRUCTURE AND PERIODIC TABLE

The current arrangement of the periodic table is a result of the repeating pattern of electron arrangement as the atomic number of the elements increases.

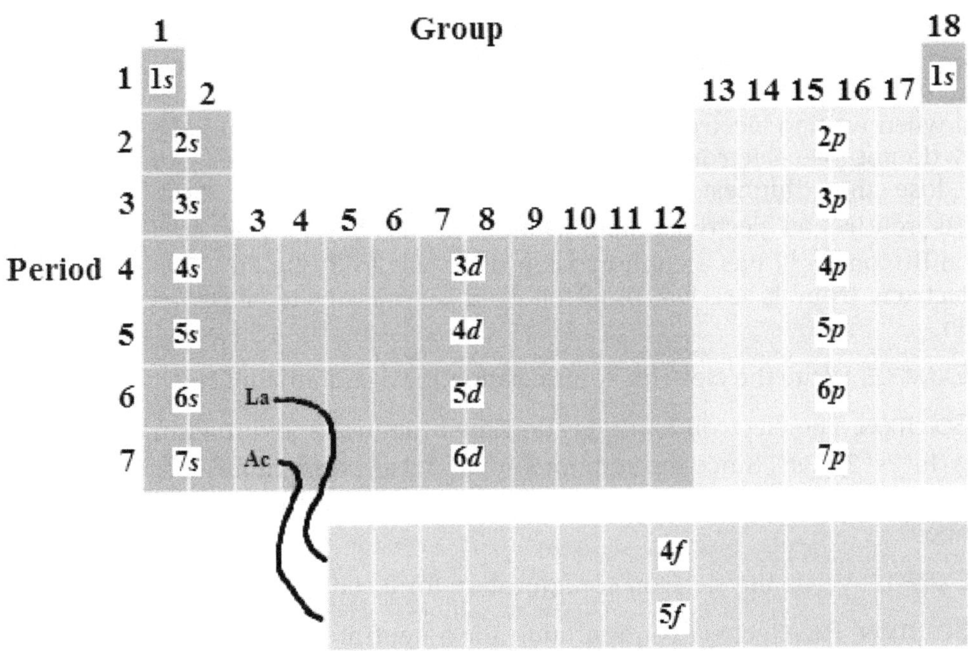

AUFBAU PRINCIPLE AND ELECTRON CONFIGURATIONS

The Aufbau principle states that electrons fill orbitals in an atom in order of increasing energy. Because it is lowest in energy, electrons enter the 1*s* orbital first, followed by the 2*s* orbital, and then the three 2*p* orbitals. The order of filling does not need to be memorized, but you do need to be able to use the periodic table to determine the order of filling. In the periodic table above, the order of filling is shown.

The electron configuration for an atom of an element represents all of the electrons in the atom, and it shows in which energy levels and orbitals the electrons reside. When electrons are in their lowest energy state (as predicted from the order of filling using the periodic table), it is referred to as ground state electron configuration. The addition of energy may cause electrons to jump to higher energy levels. This is referred to as an excited state.

EXAMPLE: Write a ground state electron configuration for oxygen.

SOLUTION: The electron configuration for oxygen is $1s^2 2s^2 2p^4$.

EXAMPLE: Write the short-hand noble gas electron configuration for germanium, Ge.

SOLUTION: The short-hand electron configuration for germanium is [Ar] $4s^2 3d^{10} 4p^2$. This is sometimes rearranged as [Ar] $3d^{10} 4s^2 4p^2$.

To write the short-hand electron configuration, write the noble gas symbol which comes in the row before the element in brackets followed by the remaining electrons. The [noble gas] represents all electrons up to that noble gas.

When atoms lose or gain electrons to become ions, their electron configuration changes to reflect the electrons lost or gained. Anions are formed when an atom gains one or more electrons. Cations are formed when an atom loses one or more electrons. Be especially careful when writing electron configurations for transition metal ions, as they do not lose electrons in the same order in which they fill. All atoms lose their highest numerical energy level electrons first. Whereas iron has an electron configuration of $1s^2 2s^2 2p^6 3s^2 3p^6 4s^2 3d^6$, the iron(III) ion, Fe^{3+}, has a configuration of $1s^2 2s^2 2p^6 3s^2 3p^6 3d^5$. The two electrons from 4*s* are lost first, followed by one of the electrons from 3*d*.

EXAMPLE: Write the electron configuration for a sodium ion, Na^+.

SOLUTION: The electron configuration for a neutral sodium atom is $1s^2 2s^2 2p^6 3s^1$. Since the sodium ion has a 1+ charge, it has lost an electron and its configuration will be $1s^2 2s^2 2p^6$ (the same as neon).

EXAMPLE: Write the electron configuration for a sulfide ion, S^{2-}.

SOLUTION: The electron configuration for a neutral sulfur atom is $1s^2 2s^2 2p^6 3s^2 3p^4$. Since the sodium ion has a 2– charge, it has gained two electrons and its configuration will be $1s^2 2s^2 2p^6 3s^2 3p^6$ (the same as argon).

VALENCE ELECTRONS

Valence electrons are the electrons in the outermost energy level or shell of an atom. In the example with oxygen, the valence electrons are in the second energy level. In the second energy level, there are 6 valence electrons, 2 in the 2*s* and 4 in the 2*p*. Coincidentally, the valence electrons of a main group element are the same as the group number in the old A-B group numbering system, except for helium which has 2 valence electrons. All of the electrons that are not in the highest energy level are known as *core* electrons. The 1*s* electrons in oxygen are core electrons.

PERIODIC TRENDS IN ATOMIC PROPERTIES

AP® Tip

When answering free-response questions on trends, follow the three C's guidelines below.

- Cannot use a trend to explain a trend
- Compare both species in the question
- Contrast both species in the question

ATOMIC RADII

Atomic radii decrease going from left to right across a period from the alkali metals to the halogens. This can be explained by an increasing effective nuclear charge (number of protons in the nucleus increases), which causes the electrons to be drawn closer to the nucleus.

Atomic radius increases down a group because of the increases in the orbital sizes in successive energy levels. Also, there is an increase in the shielding of the nucleus by the core electrons, which decreases the effective nuclear charge felt by the valence electrons.

EXAMPLE: Arrange the following atoms in order of increasing atomic radius.

C, Al, F, Si

SOLUTION: F < C < Si < Al; F and C are in the second period, which means that they have two shells or energy levels of electrons. Within a period, the radius decreases as the atomic number increases because the higher nuclear charge that results from added protons increases the attraction between the electrons and the nucleus. This means F will be smaller than C. Al and Si are in the third period and have one more energy level, so they will have a larger radius than atoms in period 2. Since Si has more protons than Al, it has a smaller radius.

IONIZATION ENERGY

Ionization energy is the energy required to remove an electron from a gaseous atom or ion.

The first ionization energy is the energy required to remove the highest energy electron (the one bound least tightly) which is one of the valence electrons.

$$X(g) + \text{energy} \rightarrow X^{1+}(g) + e^-$$

For lithium, the first ionization energy is the energy required to remove an electron from the 2*s* orbital. The configuration for lithium is $1s^22s^1$, and the valence electron is the 2*s* electron.

In general, the first ionization energy increases across a period from left to right. This trend can be related to the trend in atomic radii. As the charge of the nucleus increases, atomic size decreases and first ionization energy increases. These trends both occur because of the greater pull of the nucleus on the valence electrons. There are irregularities in this trend in ionization energy (it is these irregularities that are often seen on the AP® exam). The first ionization energy decreases from beryllium to boron because the electrons in the 2*s* orbital are, on average, closer to the nucleus than the electrons in the 2*p* orbital. This is called the penetration effect. The *s* electrons in an energy level penetrate (come closer to the nucleus) to a greater extent than do the *p* electrons in the same energy level. In all atoms except for hydrogen, the *np* orbital is slightly higher in energy than the *ns* orbital. The higher energy 2*p* electron of boron is easier to remove than one of the 2*s* electrons of beryllium.

The first ionization energy decreases from nitrogen to oxygen due to repulsion of paired electrons in the p^4 configuration of oxygen. Nitrogen has a valence electron configuration $[\text{He}]2s^22p^3$ and oxygen has a valence electron configuration of $[\text{He}]2s^22p^4$. Drawing the orbital notation for the electrons in nitrogen and oxygen as shown below can make this more clear. Each arrow represents an electron. The opposite directions of the arrows represent the opposite spins of 2 electrons sharing an orbital (lessens repulsion). According to the Pauli exclusion principle, only 2 electrons with opposite spin can occupy the same orbital.

The orbital diagram for oxygen shows unpaired electrons in 2*p*. Unpaired electrons in atoms or molecules make particles *paramagnetic*. This means that the substance is attracted to a magnet. The attraction is relatively weak in most instances. Substances with all electrons paired are said to be *diamagnetic* and are very weakly repelled from a magnetic field.

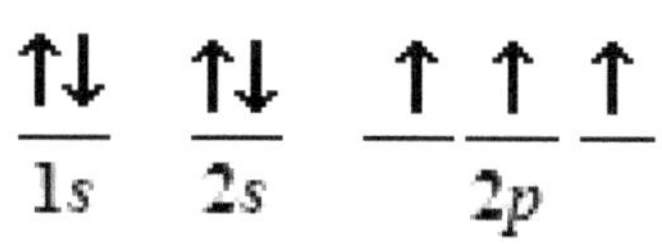

Nitrogen

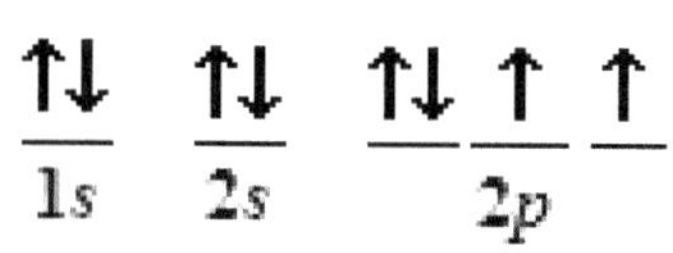

Oxygen

The PES diagrams for nitrogen and oxygen, shown on the next page, illustrate how PES data support electronic structure. Oxygen has a binding energy of 52.6 MJ/mol for removal of the 1*s* electrons, compared to 39.6 MJ/mol for the 1*s* electrons of nitrogen. This makes sense since oxygen has an additional proton in its nucleus. Note that the energy required to remove 1 mole of 2*p* electrons from oxygen is less than the energy required to remove 1 mole of 2*p* electrons from nitrogen. With a higher nuclear charge in oxygen, this is not expected. Remember, though, that oxygen has one 2*p* orbital that must contain 2 electrons. The repulsion created by these paired electrons makes the 2*p* electrons easier to remove.

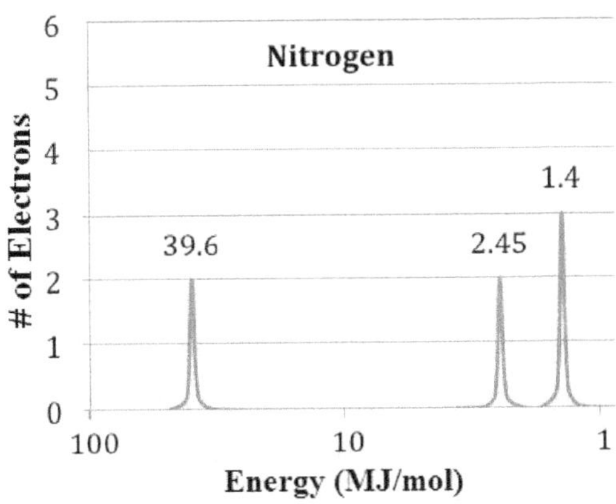

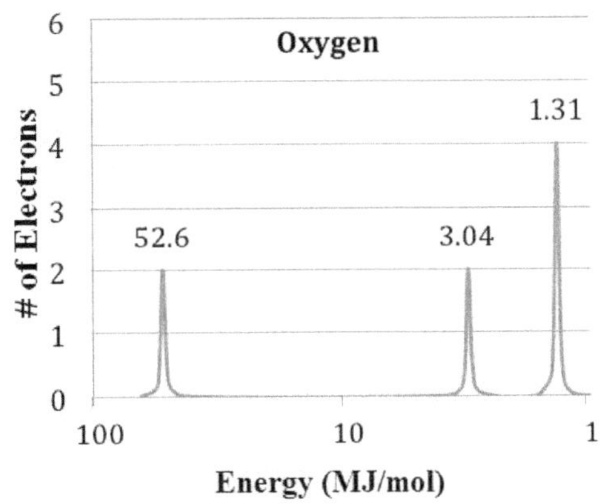

The first ionization energy decreases down a group with increasing atomic number because the valence electrons become easier to remove as the atoms become larger (added energy levels) going down a group. Shielding due to electron repulsion in core energy levels reduces the effective nuclear charge of the atoms. This lowers the attraction between the nucleus and the valence electrons which means less energy is needed to remove an electron.

The second ionization energy is the energy required to remove the second outermost electron after removal of the first electron. This value is larger than the first ionization energy. The first ionization energy removes an electron from a neutral atom. The second ionization energy removes an electron from a positive ion.

$$X^{1+}(g) + \text{energy} \rightarrow X^{2+}(g) + e^-$$

The increase in positive charge binds the electrons more tightly, and the ionization energy increases as each successive electron is removed. Note a large increase occurs in ionization energy when a core electron from a full energy level is removed.

General Increase

General Decrease

Successive Ionization Energies for Period 3 Elements

Element	IE_1	IE_2	IE_3	IE_4	IE_5	IE_6	IE_7
Na	498	4560	6910	9540	13 400	16 600	20 100
Mg	736	1445	7730	10 600	13 600	18 000	21 700
Al	577	1815	2740	11 600	15 000	18 310	23 290
Si	787	1575	3220	4350	16 100	19 800	23 800
P	1063	1890	2905	4950	6270	21 200	25 400
S	1000	2260	3375	4565	6950	8490	27 000
Cl	1255	2295	3850	5160	6560	9360	11 000
Ar	1519	2665	3945	5770	7230	8780	12 000

AP® Tip

When a question asks you to compare the ionization energies of two different elements, it is often extremely helpful to write the electron configuration and draw out the orbital notation diagram for the elements in question. This pinpoints from where the electron is removed. If the question is free-response, the electron configuration and/or orbital notation can serve as part of your justification.

EXAMPLE: Explain why the difference between the third and fourth ionization energies for aluminum is so large in comparison with the first and second and second and third ionization energies.

SOLUTION: Aluminum has the valence electron configuration [He]$3s^2 3p^1$. The first electron removed is the 3*p*, the second electron removed is a 3*s*, and the third electron removed is the remaining 3*s* electron. The fourth electron, a core electron (a 2*p*), requires much more energy for removal than the valence electrons, the 3*s* and the 3*p* electrons, because it is more tightly bound due to a higher effective nuclear charge.

EXAMPLE: Arrange the following atoms in order of increasing first ionization energy. I, Rb, Na.

SOLUTION: Na will have the smaller radius. Rb < I < Na. Na has the largest IE_1 because the electron is removed from the 2*s* sublevel. Both Rb and I have valence electrons in the fifth energy level, but I has 53 protons and a smaller radius than Rb with only 37 protons which makes I smaller than Rb. The smaller the atom, the more energy required to remove the valence electron.

ELECTRON AFFINITY

Electron affinity is the energy change associated with the addition of an electron to a gaseous atom.

$$X(g) + e^- \rightarrow X^-(g)$$

In general, electron affinity becomes increasingly negative or more exothermic as we proceed across a period with increasing atomic number. The greater the charge on the nucleus of the atom, the greater is the affinity for electrons. The addition to a noble gas, however, would require the added electron to reside in a new, higher energy level which is energetically unfavorable, so that stable, isolated X^- ions do not exist. Electron affinities are not measured for noble gases.

Electron affinities become slightly more positive and less exothermic going down a group. The electron–nucleus attraction decreases down a group due to increasing atomic size and shielding. But the added electron experiences less electron–electron repulsion, since the orbital to which the electron is being added is increasingly spread out. One exception to this trend is fluorine. The value of electron affinity for F is more positive than that of Cl due to the small size of the F atom and the greater electron–electron repulsion in the smaller 2*p* sublevel.

IONIC RADII

Cations are smaller than their parent atoms, and anions are larger than their parent atoms. The size of an ion depends on its nuclear charge, the number of electrons it contains, and the outer energy level of the atom. Positive ions are always smaller than neutral parent atoms because the nuclear charge is the same but the number of electrons has decreased, increasing the effective nuclear charge felt by each remaining electron as well as decreasing the electron–electron repulsion. When all the valence electrons are lost, the change in volume is greater because as n decreases, the radial probability is smaller so the radius is smaller.

Anions are larger than their parent atoms because the electrons gained cause the proton to electron ratio to decrease, and the effective nuclear charge felt by each electron decreases. Electron–electron repulsions also increase. The electrons spread out more to accommodate the additional electrons.

In an isoelectronic series, ions contain the same number of electrons (also the same electron configuration) and the ion with the largest atomic number will have the smallest radius.

EXAMPLE: Arrange the following ions in order of increasing size: Ca^{2+}, S^{2-}, K^+, Cl^-.

SOLUTION: All of the ions mentioned above are isoelectronic with argon, containing 18 electrons each. In order of increasing size, $Ca^{2+} < K^+ < Cl^- < S^{2-}$. Ca^{2+} is the smallest because it contains the greatest number of protons. The greater the positive charge on the nucleus, the stronger the attraction for electrons.

PART 3: MATTER ON THE MACRO SCALE

THE MOLE

The mole is defined as the number of atoms in exactly 12 g of carbon-12. Calculations involving moles utilize the conversions in the table below. Note the representative particle appropriate for each case. The mass of 1 mole of each can be termed molar mass, gram formula mass (gfm), gram atomic mass (gam, for elements), and gram molecular mass (gmm, for molecules).

1 mol of ...	is ...
monoatomic element	6.022×10^{23} atoms
molecular compound or diatomic element	6.022×10^{23} molecules
ionic compound	6.022×10^{23} formula units

MOLAR MASS

Molar mass is the mass of 1 mole of an element or compound.

EXAMPLE: The principal ore in the production of aluminum metal has a formula of $Al_2O_3 \bullet 2\ H_2O$. What is the mass in grams of 2.10×10^{24} formula units of $Al_2O_3 \bullet 2\ H_2O$?

SOLUTION: The molar mass of $Al_2O_3 \bullet 2\ H_2O$ is 138.0 g/mol. Since this is an ionic compound, 1 mol contains 6.022×10^{23} formula units.

$$2.10 \times 10^{24} \text{ formula units} \times \frac{1 \text{ mol } Al_2O_3}{6.022 \times 10^{23} \text{ formula units}} \times \frac{138.0 \text{ g}}{1 \text{ mol } Al_2O_3 \bullet 2\ H_2O}$$

$$= 481 \text{ g } Al_2O_3 \bullet 2\ H_2O$$

PERCENT COMPOSITION

The mass percentages of elements in a compound can be obtained by comparing the mass of each element present in 1 mole of the compound to the total mass of 1 mole of compound.

$$\frac{\text{Mass of element in 1 mole of compound}}{\text{Mass of 1 mole of compound}} \times 100\%$$

EXAMPLE: Calculate the percent oxygen in $Mg(NO_3)_2$.

SOLUTION: The molar mass of $Mg(NO_3)_2$ is 148.3 g/mol. There are 6 moles of oxygen in each mole of $Mg(NO_3)_2$ so the molar mass of oxygen must be multiplied by 6.

$$6 \times 16.0 = 96.0 \text{ g O} \qquad \frac{96.0 \text{ g O}}{148.3 \text{ g Mg(NO}_3)_2} \times 100 = 64.7\% \text{ O}$$

EMPIRICAL FORMULAS

EMPIRICAL FORMULA Simplest whole number ratio of the various types of atoms in a compound. For example, HO is the empirical formula for hydrogen peroxide, H_2O_2.

MOLECULAR FORMULA The formula showing the actual ratio of the elements in a compound (empirical formula) times some small whole number (may be same as empirical formula).

Determination of the empirical and/or molecular formulas of a compound from data is a common type of question found on the exam. Depending upon the type of data provided in the question, the approach may vary slightly, but the basic steps are:

1. Change mass (or given percent composition, treat it as grams) of each element to moles.
2. Set up the ratio of each element by moles.
3. Simplify the ratio by dividing each value by the smallest number of moles in the set.
4. If the simplified ratio is not in whole numbers, round each very slightly or multiply each by the same small whole number (2, 3, 4, or 5 most often) to get a whole number ratio.
5. Use the ratio as subscripts for each element. This is the empirical formula.
6. If the molecular formula is needed, the molar mass of the compound will be provided. Find the mass of the empirical formula and divide into the molar mass of the molecular formula. It should go evenly. Multiply each subscript in the empirical formula by the resulting value to determine the molecular formula.

EXAMPLE: A common artificial sweetener is found to be 57.14% C, 6.16% H, 9.52% N, and 27.18% O. The molar mass is 294.30 g. Determine the empirical and molecular formulas of this compound.

SOLUTION:

$$57.14 \text{ g C} \times \frac{1 \text{ mol C}}{12.01 \text{ g C}} = 4.758 \text{ mol C}$$

$$6.16 \text{ g H} \times \frac{1 \text{ mol H}}{1.01 \text{ g H}} = 6.10 \text{ mol H}$$

$$9.52 \text{ g N} \times \frac{1 \text{ mol N}}{14.01 \text{ g N}} = 0.680 \text{ mol N}$$

$$27.18 \text{ g O} \times \frac{1 \text{ mol O}}{16.00 \text{ g O}} = 1.699 \text{ mol O}$$

$$C:H:N:O = \frac{4.758}{0.680} : \frac{6.10}{0.680} : \frac{0.680}{0.680} : \frac{1.699}{0.680} = 7:9:1:2.5$$

$$(7:9:1:2.5) \times 2 = (14:18:2:5)$$

$$\text{Empirical formula} = C_{14}H_{18}N_2O_5$$

$$\text{Empirical formula mass} = 294.2 \text{ g/mol}$$

Since molar mass and empirical formula mass are essentially equal,

$$\text{Molecular formula} = C_{14}H_{18}N_2O_5$$

LAW OF CONSERVATION OF MASS

Mass is neither created nor destroyed in an ordinary chemical reaction. The same number and types of atoms will be found in the products as in the reactants. They are simply rearranged.

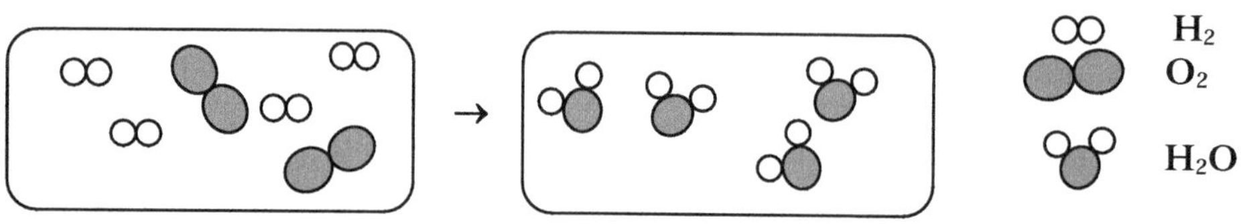

CONSERVATION OF MASS

The number of particles (atoms, formula units, molecules) in a given mass of a substance can be calculated by knowing the chemical formula of the substance and applying the mole concept. The coefficients in a balanced chemical equation indicate the relative number of particles involved in the reaction. Since the number of atoms is constant, the total mass of reactants and products must also be constant.

IDENTIFYING AND MEASURING MATTER

Atoms and molecules interact with electromagnetic radiation in a variety of ways. The most commonly used types of spectral analysis include infrared spectroscopy (IR) and ultraviolet/visible spectroscopy (UV-Vis).

INFRARED SPECTROSCOPY

IR spectroscopy involves measuring the radiation absorbed by molecules as they stretch, bend, and twist. Certain functional groups, such as –OH and –C=O, have distinctive absorbance bands on IR spectra. Each molecule produces a unique spectrum that can be used for identification purposes.

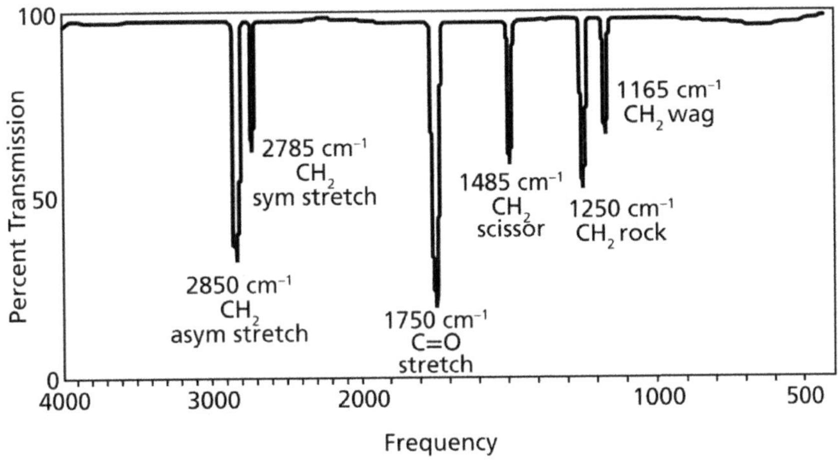

UV-VIS SPECTROSCOPY

UV-Vis spectroscopy utilizes the transition of electrons and produces diffuse bands that are generally used for quantification, not identification, of particular molecules.

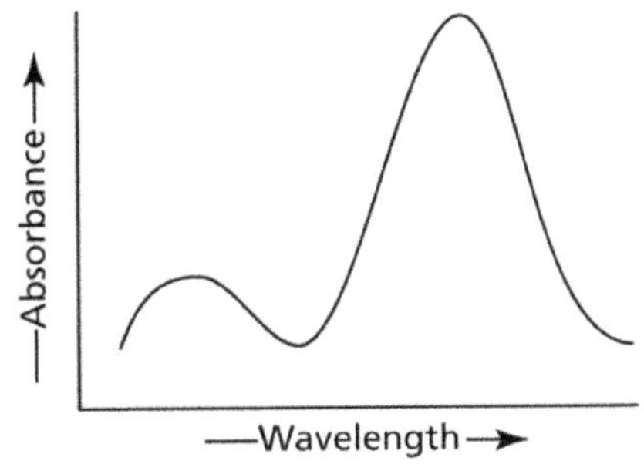

BEER'S LAW

The amount of chemical can be quantified using Beer's law (Beer-Lambert law). The absorbance of a solution at a given wavelength is directly proportional to the solution concentration.

$$A = abc$$

where A is the absorbance at the given wavelength, a is the molar absorptivity with units ($M^{-1}cm^{-1}$), b is the path length of the cell containing the chemical (usually 1 cm), and c is the concentration in molarity. Since a and b are constant, if the absorbance of a known concentration is measured, the concentration of an unknown solution can be determined from measuring the absorbance of the unknown.

When choosing a wavelength of light to use for visual spectroscopy, it is important to choose a wavelength where the sample has high absorbance. In the graph above, two peaks of absorbance are shown. The wavelength chosen for analysis should be at or close to the wavelength of maximum absorbance (λ_{max}). If transmittance is given instead of absorbance, choose the wavelength that transmits the least light as it will absorb the most.

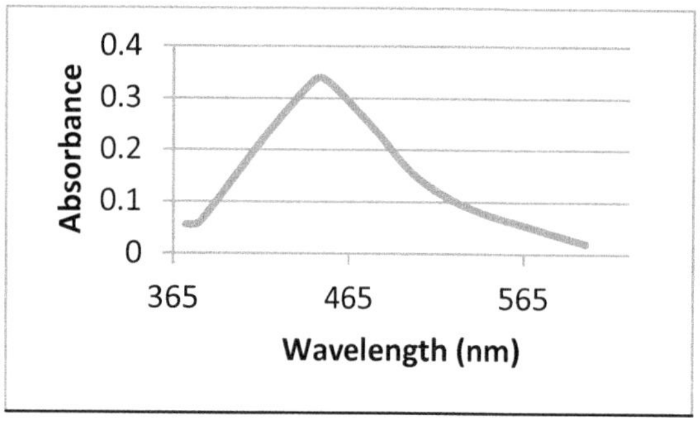

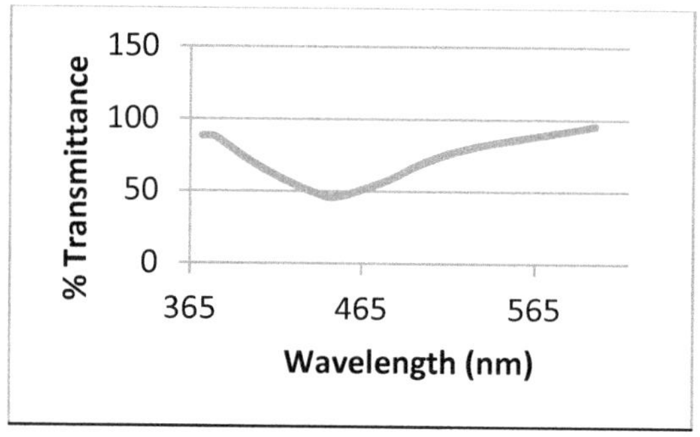

EXAMPLE: In the two graphs above, what is the approximate wavelength of light that should be used by the spectrophotometer to measure the solution?

SOLUTION: A wavelength near 450 would be best.

The graph below shows a Beer's law graph or standard curve for data representing the absorbance of light by a solution. Dilutions of known concentration are carefully prepared and their absorbance (or transmittance) is measured. If transmittance is measured, the following formula can be used to convert transmittance to absorbance: $A = -\log T$. The graph is then used with the absorbance of an unknown solution to find the concentration of the unknown solution.

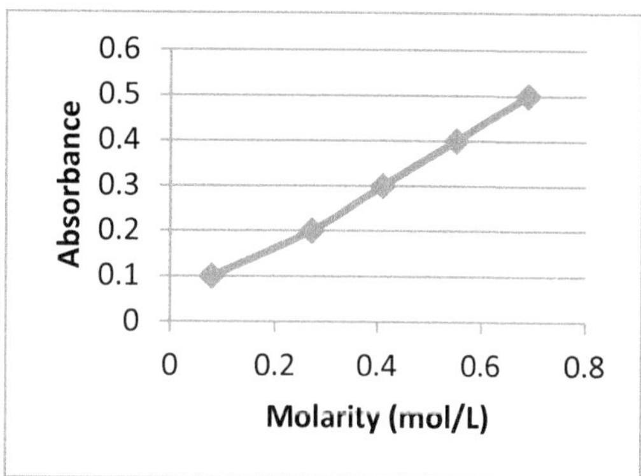

SPECTROSCOPY SUMMARY

Type of Radiation Used	Relative Energy	What It Does to the Atom/Molecule	What It Tells Us about the Atom/Molecule	Name of Spectroscopy
X-ray	↑	Removes core electrons	Measures how tightly the electrons are held by the nucleus	Photoelectron spectroscopy (PES)
Ultraviolet		Excites or removes valence electrons	Measures how tightly the electrons are held by the nucleus; Determines the identity and concentration of a molecule	PES and UV-Vis spectroscopy
Visible Light		Excites valence electrons	Determines the identity and concentration of a molecule	UV-Vis spectroscopy
Infrared		Changes the vibrations in the bonds	Identifies the functional groups within a molecule	IR (vibrational) spectroscopy
Microwave		Changes the rotations of the atoms in the bonds	Identifies the locations of hydrogen atoms within a molecule	Microwave (rotational) spectroscopy

GRAVIMETRIC ANALYSIS AND TITRATION

Analytical techniques such as gravimetric analysis and titration use conservation of mass to determine the quantity of an analyte in a solution. In gravimetric analysis, the substance of interest is selectively precipitated from a solution and the mass of the precipitate is used to determine the mass of the analyte. A titration involves the use of an indicator to determine the stoichiometric equivalence point of a reaction. Indicators are used to visually indicate the equivalence point. An indicator changes color at approximately the same point as the equivalence point. The point at which the indicator changes color is called the end point.

MULTIPLE-CHOICE QUESTIONS

No calculators can be used in this section. A periodic table and formula chart with constants are provided at the back of this book.

1st Ionization Energy	478 kJ/mol
2nd Ionization Energy	1820 kJ/mol
3rd Ionization Energy	2750 kJ/mol
4th Ionization Energy	11,600 kJ/mol

1. The table above shows the first four ionization energies for an element. Which is a correct description of the element?
   (A) The element is in Group 14 (4A) because four ionization energies are shown.
   (B) The element must be beryllium (Be) as it has 4 protons and 4 electrons.
   (C) The element must be in Group 13 (3A) because the fourth ionization energy is so much higher than the third ionization energy.
   (D) The element must be in Group 1 (1A) because the first ionization energy is lower than the successive ones.

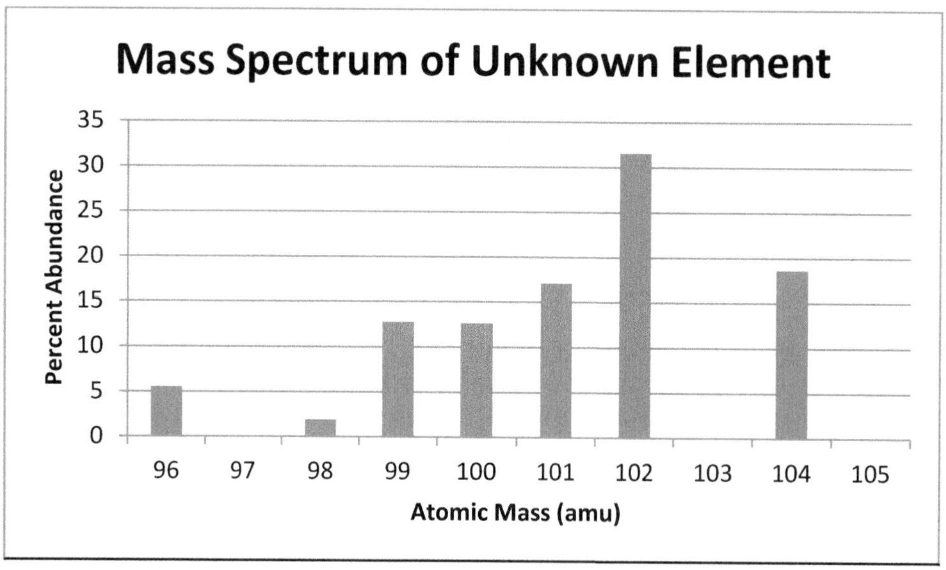

2. Name the element whose mass spectrum is shown above.
   (A) Mo
   (B) Tc
   (C) Ru
   (D) Rh

3. The formation of the most stable aluminum ion occurs when 3 electrons are removed from a neutral aluminum atom. Which statement correctly identifies the orbital location of the electron (of

the 3 removed) that requires the most energy to remove and the reason for that high energy?
(A) 3*s* because the full subshell makes it more stable.
(B) 3*s* because it experiences higher attractive forces.
(C) 3*p* because it is farther from the nucleus.
(D) 3*p* because it is the only electron in the orbital.

4. The concentration of a red-colored solution of cobalt(II) ions needs to be determined. What spectroscopic technique would be the best choice?
(A) Photoelectron spectroscopy (PES) because the electron arrangement is responsible for the color.
(B) UV-Vis spectrometry because the solution has a color.
(C) Infrared (IR) spectrometry because the solution is red in color.
(D) Mass spectrometry because knowing the mass can allow for calculation of molar concentration.

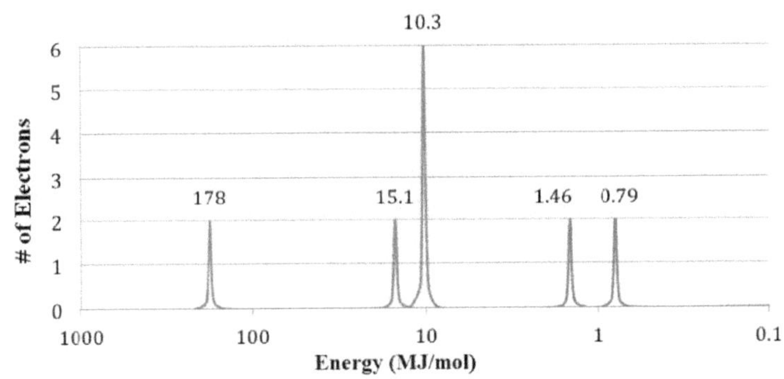

5. The complete photoelectronic spectrum of an element is shown above. Which is the correct orbital diagram showing the valence electrons for this element?

(A)

↑↓	↑↓	___	___
3*s*	3*p*	3*p*	3*p*

(B)

↑	↑	↑	↑
3*s*	3*p*	3*p*	3*p*

(C)

↑↓	↑	↓	___
3*s*	3*p*	3*p*	3*p*

(D)

↑↓	↑	↑	___
3*s*	3*p*	3*p*	3*p*

6. Glass for electronic devices needs to be more durable than typical window glass. Technology is researching various physical and chemical methods of toughening glass. In one chemical process to increase glass strength, sodium ions on the glass surface are replaced by ions of larger size by dipping the glass into a molten salt. Which type of salt should increase the glass strength?
(A) lithium nitrate
(B) magnesium nitrate
(C) potassium nitrate
(D) beryllium nitrate

7. The chemical reaction $Y_2 + 2\ Z \rightarrow 2\ ZY$ is best represented by which model?

(A)
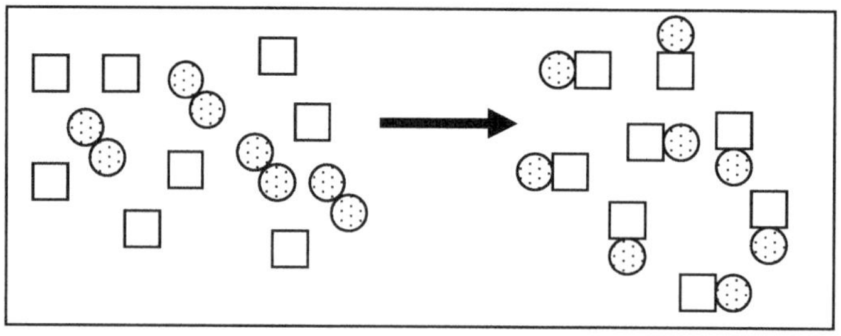

(B)
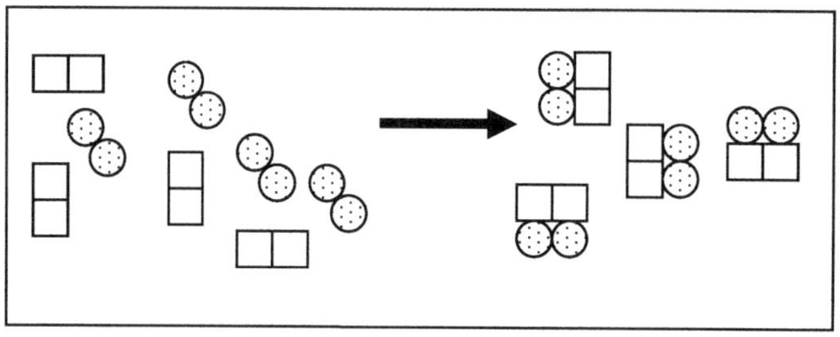

(C)
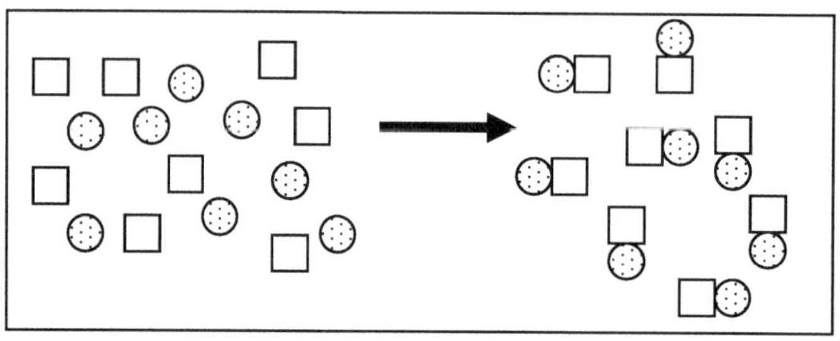

(D)
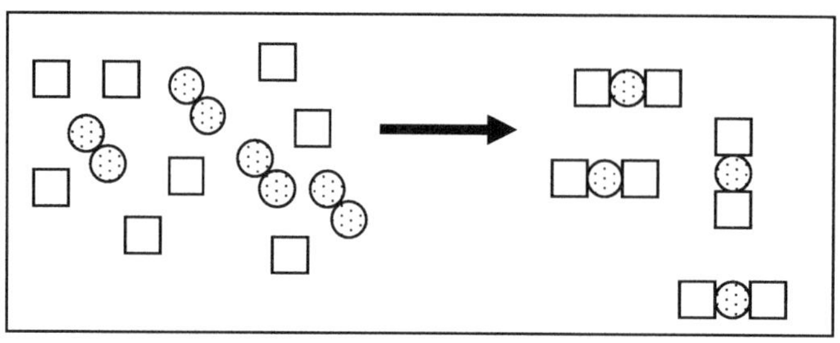

8. A student is asked to determine the concentration of a dilute solution of a weak acid, HX, by titration with a standardized solution of sodium hydroxide. Which indicator would be most appropriate for this titration?
(A) methyl yellow (pH range 2.9–4.0)
(B) methyl red (pH range 4.4–6.2)
(C) bromothymol blue (pH range 6.0–7.6)
(D) cresolphthalein (pH range 8.2–9.8)

9. The quantum mechanical atomic model replaced the Bohr atomic model because
(A) Bohr's model did not account for neutrons.
(B) the quantum mechanical model was based on mathematics.
(C) electrons act as waves as well as particles.
(D) Bohr's model could not predict the spectra for elements other than hydrogen.

10. 1.0 g of mercury is sealed in a glass tube, along with a small amount of $O_2(g)$ at 25°C. The tube is heated slightly and all of the mercury vaporizes. What mass of mercury remains in the tube?
(A) less than 1.0 g because the vapor has less mass than the liquid
(B) exactly 1.0 g because the number of atoms has not changed
(C) more than 1.0 g because the vapor occupies more space than the liquid
(D) more than 1.0 g because the mercury reacts with the oxygen

11. In 1911, Ernest Rutherford tested the atomic model existing at the time by shooting a beam of alpha particles ($^4_2$He, helium nuclei) at a very thin sheet of gold foil. He found that while most particles went straight through the foil undeflected, a very few were deflected at great angles as they passed through the foil. Why was this discovery a reason to change the atomic model?
(A) He determined that the atom contained alpha particles and was thus not indivisible as earlier models had suggested.
(B) He proved that the atom had positively charged materials as well as the negatively charged electrons proven by Thomson.
(C) He determined that the deflected alpha particles must have collided with something very dense and positively charged, discovering the nucleus.
(D) He showed that alpha particles hit very dense particles, the neutrons, and proved the existence of isotopes.

12. Silver metal reacts with sulfur in the air to form a dark gray compound that coats silver jewelry and other silver articles. Which other element would combine with silver in the same ratio as the sulfur?
(A) Br
(B) Cl
(C) N
(D) Se

13. Nitrogen and oxygen can combine in many different proportions to create different compounds. Which compound has a N:O mass ratio of 7:20?
(A) dinitrogen monoxide
(B) dinitrogen pentoxide
(C) nitrogen monoxide
(D) nitrogen dioxide

Group #	Mass of Element Q	Moles of Element Q	Increase in Mass Due to Oxygen	Moles of Oxygen
1	5.86 g	0.10	2.40 g	0.151
2	3.51 g	0.60	1.46 g	0.091
3	2.63 g	0.045	1.42 g	0.089
4	4.63 g	0.079	1.90 g	0.119

14. A laboratory experiment was performed to determine the empirical formula of a compound by burning element "Q" in the presence of oxygen to form an oxide. All student groups investigated the formation of the same compound. Partial lab data, along with mole calculations from four groups of students, are shown in the table above. Which lab group must have made an experimental error in their procedure?
(A) Group 1
(B) Group 2
(C) Group 3
(D) Group 4

15. Strontium ions can be removed from solution by reaction with an aqueous solution of sodium carbonate according to the following reaction:

$$Sr(NO_3)_2(aq) + Na_2CO_3(aq) \rightarrow SrCO_3(s) + 2NaNO_3(aq)$$

Which other aqueous solution would likely be able to precipitate the strontium ions?
(A) NaCl
(B) K_2CO_3
(C) $LiNO_3$
(D) $CaCl_2$

FREE-RESPONSE QUESTIONS

Calculators can be used in this section. A periodic table and formula chart with constants are provided at the back of this book.

1. A 1956 dime is made of a silver-copper alloy and has a mass of 2.490 g. The dime was dissolved in nitric acid and a blue solution remained. The solution was quantitatively transferred to a 100-mL volumetric flask and brought to volume with water.
   (a) A 50.00-mL portion of the solution was reacted with excess sodium chloride solution. The precipitate formed, AgCl, was filtered, washed, and dried. The mass of the precipitate was found to be 1.375 g.
      (i) Write the net ionic equation for the reaction of the silver ion with sodium chloride.
      (ii) Determine the mass of silver in the precipitate.
      (iii) What is the percent silver in the dime?

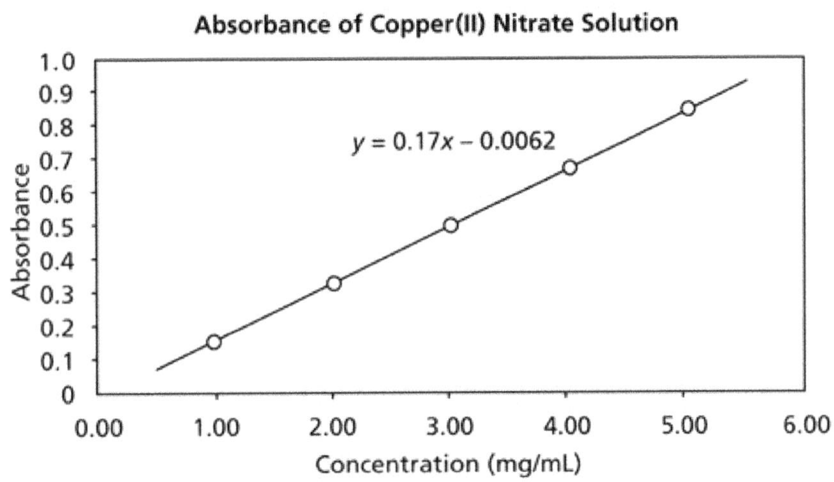

Copper(II) Nitrate Standard Curve	
Concentration (mg/mL)	Absorbance
5.00	0.843
4.00	0.672
3.00	0.507
2.00	0.336
1.00	0.161

   (b) A 1.5-mL portion of the remaining solution in the volumetric flask was analyzed with a UV-Vis spectrophotometer at 740 nm and the absorbance was 0.420. Calibration curve data and a graph are presented above.
      (i) What is the copper(II) ion concentration in the solution?
      (ii) Calculate the percent silver in the dime.

(c) The 1956 dime is known to be 90% silver.
   (i) Which method produced more accurate results? As part of your answer, calculate percent error.
   (ii) Comment on potential sources of error for the less accurate method.

2. A mysterious element was analyzed by photoelectron spectroscopy (PES) and mass spectroscopy (MS). The spectrograms are below.
   (a) Identify the mysterious element from the spectral data.
   (b) Explain how the MS supports your identification.
   (c) Explain how the PES supports your identification.
   (d) Which data (PES or MS) support that this element is paramagnetic? Justify your answer.

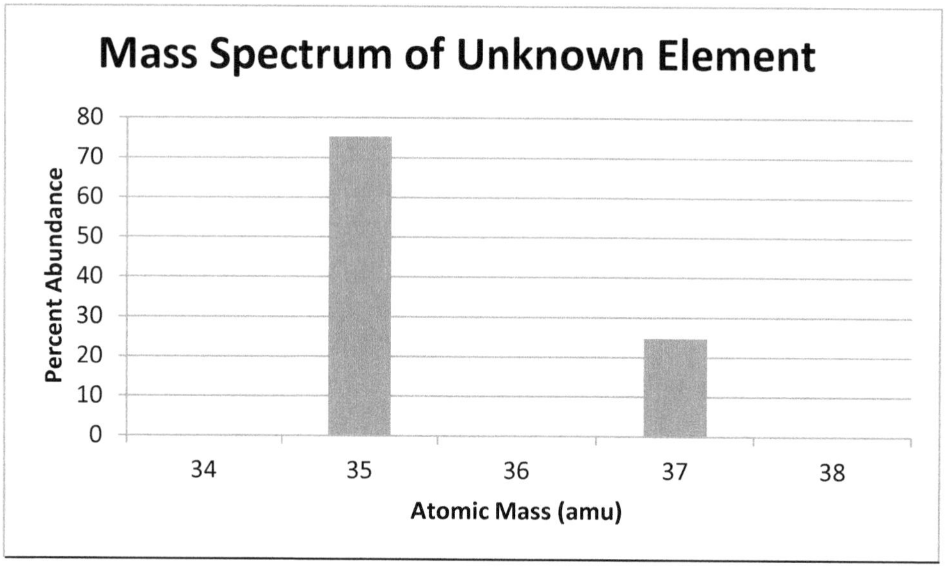

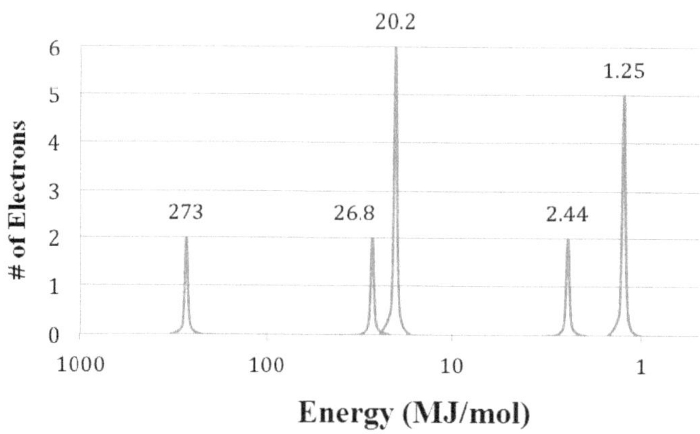

Answers

MULTIPLE-CHOICE QUESTIONS

1. C The fourth ionization energy is significantly higher than the third. This indicates that the fourth electron to be removed comes from an inner energy level. Elements in Group 13 have 3 valence electrons (LO 1.6).

2. C The average atomic mass needs to be between the masses of the lowest (96) and the highest (104) isotopes mass. Mo has an average atomic mass of 95.94 amu, Tc is 98 amu, Ru is 101.07, and Rh is 102.91. Since the average atomic mass is a weighted average, the heights of the peaks are the percentages, so the peaks of 96, 98, and 99 will approximately balance out the peak at 104. The peaks of 100 and 101 will approximately balance the 102 peak, with the mass lying closer to 101 (LO 1.14).

3. B The 3*s* electrons are on average closer to the nucleus than the 3*p* electrons so they experience a greater coulombic attraction and require more energy to remove (LO 1.5).

4. B UV-Vis spectroscopy is used to measure absorbance (directly proportional to concentration) of selected wavelengths of visible light by colored solutions. PES measures energy required to remove electrons from an atom, IR spectroscopy measures vibrations occurring in certain functional groups, and mass spectrometry separates elements and compounds by their masses (identifies masses of isotopes) (LO 1.15).

5. D The PES graph shows peaks for 1*s*, 2*s*, 2*p*, 3*s*, and 3*p*. The height of the 3*p* peak is the same as the 3*s* peak indicating that there are also 2 electrons in 3*p*. All 3*p* orbitals are degenerate (equal in energy) so 1 electron enters each orbital with parallel spins until they are all half-filled. C and D both have 2 electrons in 3*p*, but C has the 2 3*p* electrons with opposite spins instead of parallel spins (LO 1.7).

6. C The potassium ion is the only cation in the set that is larger in size than sodium. The rest are all smaller than sodium. Also, since sodium is used, another alkali metal would be an appropriate choice (LO 1.9).

7. A Z atoms are single atoms while Y must be diatomic molecules and ZY is a binary molecule. In a chemical reaction, mass and atoms must be conserved so there must be the same number and type of atoms on each side, as well as representing the particles correctly. Choice A is the only answer with equal Y and equal Z atoms on each side, as well as representing the particles correctly (LO 1.17).

8. D The titration of a weak acid with a strong base has a slightly basic equivalence point. The indicator chosen must change color in the region of approximately pH 8-10. Of these choices, only cresolphthalein changes color in that range (LO 1.20).

9. D The Bohr model of the atom is the "planetary" model with electrons in fixed orbits around the nucleus. The predicted possible energies of the electrons was not correct for elements beyond hydrogen and led to the development of the quantum mechanical model (LO 1.12).

10. B In any physical or chemical change, the total mass of the reactants and products is conserved. The mass of any element remains constant because matter is not created nor destroyed (LO 1.18).

11. C Rutherford's gold foil experiment proved the existence of the nucleus. Alpha particles are heavy (2 protons + 2 neutrons) and are positively charged. For them to be deflected, they must have hit something very dense with like charge. Since only a few were deflected, this indicated that the nucleus is very small, is positively charged, and has a high mass (LO 1.13).

12. D Chemical reactivity is strongly related to atomic structure and position in the periodic table. Sulfur is in Group 16 of the periodic table and its anions have a 2^- charge. Selenium is also in Group 16 and would also form anions with a 2^-charge. The silver compounds would be Ag_2S and Ag_2Se (LO 1.10).

13. B An N:O ratio of 7:20 would be equal to 0.5 mol N : 1.25 mol O. This is equal to a 2:5 ratio as is found in dinitrogen pentoxide (LO 1.2).

14. C A compound has a constant composition, always the same ratio by mass. Groups 1, 2, and 4 all had a 2:3 mole ratio, so all represented the same compound. Mass ratio could have been determined as well, but it would not have been as easy to calculate without a calculator. Group 3 had a 1:2 mole ratio so they must have made an experimental error (LO 1.1).

15. B Periodic trends exist in solubility. Ionic compounds containing Group 1 cations are all soluble. Compounds with Group 2 cations are less soluble. Since the precipitate contained the carbonate ion, another Group 1 carbonate is likely to precipitate the strontium ion as well (LO 1.11).

FREE-RESPONSE QUESTIONS

1. (a) (i) $Ag^+ + Cl^- \rightarrow AgCl$
   (LO 3.2)

(ii) g Ag^+ = (1.375 g AgCl)(1 mol AgCl/143.32 g AgCl) (1 mol Ag^+/1 mol AgCl) (107.87 g Ag^+/1 mol Ag^+) = 1.035 g Ag^+
(LO 1.2)

(iii) 1.035 g Ag^+ in the 50.00-mL aliquot means 2.070 g Ag^+ in the 100.00-mL sample and 2.070 g Ag in the dime.
(2.070 g Ag/2.490 g dime) × 100 = 83.13% Ag
(LO 1.19)

(b) (i) The equation of best fit, $y = 0.17x - 0.0062$, shows the relationship between absorbance and concentration of Cu^{2+}.
$x = (0.420 + 0.0062)/0.17$
$x = 2.51$ mg/mL
(LO 1.16)

(ii) $x = 0.251$ g Cu^{2+}/100 mL solution; 0.251 g Cu in the dime
(0.251 g Cu/2.490 g dime) × 100 = 10.1% Cu and 89.9% Ag
(LO 1.4)

(c) (i) The spectrometric method was more accurate with a percent error of –0.1% compared to –7.6% error in the gravimetric analysis (LO 1.3).

(ii) Potential sources of error include loss of solution during transfer for the reaction or loss of precipitate during filtration (LO 1.3).

2. (a) The element is chlorine (LO 1.14).

(b) MS shows a prominent peak at 35 amu. Chlorine's average atomic mass of 35.45 amu means the majority of the isotopes are ^{35}Cl (LO 1.2, 1.14).

(c) PES shows five occupied energy sublevels. The electrons closest to the nucleus have the highest binding energy and must be 1*s*. The height of the highest energy peak (energy 273 MJ) must correspond to 2 electrons in 1*s*. The other short peaks in the spectrum (26.8 MJ and 2.44 MJ) must also correspond to 2 electrons in *s* orbitals. The peak at 20.2 MJ is 3 times the height of the nearby peak meaning that peak must represent 6 electrons in 2*p*. The peak at 1.25 MJ is 2.5 times the height of the peak at 3*s* so that must represent 5 electrons in 3*p*. The electron configuration must be $1s^22s^22p^63s^23p^5$ (LO 1.8).

(d) PES data show that the last 5 electrons are in the 3*p* sublevel. Each 3*p* orbital holds 2 electrons so one must have a single unpaired electron (making it paramagnetic) (LO 1.6).

2

BIG IDEA 2: BONDING AND STRUCTURE

Big Idea 2

Chemical and physical properties of materials can be explained by the structure and the arrangement of atoms, ions, or molecules and the forces between them.

The chemical and physical properties of elements and compounds can be explained by the structure and arrangement of atoms, ions, or molecules and the forces between them. Big Idea 2 is divided into four chapters: Chapters 2, 3, 4, and 5. In Chapter 2, you will learn about the different types of bonds that form, the nature of those bonds, and a model that is used to predict the three-dimensional structure of covalently bonded molecules. These are *intra*molecular forces, attractions *between* atoms *within* a molecule. You will also learn about *inter*molecular forces, the attractions *between* molecules that arise as a result of these bonds. You will apply these ideas as you learn about gases (Chapter 3), liquids and solids (Chapter 4), and solutions (Chapter 5).

You should be able to:

- Identify the characteristics of ionic and covalent bonding.
- Identify the relative sizes of ions.
- Identify the effect of lattice energy on melting points of ionic compounds.
- Use electronegativity to predict the polarity of covalent bonds.
- Draw Lewis structures for molecular compounds and polyatomic ions, including resonance structures when appropriate.

- Use formal charge to identify which, of several possible, Lewis structures is the preferred structure.
- Assign molecular shapes using the VSEPR model for molecules and polyatomic ions. Assign hybrid orbitals to central atoms of molecules and polyatomic ions.
- Predict polarity of molecules, based on their shapes.
- Compare bond lengths and bond strengths in molecules.
- Identify the intermolecular forces in a substance and between substances.
- Know the difference between intramolecular forces (chemical bonds) and intermolecular forces (forces of attraction).
- Predict and explain the properties of substances based on their chemical formulas and provide explanations based on drawing representations of the substances and their intermolecular forces.

IONIC BONDING

Ionic bonding is the result of electrostatic attraction of oppositely charged ions (cations are positivity charged ions and anions are negativity charged ions). Ionic compounds form when an atom which loses electrons easily (atoms with low ionization energy) reacts with an atom that has a high affinity for electrons. Most of the time, an ionic compound is formed when a metal reacts with a nonmetal.

COULOMB'S LAW

Coulomb's law expresses the energy of interaction between a pair of ions. This is directly proportional to the charge on each ion and inversely proportional to the distance between the ions. The greater the charge on the ions, the stronger the ionic bond; if the charges are equal, then we expect a smaller ion to have a stronger ionic bond. Note that the energy will always be negative because one ion is a cation (positive) and the other an anion (negative). This formula is not on the AP® formula chart, and no calculations with it will be required. Understanding the relative values of energy with changes in charge and size is critical, however! When discussing attraction between oppositely charged ions or between electrons and the nucleus, the phrase "coulombic attraction" may be used.

$$E = (2.31 \times 10^{-19}\ \text{J} \times \text{nm})\left(\frac{Q_1 Q_2}{r}\right)$$

E is the energy of interaction between a pair of ions, in joules.

r is the distance in nanometers between ion centers.

Q_1 and Q_2 are the charges of the ions.

IONS: ELECTRON CONFIGURATION AND SIZES

PREDICTING FORMULAS OF IONIC COMPOUNDS; FORMATION OF IONIC COMPOUNDS

When a metal and a nonmetal react, each forms an ion that is isoelectronic to a noble gas (i.e., will have the same electron configuration as a noble gas). The metal loses all of its valence electrons, while the nonmetal gains electrons to complete its valence shell.

EXAMPLE: Explain how atoms of sodium and oxygen form an ionic compound.

SOLUTION: Sodium and oxygen have the following valence electron configurations:

$$\text{Na: [Ne]}3s^1 \qquad \text{O: [He]}2s^2 2p^4$$

The electronegativity of oxygen is much greater than that of sodium, so electrons are transferred from sodium to oxygen. Oxygen needs 2 electrons to fill its 2*p* valence orbitals and to achieve the electron configuration of neon. Sodium loses 1 electron to achieve the electron configuration of neon. Two sodium atoms are needed to give 1 oxygen atom the 2 electrons necessary to fill its valence orbitals. The oxide ion has an electron configuration like that of neon.

$$2\,\text{Na} \rightarrow 2\,\text{Na}^+ + 2\,e^-$$
$$\text{O} + 2\,e^- \rightarrow \text{O}^{2-}$$

To achieve an overall charge of zero, the final formula for sodium oxide will be Na_2O.

LATTICE ENERGY

Lattice energy is the energy release that occurs when separated gaseous ions are packed together to form an ionic solid. While calculations of values of lattice energy are not required, understanding relative differences in lattice energy is necessary. Lattice energy is most exothermic when small, highly charged ions are present in the compound. A difference in charge is more important than a difference in size in most lattice energies.

$$\text{X}^{m+}(g) + \text{Y}^{n-}(g) \rightarrow \text{X}_n\text{Y}_m(s) + \text{energy}$$

$$\text{Lattice energy} = k\left(\frac{Q_1 Q_2}{r}\right)$$

k is a proportionality constant dependent on the structure of the solid and the electron configuration of the ions.

r is the distance in nanometers between the ion centers.

Q_1 and Q_2 are the charges of the ions.

EXAMPLE: Which has the more exothermic lattice energy, NaCl or KCl?

SOLUTION: NaCl has the more exothermic lattice energy. Because the product $Q_1 \times Q_2$ is the same for both substances, the denominator must be considered. Na^+ is smaller than K^+, so the distance between the centers of the sodium and chloride ions is less than the distance between the potassium and chloride ions. For identical numerators, a smaller denominator produces a more exothermic lattice energy.

COVALENT BONDING

Covalent bonding involves sharing valence electrons between nuclei with similar electronegativities. A covalent or molecular compound is usually formed by bonding two or more nonmetals. There are some compounds involving metals that have covalent bonds, such as beryllium chloride, $BeCl_2$, and tin(IV) chloride, $SnCl_4$, because the electronegativities of the atoms are similar.

ELECTRONEGATIVITY

Electronegativity measures the ability of an atom in a molecule to attract shared electrons to itself. The two factors affecting electronegativity are effective nuclear charge (Z_{eff}) and electron shielding. Going from left to right across a period, there are more protons pulling on electrons (higher Z_{eff}), so the electron shells are pulled closer into the nucleus, the atomic size decreases, and the electronegativity increases. Going down a column, more electron shells are added, shielding the valence electrons from the positively charged nucleus, so the atomic size increases and the electronegativity decreases.

EXAMPLE: Place the following atoms in order of increasing electronegativity: Cs, Rb, S, Al, Sr, O.

SOLUTION: In order of increasing electronegativity: Cs < Rb < Sr < Al < S < O.

Electronegativity increases from the lower left to the upper right corner of the periodic table.

Memory Trick

To quickly recall the four most electronegative elements, in order of decreasing electronegativity, simply remember FONCl (pronounced "phone call").

POLAR BOND

The type of bond formed between 2 atoms is related to the difference in electronegativity between the 2 atoms. For identical atoms, such as a molecule of hydrogen, the electronegativity difference is zero. There is an equal sharing of electrons between the atoms, and the bond formed is a nonpolar covalent bond. When the difference in electronegativities is very small, the bond is still described as nonpolar. An important nonpolar covalent bond is the C–H bond in organic compounds.

When 2 atoms have a very large difference in electronegativity, an ionic bond forms. For example, atoms of lithium and chlorine have a large difference in electronegativity and form an ionic compound.

A polar covalent bond results when there is an unequal sharing of electrons. The difference in the electronegativities of atoms in the polar covalent bond is between that of the ionic and nonpolar bonds. In general, the farther apart 2 atoms are on the periodic table, the more polar the bond. Fluorine is the most electronegative atom, so the closer an atom is to fluorine, the more electronegative it is. An example of a polar covalent bond is HF, which has the following charge distribution.

$$^{\delta+}\text{H–F}^{\delta-}$$

EXAMPLE: Which of the following bonds will be most polar: O–F, N–F, C–F?

SOLUTION: The most polar bond will be the bond in which the difference in electronegativities is the greatest. Since you are not given electronegativities in this question, you must estimate the differences by looking at how far apart the atoms are on the periodic table. C is further from F than N or O, so the C–F bond is most polar, followed by N–F and O–F.

As a rule of thumb, bonds with an electronegativity difference less than 0.3 are usually considered to be nonpolar, and a difference greater than 1.7 is considered to be ionic. Values between 0.3 and 1.7 are considered to be polar covalent, with weakly polar covalent being closer to the 0.3 value and the strongly polar covalent being closer to the 1.7 value. The electronegative value of H is 2.1, and O has a value of 3.5. The difference is 1.4, which is why the bonds in water are considered to be strongly polar (difference of 1.4). As the difference in the electronegativity moves from a difference of 0.0 to a difference of 3.3, the percentage of ionic and covalent character changes; in fact, most bonds have characteristics of both.

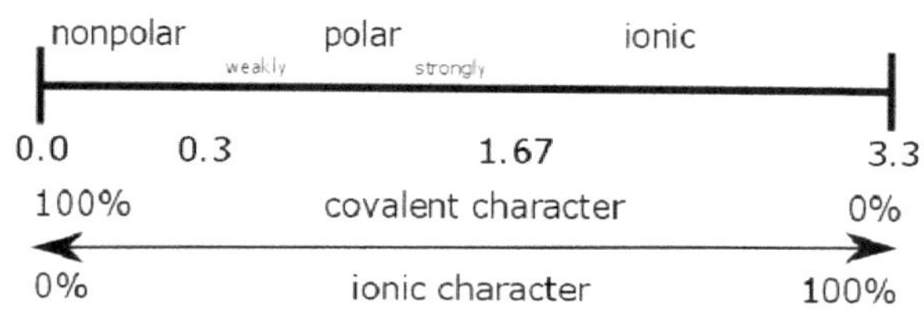

BOND POLARITY AND DIPOLE MOMENTS

When HF is placed in an electric field, the molecules tend to orient themselves with the fluoride end toward the positive pole and the hydrogen end toward the negative pole. HF is said to have a dipole moment, having two poles. An arrow points to the negative charge center, and the tail of the arrow indicates the positive center of charge.

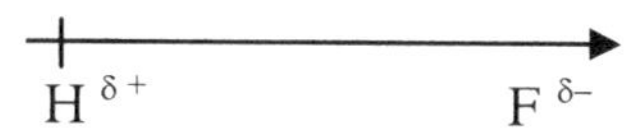

A diatomic molecule with a polar bond has a dipole moment. Molecules with more than 2 atoms can also have dipolar behavior. This is addressed in the section on molecular shapes where we look at the distribution of charge in the molecule.

MOLECULAR ORBITAL BONDING MODEL

The molecular orbital bonding model assumes that electrons occupy orbitals in a molecule in much the same way as they occupy orbitals in atoms. The electrons do not belong to any one atom. This model can be very complex for molecules consisting of two or more different elements. Molecular orbital configuration is not tested on the exam.

LOCALIZED ELECTRON BONDING MODEL

The localized electron model assumes that a molecule is composed of atoms that are bound together by sharing pairs of electrons using the atomic orbitals of the bound atoms. Lone pairs are pairs of electrons which are localized in an area around 1 atom. Bonding pairs occupy the space between atoms.

LEWIS STRUCTURES

The Lewis structure shows how the valence electrons are arranged among atoms in a molecule.

Steps for writing Lewis structures:

1. Add up the total valence electrons from each atom. For a polyatomic ion, add 1 electron for each negative charge on the anion, or take away 1 electron for each positive charge on the cation.

2. Place the least electronegative element in the middle. (Hydrogen can never be in the middle because it can form only one bond.) Draw the outer atoms around the central atom. Chemical formulas are often written in the order in which they are connected, such as HCN. The central atom is sometimes written first in a formula, when a central atom has a group of other atoms bonded to it, as in SO_4^{2-}.

3. Connect the central atom to the outer atoms using a line (or two dots) to indicate each pair of bonding electrons.

4. Arrange the remaining electrons around the atoms to follow the octet rule for Groups 14–17 and to satisfy the duet rule for hydrogen (hydrogen needs only 2 electrons to have a filled orbital).

5. Place the remaining electrons on atoms other than hydrogen to satisfy the octet rule, working your way "around" the structure and ending with the central atom.

   a. If there are not enough electrons remaining to satisfy the octet rule, and if there is a bond between C, N, O, P, and S, add a double bond for every pair of electrons you are "short" and readjust the number of electrons on each surrounding atom. A triple bond (or two double bonds) may be necessary if you need 2 extra pairs of electrons. You can only have multiple bonds between those five elements. If you are short electrons and do not have bonds between C, N, O, P, and S, then you will not be able to satisfy the octet rule (example is BF_3—see the next section).

   b. If the central atom is an element in the third period or beyond, it can exceed the octet rule.

   c. If you have the central atom being C, N, O, P, or S and one of those elements is bonded to the central atom, check to make sure that every atom has a full octet. If not, multiple bonds will be used in order for every atom to have a full octet.

6. Check that the valence electrons drawn = the total valence electrons in step 1.

EXAMPLE: Draw the Lewis structure for NF_3.

SOLUTION: Step 1:

Total valence electrons = $[1 \times N(5\ e^-)] + [3 \times F(7\ e^-)] =$

$5\ e^- + 21\ e^- = 26$ valence electrons.

Step 2: Place nitrogen in the middle and connect the fluorine atoms to nitrogen using a line to represent a single bond (1 pair of electrons).

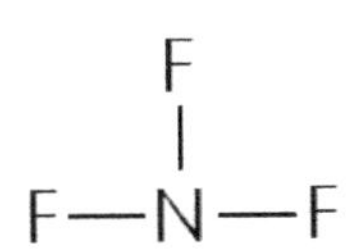

Step 3: Total the valence electrons used by counting 2 electrons for each single bond (2 × 3 single bonds = 6 electrons used). Subtract to find the number of electrons left. 26 – 6 = 20 electrons left.

Step 4: Satisfy the octet rule for each of the outer fluorine, F, atoms. Each F atom needs 6 more to complete the octet.

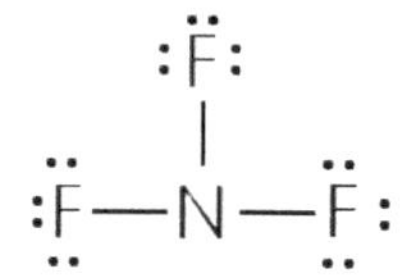

Step 5: Count electrons remaining after satisfying the outer atoms. 20 – 6(3) used = 2 electrons left. Place remaining electrons on the central atom, nitrogen, and its octet is completed.

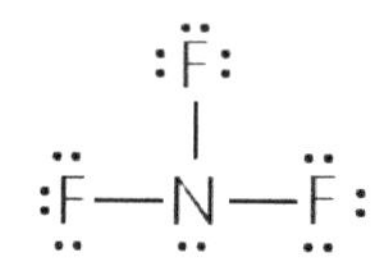

Step 6: Check that electrons drawn, 26, equals the total valence electrons in step 1.

EXCEPTIONS TO THE OCTET RULE

Elements with a Z value less than 6 often do not follow the octet rule. For example, boron tends to form compounds in which boron can have fewer than 8 electrons, such as boron trifluoride, BF_3. Boron can have 6 electrons around it.

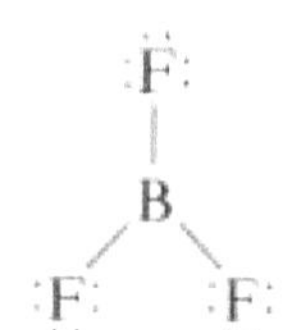

Some elements in period 3 of the periodic table and beyond can exceed the octet rule. For example, ClF_3 has 10 electrons around the central atom.

Elements in the third period have 3*s*, 3*p*, and 3*d* valence orbitals. The valence electrons in chlorine occupy the 3*s* and 3*p* orbitals. The 3*d* orbitals are empty and can hold extra electrons. Elements in the second period have only 2*s* and 2*p* orbitals. Because of the small size of these elements, exceeding the octet would cause too much electron–electron repulsion. There simply is not enough space. Also, these elements cannot have more than 8 electrons in their valence orbitals because they do not have *d* orbitals available to hold them.

RESONANCE

Resonance occurs when more than one correct Lewis structure can be drawn for a molecule or ion. This can only occur with multiple bonds, so again, with bonds that occur between C, N, O, P, and S.

EXAMPLE: Draw the Lewis structure for the nitrate ion, NO_3^-, including resonance structures.

SOLUTION: Step 1: Total valence electrons: 5 from N and 6 from each O and 1 from the negative charge on the ion = 5 + 3(6) + 1 = 24.

Step 2: Place nitrogen in the middle and connect the oxygen atoms to nitrogen.

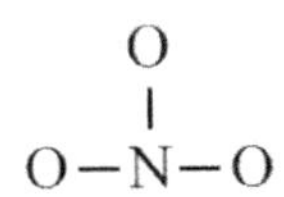

Step 3: Count valence electrons used, 2 for each single bond (2 × 3 single bonds = 6 electrons used). Subtract to find the electrons left. 24 – 6 = 18 electrons left.

Step 4: Satisfy the octet rule for each of the outer atoms of oxygen. Each O atom needs 6 more to complete the octet.

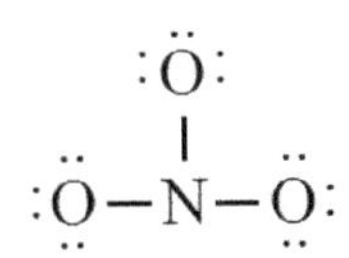

Step 5: Count electrons remaining after satisfying the outer atoms. 18 – 6(3) used = 0 electrons left. The central atom, N, still needs 2 more electrons to satisfy the octet rule. You can add a double bond for every pair of electrons you are "short."

One double bond forms between the N and the O. A double bond is made of 2 pairs of electrons. However, experimental evidence shows that the nitrate ion does not have one double bond and two single bonds. All of the three bonds are the same length, somewhere intermediate between a single bond and a double bond. The nitrate ion exists as an average of three resonance structures as pictured below.

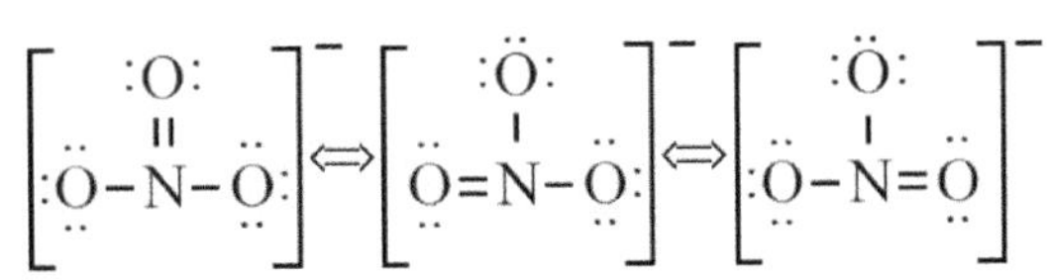

Step 6: Check that electrons drawn, 26, equals the total valence electrons in step 1.

Memory Trick

Resonance structures only occur with multiple bonds and multiple bonds only occur between CNOPS (pronounced *see nops*). Both elements in the multiple bond must be a member of CNOPS.

FORMAL CHARGE

When more than one Lewis diagram can be drawn for a species, formal charge is used to determine the most accurate Lewis structure. Formal charge is the difference between the number of valence electrons an atom has in an uncombined state and the number of valence electrons assigned to the atom when it is combined (bonded) in a molecule or ion.

The equation to calculate formal charge is shown below.

$$\text{Formal charge} = (\#\text{valence } e^-) - (\#\text{ lone } e^-) - (\frac{1}{2}\#\text{ bonded } e^-)$$

When assigning the valence electrons in a molecule we make two assumptions:

1. Electrons in the bonds are divided equally between the 2 atoms.
2. Electrons in lone pairs are entirely "owned" by the atom to which they are attached.

To evaluate different possible resonance structures, formal charge can be used. In order to determine the best resonance structure, the following rules apply:

1. Molecules with atoms that have a formal charge of zero (or as close to zero as possible) are preferred.
2. Atoms with the highest electronegativity values should be the atoms in a molecule that have a negative formal charge.
3. The total of the formal charges must equal the overall charge on the molecule (zero) or ion.

EXAMPLE: SO_4^{2-}

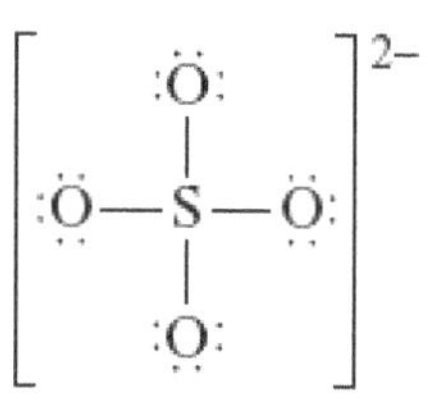

Sulfur 6 – 4 = +2
Oxygen 6 –7 = –1
2 + 4(–1) = –2

or

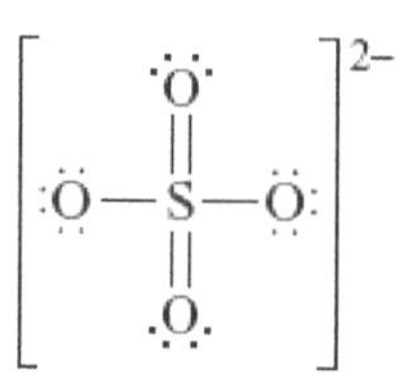

Sulfur 6 – 6 = 0
Oxygen (top and bottom O) 6 – 6 = 0
Oxygen (left and right O) 6 – 7 = –1
2(0) + 2(–1) = –2

Formal charge only needs to be considered on the AP® exam if it is specifically asked for. Either form of the sulfate ion would be acceptable on the exam unless it specifically said to minimize formal charges. The form with two double bonds minimizes formal charges.

PARAMAGNETISM AND DIAMAGNETISM

Most substances are not magnetic unless they are placed in a magnetic field. Substances that contain unpaired electrons are weakly attracted into magnetic fields and are said to be *paramagnetic*. Those whose electrons are paired are very weakly repelled by magnetic fields and are called *diamagnetic*. *Ferromagnetism* is exhibited by Fe, Co, and Ni and is much stronger than paramagnetism.

BOND ORDER

Bond order is a measure of the stability of a bond. The higher the bond order, the more stable the bond. With the localized electron model of bonding, bond order simply refers to the number of shared pairs of electrons in a bond. A single bond has a bond order of 1, a double bond has a bond order of 2, and a triple bond has a bond order of 3. A resonance structure with one single bond and one double bond would have a 1.5 bond order.

MOLECULAR STRUCTURE: THE VSEPR MODEL

The valence shell electron-pair repulsion (VSEPR) model predicts the geometries of molecules and polyatomic ions. The structure around the central atom is determined by minimizing electron-pair repulsions. Bonding and nonbonding pairs of electron are positioned as far apart as possible. Nonbonding or lone pairs of electrons require more room than bonding pairs and tend to compress the angle between bonding pairs.

Steps to apply the VSEPR model:

1. Draw the Lewis structure of the molecule.
2. Count the bonding and nonbonding electron pairs around the central atom. For structures with double or triple bonds, count the multiple bond(s) as one effective pair of electrons (or as one electron density site). If there are single electrons (as in an odd-electron molecule such as NO_2), count the odd electron as one effective pair of electrons.
3. Count the number of atoms attached to the central atom. Look in the chart below to find the arrangement that minimizes the repulsion, placing the electrons as far apart as possible. For example, BeH_2, another exception to the octet rule, has 2 electron pairs and 2 atoms attached to the central atom. If the 2 bonded pairs of electrons are placed 180° apart, the resulting linear geometry has minimal repulsion.

Molecular Shapes and Bond Angles

No. of Electron Pairs	Electron Pair Geometry (Bond Angle)	No. of Bonding Electron Pairs	No. of Lone Electron Pairs	Molecular Geometry	Example	2D Structure	Hybrid Orbital
2	Linear (180°)	2	0	Linear	BeH_2	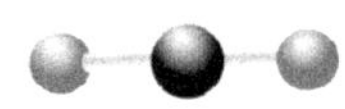	*sp*
3	Trigonal planar (120°)	3	0	Trigonal planar	CO_3^{2-}	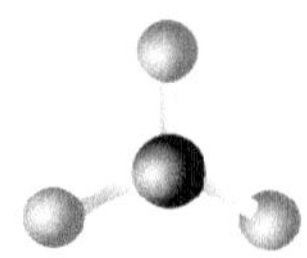	sp^2
3	Trigonal planar (<120°)	2	1	Bent	NO_2^-	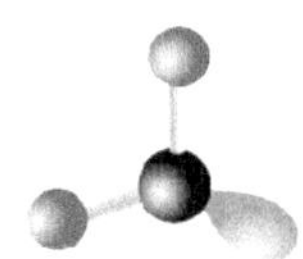	sp^2
4	Tetrahedral (109.5°)	4	0	Tetrahedral	CH_4	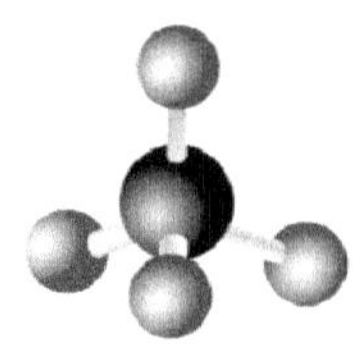	sp^3
4	Tetrahedral (<109.5°)	3	1	Trigonal pyramidal	NH_3	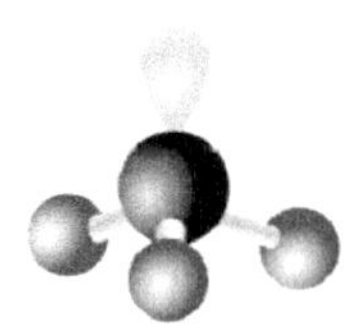	sp^3
4	Tetrahedral (<109.5°)	2	2	Bent	H_2O		sp^3
5	Trigonal bipyramidal (90°, 120°)	5	0	Trigonal bipyra-midal	PCl_5	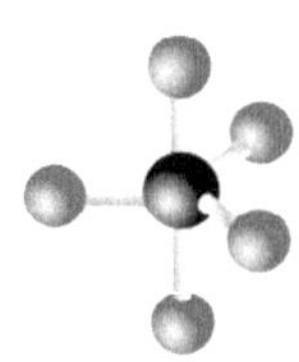	*

No. of Electron Pairs	Electron Pair Geometry (Bond Angle)	No. of Bonding Electron Pairs	No. of Lone Electron Pairs	Molecular Geometry	Example	2D Structure	Hybrid Orbital
5	Trigonal bipyramidal (<90° and <120°)	4	1	Seesaw	SF_4	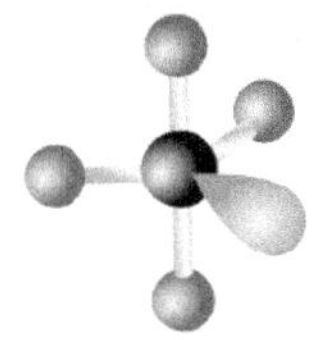	*
5	Trigonal bipyramidal (<90°)	3	2	T-shaped	BrF_3	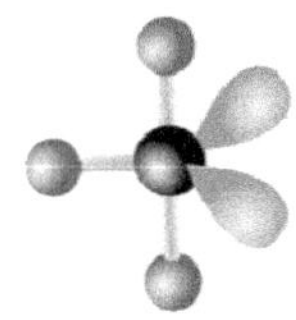	*
5	Trigonal bipyramidal (180°)	2	3	Linear	ICl_2^-	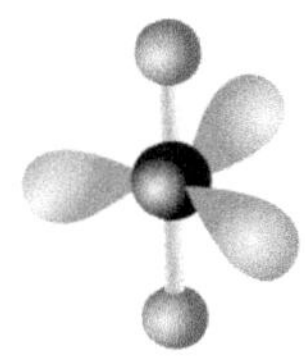	*
6	Octahedral (90°)	6	0	Octahedral	SF_6	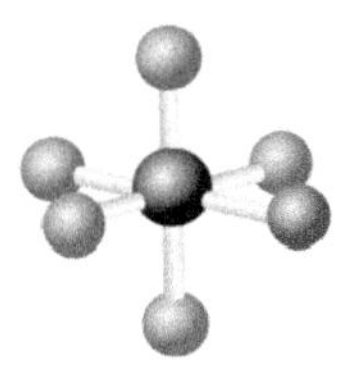	*
6	Octahedral (<90°)	5	1	Square pyramidal	BrF_5	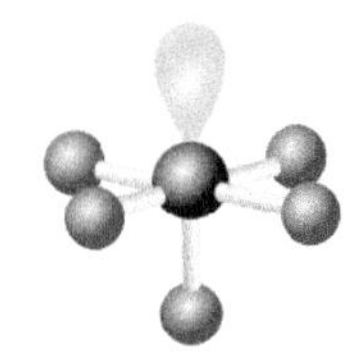	*
6	Octahedral (90°)	4	2	Square planar	ICl_4^-	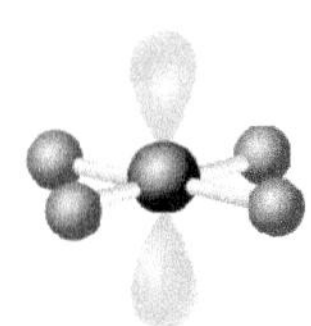	*

*Hybridization beyond sp^3 is not currently accepted by many chemists and will not be found on the AP® exam.

MOLECULAR POLARITY

If the molecule has polar bonds, the shape must be analyzed to determine whether the molecule itself is polar (has a dipole moment). The dipole moment may be given in units called debyes (D).

Symmetrical molecules with all outer atoms the same will be nonpolar because the polarity of the bonds cancels out. Nonsymmetrical molecules (bent, pyramidal, or having outer atoms that differ) will be nonpolar because the polarity of the bonds does not cancel.

Memory Trick

People **L**ike **T**o **O**pen **B**ooks is a good way to remember the five shapes that **can** be symmetrical. **P**eople = **P**lanar, **L**ike = **L**inear, **T**o = **T**etrahedral, **O**pen = **O**ctahedral, **B**ooks = trigonal **B**ipyramidal.

EXAMPLE: Predict the molecular geometry of the carbon dioxide molecule. Does this molecule have a dipole moment?

SOLUTION: First we must draw the Lewis structure for the CO_2 molecule.

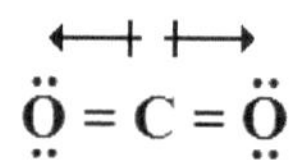

In this structure for CO_2, there are two effective pairs around the central atom (each double bond is counted as one effective pair). There are 2 atoms attached to the central atom.

According to the table, a linear arrangement is required. The electron pairs repel each other; 180° apart is the greatest angle possible.

Each C–O bond is polar, but, since the molecule is linear, the dipoles cancel out and the molecule is nonpolar.

EXAMPLE: Draw the Lewis structure for PF_5 and identify the shape and polarity.

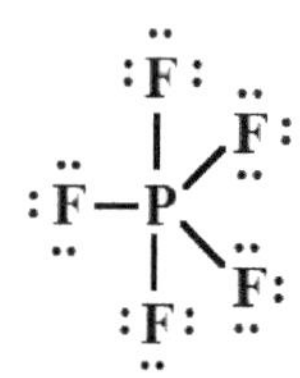

SOLUTION: PF_5 has 40 valence electrons. Connecting 5 F atoms to P, the central atom, uses 10 electrons in five single bonds. The 30 remaining electrons can be used to satisfy the octet rule for each F atom. There are 5 electron pairs around P and 5 atoms attached to P, resulting in a trigonal bipyramidal shape with bond angles of approximately 90° and 120°. The symmetrical shape causes polar bonds to cancel, resulting in a nonpolar molecule.

EXAMPLE: Draw the Lewis structure and predict the bond angles for CH_3OH.

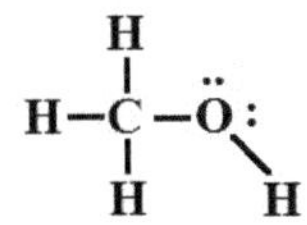

SOLUTION: This molecule has more than 1 central atom, the carbon and the oxygen. Hydrogen cannot be in the middle since it can form only one bond. Draw the correct Lewis structure and then using the rules for VSEPR, assign the molecular geometry around each central atom. The carbon atom has 4 electron pairs around it and 4 atoms attached resulting in tetrahedral geometry with approximately 109.5° bond angles. Oxygen also has 4 electron pairs around it, but only has 2 atoms attached. This leads to bent geometry around oxygen and $<109.5°$ bond angles.

EXAMPLE: Arrange the following molecules in order of increasing bond angles: H_2S, CCl_4, NF_3.

SOLUTION: First, draw the correct Lewis structure for each molecule and then assign an approximate bond angle for each.

Each of the four molecules has 4 electron pairs around the central atom, resulting in a tetrahedral arrangement of the electron pairs about the central atom and approximately a 109.5° bond angle. H_2S has 2 lone pairs on the central atom; NF_3 has 1 lone pair on the central atom. These lone pairs repel more than the bonded electron pairs and will cause the bond angle to be smaller than the expected 109.5°. The more lone pairs, the smaller the bond angle.

In order of increasing bond angles, the answer is $H_2S < NF_3 < CCl_4$.

HYBRIDIZATION

Hybridization describes the mixing of atomic orbitals to form hybrid orbitals of equal energies that share electrons during bond formation. For example, in methane (CH_4), the 2*s* orbital and three 2*p* orbitals of carbon mix together to form four sp^3 hybrid orbitals, each with the same energy. The sp^3 orbital of carbon overlaps the 1*s* orbital of hydrogen, creating a region of high electron density between the two nuclei. These overlapping orbitals form a single bond, also known as a sigma (σ) bond. Sigma bonds are very strong because the electrons are held in the region between the positively charged nuclei.

In general, the sum of the superscripts on the hybrid orbitals equals the number of effective electron pairs around the central atom. The carbon atom in CH_4 has 4 electron pairs around it. The sum of the superscripts in sp^3 equals 4.

Multiple bonds are formed by sharing electrons in the overlap of unhybridized orbitals either above and below, or in front and behind the σ bond. The carbon atoms in ethane, C_2H_4, are sp^2 hybridized. There are 3 effective electron pairs around each carbon so the sum of the superscripts is 3.The 2*s* and two of the 2*p* orbitals are hybridized to form sp^2 hybrid orbitals for each carbon atom, leaving one 2*p* orbital unused. The unhybridized $2p$ orbitals from each carbon atom overlap above and below the plane of the nuclei forming one pi (π) bond. Since the electrons in the pi bond are outside of the plane of the nuclei, this bond is weaker than a sigma bond. However, the resulting double bond consists of $1\sigma + 1\pi$ and is therefore stronger than a single bond which has only 1π bond. C_2H_4, shown below, has 5σ bonds and 1π bond. Hybridization beyond sp^3 will not be tested on the AP® exam.

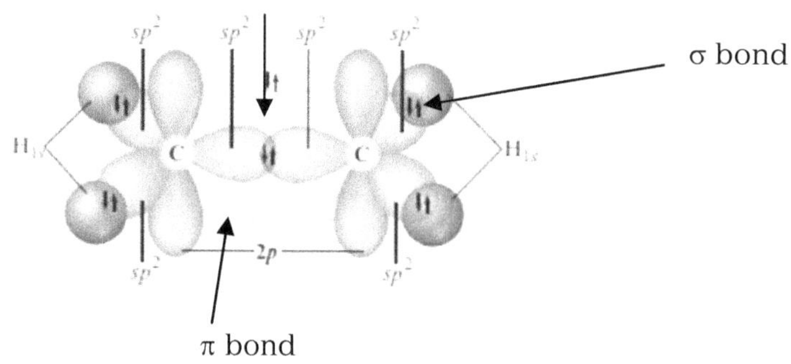

A triple bond, such as the one shown in N_2 below, consists of 1σ bond and 2π bonds.

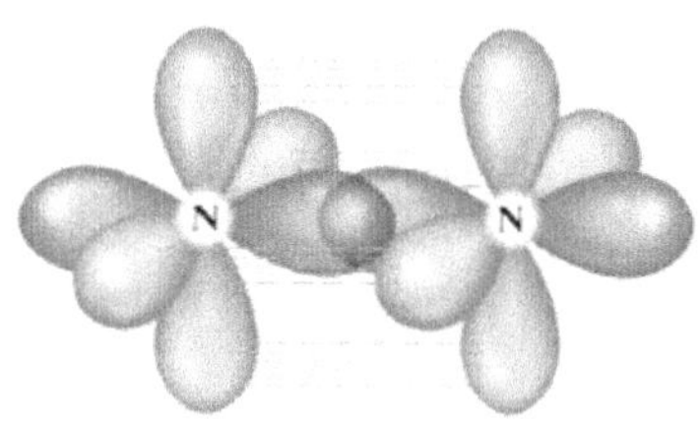

The hybrid orbitals are oriented in space according to the VSEPR model. For example, the four sp^3 orbitals have a tetrahedral arrangement (there are 4 electron groups around the central atom).

BOND LENGTH AND STRENGTH

As the number of bonds between 2 atoms increases, the bond grows shorter and stronger. The presence of pi bonds pulls the atoms closer together, while also increasing bond strength. Bond energies will be discussed in Chapter 9 of this guide.

EXAMPLE: Arrange the following molecules in order of decreasing C–C bond length: C_2H_4, C_2H_2, C_2H_6.

SOLUTION: In order of decreasing C–C bond length: C_2H_6, C_2H_4, C_2H_2.

First, draw the Lewis structure for each molecule.

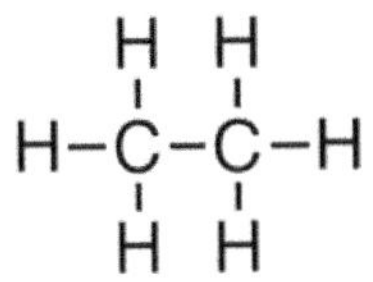

C_2H_6

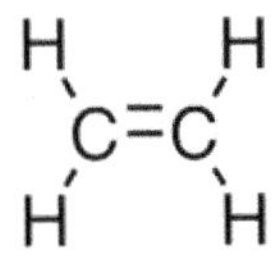

C_2H_4

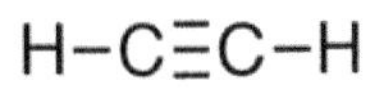

C_2H_2

The C–C triple bond is the shortest, followed by the double bond, and the single bond.

INTERMOLECULAR FORCES

Intermolecular forces (IMFs) are attractions *between* molecules, unlike intramolecular bonds which are attractions *within* a molecule. Molecules with strong intermolecular forces require more energy to separate from each other, so they will tend to have relatively high melting points, boiling points, and heats of vaporization. When in the liquid phase, they will generally have higher viscosity and higher surface tension.

DIPOLE–DIPOLE ATTRACTIONS

Polar molecules attract each other, lining up so that the positive pole of one molecule is near a negative pole of another molecule. Poles of like charges are as far apart as possible. Polar molecules generally have higher boiling points than nonpolar molecules of similar molar mass because they have dipole–dipole attractions in addition to London forces. An example of a polar molecule is SO_2.

HYDROGEN BONDING

Hydrogen bonding is an unusually strong dipole–dipole force among molecules in which hydrogen is bound to a small, highly electronegative atom such as nitrogen, oxygen, or fluorine. The hydrogen bond occurs between the hydrogen bonded to one of these electronegative atoms and a lone pair of electrons on a F, O, or N on an adjacent molecule. Hydrogen bonding can also occur within a large molecule, such as a protein—this occurs when the large molecule folds or twists around and aligns the electron pair with the H atom that is bonded to F, O, or N. This is the strongest type of intermolecular force and results in molecules with higher boiling points than molecules of similar molar mass attracted to each other by dipole–dipole forces. Examples of molecules that exhibit H-bonding include H_2O, NH_3, and HF.

Memory Trick

You can't hydrogen bond without a FON (phone).

LONDON DISPERSION FORCES

London dispersion forces (LDF) exist between all molecules and account for the boiling points of the noble gases. If there were no attractions at all between molecules, they would never liquefy. The electron cloud can experience temporary shifts that result in one side of the molecule becoming more negative than the other, causing a temporary dipole. This instantaneous dipole can induce a similar dipole on a neighboring atom.

The larger the number of electrons in the molecule, the greater the induced dipole. As the atomic number increases down a group, atoms become more polarizable, that is, the electron cloud can become polarized due to the instantaneous dipole. For example, the halogens all experience London dispersion forces, but the force becomes stronger toward the bottom of the group. F_2 and Cl_2 are found as gases in nature, Br_2 is a liquid, and I_2 is a solid due to increasing London forces as the number of electrons and size of the electron cloud increase in the group.

It is important to recognize that, particularly for organic compounds such as hydrocarbons, strength of London dispersion forces depends on the length or surface area of the molecule as much as on the number of electrons. A straight chain has more potential contact points than a branched structure with the same formula; therefore, the more contact points, the stronger the London dispersion forces will be.

EXAMPLE: Arrange C_2H_6, CH_4, and C_4H_{10} in order of increasing strength of intermolecular forces.

SOLUTION: $CH_4 < C_2H_6 < C_4H_{10}$. None of these molecules is polar, so they only exhibit London dispersion forces. The greater the number of electrons and the longer the molecule, the more polarizable the molecule becomes. C_4H_{10} has the greatest number of electrons and longest straight chain of carbons, followed by C_2H_6, and has the strongest London dispersion forces.

AP® Tip

When discussing relative strength of London dispersion forces, always mention that more electrons = more polarizeable = stronger IMF. Never discuss mass or atomic number, as it is the number of electrons that matters most!

Also, when describing intermolecular forces between polar molecules, remember that all polar molecules have both dipole–dipole and London dispersion forces; depending on the number of electrons, the London forces are often the most important part of the overall IMF.

MULTIPLE-CHOICE QUESTIONS

No calculators are to be used in this section.

1. Which of these molecules could best be represented by the ball-and-stick model shown?
(A) BeH_2
(B) CO_2
(C) H_2O
(D) SO_2

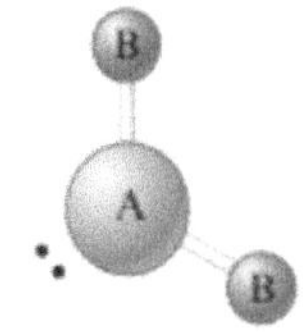

2. Which of the following substances contains both ionic and covalent bonds?
(A) NH_3
(B) CH_4
(C) NaOH
(D) C_2H_5OH

3. What type of bonding is found within a water molecule?
(A) ionic bonding
(B) polar covalent bonding
(C) nonpolar covalent bonding
(D) hydrogen bonding

4. Dinitrogen monoxide, N_2O, has two double bonds. The general structure is N=N=O. What is the formal charge on the oxygen atom in this molecule?
(A) zero
(B) positive one (+1)
(C) positive two (+2)
(D) negative one (–1)

5. The Lewis structure of which compound is best represented with resonance structures?
(A) $MgCl_2$
(B) CO_2
(C) SO_2
(D) OCl_2

6. Rank the following from weakest to strongest intermolecular force: C_2H_6, C_2H_5OH, C_2H_5Cl
(A) $C_2H_6 < C_2H_5OH < C_2H_5Cl$
(B) $C_2H_5OH < C_2H_5Cl < C_2H_6$
(C) $C_2H_5Cl < C_2H_6 < C_2H_5OH$
(D) $C_2H_6 < C_2H_5Cl < C_2H_5OH$

The following information will be used for questions 7–9.

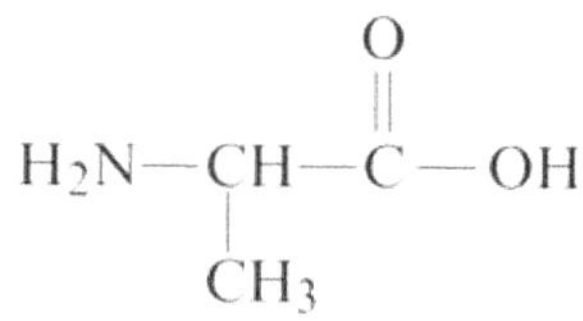

Alanine

The molecule shown above represents alanine, one of the simplest amino acids. This amino acid is considered nonessential to the human diet as it can be synthesized in the body from other compounds.

7. Which statement is correct concerning alanine?
   (A) Each carbon atom is sp^3 hybridized.
   (B) The hybridization of the nitrogen atom is sp^2.
   (C) All bond angles are approximately 109.5°.
   (D) The molecule contains 12σ and 1π bonds.

8. What is the most important intermolecular force found between neighboring alanine molecules?
   (A) covalent bonding
   (B) dipole–dipole forces
   (C) hydrogen bonding forces
   (D) London dispersion forces

9. How many unbonded pairs of electrons are found on the alanine molecule?
   (A) 0
   (B) 1
   (C) 3
   (D) 5

10. How many of the following molecules have a measureable dipole moment?
    SO_3 CCl_4 PF_3 BH_3
    (A) 1
    (B) 2
    (C) 3
    (D) 4

11. Which best describes the bonding in the cyanide ion (CN^-)?
    (A) 3σ (sigma) bonds
    (B) 2σ (sigma) bonds and 1π (pi) bond
    (C) 1σ (sigma) bond and 2π (pi) bonds
    (D) 3π (pi) bonds

12. Which statement is correct concerning hydrogen bonding?
    (A) A hydrogen bond is stronger than a covalent bond.
    (B) Hydrogen bonds can form between two parts of a single large molecule.
    (C) All bonds where electrons are shared with hydrogen are considered to be hydrogen bonds.
    (D) Dipole–dipole attractions are usually stronger than hydrogen bonds.

13. Which is correct for phosphorus tribromide?

	Hybridization of Phosphorus	Molecular Geometry
(A)	sp^3	tetrahedral
(B)	sp^3	trigonal pyramidal
(C)	sp^2	trigonal pyramidal
(D)	sp^2	trigonal planar

14. Which bond is the most polar?
    (A) F–O
    (B) H–O
    (C) Na–O
    (D) Sn–O

15. The effect of lone pairs of electrons on molecular geometry is
    (A) to push other atoms closer together because lone pairs are localized on only one nucleus, so they spread out more.
    (B) to allow the other atoms to be further apart because lone pairs take up less space than bonding pairs.
    (C) to create an aysmmetical distribution of the electrons within the atom, thus creating a polar molecule by creating dipoles.
    (D) to help to create resonance structures through the formation of pi bonds.

FREE-RESPONSE QUESTIONS

1. The element selenium is a critical dietary component for mammals. Deficiencies of selenium have been linked to many health ailments such as cancer, multiple sclerosis, and low sperm count. However, excess levels of selenium have been known to be toxic to humans and animals.
    (a) As a dietary supplement, selenium is usually found in the form of sodium selenite (Na_2SeO_3, molar mass 172.9 g/mol), a white, water-soluble, crystalline solid with a melting point of 608°C.
        (i) Is sodium selenite an ionic or molecular compound? Justify your answer using the properties given above.
        (ii) Complete the Lewis structure for the selenite ion in the box on the next page by showing all bonded and nonbonded electrons.

O Se O

O

(iii) What is the formal charge on Se in the Lewis structure that you drew?

(iv) Is the selenite ion paramagnetic or diamagnetic? Justify your answer.

(v) Using circles to represent the sodium ions and squares to represent the selenite ions, draw a bonding representation of the solid structure of sodium selenite in the box below. Use between 6 and 20 total particles.

(vi) The daily dose of this supplement is 50 μg. If the supplement is 95% pure sodium selenite, how many grams of selenium are found in 50.0 μg of a daily dose of the supplement?

(b) Selenium is commercially obtained as a byproduct of the processing of other metal ores (especially copper) which may contain copper(I) selenide (Cu_2Se, molar mass 206.1 g/mol) as an impurity. If a 100.0 kg sample of copper ore contained 0.500 g of Cu_2Se, what would be the percent selenium in the ore?

(c) Selenium has five stable isotopes. What do they have in common, and how are they different?

2. Ethylamine ($CH_3CH_2NH_2$) and dimethyl ether (CH_3OCH_3) have similar molar masses.

(a) Draw complete Lewis structures for each compound in the boxes below.

$CH_3CH_2NH_2$ CH_3OCH_3

(b) Which compound would you expect to have stronger intermolecular forces? Justify your answer.

(c) 1.000 g of one of these two compounds was burned in excess oxygen, producing 1.910 g CO_2 and 1.173 g of H_2O as the only carbon- and hydrogen-containing products. Which compound was burned? Assume a 100% yield. Support your answer with calculations.

Answers

MULTIPLE-CHOICE QUESTIONS

1. D The ball-and-stick model shows two bonds and 1 lone pair of electrons. BeH_2 and CO_2 have no lone pair, H_2O has 2 lone pairs, and SO_2 has 1 lone pair. To answer a question such as this, always count the total number of valence electrons and draw each Lewis structure (LO 2.17).

2. C NaOH is an ionic substance (metal to nonmetal). In this case, the anion is a covalently bonded polyatomic, OH^- (LO 2.17).

3. B This question is asking you to consider the nature of the bonding between hydrogen and oxygen in water (note the term *within* in the question!). An electronegativity difference (H = 2.1 and O = 3.5) of 1.4 suggests a bond of a very polar covalent nature due to the uneven sharing of the pair of electrons between hydrogen and oxygen. (*Note:* The force *between* one water molecule and another water molecule is known as a hydrogen bond.) (LO 2.17)

4. A To determine the formal charge on an atom, you need to determine the difference in the number of electrons assigned to the atom in the molecule and the number of outermost (valence) electrons on the free atom. In this case, the double bond between nitrogen and oxygen allows oxygen to "own" 2 (one-half of the 4) electrons in the bond, plus totally own the other 4 electrons. (If this is not clear to you, draw the Lewis structure for this molecule.) Since the number of electrons assigned to oxygen in the molecule and as a free atom is the same, the formal charge is zero (LO 2.21).

5. C Watch for *resonance* when it is possible to draw more than one valid Lewis structure. This tends to occur when the same kind of atoms are bonded once as a single bond and again with a multiple bond holding the same kind of atoms together. In this case, there is a single bond between sulfur and one of the oxygen atoms, and a double bond between sulfur and the other oxygen atom (LO 2.21).

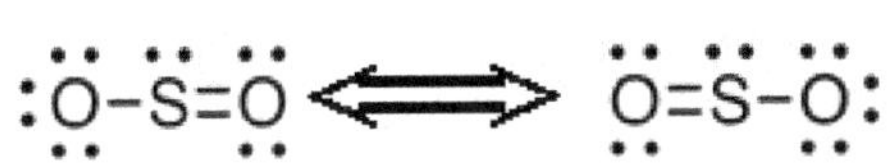

6. D Consider the intermolecular forces between molecules. All of these molecules have roughly the same molar mass, so their number of electrons will not vary significantly and their dispersion forces will be of similar strength. C_2H_6 is a nonpolar molecule with only dispersion forces. C_2H_5Cl is a polar covalent molecule and will have dipole–dipole forces as well as dispersion forces. C_2H_5OH has hydrogen bonding between molecules as well as dispersion forces. Hydrogen bonding is generally a stronger force of attraction than dipole–dipole. Dispersion forces by themselves (when # of electrons is similar) are the weakest force (LO 2.13).

7. D To answer this question, it is helpful to complete the Lewis structure by drawing in the lone pairs of electrons and also drawing bonds to the hydrogen atoms. A cannot be correct because the carbonyl carbon would be sp^3 hybridized. B cannot be correct because with its lone pair, the nitrogen atom is sp^3 hybridized. C cannot be correct because many bonds are slightly less than 109.5° and the carbonyl C bond angle is 120°. The molecule does contain exactly 12σ and 1π bonds (LO 2.21).

8. C Because of the hydrogen bonding possible both with the $-NH_2$ group (amine) and the –OH group (hydroxyl), the most important intermolecular force is hydrogen bonding (LO 2.13).

9. D Carefully draw out all lone pairs of electrons that are not shown. All elements here (other than hydrogen) follow the octet rule. You should find 1 pair on the nitrogen, and 2 pairs on each of the oxygen atoms for a total of 5 pairs (LO 2.21).

10. A In order to have a dipole moment, there must be polar covalent bonds present and an asymetrical distribution of charge. In order to determine this, you must consider both the nature of the bonding (extent of polarity) and the arrangement of these bonds (the molecular shape). If all the vectors, resulting from polar bonds, cancel each other, then there is zero dipole moment. In the case of PF_3, the shape is that of a trigonal pyramid, the vectors do not cancel each other, and the molecule is polar. SO_3 and BH_3 are trigonal planar, while CCl_4 is tetrahedral. These are all symmetrical nonpolar shapes, as long as the outside atoms are all the same (LO 2.21).

11. C The Lewis structure for the cyanide ion includes a triple bond. All bonds between atoms start with sharing a pair of electrons in overlapping orbitals. This is called a sigma bond. Multiple bonds occur when electrons in unhybridized orbitals above and below the sigma bond are shared. Each multiple bond is a pi bond, so a triple bond contains 1σ bond and 2π bonds (LO 2.21).

12. B Hydrogen bonding can occur between two different molecules or between two different regions of a very large molecule. Covalent or ionic bonds involving hydrogen are not hydrogen bonds as hydrogen bonds are a weaker attraction between the hydrogen atom (attached to a F,O, or N) on one molecule with a lone pair of electrons on a very electronegative atom (F, O, N) on

an adjacent molecule or on a distant part of the same molecule. Hydrogen bonds are stronger than dipole–dipole attractions (LO 2.13).

13. B Draw out the Lewis structure for PBr_3. The resulting structure should have 3 shared pairs and 1 unshared pair of electrons around phosphorus. This gives a trigonal pyramidal molecular geometry (molecular shape) and a hybridization of sp^3 for phosphorus (LO 2.21).

14. C The bond with the greatest polarity is the bond where the 2 atoms have the greatest electronegativity difference. This is often a metal and a nonmetal. Look at the location of the elements on the periodic table, remembering that electronegativity tends to increase up and to the right on the table. Look for the two elements that are the farthest apart (LO 2.18).

15. A Lone pairs are only localized around one nucleus, instead of being held tightly in the region between two nuclei, so electrons in lone pairs spread out and take up more space. This causes bond angles to be greater than predicted (LO 2.21).

FREE-RESPONSE QUESTIONS

1. (a) (i) Sodium selenite is an ionic compound because it consists of the sodium ion and the polyatomic selenite ion. It also has a high melting point, is soluble in water, and is a crystalline solid, all characteristic properties of ionic compounds (LO 2.22).

(ii)

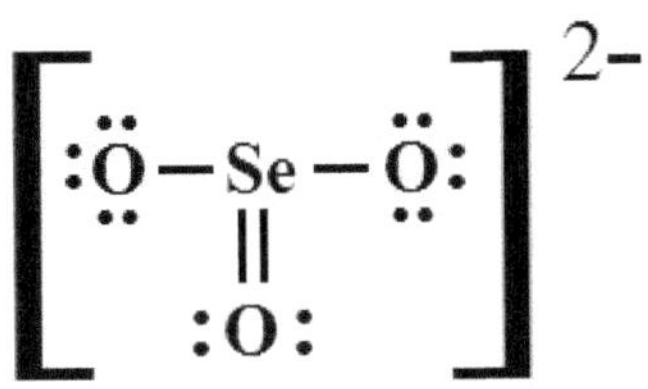

(LO 2.21)

(iii) The formal charge on Se is +2 because Se has 6 valence electrons, but has only 8/2 = 4 bonding electrons. 6 – 4 = +2 (LO 2.21).

(iv) The selenite is diamagnetic because all of its electrons are paired up. If there were odd electrons, it would be paramagnetic (LO 2.21).

(v) This needs to show a 2:1 ratio of sodium ions to selenite ions. It also needs to show a lattice structure to the compound (LO 2.23).

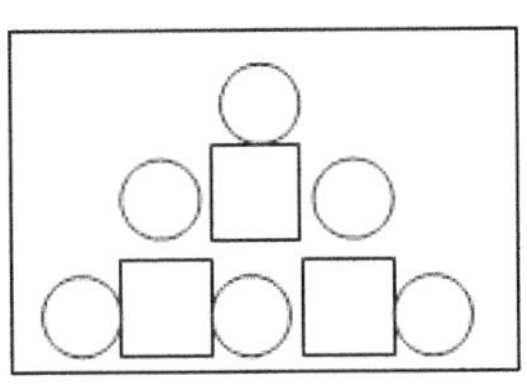

(vi) $0.95 \times 50.0\ \mu g$ supplement = $47.5\ \mu g\ Na_2SeO_3$ in a daily dose

$$47.5\ \mu g\ Na_2SeO_3 \times \frac{1\ g}{10^6 \mu g} \times \frac{1\ mol\ Na_2SeO_3}{172.9\ g\ Na_2SeO_3} \times \frac{1\ mol\ Se}{1\ mol\ Na_2SeO_3} \times \frac{79.0\ g\ Se}{1\ mol\ Se} = 2.17 \times 10^{-5} g\ Se$$

(LO 1.2)

(b)

$$0.500\ g\ Cu_2Se \times \frac{1\ mol Cu_2Se}{206.1\ g\ Cu_2Se} \times \frac{1\ mol\ Se}{1\ mol\ Cu_2Se} \times \frac{79.0\ g\ Se}{1\ mol\ Se} \times \frac{1\ kg}{1000\ g} = 1.92 \times 10^{-4}\ kg\ Se$$

$$\frac{1.92 \times 10^{-4}\ kg\ Se}{100.0\ kg\ ore} \times 100\% = 1.92 \times 10^{-4}\%\ Se$$

(LO 1.3)

(c) Each isotope of selenium has 34 protons and 34 electrons (if a neutral atom). The number of neutrons, and thus the mass number, of each isotope varies (LO 1.12).

2. (a)

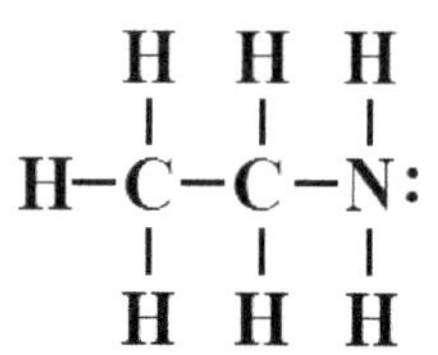

$CH_3CH_2NH_2$

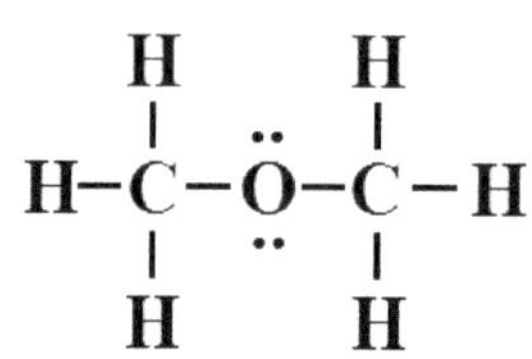

CH_3OCH_3

(LO 2.21)

(b) The ethylamine has hydrogen bonding, while the dimethyl ether has only dipole–dipole plus LDF. Hydrogen bonding is stronger than dipole–dipole and LDF (LO 2.13).

(c) $$1.910\ g\ CO_2 \times \frac{1\ mol\ CO_2}{44.01\ g} \times \frac{1\ mol\ C}{1\ mol\ CO_2} = 0.04340\ mol\ C$$

$$1.173\ g\ H_2O \times \frac{1\ mol\ H_2O}{18.02\ g\ H_2O} \times \frac{2\ mol\ H}{1\ mol\ H_2O} = 0.1302\ mol\ H$$

$$1.000\ g\ CH_3CH_2NH_2 \times \frac{1\ mol\ CH_3CH_2NH_2}{45.10\ g\ CH_3CH_2NH_2} \times \frac{2\ mol\ C}{1\ mol\ CH_3CH_2NH_2} = 0.04435\ mol\ C$$

$$1.000\ g\ CH_3OCH_3 \times \frac{1\ mol\ CH_3OCH_3}{46.08\ g\ CH_3OCH_3} \times \frac{2\ mol\ C}{1\ mol\ CH_3OCH_3} = 0.04340\ mol\ C$$

The compound is CH_3OCH_3. This could have also been proven through the determination of moles of hydrogen in each compound (LO 1.3).

3

Big Idea 2: Gases

Big Idea 2

Chemical and physical properties of materials can be explained by the structure and arrangement of atoms, ions, or molecules and the forces between them.

In this section, the laws describing gas behavior are explored. Mathematical laws relating the properties of pressure, volume, temperature, and moles of gas are considered. Densities and molar masses of ideal gases will be determined from the ideal gas law. Stoichiometric calculations will be performed for reactions involving gases. Pressures of gases in a mixture will be examined. The behavior of ideal gases will be explained by the kinetic molecular theory. Real gases will be compared to ideal gases.

Questions about gas behavior and gas laws can be either qualitative or quantitative. The free-response portion of the exam can include questions about an experiment involving gases such as the determination of the molar mass or molar volume of a gas. Essay questions about ideal or real gas behavior or the kinetic-molecular theory may also appear.

You should be able to:

- Perform calculations with the combined and ideal gas laws.
- Perform calculations with the ideal gas law to find the density or molar mass of the gas.
- Interpret or draw graphical relationships between gas variables.

- Perform stoichiometric calculations for reactions that involve gases as reactants, products, or both.
- Perform calculations with molar volume.
- Perform calculations with Dalton's law of partial pressures for a mixture of gases.
- Perform calculations for gases collected over water.
- Use the kinetic-molecular theory (KMT) and knowledge of intermolecular forces to predict ideal and nonideal gas properties.
- Draw representations of gas-phase materials that illustrate the intermolecular forces and the effects of changing physical conditions on the materials.
- Use data regarding real gases to identify deviations from ideal gas behavior and identify the molecular interactions leading to these deviations.

> **AP® Tip**
>
> You do not need to memorize the names of the individual gas laws (i.e., Dalton, Boyle's, etc.). Rather, you will need to understand the relationship between the variables and be able to apply the correct gas law for a given situation.

RELATIONSHIPS

The gas laws form a series of relationships, either direct or inverse. In a direct relationship, if one variable goes up, the other also goes up. An example would be if you work and get paid by the hour: the more hours you work, the larger the amount of your paycheck. In an inverse relationship, when one variable increases, another variable decreases. The faster you drive your car, the lower the gas mileage (miles per gallon).

Variables	Type of Relationship	Graphical Representation
Volume and Pressure	inverse	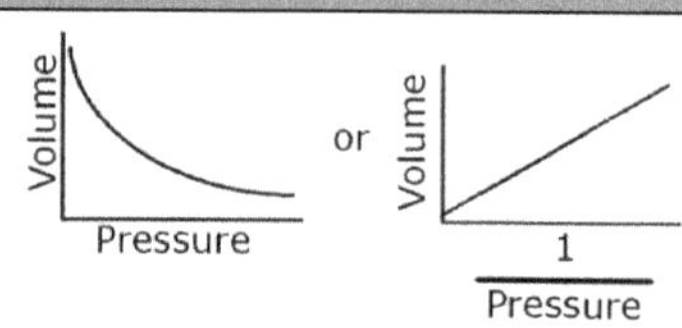
Volume and Temperature	direct	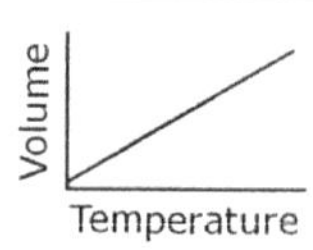

Variables	Type of Relationship	Graphical Representation
Pressure and Temperature	direct	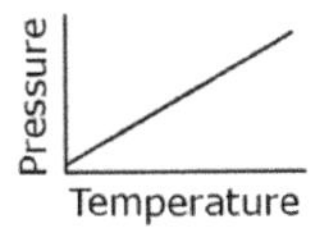
Pressure and Moles	direct	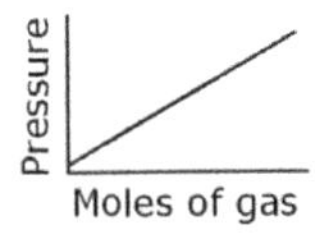

Since a calculator is not allowed in the multiple choice section, the relationship between the variables can be used to solve this problem.

EXAMPLE: If 6.6 L of gas at 315 K is heated until its temperature is 945 K, what is the new volume of the gas?

SOLUTION: Since the temperature increases by a factor of 3, and the relationship between temperature and volume is a direct relationship, the volume must also increase by a factor of 3 and the answer is 19.8 L since $3 \times 6.6 = 19.8$ L.

A visual approach to this problem might look like this:

$T_1 = 315$ K	$\times 3$ (increases by a factor of 3)	Since V and T are directly related, V must also increase by a factor of 3.	$V_1 = 6.6$ L
$T_2 = 945$ K			$V_2 = ?$

Math Trick

Think of the problem 3×6.6 as $3 \times (6 + 0.6)$ so the problem becomes $(3 \times 6) + (3 \times 0.6) = 18 + 1.8$ so the final answer is 19.8.

GAS LAWS

The table on the next page summarizes the gas laws. A typical problem involves solving for the missing variable when given a set of initial and final conditions for a gas sample. **Be sure to perform all gas calculations involving temperature using the Kelvin scale, not degrees Celsius.**

$$\text{Temperature (K)} = {}^\circ\text{C} + 273.15$$

(If the temperature is given as a whole number, then use 273. If greater precision is needed, use 273.15.)

You do not need to remember the names of the gas laws in the following table. However, one equation that is not provided on the formula chart can be very useful. The combined gas law can be used to derive several other gas laws, simply by removing the term that remains constant.

$$\frac{P_1V_1}{T_1} = \frac{P_2V_2}{T_2}$$

If the temperature remains constant, then remove the T terms and you have Boyle's law. If the pressure remains constant, remove the P terms and you have Charles' law. If the volume remains constant, remove the V terms and you have Gay-Lussac's law.

Gas Law	Equation	Definition
Boyle's law	$P_1V_1 = P_2V_2$	The product of the pressure and the volume is a constant, *k*, for a sample of gas at constant temperature.
Charles' law	$\frac{V_1}{T_1} = \frac{V_2}{T_2}$	The volume of a gas at constant pressure increases linearly with the temperature of the gas.
Avogadro's law	$\frac{V_1}{n_1} = \frac{V_2}{n_2}$	Equal volumes of gases at the same temperature and pressure contain the same number of moles of gas.
Ideal gas law	$PV = nRT$ R = universal gas constant $\frac{0.08206\ \text{L} \times \text{atm}}{\text{mol} \times \text{K}}$	A particular *state* of a gas is described by its pressure, volume, temperature, and number of moles of an ideal gas.
Dalton's law	$P_{total} = P_1 + P_2 + P_3 + \ldots$ $P_{total} = (n_{total}RT)/V$	For a mixture of gases, the sum of the pressures of the individual gases is equal to the total pressure. The total pressure depends on the total moles of gas present.

GRAPHICAL RELATIONSHIPS FOR GAS LAWS

EXAMPLE: A quantity of gas at 27.0°C is heated in a closed vessel until the pressure is doubled. To what temperature is the gas heated?

SOLUTION: Pressure is directly proportional to temperature in Kelvin units; therefore, convert the temperature to Kelvin before you do any computations. If the pressure is doubled, the temperature will also double. 27.0°C + 273.15 = 300.2 K. The answer is 600.4 K (or 327.2°C).

> **AP® Tip**
>
> On the multiple-choice section, you will likely see temperatures ending with 27 (i.e., 27°C, 127°C, 327°C, etc.). This is because when converted to Kelvin, the resulting value is a whole number (300 K, 400 K, 600 K, etc.).

GAS STOICHIOMETRY

If the pressure, volume, and temperature of an ideal gas are known, stoichiometric calculations can be performed. This is discussed in more detail in Big Idea 3.

EXAMPLE: Titanium is mined from an ore containing TiO_2, along with mineral impurities. One of the processes to determine if the ore is economically feasible to mine (usually above 5.5% titanium by mass) involves reacting the ore with bromine trifluoride and collecting the oxygen gas from the reaction to determine the percent titanium in the ore. A 1.65 kg ore sample is placed in a bath containing excess bromine trifluoride. The reaction produces 42.2 L of oxygen gas at 0.978 atm and 296 K. Is it economically feasible to mine this supply of titanium? The equation for the reaction is shown below.

$$3\ TiO_2(s) + 4\ BrF_3(l) \rightarrow 3\ TiF_4(s) + 2\ Br_2(l) + 3\ O_2(g)$$

SOLUTION:

$$PV = nRT \Rightarrow n = \frac{PV}{RT}$$

$$n = \frac{(0.978 \text{ atm})(42.2 \text{ L})}{\left(\frac{0.08206 \text{ L} \times \text{atm}}{\text{mol} \times \text{K}}\right)(296 \text{ K})} = 1.70 \text{ mol } O_2$$

$$1.70 \text{ mol } O_2 \times \frac{3 \text{ mol } TiO_2}{3 \text{ mol } O_2} \times \frac{1 \text{ mol Ti}}{1 \text{ mol TiO}} \times \frac{47.87 \text{ g}}{1 \text{ mol Ti}} = 81.4 \text{ g}$$

$$\% \text{ Ti in ore} = \frac{81.4 \text{ g}}{1650 \text{ g}} \times 100 = 4.93\%$$

Since 4.93% is less than 5.5%, it is not economically feasible to mine this sample of Ti.

MOLAR VOLUME

The conditions 0°C and 1 atm of pressure are known as standard temperature and pressure (STP). One mole of an ideal gas at STP occupies 22.414 L. The value of 22.4 L is on the chart provided with the exam. This value can be used as a conversion factor to convert between moles and volume of a gas (at STP).

EXAMPLE: What mass of helium is required to fill a 1.5-L balloon at STP?

SOLUTION:

$$1.5\text{ L} \times \frac{1\text{ mol He}}{22.4\text{ L}} \times \frac{4.0\text{ g He}}{1\text{ mol}} = 0.27\text{ g He}$$

AP® Tip

22.4 L cannot be used in problems involving molarity (M). 22.4 L/mol is *only* for a gas and *only* at STP.

MOLAR MASS OF A GAS

The molar mass of a gas can be calculated from its measured density, D. The density of a gas can also be calculated from the molar mass of the gas.

The equation relating density and molar mass can be derived from $PV = nRT$. Since n = grams/molar mass, PV = (grams/molar mass)RT.

$$\text{Molar mass} = \frac{DRT}{P} = \frac{(mass)RT}{V \times P}$$

EXAMPLE: A sample of gas with a mass of 0.800 g occupies a 256-mL flask at 100.°C and 750.0 torr. Determine the molar mass of the gas.

SOLUTION: First, convert pressure from torr to atm.

$$750.0\text{ torr} \times \frac{1.00\text{ atm}}{760.0\text{ torr}} = 0.987\text{ atm}$$

$$\text{Molar mass} = \frac{0.800\text{ g} \times \frac{0.08206\text{ L atm}}{\text{mol K}} \times 373\text{ K}}{0.256\text{ L} \times 0.987\text{ atm}} = 96.9\text{ g/mol}$$

DALTON'S LAW

For a mixture of gases in a container, the total pressure is the sum of the pressures that each gas would exert if it were alone.

$$P_{total} = P_1 + P_2 + P_3 + \ldots$$

The partial pressure of each gas can be calculated from the ideal gas law:

$$P_1 = (n_1RT)/V$$

The total pressure of the mixture, P, can be represented as:

$$P_{total} = (n_{total}RT)/V$$

The mole fraction is the ratio of the number of moles of a given component in a mixture to the total number of moles in the mixture. Mole fractions have no units and should be written in decimal form.

$$X_1 = n_1/n_{total}$$

EXAMPLE: A mixture of 1.00 g of H_2 and 1.00 g of He is placed in a 1.00-L container at 27°C. Calculate the mole fraction and partial pressure of each gas. Calculate the total pressure in the container.

SOLUTION:

$$n_{H_2} = 1.00 \text{ g } H_2 \times \frac{1 \text{ mol } H_2}{2.02 \text{ g } H_2} = 0.495 \text{ mol } H_2$$

$$n_{He} = 1.00 \text{ g He} \times \frac{1 \text{ mol He}}{4.00 \text{ g}} = 0.250 \text{ mol He}$$

Math Trick

The conversion to moles would be suitable for the multiple-choice section. The numbers make the math easy to do without a calculator—use the molar mass of hydrogen as 2.0 grams in the multiple-choice section.

$$X_{H_2} = \frac{0.495 \text{ mol}}{0.495 \text{ mol} + 0.250 \text{ mol}} = 0.664; \quad X_{He} = \frac{0.250 \text{ mol}}{0.495 \text{ mol} + 0.250 \text{ mol}} = 0.336$$

Total moles = 0.495 mol + 0.250 mol = 0.745 mol

$$P_{total} = \frac{n_{total}RT}{V} = \frac{0.745 \text{ mol} \times \frac{0.08206 \text{ L} \times \text{atm}}{\text{mol} \times \text{K}} \times 300 \text{ K}}{1.00 \text{ L}} = 18.3 \text{ atm}$$

$$P_{H_2} = X_{H_2} \times P_{total} = 0.664 \times 18.3 \text{ atm} = 12.2 \text{ atm}$$

$$P_{He} = X_{He} \times P_{total} = 0.336 \times 18.3 \text{ atm} = 6.15 \text{ atm}$$

GAS COLLECTION OVER WATER

Gases are often collected by displacement of water. A mixture of gases results due to the presence of water vapor and the gas being collected as shown to the right.

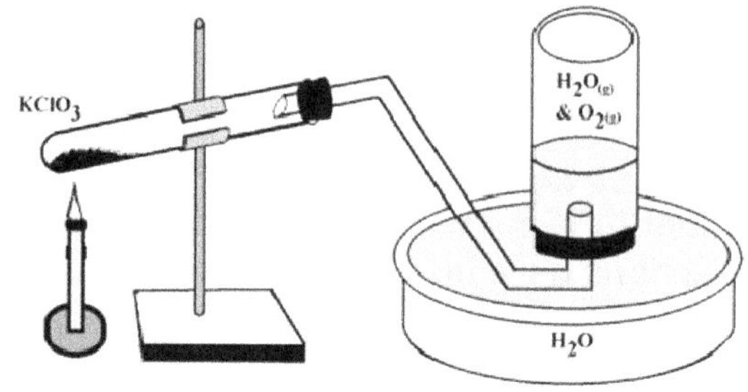

EXAMPLE: A sample with a mass of 0.986 g contains zinc and some nonreactive impurities. Excess hydrochloric acid is added and reacts with the zinc in the sample. Determine the percentage of zinc in the sample if 240.0 mL of hydrogen gas is collected over water at 30.0°C and 1.032 atm. The vapor pressure of water at this temperature is 0.042 atm.

SOLUTION: Beginning with the balanced equation:

$$Zn + 2\,HCl \rightarrow ZnCl_2 + H_2$$

One mole of zinc will produce 1 mole of hydrogen gas.

Using Dalton's law and the vapor pressure of water at the specified temperature:

$$P_{total} = P_{atm} = P_{H_2} + P_{H_2O};\ P_{H_2} = P_{atm} - P_{H_2O}$$

$$P_{H_2} = 1.032 \text{ atm} - 0.042 \text{ atm} = 0.990 \text{ atm}$$

$$n_{H_2} = \frac{P_{H_2} \times V}{RT} = \frac{0.990 \text{ atm} \times 0.240 \text{ L}}{\frac{0.08206 \text{ L atm}}{\text{mol K}} \times (30.0 + 273.15) \text{ K}} = 0.00955 \text{ mol } H_2$$

$$0.00955 \text{ mol } H_2 \times \frac{1 \text{ mol Zn}}{1 \text{ mol } H_2} \times \frac{65.4 \text{ g Zn}}{1 \text{ mol Zn}} = 0.625 \text{ g Zn}$$

$$\%\ Zn = \frac{\text{g Zn} \times 100\%}{\text{g sample}} = \frac{0.625 \text{ g}}{0.986 \text{ g}} \times 100\% = 63.4\%\ Zn$$

Note that the vapor pressure of water at various temperatures can be found in tables of known physical constants. When you solve this kind of problem, the vapor pressure of water will be provided (unless you are asked to solve for it).

KINETIC-MOLECULAR THEORY

The kinetic-molecular theory is a model that attempts to explain the properties of an ideal gas. The postulates of the kinetic-molecular theory are:

1. The volume of the individual particles of a gas can be assumed to be negligible.
2. The gas particles are in constant, random motion. The pressure exerted by the gas is due to the collisions of the gases with the walls of the container.
3. The gases are not attracted to one another.
4. The average kinetic energy of a gas is directly proportional to the Kelvin temperature.

$$(\mathrm{KE})_{\text{average}} = \frac{3}{2}RT = \frac{1}{2}mv^2$$

Note that the value of R here is 8.31 J mol^{-1} K^{-1}. The other values of R do not have the correct units for energy calculations.

EXAMPLE: Three identical sealed flasks are filled with three different gases.

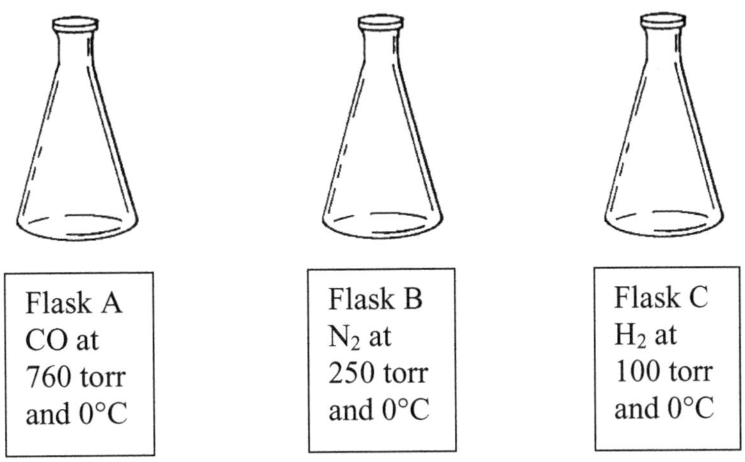

In which flask will the molecules have the greatest average kinetic energy? In which flask will the molecules have the greatest average velocity?

SOLUTION: Since temperature measures the average kinetic energy of all of the particles in a system and all three flasks are at 0°C, they will all have the same average kinetic energy. Flask C will have the greatest average velocity since hydrogen has the lowest molar mass. At any given temperature, the lightest molecules are fastest (on average).

A common method of showing gas particle velocities is with arrows (there are other ways, but the clearest to read on the test seems to be done with arrows). The length of the arrow indicates the relative speed of the molecules.

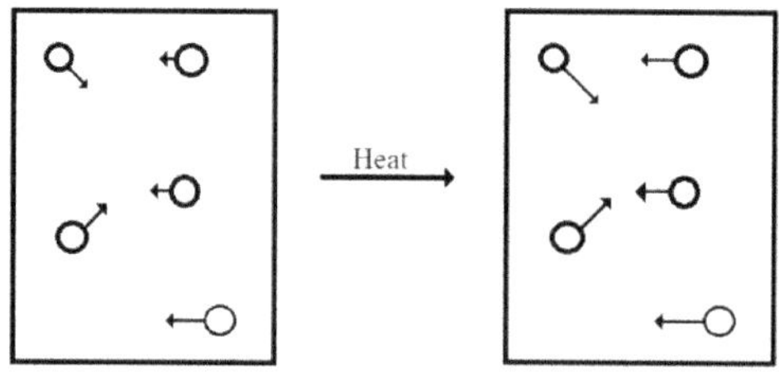

REAL GASES

The ideal gas model fails at high pressure and low temperature. Real gas molecules take up space and experience attractive forces between molecules. At high pressure, there is less empty space between molecules, and the volume of molecules becomes more significant. An ideal gas could be compressed to zero volume, but for a real gas, as the pressure doubles, the volume of empty space cannot continue to be halved. As the temperature decreases, the molecules have less kinetic energy to overcome the attractive forces between gas molecules; these attractive forces may cause the gas to condense.

The van der Waals equation modifies the assumptions of the kinetic-molecular theory to fit the behavior of real gases. The van der Waals constant *a* adjusts for the presence of intermolecular attractions; the van der Waals constant *b* adjusts for the size of the atom or molecule. Calculations involving the van der Waals equation are no longer tested; however, a qualitative understanding is certainly part of the exam.

$$[P_{obs} + a(n/V)^2] \times (V - nb) = nRT$$

EXAMPLE: Which of the following gases would you expect to deviate the most from ideal behavior: H_2, N_2, CH_4, or C_5H_{12}?

SOLUTION: An ideal gas takes up no space, and the particles have no interactions with each other. Since all of the molecules are nonpolar, the one that has the strongest IMFs will deviate the most. C_5H_{12} has the strongest London forces since it has the most polarizable electrons; it also is the largest molecule and will take up the most space.

EXAMPLE: Which of the following gases would you expect to deviate the most from ideal behavior: H_2, CO_2, CH_4, CH_3OH, or N_2?

SOLUTION: CH_3OH will deviate the most from ideal behavior. It is the only one of the five gases that is polar, so it will have strong IMFs compared to the others (especially since they all have roughly equal number of electrons, so the London forces are comparable in all five gases).

Memory Trick

Just like most students and teachers, gases behave most ideally at high temperatures (summer vacation) and low pressure (no school work)!

MULTIPLE-CHOICE QUESTIONS

No calculators are to be used in this section.

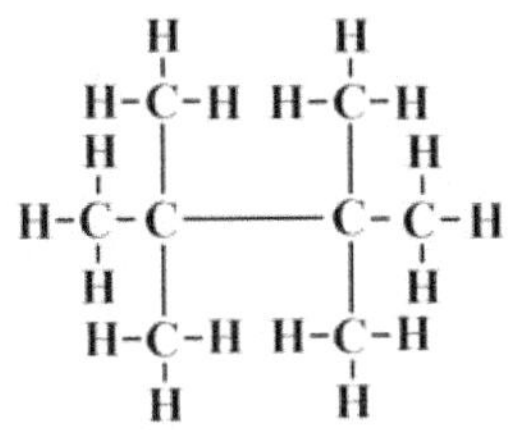

Structure 1

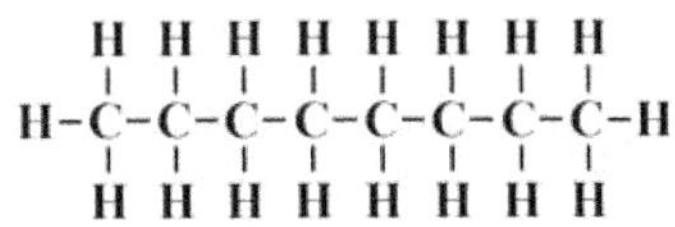

Structure 2

1. Shown above are two isomers of octane, C_8H_{18}. Assuming that they are at the same temperature and both are in the gas phase, which of the following statements is correct?
   (A) Both structures behave equally ideally as they have the same molar mass and temperature, resulting in equal average velocities. They will have the same chance of having molecular interactions.
   (B) Both structures behave equally ideally. Because they have the same number and types of atoms in their structures, they will have the same molecular volume.
   (C) Structure 1 behaves most like an ideal gas due to the branched nature of the structure resulting in a more symmetrical molecule. This results in a smaller induced dipole.
   (D) Structure 2 behaves most like an ideal gas. It has a greater induced dipole due to its linear shape causing the electron cloud to be more polarizable.

2. Under what conditions does SF_6 gas behave least like an ideal gas?
   (A) 250°C and 25 atm
   (B) 25°C and 0.225 atm
   (C) 50°C and 125 atm
   (D) 2.50°C and 0.125 atm

3. 13.1 g of an ideal gas is collected over water in a laboratory experiment. The gas has a volume of 189 mL at 30°C and 777 torr. The vapor pressure of water is 32 torr at 30°C. How would the calculated molar mass of the gas be affected if the student fails to take into account the vapor pressure of the water?
   (A) The molar mass will be too high due to the pressure used in the calculations being too low.
   (B) The molar mass will be too low due to the pressure used in the calculations being too high.
   (C) The molar mass will be too high due to the volume used in the calculations being too high.
   (D) The molar mass will be too high due to the volume used in the calculations being too low.

4. A 10.0-L cylinder contains 10.0 g NO, 40.0 g Ar, and 64.0 g O_2. What is the total pressure in the cylinder in atm at 27°C?
   (A) 0.0821 atm
   (B) 0.821 atm
   (C) 8.21 atm
   (D) 10.0 atm

5. Consider the reaction:

   $$2\ C_2H_6(g) + 7\ O_2(g) \rightarrow 4\ CO_2(g) + 6\ H_2O(g)$$

   If 6.0 g of ethane [$C_2H_6(g)$ (molar mass = 30. g/mol)] is combusted, what volume of $CO_2(g)$ will be formed at STP?
   (A) 0.20 L
   (B) 4.5 L
   (C) 9.0 L
   (D) 13 L

6. Cl_2 and F_2 react to form a gaseous product; one volume of Cl_2 reacts with three volumes of F_2 to yield two volumes of product. Assuming constant conditions of temperature and pressure, what is the formula of the product?
   (A) Cl_2F_2
   (B) ClF_2
   (C) Cl_2F
   (D) ClF_3

7. Increasing the temperature of an ideal gas from 50°C to 200°C causes the average kinetic energy to
   (A) increase by a factor of 16.
   (B) increase by a factor of 4.
   (C) increase by a factor of 2.
   (D) increase by less than a factor of 2.

8. A rigid 5.00-L vessel contains 2.00 moles of helium and 3.00 moles of hydrogen at a pressure of 10.0 atm. Maintaining a constant temperature, an additional 3.00 moles of hydrogen are added. What is the partial pressure of hydrogen gas in the vessel at the end? (Assume that the gases behave ideally.)
   (A) 6.00 atm
   (B) 10.0 atm
   (C) 12.0 atm
   (D) 20.0 atm

9. Increasing the external pressure on a gas in a flexible container at constant temperature will
   (A) increase the average kinetic energy of the gas molecules.
   (B) increase the influence of intermolecular attractions, because the molecules will be closer together.
   (C) decrease the influence of intermolecular attractions, because the molecules will be moving faster.
   (D) decrease the average kinetic energy of the gas molecules.

Questions 10–12 use the following information.

Samples of two different nonreactive gases are placed in a closed, rigid container at 273 K as shown below.

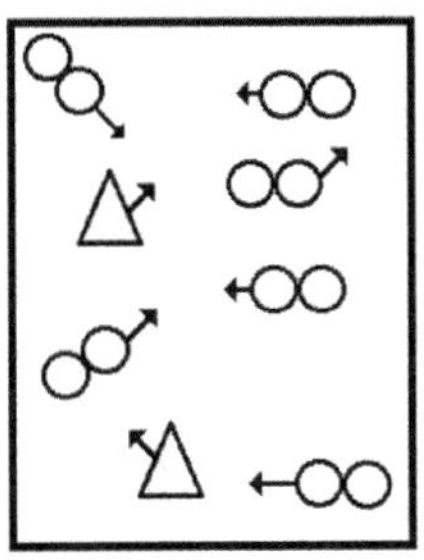

Starting Container

10. The sample consists of
   (A) two different elements.
   (B) two different compounds.
   (C) one element and one compound.
   (D) one element and two compounds.

11. If each molecule or atom shown above represents 0.1 mole and the total pressure in the container is 1200 mm Hg, what is the partial pressure of △?
   (A) 200 mm Hg
   (B) 300 mm Hg
   (C) 400 mm Hg
   (D) 600 mm Hg

12. The sample is then heated to 400 K. Which of the following containers represents the container after heating?

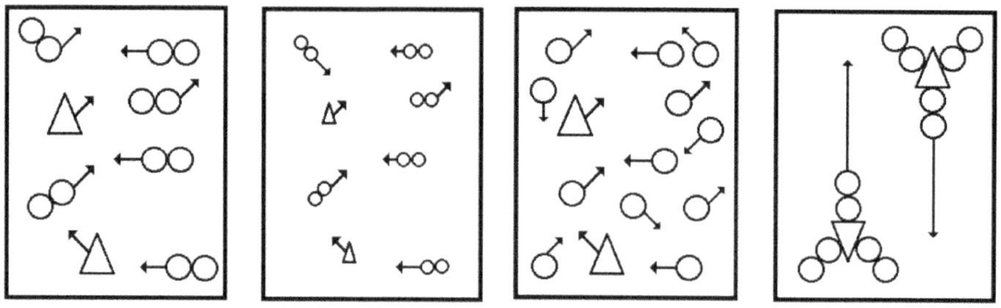

Container 1	Container 2	Container 3	Container 4
(A)	(B)	(C)	(D)

Questions 13–15 refer to the information below.

Figures 1 and 2 below show mixtures of gases in two rigid containers. The gases are identified in the key on the right. Assume that no reaction takes place between the gases and that all gases behave ideally.

Figure 1

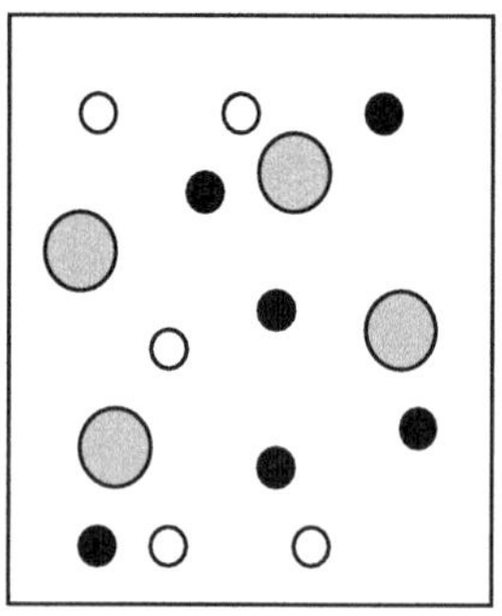

Figure 2

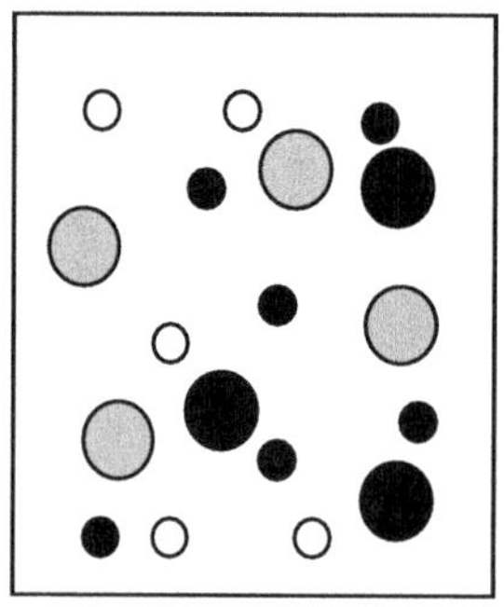

Key to Gas Symbols

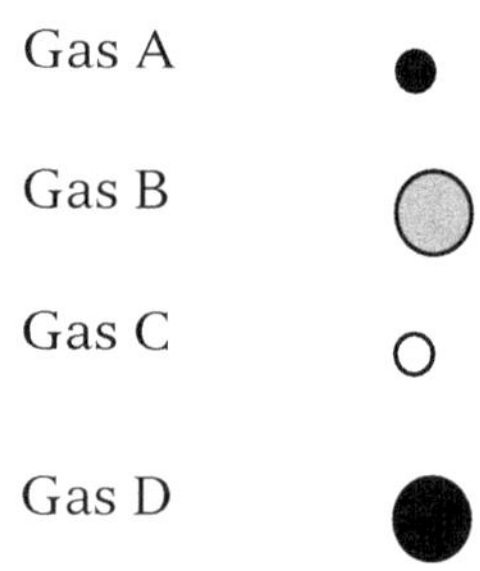

13. In Figure 1, the total pressure within the vessel is 5.00 atm. What is the partial pressure of Gas B?
(A) 0.750 atm
(B) 1.33 atm
(C) 1.60 atm
(D) 2.00 atm

14. Figure 2 represents the same container after Gas D is added to the mixture, keeping the temperature and volume constant. What happens to the partial pressure of Gas B?
    (A) P_B increases due to an increase in the number of collisions with the container's walls.
    (B) P_B decreases due to Gas D interfering with the collisions of Gas B with the container's walls.
    (C) P_B decreases due to the number of moles of all gases increasing; therefore, the mole fraction of B will decrease.
    (D) P_B stays the same, since Gas B collides with the container walls at the same rate as before.

15. The total pressure in Figure 2 is greater than 5.0 atm. Which best explains the reason for the increase in the total pressure of the system?
    (A) An increase in the number of collisions with the container's walls
    (B) An increase in the total number of moles of gas
    (C) An increase in the force of collisions with the container's walls
    (D) An increase in the number of collisions between the gas particles

FREE-RESPONSE QUESTIONS

Calculators may be used for this section.

1. Assume that two cylinders at 27°C are connected by an injection valve (a valve that only allows gas to flow in one direction), as shown below. The cylinder on the left has two electrodes, allowing a spark of electricity to flow between the cylinders and providing the activation energy for the reaction to take place. After the reaction goes to completion, the final temperature of the reaction chamber is 117°C.

   The equation for this reaction is
   $2\ H_2(g) + O_2(g) \rightarrow 2\ H_2O(g)$

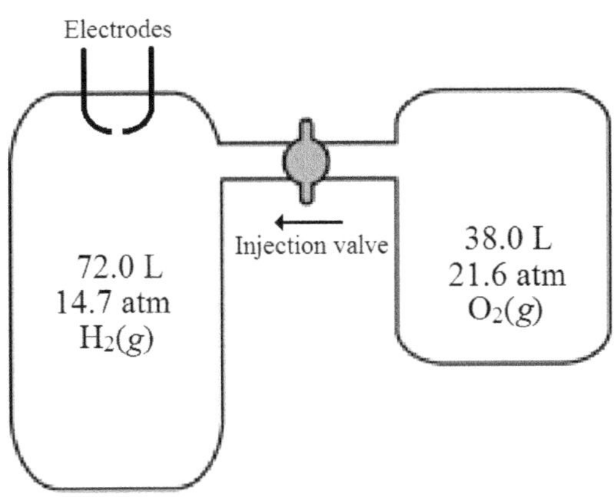

   (a) How many moles of hydrogen gas are present before the reaction takes place?

(b) Assuming that 58% of the oxygen gas is injected into the cylinder containing hydrogen, how many moles of oxygen gas remain in the cylinder on the right?
(c) After 58.0% of the oxygen is injected into the hydrogen cylinder and the mixture is sparked, the reaction goes to completion.
   (i) What is the partial pressure of the excess reactant when the reaction goes to completion?
   (ii) What is the total pressure in the cylinder when the reaction goes to completion?
(d) Which one of the two starting gases behaves most like an ideal gas? Justify your answer.
(e) The bond angle in the water molecule has been observed to be 104.5°. Explain this observation.
(f) What is the hybridization of the oxygen atom in the water molecule?
(g) Would you expect the observed pressure (P_{obs}) to be more than, less than, or equal to the calculated total pressure as determined in part (c)(ii)? Explain your reasoning.

2.

	Water	Propane
Lewis structure	H–Ö–H (O with two lone pairs)	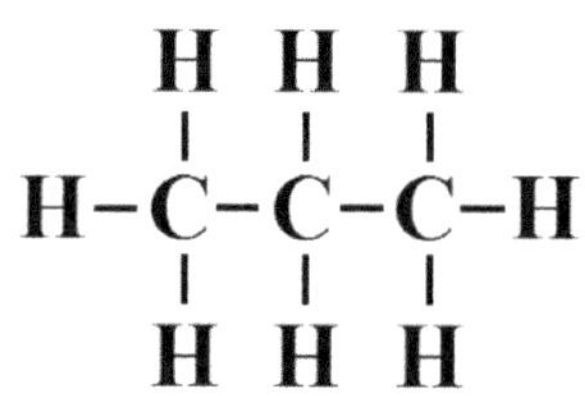
Magnetic properties	Diamagnetic	Diamagnetic
Boiling point	100. °C	–42.0°C
Molar mass	18.02 g/mol	44.1 g/mol

(a) One mole of water and 1 mole of propane are placed in separate, closed 1-L containers and both containers are heated to 110°C. After heating, the pressure in the water vapor container is found to be less than the pressure in the propane container. Explain this observation.
(b) What is the density of water vapor at 0°C and 1 atm pressure?

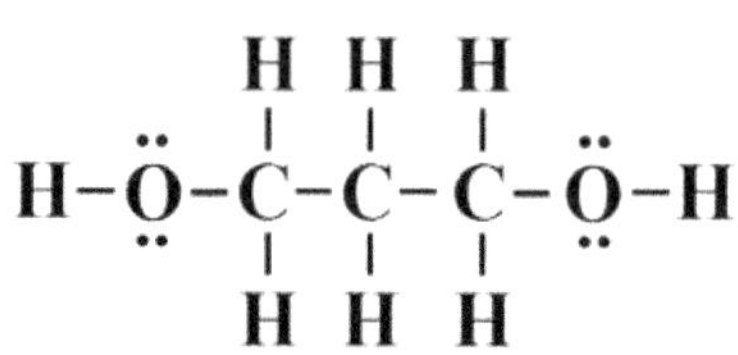

Structure A

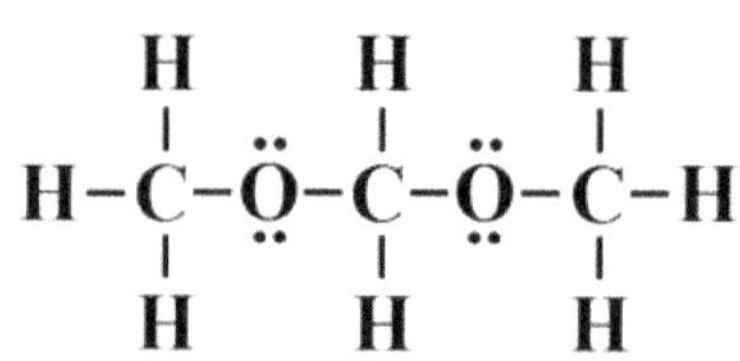

Structure B

(c) The two structures shown above both have the molecular formula $C_3H_8O_2$. If each were in the gaseous state, which would be expected to behave most ideally? Justify your answer.

Answers

MULTIPLE-CHOICE QUESTIONS

1. C Ideal gas particles have no interactions with each other. The longer a molecule, the more polarizable the electron cloud, so the stronger the London dispersion forces are. Since Structure 1 is a more compact molecule, the electron cloud is not as polarizable as Structure 2; therefore, Structure 1 will have weaker dispersion forces (LO 2.4).

2. C Ideal gas molecules have negligible volume and no interactions with each other. The conditions that would increase the interactions of the SF_6 particles the most would cause it to behave the least like an ideal gas. High pressure would push the molecules closer together which both increases interactions and makes the volume of the gas molecules less negligible. The lower the temperature, the slower the particles move and the more likely they are to interact (LO 2.12).

3. B The total pressure = 777 torr. Of this, 32 torr is due to the water vapor; hence, 777 – 32 = 745 torr of pressure is due to the dry gas and this value should have been used in the calculations. Molar mass is g/mol, so 13.1 g is divided by the number of moles. Using $PV = nRT$, the number of moles would be equal to PV/RT; therefore, using a larger number for P than should have been used would result in a larger number of moles and a smaller molar mass (LO 2.6).

4. C Use $P_{total} = (n_{total}RT)/V$ to determine the pressure. From the mass of each of the gases, you can find 0.333 mole of NO, 1.00 mole of Ar, and 2.00 moles of O_2 to give a total number of moles (n) of 3.33 moles. The trick here is to think of 3.33 as the fraction $\frac{10}{3}$.

   Therefore,

$$P = \frac{\left(\frac{10}{3}\right) \times R \times 300K}{10.0L} = 100R.$$

   Note: You do not need a calculator for this math (LO 2.6).

5. C

$$6.0gC_2H_6 \times \frac{1 \text{ mol } C_2H_6}{30.gC_2H_6} \times \frac{4 \text{ mol } CO_2}{2 \text{ mol } C_2H_6} \times \frac{22.4L\ CO_2}{1 \text{ mol } CO_2}$$

$$= \frac{1}{5} \times 2 \times 22.4 \cong \frac{1}{5} \times 45 = 9$$

   (LO 2.6)

6. D This problem assumes you understand Avogadro's law, "At constant temperature and pressure, the volume of a gas is directly proportional to the number of moles of the gas." To apply that to this problem, 1 volume Cl_2 + 3 volumes $F_2 \rightarrow$ 2 volumes Cl_xF_{3x}. The simplest formula for the product becomes ClF_3 (LO 2.6).

7. D Be careful here to note the difference between the types of temperature scales and what they mean. Even though the Celsius temperature is increased by a factor of 4, the average kinetic energy is proportional to the Kelvin temperature. In this case, the temperature increases from (50 + 273) K to (200 + 273) K, a ratio that isn't even doubled, so the temperature increases by a factor of less than 2 (LO 2.6).

8. C The initial partial pressure of the hydrogen is found using $P_{H_2} = X_{H_2} \times P_{total} = (3.0/5.0) \times 10 \text{ atm} = 6.0 \text{ atm}$. The amount of hydrogen present is doubled from 3.0 moles to 6.0 moles; therefore, the partial pressure will increase proportionally to 12.0 atm. According to KMT, the volume of the particles is negligible and the particles do not interact with each other, so it does not matter what other gases are added or what other gases are present (LO 2.4).

9. B Since volume and pressure are inversely proportional, the higher the external pressure, the smaller the flexible container will be. Since the space is smaller, the intermolecular forces are more significant when molecules are closer together (LO 2.4).

10. A Each circle of the two-circle combination is identical, so that would indicate a diatomic element, so that is not a compound. Diatomic molecules need to have two different elements to be considered a compound. The triangle is not made out of smaller shapes, so that would indicate that it is also an element (LO 2.5).

11. B The partial pressure = mole fraction × total pressure. In this case, partial pressure of gas triangle = [0.2/(0.2 + 0.6)](1200mm Hg) = (¼)(1200 mm Hg) = 300 mm Hg (LO 2.4).

12. A Since the sample is being heated, the sample's average kinetic energy should increase. As the average kinetic energy increases, each particle should have more kinetic energy, which is indicated by a longer arrow (LO 2.5).

13. B The partial pressure = mole fraction × total pressure. In this case, partial pressure of Gas B = (4/15)(5.00 atm) = 1.33 atm (LO 2.4).

14. D Since the amount of Gas B remains the same, there would be no change in the partial pressure of this gas. The same number of particles is hitting the walls of the container with the same force, so the pressure must be the same (LO 2.4).

15. A Since the total number of moles of gas in the vessel has now increased, the total pressure will also increase; there are more

collisions with the walls of the container by the increased number of gas particles (LO 2.4).

FREE-RESPONSE QUESTIONS

1. (a)

$$PV = nRT \Rightarrow n = \frac{PV}{RT} \Rightarrow n = \frac{(14.7\text{ atm})(72.0\text{ L})}{\left(\frac{0.08206\text{ L} \times \text{atm}}{\text{mol} \times \text{K}}\right)(27 + 273\text{ K})} = 43.0\text{ mol H}_2$$

(LO 2.6)

(b)

$$PV = nRT \Rightarrow n = \frac{PV}{RT} \Rightarrow n = \frac{(21.6\text{ atm})(38.0\text{ L})}{\left(\frac{0.08206\text{ L} \times \text{atm}}{\text{mol} \times \text{K}}\right)(27 + 273\text{ K})} = 33.0\text{ mol}$$

58.0% of 33.3 mol = 19.3 mol O_2 injected, so 14.0 mol O_2 are left in cylinder (LO 2.6).

(c) (i) For this part of the problem, two different techniques will be shown.

Since we will need the moles of water produced in this reaction and it is a limiting reagent problem, we will convert both starting amounts into the product.

$$19.3\text{ mol O}_2 \times \frac{2\text{ mol H}_2\text{O}}{1\text{ mol O}_2} = 38.6\text{ mol H}_2\text{O} \quad \textit{and}$$

$$43.0\text{ mol H}_2 \times \frac{2\text{ mol H}_2\text{O}}{1\text{ mol H}_2} = 43.0\text{ mol H}_2\text{O}$$

Since 38.6 moles is less than 43.0 moles, there are only 38.6 moles of H_2O produced and the oxygen is the limiting reagent (may also be called limiting reactant). To determine the amount of excess reactant, use the limiting reactant and do a stoichiometry problem.

$$19.3\text{ mol O}_2 \times \frac{2\text{ mol H}_2\text{O}}{1\text{ mol O}_2} = 38.6\text{ mol H}_2 \Rightarrow$$

$$43.0\text{ mol H}_2 - 38.6\text{ mol H}_2 = 4.4\text{ mol H}_2\text{ excess}$$

Reaction	$2\ H_2$ +	O_2 →	$2\ H_2O$
Before	43.0 mol	19.3 mol	0
Change	$-2x$	$-x$	$+2x$
After	$43.0 - 2x$	$19.3 - x$	$2x$

The second technique is using a BCA table. If H_2 is limiting, $2x$ will be 43.0 and x will be 21.5. That is impossible as the remaining amount of O_2 would be negative. O_2 must be limiting and x must be 19.3 mol. That results in $2x = 38.6$ mol and $43.0 - 38.6 = 4.4$ mol of excess H_2 remaining.

$$PV = nRT \Rightarrow P = \frac{nRT}{V}$$

$$= \frac{(4.4 \text{ mol } H_2)\left(\frac{0.08206 \text{ L} \times \text{atm}}{\text{mol} \times \text{K}}\right)(117 + 273\text{K})}{72.0 \text{ L}} = 2.0 \text{ atm } H_2$$

(LO 1.4)

(ii) Total moles of gas would be 4.4 mol H_2 + 38.6 mol H_2O = 43.0 mol total gas.

$$PV = nRT \Rightarrow P = \frac{nRT}{V}$$

$$= \frac{(43.0 \text{ mol gas})\left(\frac{0.08206 \text{ L} \times \text{atm}}{\text{mol} \times \text{K}}\right)(117 + 273\text{K})}{72.0 \text{ L}} = 19.1 \text{ atm}$$

(LO 2.6)

(d) Since both gases are nonpolar and only have London dispersion forces as the IMF, the one with more polarizable electrons will deviate the most from ideal behavior and that will be O_2 (LO 2.4).

(e) There are four sites of electron density around the oxygen atom resulting in a tetrahedral structure with bond angles of approximately 109.5°. Water has 2 bonded and 2 nonbonded (lone) pairs of electrons, which results in a bent structure. The nonbonded pair of electrons is pulled more toward the central atom, causing more repulsion between the nonbonded pairs of electrons and the bonded pairs of electrons. This results in a slightly smaller bond angle (~104.5°) (LO 2.21).

(f) sp^3 (LO 2.21)

(g) Water's major IMF is hydrogen bonding. Since we are just above the boiling point of water (117°C) and at a moderate pressure (19. atm), the intermolecular forces will be significant and the observed pressure will be less than that calculated in (c)(ii) (LO 2.4).

2. (a) Water molecules are polar and attracted to each other by hydrogen bonding. At temperatures just above water's boiling point, these attractions will be more important. Because of this attraction, the molecules will hit the container walls less frequently and the pressure will be less than predicted by the ideal gas law. Nonpolar propane molecules only experience London dispersion forces of attraction which are weak and easily overcome, especially since this sample is well above its boiling point (LO 2.4).

(b) Density is mass/volume. The density of an object is independent of the amount of material, so you can choose any amount you want. Since we are at STP conditions and you know that 1 mole of any gas at STP is 22.4 L and the mass of 1 mole of water is 18.02 g, the density must be 18.02 g/22.4 L = 0.804 g/L (LO 2.6).

(c) Structure B will behave the most like an ideal gas. Structure A has two O–H bonds, so particles will have a large amount of intermolecular interactions (via hydrogen bonding, dipole–dipole, and London dispersion forces). Structure B has weaker intermolecular interactions (it exhibits dipole–dipole interactions plus London dispersion forces) and will behave more ideally (LO 2.4).

4

BIG IDEA 2: LIQUIDS AND SOLIDS

Big Idea 2

Chemical and physical properties of materials can be explained by the structure and the arrangement of atoms, ions, or molecules and the forces between them.

The forces that hold together solids and liquids, the condensed states of matter, are similar. The effect of these forces on properties such as surface tension and vapor pressure will be reviewed. The bonding models, structure, and properties of liquids and solids will be discussed. Changes in state from solid to liquid to gas will also be considered.

You should be able to:

- Describe the intermolecular forces that account for observed properties of liquids such as surface tension, capillary action, viscosity, vapor pressure, and boiling point.
- Draw or interpret diagrams that illustrate the observed differences between solids and liquids and identify the interparticle forces (also known as intermolecular forces) that account for those differences.
- Describe the differences in the structures and interparticle forces present in solids (molecular, covalent network solids, ionic, and metals) and how these affect properties such as melting point, malleability, and electrical conductivity.
- Use your understanding of the types of bonding present in solids (molecular, covalent network solids, ionic, and metals)

to draw or interpret pictures that illustrate the observed differences in the types of bonding and identify the interparticle forces that account for those differences.
- Describe the structure and properties of metal alloys.

INTERMOLECULAR FORCES

Intermolecular forces (IMF) were introduced in Chapter 3. In this chapter, you will see the effect of intermolecular forces on properties of liquids and solids such as melting point, boiling point, surface tension, viscosity, and vapor pressure. When molecular compounds such as water undergo changes of state, intermolecular forces are involved, rather than the intramolecular forces (covalent bonds). When ionic compounds such as sodium chloride change state, ionic bonds are broken.

PROPERTIES OF LIQUIDS

SURFACE TENSION

↑IMF = ↑Surface Tension

Molecules at the surface of a liquid are pulled inward by the molecules beneath them. The surface tension of a liquid is the resistance of a liquid to increasing its surface area. Liquids with strong intermolecular forces tend to have high surface tensions.

EXAMPLE: Equal-sized drops of Hg, Br_2, and H_2O are placed onto the same glass surface. Which of the three liquids will have the most spherical shape drop?

SOLUTION: Hg drops will have the most spherical shape due to the metallic bonding which occurs between mercury atoms. Metallic bonding is stronger than both the hydrogen bonding in water and the London dispersion forces of bromine.

CAPILLARY ACTION

The rise of liquids up very narrow tubes is called capillary action. Water and dissolved nutrients move upward through plants via capillary action.

Adhesive forces, the attractions between a liquid and its container, cause the liquid to creep up the walls of the container. Adhesive forces are opposed by *cohesive forces,* the intermolecular forces in the liquid that try to decrease the liquid's surface area. Molecules with strong intermolecular attractions will have stronger cohesive forces than adhesive forces.

EXAMPLE: Two different liquids are placed in separate glass tubes. Liquid A forms a concave meniscus. Liquid B forms a convex meniscus. Why do the meniscuses of Liquids A and B have two different shapes? Suggest a possible identity for each liquid.

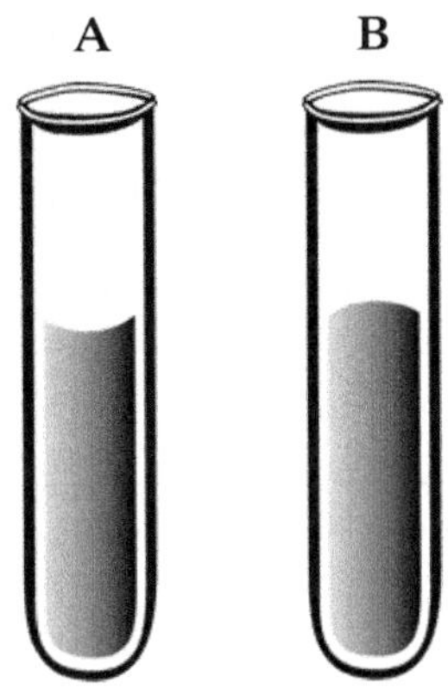

SOLUTION: Liquid A's adhesive forces are stronger than its cohesive forces, as in water. Liquid B's cohesive forces are stronger than its adhesive forces, as in mercury.

VISCOSITY

↑IMF = ↑Viscosity

Viscosity is a measure of a liquid's resistance to flow. The greater its resistance, the more slowly it will flow. Molasses, motor oil, and glycerol are viscous liquids. Highly viscous liquids have large intermolecular forces. More complex molecules are more viscous because they tangle up.

Viscosity decreases with increasing temperature because as temperature (average kinetic energy) increases, more molecules will have the required kinetic energy to overcome the intermolecular forces of the liquid.

STRUCTURE AND TYPES OF SOLIDS

CRYSTALLINE SOLIDS

Crystalline solids are classified according to the type of particle that occupies the lattice points: atoms, molecules, or ions. The four types of crystalline solids are molecular, ionic, metallic, and network covalent. The table on the next page describes each of these solids.

A lattice is a three-dimensional system of points designating the points of the components (atoms, molecules, or ions) that make up the substance. A unit cell is the smallest repeating unit of a lattice.

Type of Solid	Molecular	Ionic	Atomic	Metallic	Network Covalent
Examples	$C_6H_{12}O_6$	NaCl	Ar	Cu	C (diamond)
Particles occupying lattice points	Molecules	Ions	Atoms	Cations	Atoms
Bonds or forces between lattice points	Dispersion forces, dipole–dipole, hydrogen bonding	Ionic	Dispersion forces	Delocalized nondirectional metallic	Covalent
Melting points	Low	High	Low	High	Very high
Electrical conductivity as solid	None	None	None	Yes	None (other than Si)
Electrical conductivity when melted	None	Yes	None	Yes	None
Other properties		Brittle		Malleable and ductile	Very hard

AMORPHOUS SOLIDS

Amorphous solids, such as rubber, plastics, and glass, have particles with no orderly structure.

STRUCTURE AND BONDING IN METALS

Metallic bonding is characterized as an array of positively charged nuclei surrounded by a "sea" of mobile valence electrons. It is said that the bonding is nondirectional; the metallic atoms can easily slide over each other. This model explains the observed properties of metals: they are malleable (can be formed into thin sheets), conductive, and ductile (can be pulled into wires).

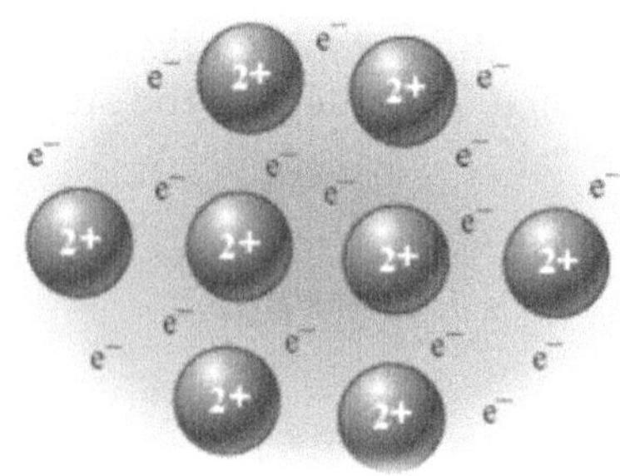

ALLOYS

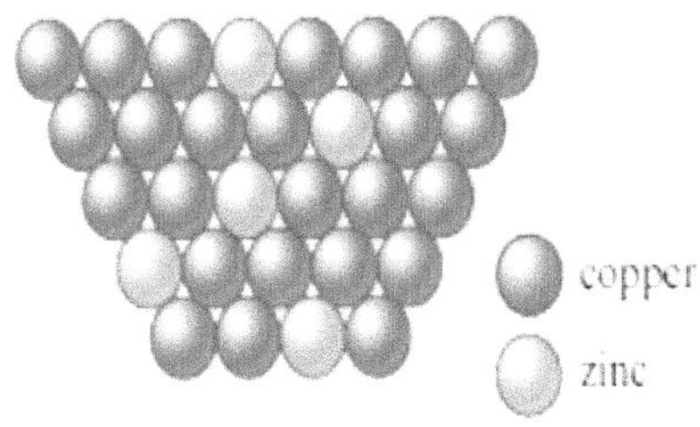

Substitutional Alloy

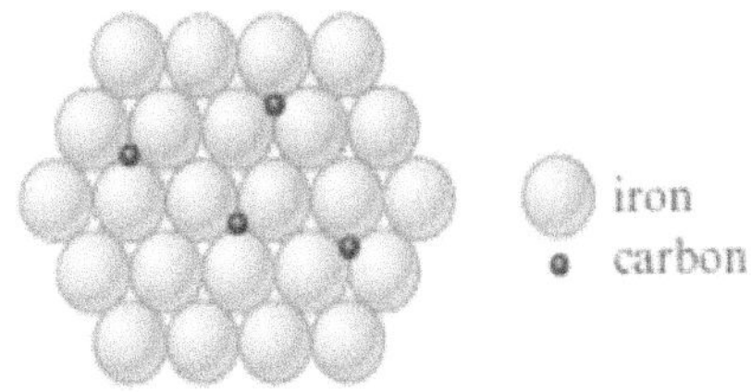

Interstitial Alloy

Alloys are mixtures of metallic elements (and sometimes small nonmetallic elements, such as carbon) and have metallic properties. Substitutional alloys, like brass and sterling silver, have host metal atoms replaced by atoms of similar size. Brass is a substitutional alloy made of copper and zinc. Interstitial alloys have smaller atoms occupying the empty spaces between the metal atoms within the structure. The presence of the smaller atoms changes the properties of the host making it harder and stronger. Steel is an interstitial alloy containing iron and carbon.

SEMICONDUCTORS

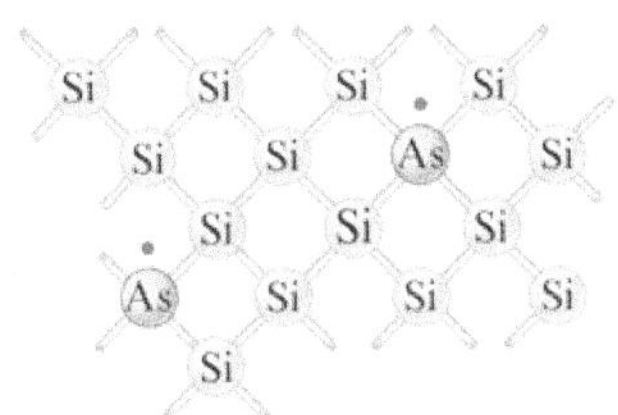

n-Type Semiconductor

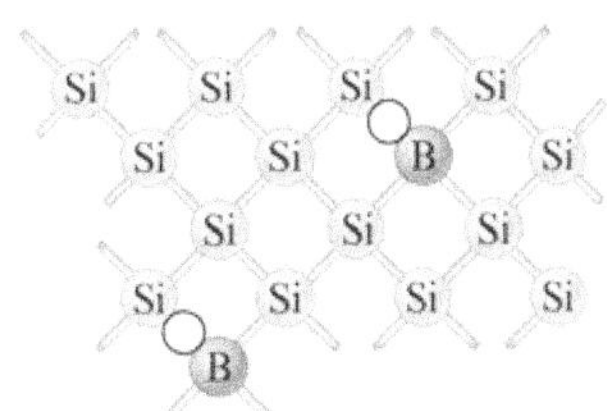

p-Type Semiconductor

Elemental silicon has the same structure as diamond, but it is able to conduct electricity weakly at room temperature and more strongly at higher temperatures, making it a *semiconductor*. Most metals have decreased conductivity at higher temps.

Doping is a process that increases the conductivity of silicon. Very few of the Si atoms are replaced by other atoms such as arsenic.

There are two main types of semiconductors:

- *n-type semiconductor* (*n* stands for negative): The atom that is substituted for silicon has more valence electrons than silicon. The extra electron(s) are free to move throughout the structure and help it conduct electricity. Silicon with a few arsenic atoms substituted makes an *n*-type semiconductor.
- *p-type semiconductor* (*p* stands for positive): The atom being substituted in has fewer electrons than silicon. An example would be boron. This creates a "hole" where the missing electron would be. Mobile electrons can use that hole as they travel through the structure to conduct electricity.

CHANGES IN STATE

MELTING POINT

↑IMF = ↑MP

↑IMF = ↑ΔH_{fus}

The *normal melting point* of a substance is the temperature at which the particles of a solid overcome their intermolecular forces to flow as a liquid at 1 atm of pressure. At this temperature, the vapor pressures of the solid and liquid are equal.

The energy required to overcome some of the attractive forces to melt a substance at its normal melting point is the *heat of fusion,* ΔH_{fus}. The stronger the intermolecular/interparticle forces that a substance has, the higher the melting point and the higher the heat of fusion.

VAPOR PRESSURE

↑IMF = ↓VP

↑IMF = ↑ΔH_{vap}

In the process of *vaporization* (evaporation), liquid molecules leave the surface and form a vapor. The heat of vaporization (enthalpy of vaporization, ΔH_{vap}) is the energy required to vaporize 1 mole of a liquid at its normal boiling point (1 atm). This quantity is endothermic because energy is required to overcome the intermolecular forces. In the reverse process, condensation, energy is released as the IMFs form between the particles when the gas molecules get close enough to interact with each other.

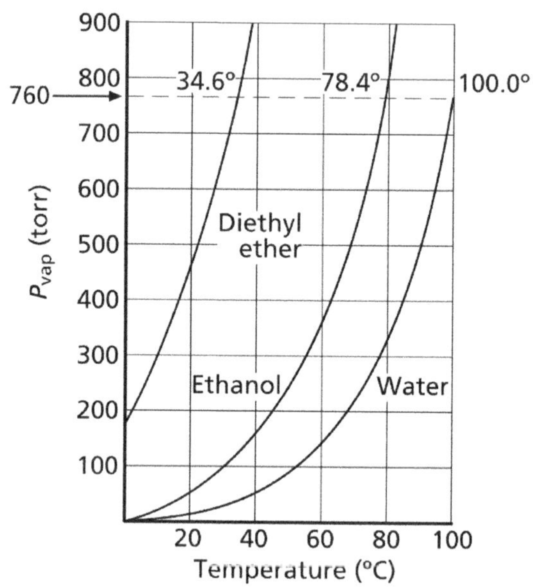

In a closed container, a liquid and its vapor are at equilibrium when the rate of evaporation equals the rate of condensation. The equilibrium vapor pressure is the pressure above the liquid from the molecules of the liquid in the gas phase.

Volatile liquids are liquids with high vapor pressures. Strong intermolecular forces result in lower vapor pressure. Weak intermolecular forces result in higher vapor pressure because the liquid molecules are not held together as strongly and readily escape from the surface.

The vapor pressure of a liquid increases with increasing temperature. At higher temperatures, more molecules will have sufficient kinetic energy to overcome intermolecular forces and break free from the surface of the liquid.

EXAMPLE: Explain why diethyl ether, $CH_3CH_2OCH_2CH_3$, is more volatile than ethanol, CH_3CH_2OH, at 25°C.

SOLUTION: The dipole forces in diethyl ether are weaker than the hydrogen bonding in ethanol. The weaker the intermolecular forces, the higher the liquid's vapor pressure and volatility.

BOILING POINT

↑IMF = ↑BP

The *normal boiling point* of a liquid is the temperature at which the vapor pressure of the liquid is equal to standard atmospheric pressure (1 atm = 760 mm Hg). The energy required to overcome the intermolecular forces to vaporize a substance is the heat of vaporization.

The reverse process, condensation, requires the release of energy as new intermolecular forces are formed. The energy released is the heat of condensation.

EXAMPLE: Using the vapor pressure graph on the previous page, determine the normal boiling points of diethyl ether and ethanol.

SOLUTION: The normal boiling points for diethyl ether and ethanol are approximately 34.6°C and 78.4°C, respectively. The normal boiling point occurs at 1 atm and can be found by interpolating the vapor pressure curves.

When a solid goes directly to a gas without going through the liquid phase, the change is called *sublimation*. The energy involved is the heat of sublimation, ΔH_{sub}. The reverse process is called *deposition*.

HEATING CURVES

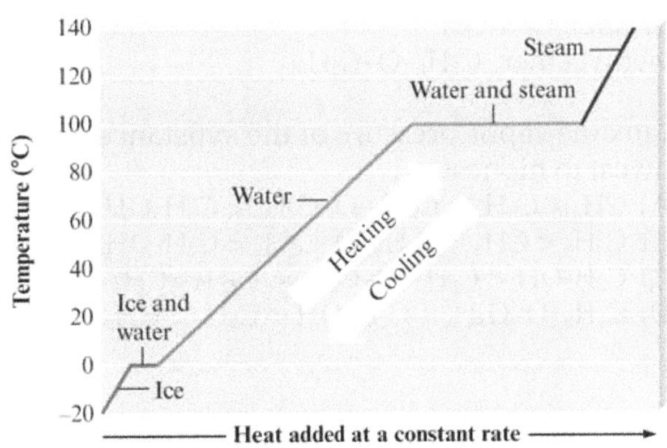

Chapter 9 of this book (Big Idea 5) includes a discussion of the quantitative aspects of the heating curve. Energy is added at a constant rate to a solid and a plot of temperature vs. time is drawn. The first plateau on the heating curve represents the melting point of the substance. The second plateau on the heating curve represents the boiling point of the substance. The temperature does not change during melting or boiling because the energy added is used to overcome intermolecular forces.

Usually, the plateau for vaporization is longer in duration than the melting plateau because it takes more energy to separate the molecules from a liquid into a gaseous state than from a solid into a liquid state. This greater energy is related to the fact that when changing from a solid to a liquid, only some of the intermolecular forces are overcome, but changing from a liquid to a gas, all of the intermolecular forces must be overcome.

MULTIPLE-CHOICE QUESTIONS

No calculators are to be used in this section.

1. Which of the following diagrams best shows the structure of the most common compound formed from K and Cl?

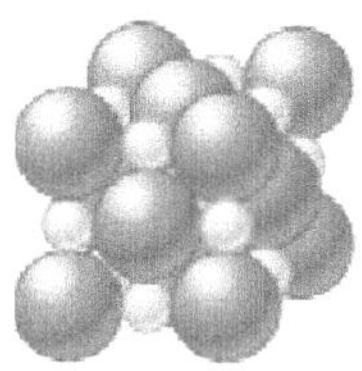
(A)

(B)

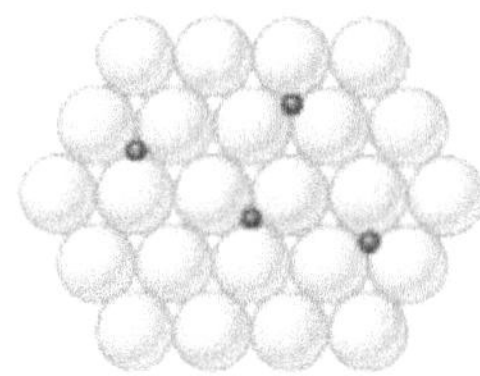
(C)

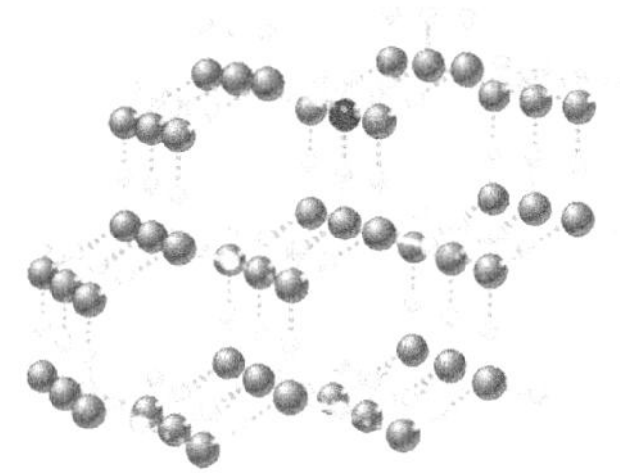
(D)

2. In some types of plastic graduated cylinders, water shows no visible meniscus. Which is an explanation for this observation?
   (A) Water has stronger cohesive forces for itself than adhesive forces for the plastic.
   (B) Water has stronger adhesive forces for the plastic than cohesive forces for itself.
   (C) Water has no adhesive forces for the plastic and strong cohesive forces for itself.
   (D) Water has equally strong cohesive forces for itself and adhesive forces for the plastic.

Methane, CH_4
Propane, C_3H_8
Ethanol, C_2H_5OH
Diethyl ether, $C_2H_5{-}O{-}C_2H_5$

3. Rank the vapor pressure of the substances shown above from lowest to highest.
   (A) $CH_4 < C_3H_8 < C_2H_5{-}O{-}C_2H_5 < C_2H_5OH$
   (B) $C_3H_8 < CH_4 < C_2H_5{-}O{-}C_2H_5 < C_2H_5OH$
   (C) $C_2H_5OH < C_2H_5{-}O{-}C_2H_5 < C_3H_8 < CH_4$
   (D) $C_2H_5{-}O{-}C_2H_5 < C_2H_5OH < C_3H_8 < CH_4$

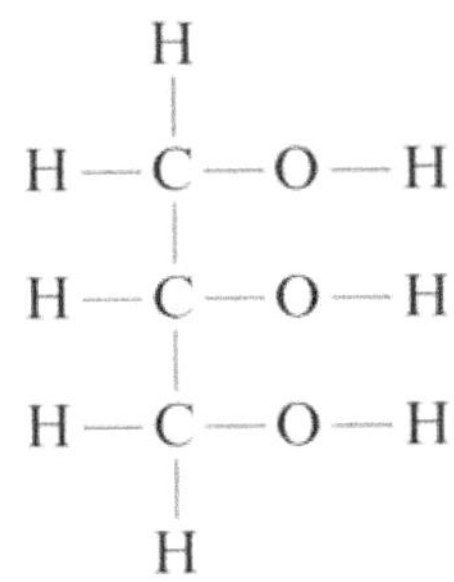

Glycerol

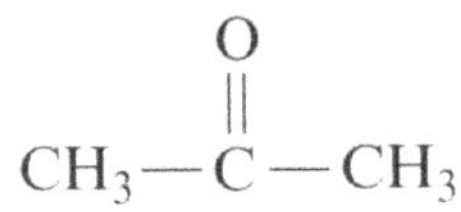

Acetone

4. The viscosities of four liquids (glycerol, water, acetone, and hexane, C_6H_{14}) are compared by adding 800 mL of each liquid to different 1000-mL graduated cylinders and then dropping a steel ball of uniform size and mass into each. The time required for the ball to reach the bottom of the cylinder is measured and recorded. The structures of glycerol and acetone are shown above. Through which liquid did the ball travel for the longest time?
(A) glycerol
(B) water
(C) acetone
(D) hexane

Element	Electron Configuration	Atomic Radii (pm)	Density (g/cm³)
Gold	$[Xe]4f^{14}5d^{10}6s^{1}$	135	19.3
Sulfur	$[Ne]3s^{2}3p^{4}$	104	2.07
Germanium	$[Ar]3d^{10}4s^{2}4p^{2}$	122	5.32

5. Gold can easily be flattened into a sheet that is only a few atoms thick (gold foil) and is used in many decorative applications. Solid sulfur and germanium are brittle and cannot be formed into thin sheets. Some properties of these three elements are given in the table above. What property of gold allows it to form such a thin layer without shattering?
(A) A high density that results in a very strong material
(B) Large atomic radii that lend stability
(C) Valence electrons that are mobile
(D) Presence of *f* subshell electrons that create an exception to the Aufbau principle

Questions 6 and 7 refer to the following.

Noble Gas	Normal Boiling Point
Helium	4 K
Neon	27 K
Argon	87 K
Krypton	122 K
Xenon	167 K
Radon	212 K

6. The boiling points of the noble gases are shown in the table above. The trend observed when going down the group of noble gases on the periodic table is due mainly to a(n)
(A) increase in hydrogen bonding.
(B) decrease in dipole–dipole forces.
(C) increase in the molar mass.
(D) increase in the number of polarizable electrons.

7. At a temperature of 200 K, radon has a density of 4.4 g/cm^3. Which of the following is correct concerning radon at 200 K?
(A) Radon is a gas with atoms spaced far apart. It takes the shape and volume of its container.
(B) Radon remains a gas with atoms close together. It takes the shape but not the volume of its container.
(C) Radon is a liquid with atoms spaced far apart. It takes the shape and volume of its container.
(D) Radon is a liquid with atoms close together. It takes the shape but not the volume of its container.

8. A sample of the material represented in the drawing above is brittle and shatters when hit with a hammer. Which answer best explains this lack of malleability?
(A) The compound is molecular. The covalent bonds are strong and do not hold the atoms tightly, not allowing them to spread apart.
(B) The sample is not a compound, but a mixture of metals (an alloy). While pure metals are malleable, alloys are not.
(C) The compound is ionic and made up of ions of opposite charges. The opposite charges repel each other when forced together.
(D) The sample is a covalent network solid. Each atom is covalently bonded to multiple other atoms, essentially forming a giant molecule.

9. Neon, ammonia, carbon monoxide, and methylamine (CH_3NH_2) all have low molar masses. Place these substances in order of increasing boiling points.
(A) $Ne < CO < NH_3 < CH_3NH_2$
(B) $NH_3 < CO < CH_3NH_2 < Ne$
(C) $Ne < CO < CH_3NH_2 < NH_3$
(D) $CH_3NH_2 < CO < NH_3 < Ne$

10. Which of these compounds would be expected to have the lowest melting point?
(A) NaCl
(B) KCl
(C) $MgCl_2$
(D) $CaCl_2$

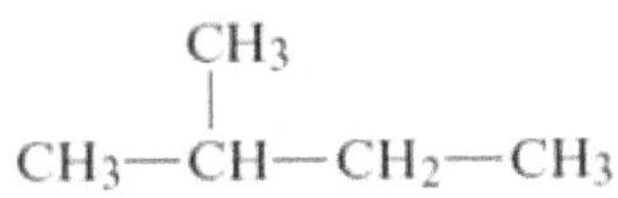

2-methylbutane

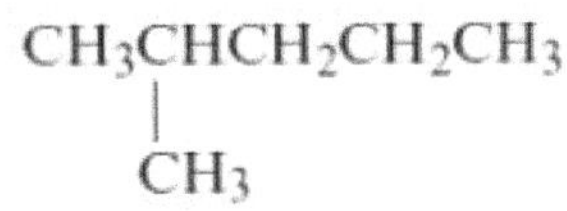

2-methylpentane

11. The structures of two hydrocarbons are shown above. Which statement correctly gives the compound with the higher boiling point and the correct justification?
(A) 2-methylbutane has the higher boiling point because it has a smaller size. This allows the molecules of 2-methylbutane to attract each other more strongly than the molecules of 2-methylpentane.
(B) 2-methylbutane has the higher boiling point because of its lower molar mass, resulting in a greater molecular velocity.
(C) 2-methylpentane has the higher boiling point because it has a higher molar mass, causing the molecules to move more slowly and attract each other more.
(D) 2-methylpentane has the higher boiling point because it has more electrons and is more polarizable.

12. When sodium chloride (NaCl) crystals are placed in a flame, the NaCl is vaporized and the flame color turns bright yellow-orange. Which particle view diagram best reflects the state of the NaCl after vaporization?

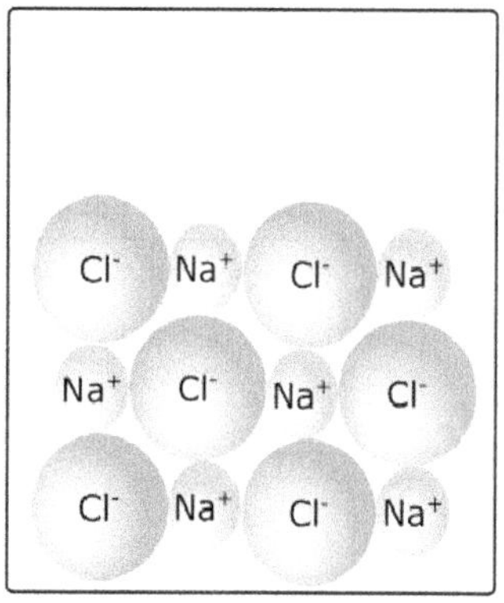

(A)

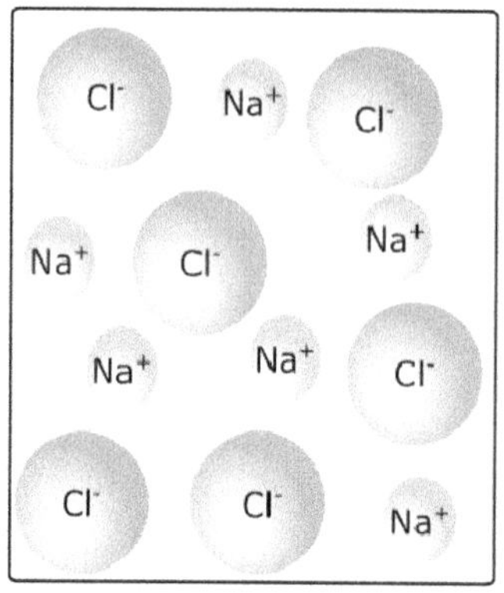

(B)

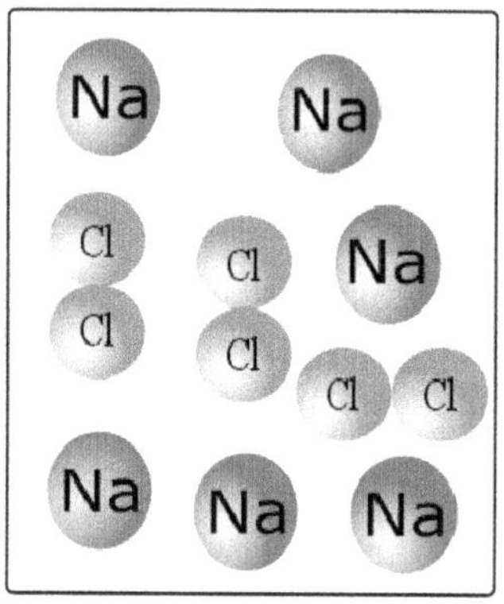

(C)

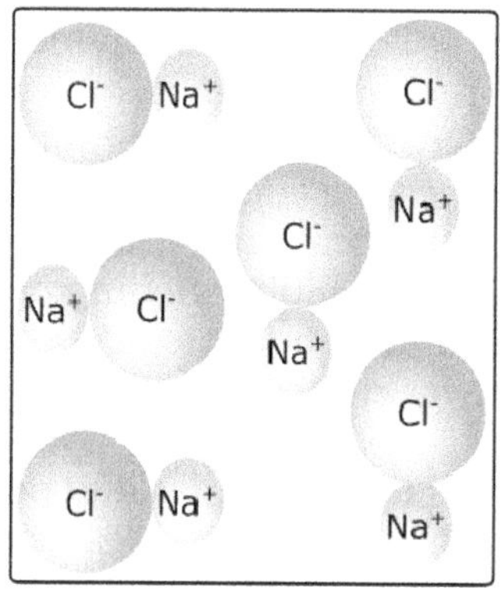

(D)

Test	Results
Electrical conductivity as a solid	Does not conduct electricity
Melting point	Low melting point
Solubility	Insoluble in water
Appearance	Reflects some light

13. A student is conducting a laboratory investigation to determine the type of bonding in an unknown substance. The student's data table is shown above. Which type of compound is this most likely to be?
(A) metal
(B) ionic solid
(C) molecular solid
(D) network covalent solid

14. The fact that metals are conductive is best explained by which of the following statements?
    (A) Metal atoms are malleable. Electrons can be "squeezed" out of them.
    (B) Metal nuclei reside in a "sea" of delocalized electrons that are free to move.
    (C) Metals have *d*-block electrons that can contract, allowing the electrons to conduct electricity.
    (D) Metals can form more than one oxidation state, allowing the excess electrons to conduct electricity.

15. Which of these compounds is most likely the identity of a colorless, crystalline solid that has the following properties?
    - Melting point of 122°C
    - Solubility in water = 3 g/100 g at 25°C
    - Solubility in ethanol (C_2H_5OH) = 57 g/100 g at 25°C
    - Does not conduct electricity as a solid

    (A) $Fe(s)$
    (B) $SiC(s)$
    (C) $C_6H_5COOH(s)$
    (D) $KNO_3(s)$

FREE-RESPONSE QUESTIONS

Calculators and equation tables may be used.

1. Four solids were tested in the laboratory, and the results of various tests are shown below.

	Solid A	Solid B	Solid C	Solid D
Solid conducts electricity?	Yes	No	No	No
Soluble in water?	No	Yes	No	Yes
Aqueous solution conducts electricity?	Not soluble	Yes	Not soluble	No
Relative melting point	High	High	Very high	Low

   (a) Identify the type of bonding present in each solid based on these data. Carefully explain how the data lead to your conclusions.
   (i) Solid A
   (ii) Solid B
   (iii) Solid C
   (iv) Solid D

(b) Draw a particle view representation that shows the structure of each type of solid.

Solid A

Solid B

Solid C

Solid D

(c) Alloys are mixtures of elements made to produce the desired properties.

(i) A substitutional alloy is formed when a metallic element of similar atomic size replaces another element in its crystal structure. Draw a representation of a substitutional alloy of 75% silver and 25% copper in the box below.

(ii) A sheet of shiny metallic orange-colored copper (density 8.96 g/cm^3) was electroplated with metallic silver-colored zinc (density 7.14 g/cm^3). The resulting multilayered material which was copper colored on the inside and silver colored on the outside was then exposed to high heat. The material became a uniform bright gold color throughout and had a density of 8.55 g/cm^3. The new material conducted electricity and was malleable. Was an alloy formed? Justify your answer in terms of the properties and structure of the new material.

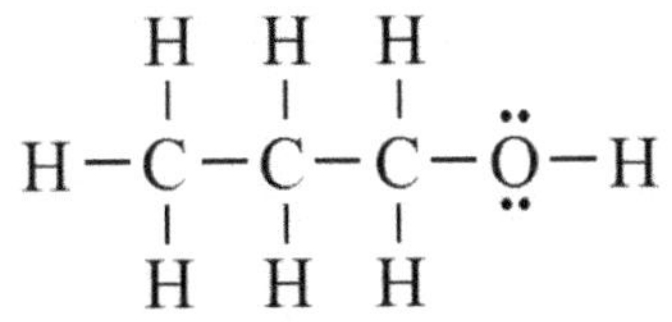

1-propanol

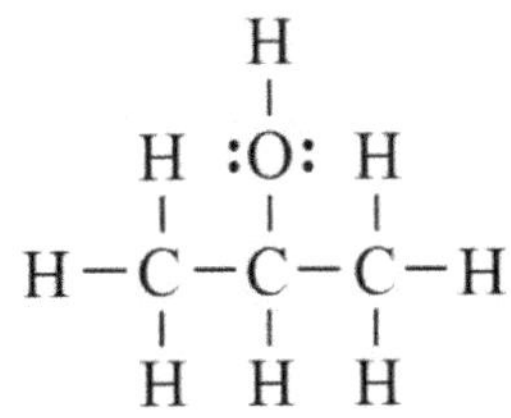

2-propanol

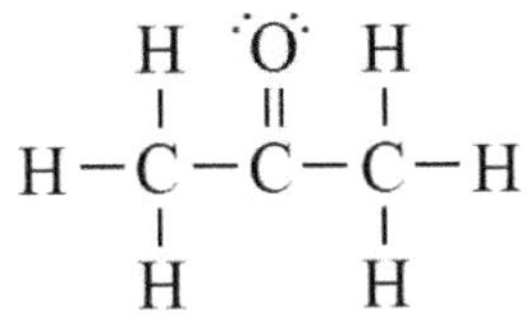

2-propanone

2. Three different compounds of carbon, hydrogen, and oxygen are shown above.
   (a) Would 2-propanol or 2-propanone have a higher vapor pressure? Justify your answer in terms of molecular structure and intermolecular forces.
   (b) 1-propanol boils at 97°C, while 2-propanol boils at 87.5°C.
      (i) Which of these two compounds has stronger intermolecular forces?
      (ii) Account for the difference in boiling points between 1-propanol and 2-propanol in terms of intermolecular forces.

Answers

MULTIPLE-CHOICE QUESTIONS

1. A Potassium and chlorine (metal plus nonmetal) would combine to form an ionic compound. The structure of an ionic compound is a lattice of alternating anions and cations (LO 2.24, 2.28, 2.30, 2.32).

2. D The shape of a meniscus is determined by the relative attractions of the liquid molecules for themselves and for the substance making up the container. In a glass cylinder, the meniscus of water would be concave as the water has stronger attractions for itself than for the glass. If the meniscus is flat, it indicates that the attractions are equal (LO 2.16).

3. C Examining the IMFs will indicate that only the alcohol (C_2H_5OH) has an exposed –OH, suggesting hydrogen bonding. C_2H_5–O–C_2H_5 is a polar molecule with dipole–dipole attractions as well as

London dispersion forces. The other two have only weaker London dispersion forces, which are greater in the larger C_3H_8 molecules that have more polarizable electrons. Realize also that the higher the IMF, the lower the vapor pressure (LO 2.16).

4. A The resistance to flow of any fluid is called viscosity. As you would predict, liquids with high viscosity (e.g., maple syrup) have large intermolecular forces. Of these four compounds, glycerol has the strongest IMFs as it has extensive hydrogen bonding (LO 2.16).

5. C Gold is a metal. A metal has mobile valence electrons that allow it to conduct electricity and heat and also to be hammered into a thin sheet (malleability). Sulfur is a nonmetal, and germanium is a metalloid (LO 2.20).

6. D Noble gases are all nonpolar. They do have London dispersion forces that increase in strength with increasing numbers of electrons, since the additional electrons are more polarizable. As you go down the group of noble gases, the number of electrons that each atom has increases because they each have an extra energy level of electrons (LO 2.11).

7. D The boiling point of Argon is 212 K. Since the temperature is 200 K, we know that it is not yet a gas. Radon has a density very similar to that of liquid water at about 377 K. This indicates that Argon has properties of a liquid, not of a gas (LO 2.3).

8. C The structure shows a crystal lattice that indicates an ionic compound. Ionic compounds shatter when the crystal is hit by a hammer because the electrons are not mobile and ions of opposite charge collide with each other. The resulting repulsion causes the crystal to shatter (LO 2.19).

9. A The general strengths of intermolecular forces in order are London dispersion forces (the weakest), then dipole–dipole, then hydrogen bonding, and then ionic forces (the strongest). The stronger the IMF, the higher the boiling point. Neon has only LDF; CO has dipole–dipole plus LDF; NH_3 has hydrogen bonding, dipole–dipole, and LDF. CH_3NH_2 has hydrogen bonding, dipole–dipole, and LDF, but it has more extensive LDF because it has more electrons and is thus more polarizable (LO 2.16).

10. B Weaker IMF = Lower boiling point. These compounds are all ionic, so all have ionic forces. The weakest ionic attraction will be between the largest ions with the lowest charges. This eliminates C and D as calcium and magnesium both have a 2+ charge. Potassium is larger than sodium so the potassium ion and the chloride ion will be further apart and it will not take as much energy to separate them (LO 2.14).

11. D The only difference in the two compounds is that 2-methylpentane has one extra CH_2. This results in more electrons (thus more polarizable) and stronger LDF. The structure with the highest boiling point will be the one with the strongest IMF. This will be 2-methylpentane (LO 2.11).

12. B When an ionic solid is placed into a flame, it is vaporized and the energy causes the ions to separate and become gaseous. No chemical reaction takes place (LO 2.3).

13. C Since the solid does not conduct electricity, a metal is ruled out. A low melting point indicates a molecular solid. It is not soluble in water which indicates that it may be a nonpolar molecular solid (LO 2.22).

14. B The "electron sea" model was developed to explain metal conductivity of heat and electricity. The metal cations are in a sea of mobile valence electrons (also called delocalized electrons). If 1 mole of electrons enters one end of a metal wire, 1 mole of electrons would exit the other end (LO 2.26).

15. C A melting point of 122°C is relatively low (can't be SiC, a network solid, or KNO_3, an ionic compound). This compound is more soluble in ethanol than in water (indicates more nonpolar than polar structure). It does not conduct electricity as a solid (can't be a metal) (LO 2.1).

FREE-RESPONSE QUESTIONS

1. (a) (i) Solid A is most likely metallic. The delocalized "sea of electrons" results in high conductivity as a solid. Lack of solubility in water indicates that the intermolecular forces are very strong (LO 2.22).

   (ii) Solid B is most likely ionic. Ionic solids are not conductive because their electrons are localized around their respective nuclei. Many ionic solids are soluble in water; when they dissolve, they split up into cations and anions which are attracted to the polar water molecules. The separation of charges leads to formation of a conductive solution (LO 2.22).

   (iii) Solid C is most likely network covalent, such as SiO_2 or diamond. These solids are not particularly conductive because the electrons are localized in the covalent bonds between the atoms. They are also not soluble in polar solvents because the molecules themselves have very strong covalent bonds to each other (LO 2.22).

   (iv) Solid D is most likely molecular. Like network covalent solids, the electrons are localized in the covalent bonds between the atoms. However, many are soluble in water because they are attracted to each other primarily by London dispersion forces, which can be overcome by the attraction of polar parts of the molecules to polar water (LO 2.22).

(b)

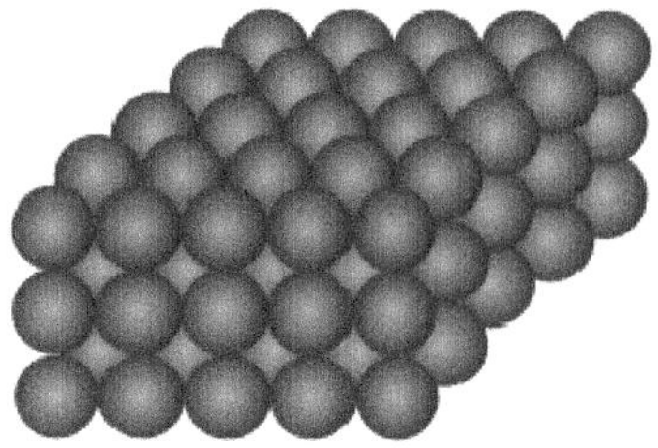

Solid A (LO 2.23)

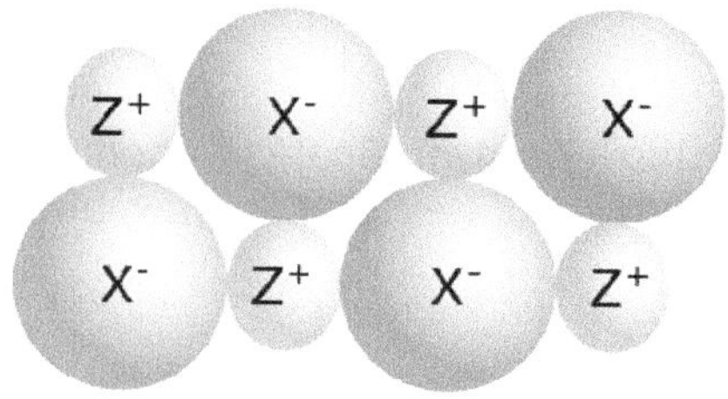

Solid B (LO 2.27)

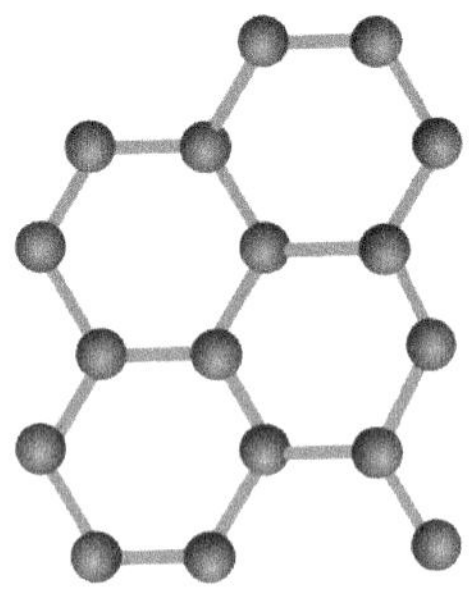

Solid C (LO 2.29)

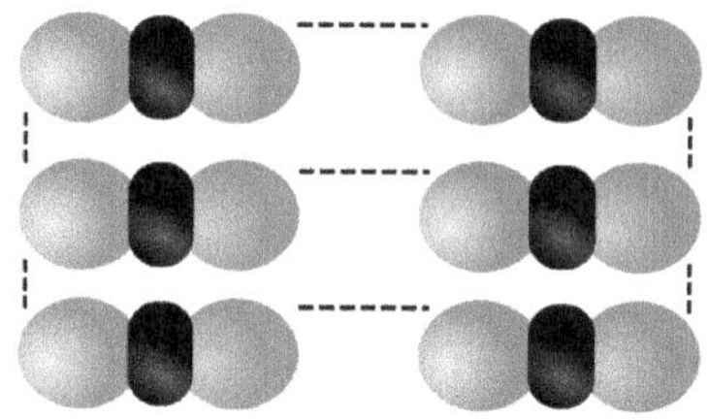

Solid D (LO 2.31)

(c) (i)

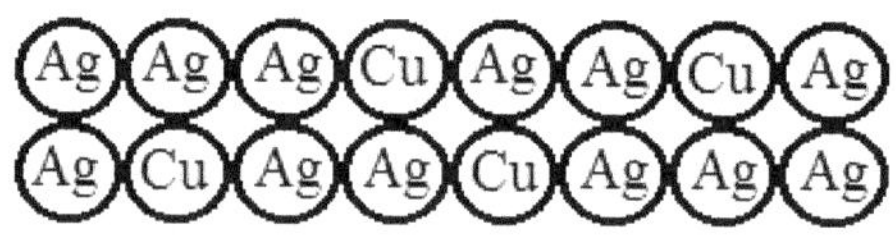

(LO 2.27)

(ii) A substitutional alloy was formed between two similar sized metals. The properties of the alloy are metallic, and the density of the alloy is in between the densities of the two metals that make it up. The appearance of the alloy is uniform and is different than that of either metal (LO 2.25).

2. (a) 2-propanone would have the higher vapor pressure. High vapor pressure results from weak intermolecular forces. 2-propanone is a slightly polar molecule with only dipole–dipole forces and LDF. 2-propanol has stronger hydrogen bonding between molecules and would have a lower vapor pressure (LO 2.16).

(b) (i) The compound with the higher boiling point (1-propanol) would have the stronger intermolecular forces (LO 2.13).

(ii) Both compounds have hydrogen bonding. The hydroxyl group on the end of the molecule in 1-propanol, along with its more linear shape, allows greater interaction between adjacent molecules and thus stronger IMF (LO 2.16).

5

Big Idea 2: Solutions

Big Idea 2

Chemical and physical properties of materials can be explained by the structure and arrangement of atoms, ions, or molecules and the forces between them.

In this section, the properties of solutions will be reviewed. Solution composition and factors affecting solubility will be covered.

You should be able to:

- Perform calculations with solution concentrations (molarity) and solve dilution problems.
- Explain how the attractions between ions and solvents affect factors determining the solubility of ionic and molecular compounds in water and other solvents.
- Draw or explain representations of solutions that illustrate the interactions between solute and solvent.
- Design or explain the results of separation experiments (chromatography, filtration, distillation) based on the relative strengths of interactions among and between the components.

SOLUTIONS AND THEIR COMPOSITIONS

TYPES OF SOLUTIONS

A solution is a homogeneous mixture and can involve any combination of the three states of matter. There are nine different types of solutions. In this chapter, the main focus will be on aqueous solutions.

The substance being dissolved is the *solute*. When referring to a liquid–liquid or gas–gas solution, the *solvent* is the substance present in the largest amount.

> **AP® Tip**
>
> On the AP® exam, all solution volumes are considered to be additive. If you pour 75.0 mL of solution A into 50.0 mL of solution B, your final volume will be 125.0 mL.

MOLARITY

The molarity, *M*, of a solution (its concentration) is the number of moles of solute per liter of solution. The unit *M* is pronounced molar.

EXAMPLE: A solution is prepared by mixing 30.0 mL of acetone (C_3H_6O, d = 0.791 g/mL) with 65.0 mL of water (d = 1.00 g/mL). Assuming that the volumes of the liquids are additive, calculate the molarity of the acetone in the solution.

SOLUTION: Acetone is the solute because less of it is used. First, find the moles of acetone.

$$30.0\text{ mL} \times \frac{0.791\text{ g}}{\text{mL}} \times \frac{1\text{ mol } C_3H_6O}{58.0\text{ g}} = 0.409\text{ mol } C_3H_6O$$

The new solution volume is 95.0 mL, or 0.0950 L. Use this to find the molarity.

$$\frac{0.409\text{ mol } C_3H_6O}{0.0950\text{ L}} = 4.31\ M$$

SOLUTION PREPARATION

When preparing a solution, the desired accuracy must be considered when choosing measuring equipment. If a 1.00 *M* solution is needed, it is best to use a volumetric flask to contain the solution and a pipet (if liquid) or an analytical balance (if solid) to measure the solute. If a roughly 1 *M* solution is needed, beakers (to hold the solution) and graduated cylinders to measure both solute (if liquid) and solvent will be fine.

When preparing a solution involving a solid solute, fill the volumetric flask approximately halfway with purified water, add the carefully massed solid, stopper, and shake until the solute is dissolved. Remove the stopper and carefully add additional water, pausing to mix once or twice until the volume reaches the engraved line on the volumetric flask. Replace the stopper and mix thoroughly.

When preparing a solution involving a liquid solute or when diluting a more concentrated solution, add some purified water to the flask (allowing plenty of space for the solute or concentrated solution to be added), add the solute or concentrated solution, stopper, and mix well. Continue adding water with mixing until the volume reaches the engraved line on the volumetric flask. Replace the stopper and mix thoroughly.

EXAMPLE: Prepare 2.00 L of 0.250 *M* NaOH from solid NaOH.

SOLUTION:

$$2.00\ \text{L} \times \frac{0.250\ \text{mol NaOH}}{1\ \text{L}} \times \frac{40.00\ \text{g NaOH}}{1\ \text{mol NaOH}} = 20.0\ \text{g NaOH}$$

Place 20.0 g NaOH in a 2-L volumetric flask, add water to dissolve the NaOH, and fill to the mark with distilled water, mixing several times along the way.

Another way to prepare a solution of a specific molarity is to dilute a more concentrated solution, using the relationship $M_1V_1 = M_2V_2$:

EXAMPLE: Prepare 2.00 L of 0.250 *M* NaOH from 1.00 *M* NaOH.

SOLUTION: $1.00\ M\ (V_1) = (0.250\ M \times 2.00\ \text{L}) \qquad V_1 = 0.500\ \text{L}$

Add 500. mL of 1.00 *M* NaOH stock solution to a 2-L volumetric flask; add deionized or distilled water in several increments with mixing until the flask is filled exactly to the mark on the neck of the flask. Solution volumes are not always additive. Therefore, it would not be correct to say that 1500 mL of water should be added to 500 mL of NaOH.

Note: A common error is to use the $M_1V_1 = M_2V_2$ formula for a chemical reaction (such as in a titration). This often results in the incorrect answer because the mole ratio is not always 1 to 1.

FACTORS AFFECTING SOLUBILITY

The formation of a liquid solution begins by separating the solute into its individual components. This process always requires energy; therefore, energy flows into the system from the surroundings, and ΔH is positive. Next, the solvent's intermolecular forces must be overcome to make room for the solute. This also always requires energy; therefore, energy flows into the system from the surroundings, and ΔH is positive. Finally, the solute and solvent interact to form the solution. This process always releases energy; therefore, energy flows from the system to the surroundings, and ΔH is negative.

STRUCTURAL EFFECTS

Solubility is favored if the solute and solvent have similar intermolecular forces, as determined by their structure. "Like dissolves like" is a general observation for solubility, but is ***not a reason*** for it. Substances with similar types of intermolecular forces tend to dissolve in each other.

Polar solvents tend to dissolve polar or ionic solutes. Nonpolar solvents tend to dissolve nonpolar solutes. When asked about reasons for solubility, discuss the strengths of the forces between the solute particles, between the solvent particles, and between the solute and the solvent (i.e., no credit is earned for "like dissolves like"). The table below summarizes the solubility of different types of solutes in different types of solvents. *Note:* If solute and solvent can form hydrogen bonds with each other, the force of attraction is hydrogen bonding.

Type of Solute	Type of Solvent	Solubility	Attraction between Solute and Solvent	Example
Ionic	Polar	Usually soluble	Ion–dipole	$LiCl$ in H_2O
Polar	Polar	Soluble (miscible)	Dipole–dipole or hydrogen bonding	CH_3OH in H_2O
Nonpolar	Polar	Insoluble (immiscible)	Induced dipole–dipole	C_6H_{14} in H_2O
Nonpolar	Nonpolar	Soluble (miscible)	Induced dipole–induced dipole	C_6H_{14} in CCl_4

AP® Tip

Simply stating "like dissolves like" will not get you credit on the exam. You must elaborate as to the attractions between solute particles, solvent particles, and solute particles with solvent particles!

ENERGIES INVOLVED IN SOLUTION FORMATION

The solubility of a substance is governed by the overall enthalpy and entropy changes involved in the process. While enthalpy and entropy will be discussed to a much greater extent in Big Idea 5, their role in solution formation is introduced here.

STEP 1: BREAKING UP THE SOLUTE

The energy required to overcome the attraction between the molecules or ions is determined by the interparticle forces holding them together. This step always requires energy (is endothermic). Ionic solutes require a large amount of energy, while nonpolar compounds require much less energy.

STEP 2: BREAKING UP THE SOLVENT

The energy required to overcome the attraction between the solvent molecules is also always endothermic. Solvents with extensive hydrogen bonding require a large amount of energy, while nonpolar solvents require much less energy.

STEP 3: INTERACTION BETWEEN SOLUTE AND SOLVENT

There is always some degree of attraction between the solute and the solvent (Oil and water do not repel each other!). This attraction releases energy (is exothermic).

The sum of the energy values for the three steps determines whether the dissolving process is endothermic or exothermic. Most often, the process is endothermic. Usually, the dissolving process involves an increase in the entropy of the system as the particles become more dispersed. As will be discussed in Big Idea 5, the combination of energy (enthalpy) and disorder (entropy), along with temperature, determines whether a process occurs (is thermodynamically favored).

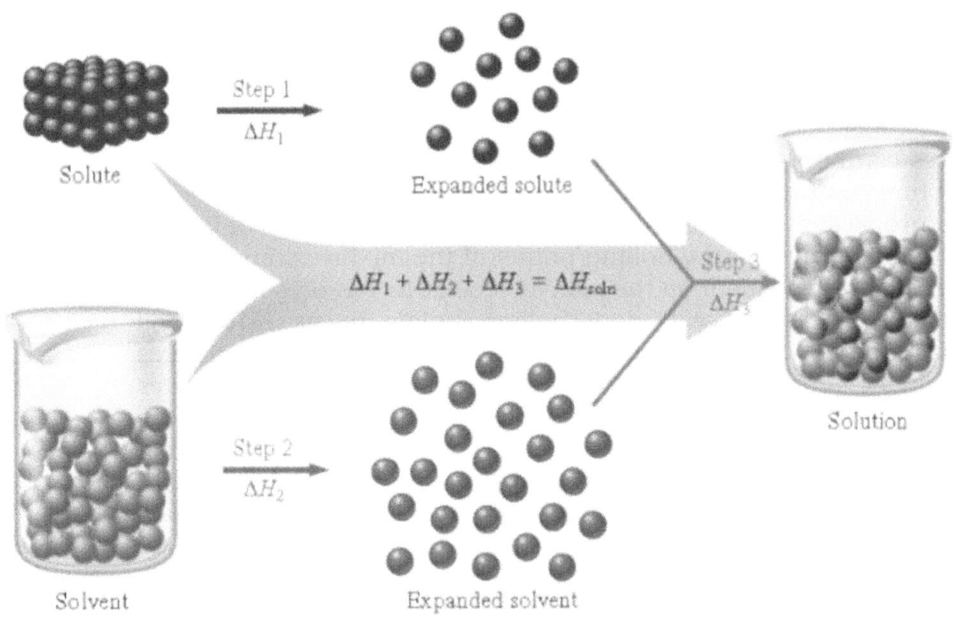

EXAMPLE: Discuss the solubility of each of the following solutes in hexane (C_6H_{14}):

- ammonium nitrate (NH_4NO_3)
- 1-propanol ($CH_3CH_2CH_2OH$)
- pentane ($CH_3CH_2CH_2CH_2CH_3$).

Explain why each solute will or will not dissolve in the solvent. Name the predominant solute–solvent attraction in each case.

SOLUTION: First, determine the polarity of the substances involved. C_6H_{14} is a long, linear hydrocarbon, so it is nonpolar. The energy required to break apart the solvent will be low in all cases. Ammonium nitrate is an ionic compound; it will not dissolve in a nonpolar solvent because it takes a large amount of energy to separate the ions from each other (breaking apart the solute) and there is insufficient attraction between the ions and the solvent. The attraction between NH_4NO_3 and C_6H_{14} is ion–induced dipole.

1-propanol contains a polar hydroxyl (–OH) group and a nonpolar hydrocarbon portion. It will be only slightly soluble in C_6H_{14} because it takes a large amount of energy to break the hydrogen bonds between 1-propanol molecules (breaking apart the solute). The attraction between the 1-propanol and C_6H_{14} will be low and is termed dipole–induced dipole.

Pentane is a nonpolar hydrocarbon chain. It will be soluble in C_6H_{14} because it takes very little energy to break apart the solute and the solvent. Interaction between solute and solvent is also weak. The attraction between C_5H_{12} and C_6H_{14} is induced dipole–induced dipole (LDF).

SEPARATION OF MIXTURES

Mixtures can be separated by various means using the properties of the mixture components. Magnetism, solubility, and polarity are some of the properties that can be used. Distillation separates mixtures based on differences in the boiling points of the mixture components. This process is dependent on the intermolecular forces present in each compound. Substances with low boiling points (weak IMFs) will vaporize first and can be condensed and collected, leaving substances with higher boiling points behind. Chromatography separates mixtures based on the relative attractions of the components for the solvent, the mobile phase and the stationary phase. Substances that are very soluble in the mobile phase will travel the farthest. Those with greater attraction to the stationary phase will travel little, if at all. This same principle applies to paper chromatography, gas chromatography, and thin-layer chromatography.

MULTIPLE-CHOICE QUESTIONS

No calculators may be used on this part of the exam.

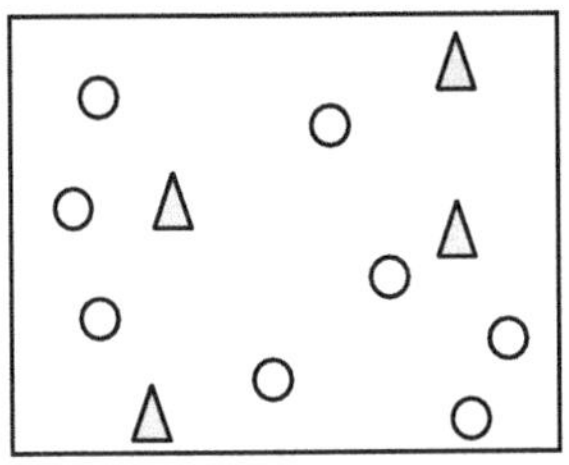

O sodium ion

Δ sulfate ion

1. If the box shown above represents a 2.50-L volume and each ion shown represents 1 mole of ions, what is the molarity of the sodium sulfate (Na_2SO_4) solution? Water molecules are not shown for clarity.
(A) 1.6 *M*
(B) 3.2 *M*
(C) 4.0 *M*
(D) 8.0 *M*

2. Potassium nitrate (KNO_3) is a salt used in the preservation of meats. When potassium nitrate dissolves in water, what is the interaction between the potassium ion and the water?
(A) Hydrogen bonding
(B) Ion–hydrogen bonding
(C) Ion–ion forces
(D) Ion–dipole forces

3. Hydrochloric acid is formed when hydrogen chloride gas dissolves in water, and hydrofluoric acid is formed when hydrogen fluoride gas dissolves in water. Which is the predominant solute–solvent interaction present in each case after the solution is formed?

	HCl(*aq*)	HF(*aq*)
(A)	Ion–dipole	Hydrogen bonding
(B)	Dipole–dipole	Hydrogen bonding
(C)	Dipole–dipole	Dipole–dipole
(D)	Ion–dipole	Dipole–dipole

4. A mixture of propanol (C_3H_7OH) and hexanol ($C_6H_{13}OH$) can be separated by distillation. Which best explains why the two alcohols can be separated by this means?
   (A) The compounds are separated based on different boiling points due to their molar mass differences.
   (B) The compounds are separated based on different boiling points due to the differences in their London dispersion forces.
   (C) The compounds are separated based on their differing attractions for the mobile phase due to differing amounts of hydrogen bonding.
   (D) The compounds are separated based on their differing attractions for the mobile phase because of different polarities.

Questions 5–6 refer to the information below.

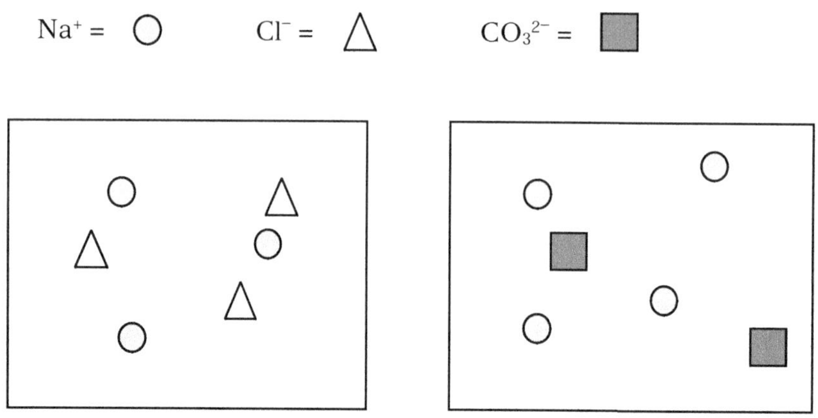

Container 1 (NaCl) **Container 2 (Na_2CO_3)**

The containers above each represent a volume of 1.0 L of solution. Each particle shown represents one-tenth (0.10) mole of ion in solution.

5. What are the concentrations of each solution?

	NaCl	Na_2CO_3
(A)	0.30 *M*	0.20 *M*
(B)	0.30 *M*	0.40 *M*
(C)	0.60 *M*	0.30 *M*
(D)	0.60 *M*	0.60 *M*

6. If the contents of Containers 1 and 2 were combined and mixed, what is the new sodium ion concentration? Assume that the volumes are additive.
   (A) 0.25 *M*
   (B) 0.35 *M*
   (C) 0.50 *M*
   (D) 0.70 *M*

7. A paper chromatography experiment is set up to investigate the properties of two colored solutions. The two solutions are 1*M* $CuCl_2(aq)$ (molar mass = 135 g/mol), with a blue color, and 1*M* beta carotene ($C_{40}H_{56}$, molar mass = 537 g/mol) dissolved in CH_3OH, with an orange color. Drops of each test solution are placed on separate chromatography paper strips and allowed to dry. Only the ends of the papers are lowered into the acetone (C_3H_6O) solvent (drops of samples remained above the level of the solvent), and the solvent is allowed to travel up the paper strip, carrying the samples with it to varying degrees. Which of the samples would be expected to travel the *smallest* distance up the strip? Why?
   (A) $CuCl_2(aq)$ because its charged particles repel the acetone solvent.
   (B) $CuCl_2(aq)$ because its ions have the least attraction for the acetone solvent.
   (C) Carotene, because its molar mass is very large.
   (D) Carotene, because it is a large, nonpolar molecule.

8. Which of the following pairs of compounds will have two separate phases (identifiable layers) when mixed?
   (A) Sodium chloride and water
   (B) Sodium chloride and hexane (C_6H_{14})
   (C) Water and ethanol (C_2H_5OH)
   (D) Water and sucrose ($C_{12}H_{22}O_{11}$)

9. A solution of ethanol (C_2H_5OH) dissolved in water is left open to the atmosphere for one week. Which statement is most correct?
   (A) The concentration of ethanol does not change because the hydrogen bonds between ethanol and water are strong.
   (B) The concentration of ethanol increases because water is more volatile than ethanol.
   (C) The concentration of ethanol decreases because ethanol is more volatile than water.
   (D) The solution changes in volume, but not concentration, because the two substances evaporate at an equal rate.

10. A small amount of copper is added to silver to form sterling silver which exhibits metallic properties, such as electrical conductivity and malleability. This is a
   (A) chemical reaction because the metal atoms bond with each other to form a new compound.
   (B) chemical reaction because the alloy has different properties than its constituents.
   (C) mixture, with silver as the solvent and copper as the solute.
   (D) mixture, with copper as the solvent and silver as the solute.

11. Which diagram most correctly shows the *predominant* interactions present in an aqueous solution of potassium iodide (KI)?

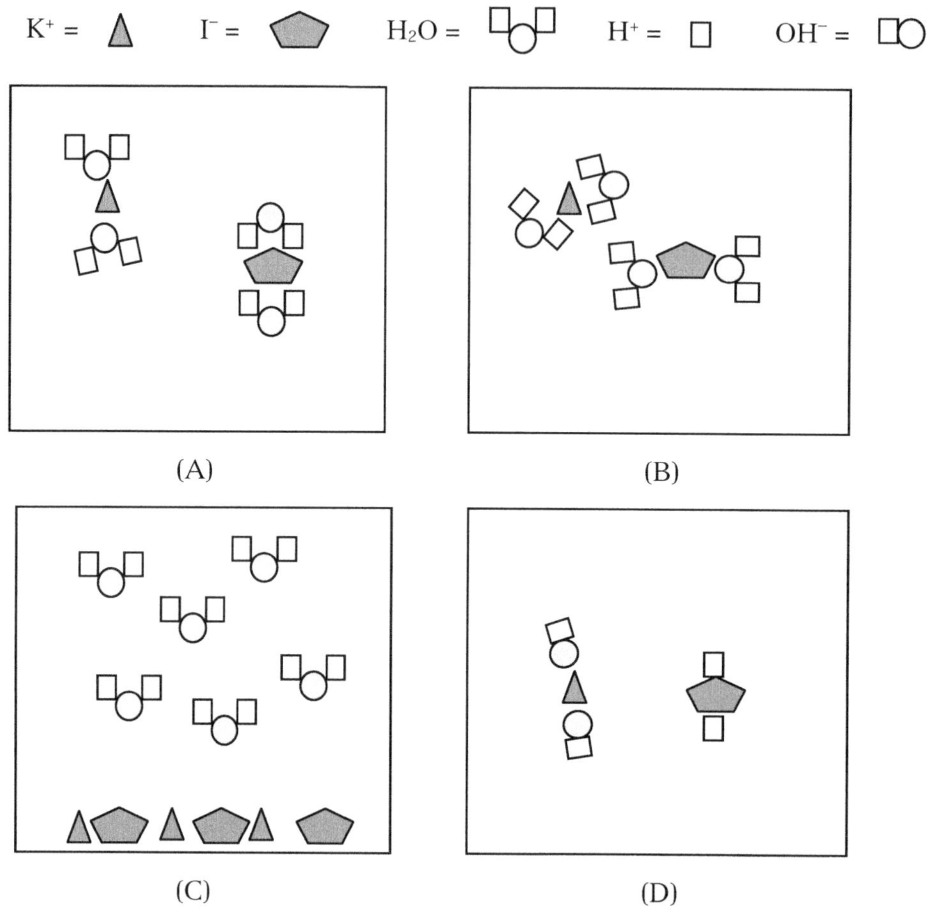

12. Which equation for the dissolving of solid sodium sulfate in water is written correctly?
(A) $Na_2SO_4(s) \rightarrow Na_2^{2+}(aq) + SO_4^{2-}(aq)$
(B) $Na_2SO_4(s) \rightarrow 2\ Na^+(aq) + SO_4^{2-}(aq)$
(C) $Na_2SO_4(s) \rightarrow 2\ Na^+(aq) + S^{2-}(aq) + 4\ O^{2-}(aq)$
(D) $Na_2SO_4(s) \rightarrow Na_2^{2+}(aq) + S^{2-}(aq) + 4\ O^{2-}(aq)$

13. Which solute would dissolve to the greatest extent in carbon tetrachloride, CCl_4?
(A) SrF_2
(B) CH_3OH
(C) CH_2O
(D) NaCl

Compound	Boiling Point (°C)
C_6H_{14}	68.0
CH_3OH	64.7
$CHCl_3$	61.2

14. The boiling points of three organic compounds are shown in the table above. A mixture of the three compounds is to be separated using a distillation process. Which statement is the best explanation for why hexane (C_6H_{14}) would be the last compound to remain in the distillation flask?
(A) Hexane has the greatest number of atoms per molecule.
(B) Hexane has the most extensive hydrogen bonding.
(C) Hexane has the most polar bonds.
(D) Hexane is a long molecule with many electrons.

15. Which of the following descriptions of the energies involved in the formation of a solution is correct?

	Energy Change with Solute Breakup	Energy Change with Solvent Breakup	Energy Change with Solute–Solvent Interaction
(A)	Endothermic	Endothermic	Exothermic
(B)	Endothermic	Exothermic	Exothermic
(C)	Exothermic	Endothermic	Endothermic
(D)	Exothermic	Exothermic	Endothermic

FREE-RESPONSE QUESTIONS

1. Students are given unknown samples A, B, C, and D of four different clear, colorless pure liquids or aqueous solutions and told that they are to determine their identities. The four possibilities are:

- hexane (C_6H_{14})
- ethanol (C_2H_5OH)
- glycerol [$CH_2(OH)CH(OH)CH_2(OH)$]
- 1 M NaCl(aq)

(a) The student decides to use an electrical conductivity tester to see if any of the samples conduct electricity. Sample A is found to be the only one that conducts electricity.
(i) What is the likely identity of Sample A?
(ii) Why is it able to conduct electricity?

(b) Sample B is found to have a much greater viscosity than the other samples.
  (i) What is the likely identity of Sample B?
  (ii) Why does it have great viscosity?

(c) The student determines the boiling point of the two remaining samples. Sample C has a lower boiling point than Sample D. The boiling points of Sample C and D are both below 100°C. Which sample, C or D, has the lowest intermolecular attraction? Explain.

(d) Samples C and D are each added separately to pure water to test for solubility. It is found that Sample D is very soluble in water, while Sample C is not soluble. What is the identity of Sample C?

(e) The molar mass of sodium chloride is 58.44 g/mol. If 100.00 mL of 1.00 *M* NaCl solution is needed for a laboratory experiment, describe how the solution should be prepared from solid NaCl. Include any necessary equipment and show calculations.

(f) Glycerol has a structure containing three hydroxyl (–OH) groups, each on a different carbon atom. Draw the complete Lewis structure for glycerol in the box below.

2. Sodium chloride is quite soluble in water but is only slightly soluble in ethanol. In the boxes below, show how solvent molecules would be oriented towards this solute and give the interparticle forces present between solute and solvent in each case.
  (a) Sodium chloride in water
    (i) Draw at least 4 water molecules in correct orientation with the solute.

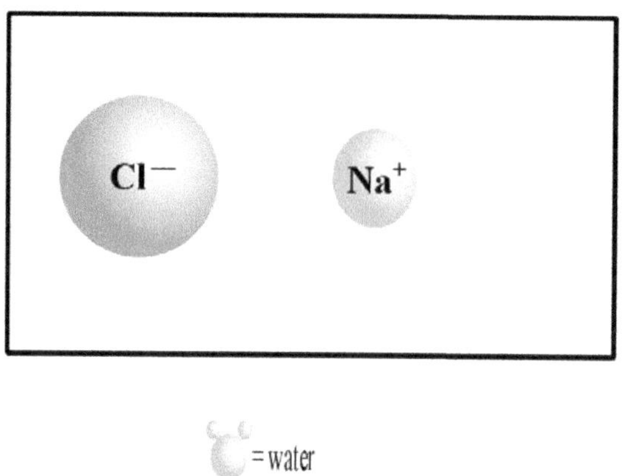

    (ii) What interparticle attraction is present between the solute and solvent?

(b) Sodium chloride in ethanol
  (i) Draw (and label) 1 sodium ion and 1 chloride ion in correct orientation with the ethanol solvent. Use a dotted line to clearly show the attraction.

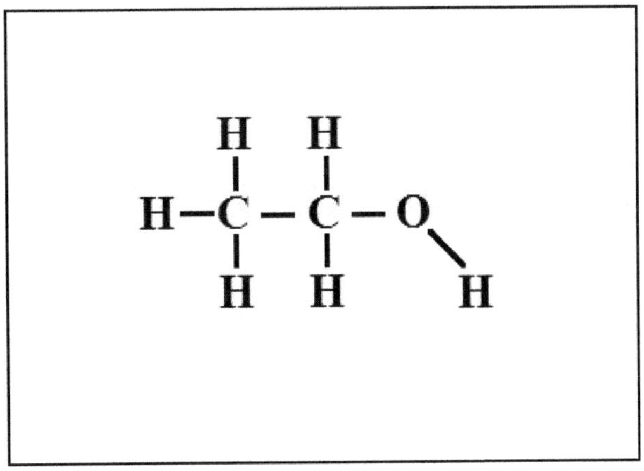

  (ii) What interparticle attraction is present between the solute and solvent?

Answers

MULTIPLE-CHOICE QUESTIONS

1. A The graphic shows 8 moles of sodium ions and 4 moles of sulfate ions. This represents 4 moles of sodium sulfate (Na_2SO_4) in 2.5 L solution. 4/2.5 = 1.6 *M* (LO 2.9).

2. D The potassium ion will be attracted to the oxygen end of the water molecule through ion–dipole attraction (LO 2.15).

3. A Hydrochloric acid is a strong electrolyte and ionizes completely, while hydrofluoric acid is a weak electrolyte and only ionizes slightly. The hydrogen (or hydronium) ion from HCl would be attracted to the oxygen end of water, and the chloride ion would be attracted to the hydrogen end of water in ion–dipole attractions. The fluorine end of the hydrogen fluoride molecule will form a hydrogen bond with the hydrogen end of a water molecule. The hydrogen end of the HF molecule will form a hydrogen bond with the oxygen end of a water molecule (LO 2.2).

4. B In distillation, solutions are heated so that components are separated based on their boiling points, with the most volatile substances evaporating first. Substances with stronger intermolecular forces have higher boiling points. Remember that molar mass is never used as a basis for determining London dispersion forces (LO 2.10).

5. A In Container 1, 3 Na^+ ions and 3 Cl^- ions are shown. Since each particle represents 0.10 mol and the container volume is 1 L, the molarity of NaCl is 0.30 *M*. In Container 2, 4 Na^+ ions and 2 CO_3^{2-} ions are shown, representing 0.40 mol of Na^+ and 0.20 mol of CO_3^{2-}. Since sodium carbonate is Na_2CO_3, the molarity of the solution is 0.20 *M* (LO 2.9).

6. B Container 1 had 0.30 mol of Na^+, and Container 2 had 0.40 mol of Na^+. The total volume of the two containers is 2 L.

$$0.30 + 0.40 = 0.70 \text{ mol Na}^+ \quad 0.70 \text{ mol/2 L} = 0.35\ M$$

(LO 2.9)

7. B The component with the highest affinity for the mobile phase will travel farthest. Since acetone, a nonpolar substance, is the mobile phase, the solute with the least attraction for acetone will move the least. $CuCl_2$ is an ionic substance and would have minimal attraction to the nonpolar acetone (LO 2.7).

8. B When the solute and solvent do not have strong attractions for each other, they are said to be immiscible and will not uniformly mix. Sodium chloride is ionic so it will not have strong attractions for the nonpolar solvent hexane. The NaCl would remain undissolved at the bottom of the container (LO 2.15).

9. C Both liquids will evaporate, but ethanol will evaporate more rapidly due to its higher vapor pressure. Water can form 2 hydrogen bonds per molecule, but ethanol can only form 1. Because ethanol has weaker intermolecular forces than water, less energy is required to vaporize the ethanol, leading to a greater rate of evaporation, lowering the ethanol concentration. The total volume of solution will decrease, while the concentration of ethanol decreases faster than the concentration of water (LO 2.16).

10. C The two metals combine to form an alloy. Copper is present in the smallest amount; so it is the solute, and silver is the solvent (LO 2.26).

11. A Water is a polar molecule with a positive hydrogen end and a negative oxygen end. Potassium iodide (KI) is a water soluble ionic compound. The ions will separate in water with the positively charged potassium ions being attracted to and surrounded by the oxygen ends of several water molecules. The negative iodide ions will be attracted to and surrounded by the hydrogen ends of several water molecules in solution (LO 2.15).

12. B When dissolved in water, Na_2SO_4 will break into its ions. The cation is Na^+, and the anion is sulfate, SO_4^{2-} (LO 3.1).

13. C You may need to draw Lewis structures to determine the polarity of the various substances. Carbon tetrachloride, CCl_4, is a nonpolar solvent. Nonpolar solutes as well as slightly polar solutes will dissolve in CCl_4 because little energy is required to break apart the solute. CH_2O is slightly polar, CH_3OH is polar with hydrogen

bonding, and NaCl and SrF_2 are both ionic with strong ionic bonds. A large amount of energy would be required to break apart the hydrogen bonds between adjacent CH_3OH molecules or to separate the ions in the ionic compounds. These would only be soluble in a very polar solvent such as water where the attractions between solute and solvent are also strong (LO 2.15).

14. D The boiling point of a substance is determined by the strength of its intermolecular forces. The last substance to remain in the flask would be the one with the highest boiling point. The data table shows us that the boiling point of hexane is the highest. This must be due to its large number of polarizable electrons and long, linear shape causing increased dispersion forces. Without the data in the table, we might incorrectly conclude that the CH_3OH would have the highest boiling point because of its hydrogen bonding, with the polar $CHCl_3$ having the second highest boiling point and the nonpolar C_6H_{14} having the lowest boiling point (LO 2.10).

15. A Energy must always be added (endothermic) to break bonds or interactions between the solute particles and between the solvent particles. When the solute and solvent interact, new interactions form and energy is released (exothermic) (LO 5.8).

FREE-RESPONSE QUESTIONS

1. (a) (i) The NaCl(*aq*) would be the only sample to conduct electricity so it is Sample A (LO 2.1, 2.19, 2.22).
   (ii) NaCl is ionic. When dissolved in water, the ions are free to move and carry an electrical charge (LO 2.19).
   (b) (i) Glycerol has the highest viscosity so it is Sample B (LO 2.16, 2.22).
   (ii) Viscosity is resistance to flow and is caused by strong intermolecular forces. Glycerol has three hydroxyl groups so it has extensive hydrogen bonding and is a syrupy liquid (LO 2.16).
   (c) Boiling point increases with increasing strength of intermolecular forces. Sample C has the lowest boiling point because it has the weakest IMFs (LO 2.16).
   (d) Very polar and ionic substances are soluble in water. Since the two remaining samples are hexane and ethanol, the ethanol (having hydrogen bonding) would be soluble in water and nonpolar hexane would be insoluble. Hexane must be Sample C (LO 2.16, 2.22).
   (e) $0.1000\ \text{L} \times \frac{1.00\ \text{mol}}{1\ \text{L}} \times \frac{58.44\ \text{g}}{1\ \text{mol NaCl}} = 5.844\ \text{g NaCl}$

   Fill a 100-mL volumetric flask approximately half full with distilled water. Mass out 5.844 g NaCl and add to the flask. Swirl until the NaCl is dissolved. Add additional water, with occasional mixing, until the level of solution reaches the engraved line on the volumetric flask. Stopper and shake (LO 1.4).

(f)

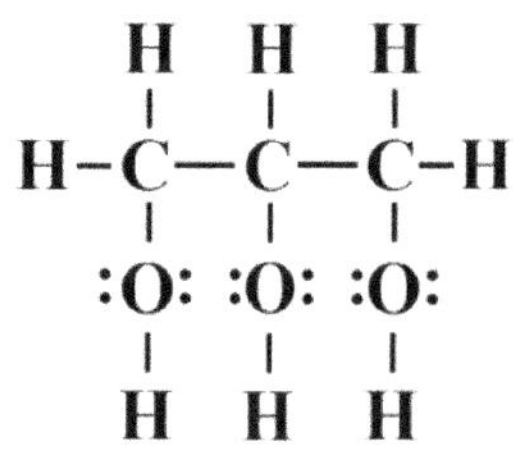

(LO 2.21)

2. (a) (i) (LO 2.8)

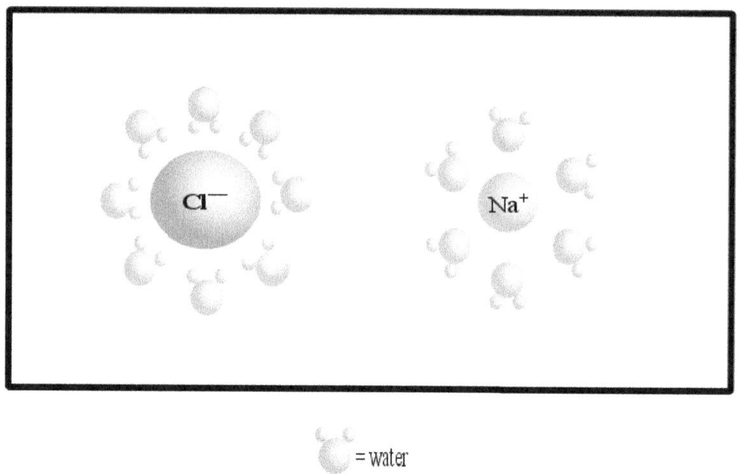

(ii) Water is very polar, but it cannot hydrogen bond with sodium or chloride ions. The attraction between the negative oxygen of water and Na^+, and between water's positive hydrogen atoms and Cl^-, is ion–dipole attraction (LO 2.15).

(b) (i)

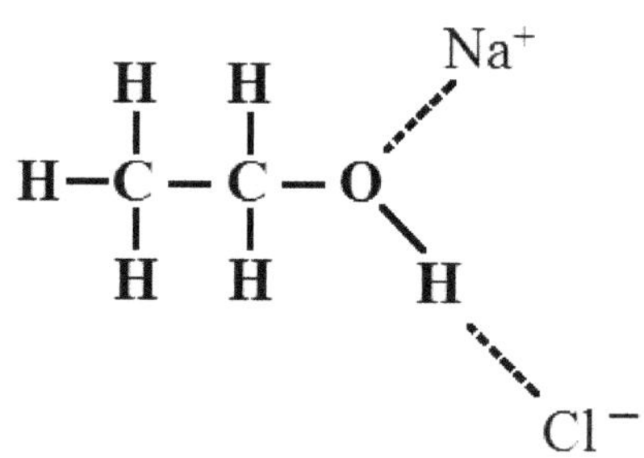

(LO 2.8)

(ii) Ethanol has a nonpolar end and a polar end and is overall less polar than water. NaCl is less soluble in ethanol than it is in water. This attraction is called ion–dipole (LO 2.15).

6

BIG IDEA 3: CHEMICAL REACTIONS AND STOICHIOMETRY

Big Idea 3

Changes in matter involve the rearrangement and/or reorganization of atoms and/or the transfer of electrons.

The two main kinds of changes studied in chemistry are phase changes, which are physical changes (melting, evaporating, freezing), and rearrangement of atoms, which involves chemical changes. Both processes are accompanied either by the release or absorption of energy. In Chapter 6, you learn about the types of reactions, balancing equations, predicting products of chemical reactions, and writing complete and net ionic equations. You will also learn about stoichiometry—calculations relating amounts of reactants and products. Chapter 7 focuses on electrochemistry, a specific application of oxidation–reduction processes. You need to have a qualitative understanding of the energy changes associated with a reaction or process, but specific calculations are covered in Big Idea 5.

You should be able to:

- Classify reactions by type.
- Write balanced molecular equations, complete ionic equations and net ionic equations, and determine which equation would be appropriate for a given situation.

- Predict if a precipitate will form (the only solubility rule you need to memorize is that all Na^+, K^+, NH_4^+, and NO_3^- salts are soluble).
- Identify compounds as Brønsted–Lowry acids and bases based on evidence of proton transfer in the balanced reaction equation.
- Identify a reaction as oxidation–reduction based on evidence of electron transfer and balance the reaction equation.
- Predict products of common reactions, given the chemical names or formulas of the reactants.
- Write balanced chemical equations and predict the amount of product formed from a given mass of reactant or the amount of reactant required to produce a desired amount of product.
- Identify limiting reactants and calculate the amount of product formed when given the amounts of all of the reactants present.
- Calculate the percent yield of a reaction.
- For reactions in solution, given the molarity and the volume of the reactants, calculate the amount of product produced or the amount of reactant required to react.
- Use data from a redox or acid–base titration, or describe an experiment using titration, to find the concentration of a solution.

AP® Tip

Learn the most common charges of transition metal ions and the physical states of elements at room temperature (i.e., if calcium is placed into water, write as Ca, not Ca^{2+}). Write solids, liquids, and gases in molecular form (i.e., write hydrogen chloride gas as HCl, not as separate ions).

EVIDENCE OF A CHEMICAL REACTION

You should be able to recognize macroscopic evidence that a chemical reaction has occurred, including:

- Releasing heat or light (exothermic) or cooling off (endothermic)
- Formation of a solid when two aqueous solutions are combined (precipitation)
- Color change, usually because of an indicator in an acid–base reaction or of the formation of products differently colored than the reactants
- Gas formation, usually in acid–base reactions and in oxidation–reduction reactions

Physical changes also may involve the formation of a solid (freezing), production of a gas (boiling), loss or gain of heat (phase changes), and color changes (rarely).

Differentiating between a physical and chemical change is not always easy. Some changes involve both and may be termed "ambiguous change." A common example of this is the dissolving of many solutes in solvents. Dissolving is generally considered a physical change because the solute can be recovered by evaporation. However, the intermolecular forces of attraction that are formed between solute and solvent are chemical, to some degree.

DESCRIBING REACTIONS IN AQUEOUS SOLUTION

1. MOLECULAR EQUATION

A molecular equation gives the overall reaction and provides information needed for stoichiometry. Including the states of matter in the equation is helpful for knowing when to ionize compounds when ionic equations are written.

$$2\ KCl(aq) + Pb(NO_3)_2(aq) \rightarrow PbCl_2(s) + 2\ KNO_3(aq)$$

2. COMPLETE IONIC EQUATION

The complete ionic equation gives the equation including all the ions in solution. Any compound that is in solution in the form of ions should be written in its ionic form. All solids, pure liquids, and weak electrolytes should be left together. For example,

$$2\ K^+(aq) + 2\ Cl^-(aq) + Pb^{2+}(aq) + 2\ NO_3^-(aq) \rightarrow PbCl_2(s) + 2\ K^+(aq) + 2\ NO_3^-(aq)$$

This is generally a helpful step toward writing the net ionic equation but is usually not required to be shown.

3. NET IONIC EQUATION

When writing a net ionic equation, eliminate the ions that are the same on both the reactant and product sides; these are called "spectator ions." Only those species that undergo a chemical change are included. The equation must still be balanced for mass and for charge!

$$Pb^{2+}(aq) + 2\ Cl^-(aq) \rightarrow PbCl_2(s)$$

TYPES OF REACTIONS

Classification of chemical reactions by type allows the prediction of products, if given the reactants.

Main Reaction Types
Precipitation (a type of double-replacement)
Acid–base (a type of double-replacement)
Redox (oxidation–reduction), including combustion and single-replacement
Synthesis/decomposition (may also be redox)

PRECIPITATION REACTIONS

Precipitation reactions involve the formation of one or two solids when two aqueous solutions are mixed. One or both products are insoluble because the ions involved have greater attraction for each other than they do for water.

PREDICTING PRECIPITATES

When two solutions of ionic compounds are mixed, their anions may exchange to form new products. If no solid (or molecular compound or gas) is formed, there is no chemical reaction! The only solubility rule you must memorize is that all Na^+, K^+, NH_4^+, and NO_3^- salts are soluble. You can deduce other rules from the context of the question. Lab data or K_{sp} values may be given that provide information as to the solubility of a substance.

EXAMPLE: Write the molecular, complete ionic, and net ionic equations when solutions of calcium chloride and sodium carbonate are mixed and a solid product is formed.

SOLUTION:

1. Molecular Equation:

$$CaCl_2(aq) + Na_2CO_3(aq) \rightarrow CaCO_3(s) + 2\ NaCl(aq)$$

Calcium carbonate must be the solid because all sodium compounds are soluble and there needs to be reaction!

2. Complete Ionic Equation:

$$Ca^{2+}(aq) + 2\ Cl^-(aq) + 2\ Na^+(aq) + CO_3^{2-}(aq) \rightarrow$$

$$CaCO_3(s) + 2\ Na^+(aq) + 2\ Cl^-(aq)$$

3. Net Ionic Equation:

$$Ca^{2+} + CO_3^{2-} \rightarrow CaCO_3(s)$$

Be sure to write the equation in whatever form is requested by the AP® exam. When in doubt, write the molecular equation with coefficients reduced to the smallest whole-number ratio.

ACID–BASE THEORIES: ARRHENIUS AND BRØNSTED–LOWRY

To work with acid–base reactions, you must be able to identify acids and bases.

The first definition of acids and bases was the *Arrhenius theory*, which states that, in aqueous solution, acids produce hydrogen ions and bases produce hydroxide ions.

The *Brønsted–Lowry* theory is used most often in this course. It says that an *acid* is a proton (H^+) donor and a *base* is a proton acceptor. It is more general than the Arrhenius theory, in which only hydroxide bases are considered. When an acid loses a hydrogen ion, the product formed is called the *conjugate base*. When a base gains a hydrogen ion, the product is called the *conjugate acid*. These are termed conjugate acid–base pairs, and they differ by only one hydrogen ion. For example:

$HC_2H_3O_2$	+	$NaOH$	→	H_2O	+	$NaC_2H_3O_2$
Acid		*Base*		*Conjugate Acid*		*Conjugate Base*

NH_3	+	H_2O	⇄	NH_4^+	+	OH^-
Base		*Acid*		*Conjugate Acid*		*Conjugate Base*

ACID–BASE REACTIONS

$$\text{Acid} + \text{Base} \rightarrow \text{Salt} + H_2O$$

Memorize the list of strong acids and strong bases below. All others may be considered as weak acids or bases for this course, unless otherwise instructed.

Strong Acids	Strong Bases
HCl, HBr, HI	Group I Hydroxides (LiOH, NaOH, ...)
H_2SO_4 (HSO_4^- = weak)	Group II Hydroxides [$Mg(OH)_2$, $Ba(OH)_2$, ...]
HNO_3	
$HClO_4$	

STRONG ACID–STRONG BASE REACTIONS

Strong acids and bases are assumed to be 100% ionized when dissolved in aqueous solution; write all reactants in ionic form, other than Group 2 hydroxides, such as $Mg(OH)_2$, that have very low solubilities. After combining the hydrogen ion from the acid with the hydroxide ion from the base to make water, the remaining cation and anion make up the salt, which in most cases will be soluble.

EXAMPLE: Solutions of hydrochloric acid and sodium hydroxide are mixed.

SOLUTION:

1. $HCl(aq) + NaOH(aq) \rightarrow NaCl(aq) + H_2O(l)$

2. $H^+(aq) + Cl^-(aq) + Na^+(aq) + OH^-(aq) \rightarrow$
$Na^+(aq) + Cl^-(aq) + H_2O(l)$

3. $H^+(aq) + OH^-(aq) \rightarrow H_2O(l)$

Note: For most strong acid–strong base reactions, this will be the net ionic equation, but make sure that the salt formed is soluble.

WEAK ACID–STRONG BASE REACTIONS (OR STRONG ACID–WEAK BASE)

Weak acids and bases ionize only slightly and are classified as weak electrolytes. Write the formula for the weak acid or weak base in molecular form (do not separate into ions!). Physical states are not required, but are often helpful.

EXAMPLE: Write the net ionic equation for the reaction when equimolar solutions of acetic acid and sodium hydroxide are mixed.

SOLUTION:

1. $HC_2H_3O_2(aq) + NaOH(aq) \rightarrow NaC_2H_3O_2(aq) + H_2O(l)$

2. $HC_2H_3O_2(aq) + Na^+(aq) + OH^-(aq) \rightarrow$
$Na^+(aq) + C_2H_3O_2^-(aq) + H_2O(l)$

3. $HC_2H_3O_2(aq) + OH^-(aq) \rightarrow C_2H_3O_2^-(aq) + H_2O(l)$

EXAMPLE: Equimolar solutions of barium hydroxide and sulfuric acid are mixed and a precipitate is formed.

SOLUTION:

1. $Ba(OH)_2(aq) + H_2SO_4(aq) \rightarrow BaSO_4(s) + 2\ H_2O(l)$

Barium hydroxide is only slightly soluble, but since it is a solution, the dissolved ions are involved and it should be written as separate ions. The first hydrogen ion completely ionizes from H_2SO_4 in all cases, but the second does not ionize completely until it reacts with a strong base. The $BaSO_4$ that forms is insoluble and is thus left together. There are no spectator ions to cancel. Therefore, the net ionic equation for the reaction above is

2. and 3. $Ba^{2+}(aq) + 2\ OH^{-}(aq) + H^{+}(aq) + HSO_4^{-}(aq) \rightarrow$

$BaSO_4(s) + 2\ H_2O(l)$

Often, gases are produced in acid–base reactions when one of the products decomposes. If a metallic carbonate reacts with an acid, the carbonic acid produced is immediately decomposed into water and carbon dioxide gas (bubbling would be observed).

$$Na_2CO_3(aq) + 2\ HCl(aq) \rightarrow 2\ NaCl(aq) + \cancel{H_2CO_3(aq)}\ H_2O(l) + CO_2(g)$$

The net ionic equation for the reaction would be:

$$CO_3^{2-} + 2\ H^{+} \rightarrow H_2O + CO_2$$

When sulfurous acid (H_2SO_3) is produced in a double-replacement reaction, it breaks down into water and SO_2. If ammonium hydroxide (NH_4OH) appears to be produced in a double-replacement reaction, the products are actually NH_3 and H_2O.

OXIDATION–REDUCTION REACTIONS

Oxidation–reduction (redox) reactions involve the transfer of electrons from the atom(s) being oxidized to the atom(s) being reduced. Skills required for this section include the assignment of oxidation numbers and the identification of oxidation–reduction reactions. This category includes combustion and most synthesis and decomposition reactions, so check for electron transfer when classifying these kinds of reactions.

In a redox reaction, you need to be able to identify the atoms that are oxidized and the atoms that are reduced.

Memory Trick

"**OIL RIG**": **O**xidation **I**s **L**oss of electrons and **R**eduction **I**s **G**ain of electrons.

Or "**LEO** the lion goes **GER**!" LEO: **L**ose **E**lectrons = **O**xidation and GER: **G**ain **E**lectrons = **R**eduction.

Rules for Assigning Oxidation States (also called Oxidation Numbers)	
The oxidation state of ...	**Examples**
A free element in the ground state is zero.	Na(*s*), O_2(*g*)
A monatomic ion is the same as its charge.	Na^+
Oxygen is usually –2 in its compounds. Exception: peroxides (containing O_2^{2-}, ex. Na_2O_2) in which oxygen is –1.	H_2O, CO_2
Hydrogen is +1 in its covalent compounds. (Hydrogen is –1 in metallic hydrides, ex. NaH).	H_2O, NH_3
For an electrically neutral compound, the sum of the oxidation states must be zero.	$KMnO_4$ Solution: K = +1, O = –2 +1 + Mn + 4(–2) = 0 –7 + Mn = 0; Mn = +7
For an ionic species, the sum of the oxidation states must equal the overall charge. *Note:* While ionic charges have the +/– sign after the value, oxidation states have the +/– sign before the value.	Ex: $Cr_2O_7^{2-}$ Solution: O = –2, 2 Cr + 7(–2) = –2; 2 Cr = +12; Cr = +6

EXAMPLE: Is the following reaction an oxidation–reduction reaction? If it is a redox reaction, identify which atoms are oxidized and which atoms are reduced.

$$Zn + Cu(NO_3)_2 \rightarrow Cu + Zn(NO_3)_2$$

SOLUTION: Assign oxidation states to each atom. If no element changes oxidation state, it is not a redox reaction.

$$\overset{0}{Zn} + \overset{+2\ +5\ -2}{Cu(NO_3)_2} \rightarrow \overset{0}{Cu} + \overset{+2\ +5\ -2}{Zn(NO_3)_2}$$

Zinc goes from Zn to Zn^{2+}. (It is oxidized, its oxidation number goes up by 2, and it loses 2 electrons.) Copper goes from Cu^{2+} to Cu. (It is reduced, its oxidation number goes down by 2, and it gains 2 electrons.) This is an oxidation–reduction reaction.

EXAMPLE: Is the following reaction an oxidation-reduction reaction? If it is, identify which atoms are oxidized and which atoms are reduced.

$$CaCO_3 \rightarrow CaO + CO_2$$

SOLUTION:

$$\overset{+2\ +4\ -2}{CaCO_3} \rightarrow \overset{+2\ -2}{CaO} + \overset{+4\ -2}{CO_2}$$

Decomposition is usually oxidation–reduction, but this one is not. No oxidation numbers change, so no electrons are transferred.

STEPS FOR BALANCING OXIDATION–REDUCTION REACTIONS

Some very simple redox reactions, such as single-replacement, synthesis, and decomposition, can be balanced in the normal method, by inspection. Many, however, are more complex and require the half-reaction method of balancing. Always check both mass and charge to ensure a balanced equation.

BALANCE OXIDATION–REDUCTION REACTIONS IN ACIDIC SOLUTION USING THE HALF-REACTION METHOD

EXAMPLE: $Cr_2O_7^{2-} + Cl^- \rightarrow Cr^{3+} + Cl_2$ (acidic solution)

STEP 1: *Write separate half-reactions:*

$Cr_2O_7^{2-} \rightarrow Cr^{3+}$ $\qquad$ $Cl^- \rightarrow Cl_2$

STEP 2: *Balance all atoms except H and O:*

$Cr_2O_7^{2-} \rightarrow 2\ Cr^{3+}$ $\qquad$ $2\ Cl^- \rightarrow Cl_2$

STEP 3: *Balance oxygen using water:*

$Cr_2O_7^{2-} \rightarrow 2\ Cr^{3+} + 7\ H_2O$ $\qquad$ $2\ Cl^- \rightarrow Cl_2$

STEP 4: *Balance hydrogen with H^+:*

$14\ H^+ + Cr_2O_7^{2-} \rightarrow 2\ Cr^{3+} + 7\ H_2O$ $\qquad$ $2\ Cl^- \rightarrow Cl_2$

STEP 5: *Balance charge using electrons:*

$6\ e^- + 14\ H^+ + Cr_2O_7^{2-} \rightarrow 2\ Cr^{3+} + 7\ H_2O$ $\qquad$ $2\ Cl^- \rightarrow Cl_2 + 2\ e^-$

STEP 6: *Equalize electron transfer.* Multiply each reaction by numbers that will allow both reactions to have the same number of electrons exchanged:

$6\ e^- + 14\ H^+ + Cr_2O_7^{2-} \rightarrow 2\ Cr^{3+} + 7\ H_2O$ $\qquad$ $6\ Cl^- \rightarrow 3\ Cl_2 + 6\ e^-$

Note: Second reaction is multiplied by 3.

STEP 7: *Add the two half-reactions,* cancelling out all the electrons and the species that appear on both sides of the equation.

$$14\ H^+ + Cr_2O_7^{2-} + 6\ Cl^- \rightarrow 2\ Cr^{3+} + 7\ H_2O + 3\ Cl_2$$

Sum of Charges: $+14 - 2 - 6 = +6 \rightarrow +6 + 0 + 0 = +6$

STEP 8: *Double-check* that there is the same number of each kind of atom on both sides and that the sums of all charges are the same on both sides.

BALANCE OXIDATION–REDUCTION REACTIONS IN BASIC SOLUTION USING THE HALF-REACTION METHOD

Repeat steps 1–6 above. After step 6: Add the same number of OH^- to both sides of the equation to cancel out any H^+. Form H_2O on the side containing both H^+ and OH^- ions. Eliminate any extra H_2O appearing on both sides. Continue with step 7.

EXAMPLE: $NO_2^- + Al \rightarrow NH_3 + AlO_2^-$ (basic solution)

STEP 1: *Write separate half-reactions:*

$Al \rightarrow AlO_2^-$ $\qquad$ $NO_2^- \rightarrow NH_3$

STEP 2: *All atoms except H and O are already balanced.*

STEP 3: *Balance O with H_2O:*

$2\,H_2O + Al \rightarrow AlO_2^-$ $\qquad$ $NO_2^- \rightarrow NH_3 + 2\,H_2O$

STEP 4: *Balance H with H^+:*

$2\,H_2O + Al \rightarrow AlO_2^- + 4\,H^+$

$7\,H^+ + NO_2^- \rightarrow NH_3 + 2\,H_2O$

STEP 5: *Balance charge using electrons:*

$2\,H_2O + Al \rightarrow AlO_2^- + 4\,H^+ + 3\,e^-$

$6\,e^- + 7\,H^+ + NO_2^- \rightarrow NH_3 + 2\,H_2O$

STEP 6: *Equalize electron transfer:*

$(2\,H_2O + Al \rightarrow AlO_2^- + 4\,H^+ + 3\,e^-) \times 2$

$6\,e^- + 7\,H^+ + NO_2^- \rightarrow NH_3 + 2\,H_2O$

STEP 7: *Add the two reactions and cancel:*

$NO_2^- + 2\,H_2O + 2\,Al \rightarrow 2\,AlO_2^- + H^+ + NH_3$

STEP 8: *New step for basic solution: Add OH^- to both sides equal to the number of H^+ ions:*

$OH^- + NO_2^- + 2\,H_2O + 2\,Al \rightarrow 2\,AlO_2^- + H^+ + NH_3 + OH^-$

Combine the hydrogen and hydroxide ions to form water:

$$OH^- + NO_2^- + 2\ H_2O + 2\ Al \rightarrow 2\ AlO_2^- + NH_3 + H_2O$$

Cancel extra water: $OH^- + NO_2^- + H_2O + 2\ Al \rightarrow 2\ AlO_2^- + NH_3$

STEP 9: *Double-check mass and charge balance.*

ELECTRON TRANSFER TO OXYGEN (INCLUDING COMBUSTION)

Combustion is the name given to the complete oxidation of any organic compound containing C, H, and/or O to yield CO_2, H_2O, and heat. Other elements and compounds undergo similar reactions with oxygen, and all are oxidation–reduction reactions.

$$C_2H_4(g) + 3\ O_2(g) \rightarrow 2\ CO_2(g) + 2\ H_2O(g)$$

$$SiH_4(g) + 2\ O_2(g) \rightarrow SiO_2(s) + 2\ H_2O(g)$$

$$2\ Mg(s) + O_2(g) \rightarrow 2\ MgO(s)$$

SYNTHESIS/DECOMPOSITION

SYNTHESIS (ALSO KNOWN AS COMBINATION OR ADDITION)

In synthesis, 2 atoms and/or molecules combine to form a new compound. These are redox reactions if electron transfer occurs. It is especially important to know your ionic charges to predict the correct formulas for products.

GENERAL RULES FOR SYNTHESIS REACTIONS:

Two or more elements or compounds combine to form a single product.

- A Group 1 or 2 metal may combine with a nonmetal to make a salt.

$$6\ Li(s) + N_2(g) \rightarrow 2\ Li_3N(s)$$

- Two nonmetals may combine to form a molecular compound.

$$C(s) + O_2(g) \rightarrow CO_2(g)$$

Two compounds combine to form a single product.

- A metallic oxide plus carbon dioxide yields a metallic carbonate.

$$CaO(s) + CO_2(g) \rightarrow CaCO_3(s)$$

- A metallic oxide plus sulfur dioxide yields a metallic sulfite.

$$CaO(s) + SO_2(g) \rightarrow CaSO_3(s)$$

- A metallic oxide plus water yields a metallic hydroxide.

$$MgO(s) + H_2O(l) \rightarrow Mg(OH)_2(s)$$

- A nonmetallic oxide plus water yields an acid.

$$CO_2(g) + H_2O(l) \rightarrow H_2CO_3(aq)$$

DECOMPOSITION

Decomposition is the opposite of synthesis, with one reactant forming two or more products. Some decomposition reactions are redox.

GENERAL RULES FOR DECOMPOSITION REACTIONS:

A compound may break down to produce two elements.

- When electrolyzed, a molten salt decomposes to its elements.

$$2\ NaCl(l) \rightarrow 2\ Na(s) + Cl_2(g)$$

A compound may break down to produce an element and a compound.

- Hydrogen peroxide decomposes into water and oxygen.

$$2\ H_2O_2(aq) \rightarrow 2\ H_2O(l) + O_2(g)$$

- Metal peroxides decompose into the metal oxide and oxygen gas.

$$2\ BaO_2(s) \rightarrow 2\ BaO(s) + O_2(g)$$

- Metallic chlorates break down to yield metallic chlorides and oxygen.

$$2\ KClO_3(s) \rightarrow 2\ KCl(s) + 3\ O_2(g)$$

A compound may break down to produce two compounds.

- Metallic carbonates break down to yield metallic oxides and carbon dioxide.

$$CaCO_3(s) \rightarrow CaO(s) + CO_2(g)$$

- Metallic sulfites break down to yield metallic oxides and sulfur dioxide.

$$CaSO_3(s) \rightarrow CaO(s) + SO_2(g)$$

- Sulfurous acid decomposes into water and sulfur dioxide.

$$H_2SO_3(aq) \rightarrow H_2O(l) + SO_2(g)$$

- Carbonic acid decomposes into water and carbon dioxide.

$$H_2CO_3(aq) \rightarrow H_2O(l) + CO_2(g)$$

- Hydrated salts decompose into the salt and water.

$$CuSO_4 \bullet 5\ H_2O(s) \rightarrow CuSO_4(s) + 5\ H_2O(l)$$

SINGLE-REPLACEMENT REACTIONS

In a single-replacement reaction, one element displaces another in a compound. One element is oxidized and another is reduced.

$$Q_2 + 2\ RT \rightarrow 2\ RQ + T_2$$

Active nonmetals replace less active nonmetals from their compounds in aqueous solution. Each halogen will displace heavier halogens from their binary salts.

EXAMPLES:

$$Cl_2(g) + 2\ KI(aq) \rightarrow I_2(s) + 2\ KCl(aq)$$

$$I_2(s) + KCl(aq) \rightarrow \text{no reaction}$$

ACTIVITY SERIES OF HALOGENS	
Decreasing Activity ↓	Fluorine
	Chlorine
	Bromine
	Iodine

Active metals replace less active metals or hydrogen from their compounds in aqueous solution. A general idea of the order of the activity series is helpful for determining reactivity. There is no need to memorize it! Simply remember that the most active metals are found in Groups 1 and 2 of the periodic table, followed by the transition metals, and finally by the "jewelry" or "precious" metals (Ag, Au, Pt, Hg, Cu) on the bottom (least reactive). A reduction potential table as discussed in the following chapter is a more detailed version of the activity series.

ACTIVITY SERIES OF METALS	
Decreasing Activity ↓	Lithium
	Potassium
	Barium
	Calcium
	Sodium
	Magnesium
	Aluminum
	Zinc
	Iron
	Cadmium
	Nickel
	Tin
	Lead
	Hydrogen (a nonmetal)*
	Copper
	Mercury
	Silver
	Gold
	Platinum

*Only metals from Li to Na will displace hydrogen from water (i.e., they will react with water). All metals above hydrogen will displace hydrogen from acids.

EXAMPLES:

$$3\ Mg(s) + 2\ FeCl_3(aq) \rightarrow 2\ Fe(s) + 3\ MgCl_2(aq)$$

$$2\ Na(s) + 2\ H_2O(l) \rightarrow 2\ NaOH(aq) + H_2(g)$$

$$Fe(s) + H_2O(l) \rightarrow \text{no reaction}$$

AP® Tip

For single- and double-replacement reactions, write water as HOH. This will help you get the correct products and helps in balancing the equation.

CONVERSIONS INVOLVED IN STOICHIOMETRIC CALCULATIONS

Type of Conversion	Example
g → mol (using molar mass)	$\frac{18.0\ g}{1\ mol\ H_2O}$
mol → mol (using mole ratio – coefficients from balanced chemical equation)	$\frac{1\ mol\ O_2}{2\ mol\ H_2O}$
mol → L of solution (using molarity which is a solution's concentration equal to the number of moles of solute in 1 L of solution)	$\frac{6\ mol\ HCl}{1\ L}$
mol → L of gas at STP (1 mole of any gas at 0°C and 1 atm pressure has a volume of 22.4 L)	$\frac{1\ mol\ H_2}{22.4\ L}$

EXAMPLE: 25.0 g of propane gas (C_3H_8) is burned in excess oxygen. Determine the mass of water vapor formed.

SOLUTION:

STEP 1: *Write a balanced equation.*

STEP 2: *Convert the mass of propane into moles of propane* using the molar mass of propane.

STEP 3: *Convert the moles of propane into moles of water* using the mole ratio from the balanced equation.

STEP 4: *Convert moles of water into mass of water* using the molar mass of water.

(Steps 2–4 can be done in a single dimensional analysis problem.)

$$C_3H_8 + 5\ O_2 \rightarrow 3\ CO_2 + 4\ H_2O$$

$$25.0\ g\ C_3H_8 \times \frac{1\ mol\ C_3H_8}{44.11\ g\ C_3H_8} \times \frac{4\ mol\ H_2O}{1\ mol\ C_3H_8} \times \frac{18.02\ g\ H_2O}{1\ mol\ H_2O} = 40.9\ g\ H_2O$$

EXAMPLE: Oxygen gas can be obtained in the laboratory by the decomposition of potassium chlorate ($KClO_3$). If 2.25 L of oxygen gas, measured at STP, is required for a laboratory test, what mass of $KClO_3$ would be needed to prepare this oxygen? Assume a 100% yield.

SOLUTION:

STEP 1: *Write a balanced equation.*

STEP 2: *Convert the volume of oxygen gas into moles of oxygen gas* by using the molar volume, 22.4 L.

STEP 3: *Convert moles of oxygen gas into moles of potassium chlorate* by using the mole ratio.

STEP 4: *Convert moles of potassium chlorate into grams of potassium chlorate* by using the molar mass of $KClO_3$.

(Steps 2–4 can be done in a single dimensional analysis problem.)

$$2\ KClO_3 \rightarrow 2\ KCl + 3\ O_2$$

$$2.25\ L\ O_2 \times \frac{1\ mol\ O_2}{22.4\ L\ O_2} \times \frac{2\ mol\ KClO_3}{3\ mol\ O_2} \times \frac{122.55\ g\ KClO_3}{1\ mol\ KClO_3} = 8.21\ g\ KClO_3$$

DETERMINATION OF EMPIRICAL FORMULA BY COMBUSTION ANALYSIS

You learned about empirical formulas in Chapter 1. A standard way of determining the empirical formula of a compound is by combustion analysis. In this procedure, a small sample of the substance is burned in excess oxygen, and the mass of resulting products, which may be solid or gaseous oxides, is measured.

If the compound is a hydrocarbon (only contains hydrogen and carbon), the masses of CO_2 and H_2O can be converted directly to moles of carbon and moles of hydrogen using stoichiometry. The mole ratio and empirical formula can then be determined. If oxygen is also present, the amount of oxygen will have to be found indirectly as oxygen must also be added to the compound for complete combustion. In this case, the masses of CO_2 and H_2O must be converted to grams of carbon and grams of hydrogen. The masses of carbon and hydrogen can then be subtracted from the original mass of the compound to determine the mass of oxygen present in the compound. Once the masses of all elements are known, convert the grams to moles and find the mole ratio and empirical formula.

EXAMPLE: A compound contains only carbon, hydrogen, and oxygen. Combustion of 1.068 g of the compound yields 1.601 g CO_2 and 0.437 g of H_2O. What is the empirical formula of the compound?

SOLUTION: Assume that all the carbon in the compound is converted to CO_2 and that all of the hydrogen in the compound is converted to H_2O. Determine the masses of carbon and hydrogen present in the 1.068 g sample.

$$1.601 \text{ g } CO_2 \times \frac{1 \text{ mol } CO_2}{44.01 \text{ g } CO_2} \times \frac{1 \text{ mol C}}{1 \text{ mol } CO_2} \times \frac{12.01 \text{ g C}}{1 \text{ mol C}} = 0.4369 \text{ g C}$$

$$0.437 \text{ g } H_2O \times \frac{1 \text{ mol } H_2O}{18.02 \text{ g } H_2O} \times \frac{2 \text{ mol H}}{1 \text{ mol } H_2O} \times \frac{1.01 \text{ g H}}{1 \text{ mol H}} = 0.0490 \text{ g H}$$

$$1.068 \text{ g cmpd} - (0.4369 \text{ g C} + 0.0490 \text{ g H}) = 0.582 \text{ g O}$$

$$0.4369 \text{ g C} \times \frac{1 \text{ mol C}}{12.01 \text{ g C}} = 0.03638 \text{ mol C}$$

$$0.0489 \text{ g H} \times \frac{1 \text{ mol H}}{1.01 \text{ g H}} = 0.0484 \text{ mol H}$$

$$0.582 \text{ g O} \times \frac{1 \text{ mol O}}{16.00 \text{ g O}} = 0.0364 \text{ mol O}$$

$$\text{C:H:O} \quad \frac{0.03638}{0.0368} : \frac{0.0484}{0.0368} : \frac{0.0364}{0.03638}$$

simplifies to 3:4:3 $C_3H_4O_3$

SOLUTION STOICHIOMETRY

Solution stoichiometry involves calculations for reactions that occur in aqueous solution. The moles of product or reactant are determined by multiplying concentration by volume, being careful to keep units consistent.

EXAMPLE: Mercury(II) ions can be removed from water by precipitation with sodium sulfide. What volume of 0.100 *M* Na_2S must be added to 100.0 mL of a 0.150 *M* $HgCl_2$ solution to remove all of the Hg^{2+} as HgS(*s*)?

SOLUTION: First, write a balanced molecular equation.

$$HgCl_2 + Na_2S \rightarrow HgS + 2\,NaCl$$

Find the volume of Na_2S solution required from the volume and molarity of $HgCl_2$.

$$0.1000 \text{ L } HgCl_2 \times \frac{0.150 \text{ mol } HgCl_2}{1 \text{ L}} \times \frac{1 \text{ mol } Na_2S}{1 \text{ mol } HgCl_2} \times \frac{1 \text{ L}}{0.100 \text{ mol } Na_2S} = 0.150 \text{ L } Na_2S$$

TITRATION

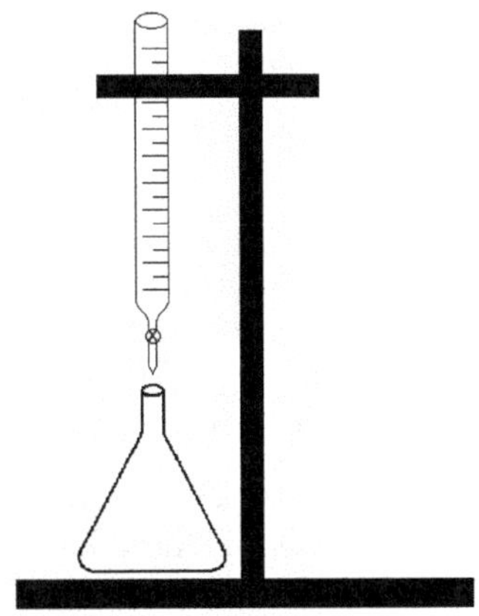

A titration is a laboratory procedure that can be used to determine the concentration of an unknown solution. A solution of known concentration (the titrant) is slowly added to the solution of unknown concentration until it completely reacts with the unknown solution (equivalence point). An indicator that changes color at or very near to the equivalence point may be used, or one of the reactants may change color when it is completely consumed. The calculations involve stoichiometric principles.

The example here is a redox titration. Acid–base titrations will be discussed further in Big Idea 6, but the principle and calculations are the same.

Changing the oxidation state of a substance often results in a color change that can be used to identify the endpoint of a titration, without the addition of an indicator. For example, the following reaction can be used to find the concentration of Fe^{2+} in solution:

$$8\ H^+(aq) + 5\ Fe^{2+}(aq) + \underset{\text{purple}}{MnO_4^-(aq)} \rightarrow 5\ Fe^{3+}(aq) + \underset{\text{colorless}}{Mn^{2+}(aq)} + 4\ H_2O(l)$$

Permanganate solution of known concentration is added dropwise, with a buret, to the Fe^{2+} solution. The solution will initially be colorless because the ions are reacting. When the solution stays purple, there is no Fe^{2+} remaining to react with MnO_4^-. This is the equivalence point of the reaction, when enough moles of titrant have been added to react exactly with the moles of the substance being titrated. For this reaction, at equivalence point:

$$\text{mol } MnO_4^- = 5 \times \text{mol } Fe^{2+}$$

EXAMPLE: A 50.0-mL solution of Fe^{2+} is titrated with 0.0020 *M* $KMnO_4$. Find the concentration of Fe^{2+} if 20.0 mL of $KMnO_4$ is required.

SOLUTION: First, write the balanced equation for the reaction, which in this example is given above. Then use the volume and molarity of the $KMnO_4$ and the mole ratio from the balanced equation to calculate the moles of Fe^{2+} in solution.

$$0.0200 \text{ L KMnO}_4 \times \frac{0.0020 \text{ mol KMnO}_4}{1 \text{ L}} \times \frac{5 \text{ mol Fe}^{2+}}{1 \text{ mol KMnO}_4} = 2.0 \times 10^{-4} \text{ mol Fe}^{2+}$$

The concentration of Fe^{2+} is:

$$\frac{2.0\times10^{-4} \text{ mol Fe}^{2+}}{0.050 \text{ L}} = 4.0 \times10^{-3}\ M \text{ Fe}^{2+}$$

Remember to use the initial volume of solution when determining the molarity!

LIMITING REACTANT

In a limiting reactant problem, the amounts of all reactants are given and the amount of product is to be determined. The limiting reactant is completely consumed when the reaction goes to completion. It determines how much product is formed.

EXAMPLE: What mass of precipitate can be produced when 50.0 mL of 0.200 *M* $Al(NO_3)_3$ is added to 200.0 mL of 0.100 *M* KOH?

SOLUTION: Always begin with a balanced molecular equation.

$$\text{Al(NO}_3)_3(aq) + 3 \text{ KOH}(aq) \rightarrow \text{Al(OH)}_3(s) + 3 \text{ KNO}_3(aq)$$

The precipitate must be the aluminum hydroxide as all nitrate compounds are soluble. For each reactant, determine the amount of a designated product produced (in mol or g).

$$0.0500 \text{ L} \times \frac{0.200 \text{ mol Al(NO}_3)_3}{1 \text{ L}} \times \frac{1 \text{ mol Al(OH)}_3}{1 \text{ mol Al(NO}_3)_3} = 0.0100 \text{ mol Al(OH)}_3$$

$$0.2000 \text{ L} \times \frac{0.100 \text{ mol KOH}}{1 \text{ L}} \times \frac{1 \text{ mol Al(OH)}_3}{3 \text{ mol KOH}} = 0.00667 \text{ mol Al(OH)}_3$$

The limiting reactant produces the least amount of product. The limiting reactant in this case is KOH. It produces only 0.00667 mole of product, which is less than the 0.0100 mole of product produced by $Al(NO_3)_3$.

The mass of precipitate produced is:

$$0.00667 \text{ mol Al(OH)}_3 \times \frac{78.00 \text{ g Al(OH)}_3}{1 \text{ mol Al(OH)}_3} = 0.520 \text{ g Al(OH)}_3$$

BCA TABLE

An alternative method that can be used to help solve limiting reagent problems is a BCA table. BCA stands for **B**efore (initial concentrations in moles), **C**hange (in concentration in moles), and **A**fter (final concentration in moles). First, convert the given amounts of reactants to moles and enter the moles into the chart. Then, analyze the mole ratio to determine which reactant (if any) will run out first (limiting reactant). Take its concentration down to zero and determine what is left at the end.

Using the previous example, note that 3 moles of KOH are required for every 1 mole of $Al(NO_3)_3$. We don't have 3 times as much KOH so it is limiting. If all of the KOH reacts, 0.0200 mol/3 = 0.00667 mol of the $Al(NO_3)_3$ will react.

	$Al(NO_3)_3$ +	3 KOH →	$Al(OH)_3$ +	3 KNO_3
Before	0.0100 mol	0.0200 mol	0	0
Change	–0.00667 mol	0.0200 mol	+0.00667 mol	+0.0200 mol
After	0.00333 mol	0	0.00667 mol	0.0200 mol

When the entire final row is filled in, reread the question to see exactly what it was asking for and use the information in the table to finish the problem. This example asked for the mass of precipitate, so 0.00667 mol $Al(OH)_3$ × 78.00 g = 0.520 g $Al(OH)_3$. If you choose the limiting reagent incorrectly, one of the values of the reactants will be negative, which is impossible to have, and you will need to select the other reactant as the limiting reagent.

PERCENT YIELD

The *percent yield* of a reaction is the actual yield of a product as a percentage of the theoretical yield. The *actual yield* is the amount of product obtained in an experiment. The *theoretical yield* is the amount of product calculated from the amounts of reactants used. The percent yield is generally less than 100% because the reaction doesn't go to completion, because of side reactions that occur without generating desired product, and because of difficulties in collecting the entire product. Yields greater than 100% are evidence of some contamination or error in calculation or measurement.

EXAMPLE: If the reaction above has an 85.3% yield of precipitate, how much aluminum hydroxide is produced?

SOLUTION:

$$\text{Percent Yield} = \frac{\text{Actual Yield}}{\text{Theoretical Yield}} \times 100\%$$

$$\text{Actual Yield} = 0.853 \times 0.520 \text{ g} = 0.444 \text{ g } Al(OH)_3$$

MULTIPLE-CHOICE QUESTIONS

No calculators are to be used in this section.

1. Aqueous solutions of $CaCl_2$ and Na_2CO_3 are mixed, forming a precipitate. What is the net ionic equation for the reaction?
   (A) $Ca^{2+}(aq) + CO_3^{2-}(aq) \rightarrow CaCO_3(s)$
   (B) $2\ Na^{+}(aq) + 2\ Cl^{-}(aq) \rightarrow 2\ NaCl(s)$
   (C) $Ca^{2+}(aq) + CO_3^{2-}(aq) + 2\ Na^{+}(aq) + 2\ Cl^{-}(aq) \rightarrow$
   $CaCO_3(s) + 2\ NaCl(aq)$
   (D) $Ca^{2+}(aq) + 2\ CO_3^{-}(aq) \rightarrow Ca(CO_3)_2(s)$

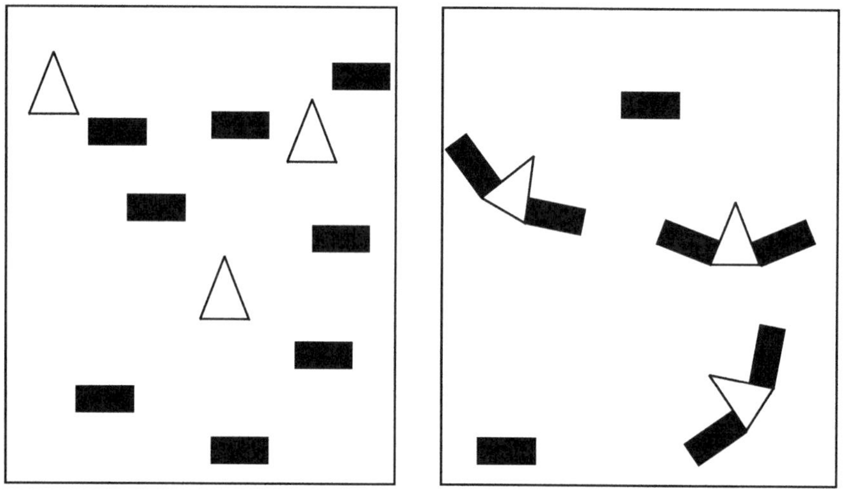

Reactants Products

2. The boxes above represent the reactants and products of a reaction. Excess reactants may be present. Which balanced equation best represents the reaction?
   (A) 8 ▬ + 3 Δ → 3 Δ▬$_2$ + 2 ▬
   (B) 8 ▬ + 3Δ → 3 Δ + 8 ▬
   (C) 6 ▬ + 3Δ → 3 Δ▬$_2$
   (D) 2 ▬ + Δ → Δ▬$_2$

3. If a solution of sodium bicarbonate is mixed with a solution containing an equal number of moles of nitric acid, sodium nitrate, water, and carbon dioxide are produced. What is the net ionic equation representing the reaction?
   (A) $NaHCO_3(aq) + HNO_3(aq) \rightarrow NaNO_3(aq) + H_2O(l) + CO_2(g)$
   (B) $HCO_3^{-}(aq) + HNO_3(aq) \rightarrow NO_3^{-}(aq) + H_2O(l) + CO_2(g)$
   (C) $Na^{+}(aq) + HCO_3^{-}(aq) + HNO_3(aq) \rightarrow$
   $Na^{+}(aq) + NO_3^{-}(aq) + H_2O(l) + CO_2(g)$
   (D) $HCO_3^{-}(aq) + H^{+}(aq) \rightarrow H_2O(l) + CO_2(g)$

4. Which of the following reactions involves the transfer of electrons from one species to another?
(A) $Ca(OH)_2(aq) + 2\ HNO_3(aq) \rightarrow Ca(NO_3)_2(aq) + 2\ H_2O(l)$
(B) $Fe_2O_3(s) + 3\ CO(g) \rightarrow 2\ Fe(s) + 3\ CO_2(g)$
(C) $CuBr_2(aq) + 2\ NaOH(aq) \rightarrow Cu(OH)_2(s) + 2\ NaBr(aq)$
(D) $2\ NaHCO_3(s) \rightarrow Na_2CO_3(s) + H_2O(l) + CO_2(g)$

The following information refers to Questions 5–6.

Many different types of vinegars are used in the preparation of food. These vinegars are made from a variety of fruits and grains and each has a different concentration of acetic acid, $HC_2H_3O_2$ (molar mass 60.1 g/mol). In order to determine the concentration of acetic acid, the vinegar may be titrated with standardized sodium hydroxide solution (with the presence of an appropriate indicator). The reaction for the neutralization of acetic acid by sodium hydroxide is shown below.

$$NaOH + HC_2H_3O_2 \rightarrow NaC_2H_3O_2 + H_2O$$

5. Which of these is an acid/conjugate base pair for the neutralization reaction of acetic acid?
(A) $HC_2H_3O_2/H_2O$
(B) $HC_2H_3O_2/NaOH$
(C) $HC_2H_3O_2/NaC_2H_3O_2$
(D) $NaC_2H_3O_2/NaOH$

6. If a 10.0-mL sample of a rice vinegar (density 1.00 g/mL) requires 25.0 mL of 0.30 *M* NaOH for complete reaction, what is the % acetic acid in the rice vinegar?
(A) 0.75%
(B) 1.5%
(C) 3.0%
(D) 4.5%

7. How many moles of electrons are transferred between the substance being oxidized and the substance being reduced in the reaction given below?

$$4\ NH_3(g) + 5\ O_2(g) \rightarrow 4\ NO(g) + 6\ H_2O(l)$$

(A) 5
(B) 10
(C) 16
(D) 20

8. When 120 g of NaOH(*s*) is added to 1 L of water and allowed to dissolve, it is observed that the solution increases in temperature and the total volume of solution is less than 1 L. Which description best describes what is happening?
   (A) A chemical change is occurring because heat is released and the solution is colorless while the solid was white.
   (B) A change in the nucleus is occurring because energy is being released.
   (C) A physical change is occurring because the chemical composition is still NaOH. Solid NaOH could be obtained by evaporation.
   (D) An ambiguous change is occurring because, while dissolving is a physical change, the new attractions being formed between the ions and water molecules are chemical changes.

9. When a piece of copper wire is immersed in a beaker containing 200 ml of 0.05 *M* $AgNO_3$ solution for 10 minutes, shiny crystals appear on the copper wire. The solution also becomes slightly blue in color. Which is the correct equation for the oxidation half-reaction occurring in this situation?
   (A) $Cu^{2+} + 2\,e^- \rightarrow Cu$
   (B) $Cu \rightarrow Cu^{2+} + 2\,e^-$
   (C) $Ag^+ + e^- \rightarrow Ag$
   (D) $Ag \rightarrow Ag^+ + e^-$

Mass of evaporating dish	25.00 g
Mass of evaporating dish plus magnesium strip	27.43 g
Mass of evaporating dish plus product	28.50 g

10. In the laboratory, a student wanted to confirm the identity of the product formed when several shiny silver-colored strips of magnesium metal were placed in an evaporating dish and ignited, resulting in a dull gray powder with no metallic-appearing material remaining. The data from the experiment are shown in the table above. The student was surprised by the results as they expected the product to be MgO. How can the results best be interpreted?
    (A) The magnesium strip was not completely oxidized.
    (B) Some of the product was lost into the air during the combustion process.
    (C) The product was actually MgO_2, not MgO.
    (D) The final product was allowed to sit out in the humid room for 1 hour before massing.

Questions 11–12 refer to the information below.

The amount of oxalic acid in a food product can be found by an oxidation–reduction titration using potassium permanganate. The oxalic acid is dissolved in an acidic solution prior to the titration. As the purple-colored permanganate solution is added to the oxalate solution, it is converted to the colorless Mn^{2+} ion and a faint pink color is present at the equivalence point. The reaction occurring is:

$$2\ MnO_4^-(aq) + 6\ H^+(aq) + 5\ H_2C_2O_4(aq) \rightarrow 2\ Mn^{2+}(aq) + 8\ H_2O(l) + 10\ CO_2(g)$$

11. Which is correct concerning the reaction?
    (A) Carbon is oxidized from +2 to +4.
    (B) Carbon is reduced from +3 to +4.
    (C) Manganese is reduced from +7 to +2.
    (D) Manganese is oxidized from +4 to +2.

12. If 0.0300 L of 0.0500 *M* $KMnO_4$ solution was required to reach the equivalence point of the titration, how many moles of $H_2C_2O_4$ were present?
    (A) 1.50×10^{-3}
    (B) 7.50×10^{-3}
    (C) 3.75×10^{-3}
    (D) 1.25×10^{-1}

13. How many moles of barium sulfide, BaS, will form when 60.0 mL of 1.0 *M* $Ba(NO_3)_2$ are mixed with 25.0 mL of 0.80 *M* K_2S solution to form barium sulfide solid?
    (A) 0.020 mol
    (B) 0.040 mol
    (C) 0.060 mol
    (D) 0.10 mol

14. Consider the neutralization reaction:
    $Mg(OH)_2(s) + 2\ HCl(aq) \rightarrow MgCl_2(aq) + 2\ H_2O(l)$. What volume of 5.00 *M* HCl is required to react completely with 5.83 g of $Mg(OH)_2$ (molar mass of $Mg(OH)_2$ = 58.3 g/mol)?
    (A) 10.0 mL
    (B) 30.0 mL
    (C) 40.0 mL
    (D) 50.0 mL

15. When 16 g of methane (CH_4) and 32 g of oxygen (O_2) reacted to produce carbon dioxide and water, 5.6 L of carbon dioxide gas was produced at STP. Calculate the percent yield of carbon dioxide in this reaction.
    (A) 5.0%
    (B) 10%
    (C) 25%
    (D) 50%

FREE-RESPONSE QUESTIONS

1. Two dilute aqueous solutions are prepared: one of $Cu(NO_3)_2$ and one of KOH.
   (a) Using the key below, draw particle-level representations of the two solutions, assuming that one beaker contains 3 formula units of $Cu(NO_3)_2$ and the other beaker contains 4 formula units of KOH. Water molecules do not need to be shown.

   Cu^{2+} = □ NO_3^- = ○ K^+ = △ OH^- = ♡

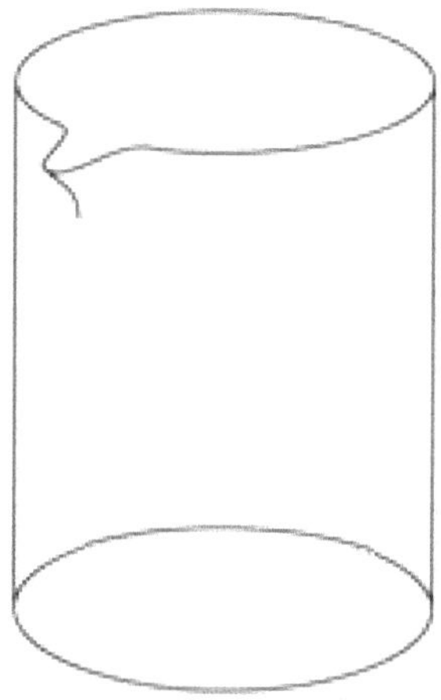

$Cu(NO_3)_2$

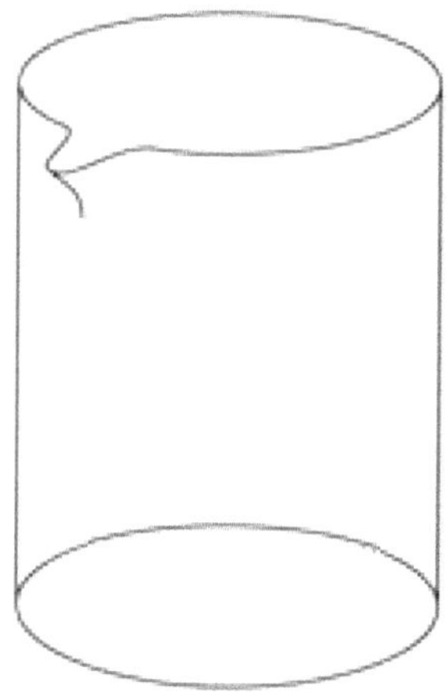

KOH

   (b) $Cu(NO_3)_2$ and KOH react chemically when combined.
      (i) Draw a molecular representation of the solution that results when the contents of the two beakers are combined and mixed in the third beaker shown below. One of the products is considered to be insoluble in water. Include the correct number of formula units or ions and the correct amounts and kinds of ions remaining. No water molecules need to be shown.

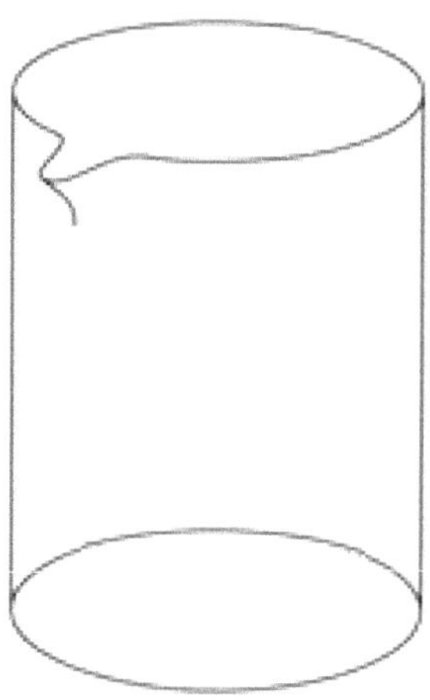

      (ii) Write the balanced net ionic equation for the reaction.

(c) Now, consider that the same two solutions have 0.10 *M* concentrations and that 100 mL of each solution is combined.

(i) The temperatures of the individual solutions are measured before they are poured together. Soon after the two solutions are combined and mixed, the temperature is recorded again. The final temperature is found to be significantly higher than the temperatures of the individual solutions. Complete the reaction diagram below to show this energy change.

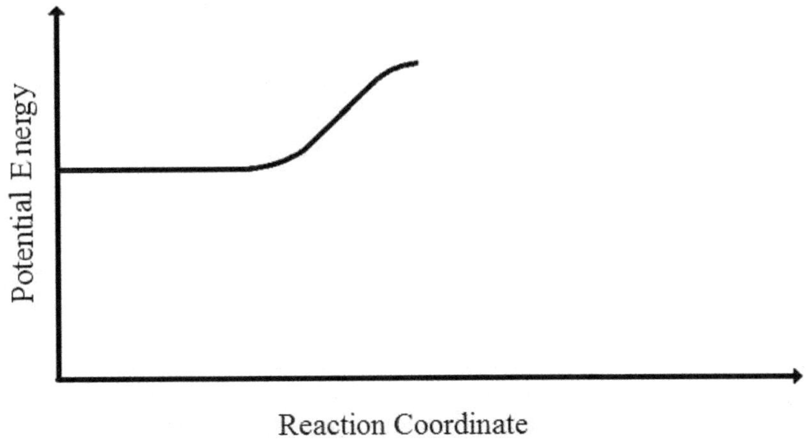

(ii) The solution is decanted, leaving the precipitate in the beaker. After the precipitate is rinsed several times with distilled water and 50 mL of distilled water is added, the mixture is heated. The precipitate decomposes to form a black solid that is an oxide of copper. The remaining water is decanted, and the oxide of copper remaining in the beaker is dried. 50 mL of 3 *M* HCl is added to dissolve the copper.Very small grains of zinc metal are added until no further gas production is observed and the solution has lost its blue color, leaving only solid copper, which is then rinsed and dried. List the data that you would need to collect to determine the empirical formula of the oxide of copper.

(iii) Show sample calculation setups for the determination of the empirical formula of the copper oxide. (No actual calculations are required.)

2. Blood alcohol (C_2H_5OH) level can be determined by titrating a sample of blood plasma with an acidic potassium dichromate solution. The *unbalanced* equation for the reaction is:

$$Cr_2O_7^{2-}(aq) + C_2H_5OH(aq) \rightarrow Cr^{3+}(aq) + CO_2(g)$$

(a) Identify which species is oxidized and which is reduced.

(b) Balance the equation, using smallest whole-number coefficients.

(c) If a 2.00-mL sample of plasma reacts completely with 3.24 mL of 0.0432 *M* $K_2Cr_2O_7$, what is the concentration of alcohol in the serum in mol/L?

Answers

MULTIPLE-CHOICE QUESTIONS

1. A The complete molecular equation for the reaction is: $CaCl_2(aq)$ + $Na_2CO_3(aq) \rightarrow CaCO_3(s) + 2\ NaCl(aq)$. All sodium compounds are soluble so the precipitate is $CaCO_3$. All of the aqueous solutions in this reaction involve strong electrolytes so they must be ionized. The chloride ions and sodium ions show up on both sides of the reaction so they are spectator ions and are cancelled out (LO 3.2).

2. D The reactants are 8 rectangles and 3 triangles. The products are 2 rectangles along with 3 molecules consisting of 1 triangle plus 2 rectangles each. This indicates that 2 rectangles were excess.

   6 ■ + 3△ → 3△■ $_2$ which can be reduced down to 2 ■ + △ → △■ $_2$

   (LO 3.1)

3. D Because sodium bicarbonate and sodium nitrate are both strong electrolytes (salts), they are written as ions in aqueous solution. Nitric acid is a strong acid and is completely ionized in aqueous solution. Because water and carbon dioxide are molecular species and have very little ionization occurring, they stay written in the molecular form. The two spectator ions, Na^+ and NO_3^-, are cancelled from both sides of the equation (LO 3.2).

4. B To answer this question, assign oxidation numbers to each atom in each equation. If an oxidation number changes, that reaction is a redox reaction and will involve the transfer of electrons. In reaction B, the oxidation number of iron goes from +3 on the reactant side to 0 on the product side, while the oxidation number of carbon increases from +2 to +4 (LO 3.8).

5. C The formula of an acid and its conjugate base will differ by a single hydrogen ion. Acetic acid and sodium acetate (the acetate ion) are a conjugate acid/base pair (LO 3.7).

6. D 0.025 L × 0.30 *M* NaOH = 0.0075 mol of NaOH. Acetic acid (CH_3COOH) reacts in a 1 to 1 ratio with NaOH, so the moles of acetic acid present are also 0.0075. 0.0075 mol CH_3COOH × 60.1 g/mol = (0.45 g CH_3COOH/10 g solution) × 100 = 4.5% acetic acid in the vinegar sample (LO 1.20).

7. D The change in oxidation number of the nitrogen is from –3 in the NH_3 to +2 in the nitrogen monoxide molecule. This requires 5 electrons/nitrogen atom, or 5 moles electrons/moles nitrogen. Since there are 4 moles of nitrogen atoms, the total moles of electrons needed are 4 × 5 or 20 moles of electrons (LO 3.8).

8. D The dissolving process is a physical change. However, the new interactions formed in the solution (evidenced by the temperature change and by the decreased volume of the solution) may be

considered chemical. Since both are present, it is best termed an ambiguous change (LO 3.10).

9. B The reaction described is $Cu + 2\ AgNO_3 \rightarrow Cu(NO_3)_2 + 2\ Ag$. Copper is being oxidized from +0 to +2 oxidation state and is losing 2 electrons in the process (LO 3.8).

10. B 27.43 − 25.00 = 2.43 g Mg   2.43 g Mg = 0.100 mol Mg

28.50 g − 27.43 g = 1.07 g O

1.07 g O is significantly less than 0.1 mole of oxygen. The magnesium strip must have completely oxidized because it no longer appeared shiny. The product could not be MgO_2 as that would require 0.2 mole oxygen per 0.1 mole Mg. If the product had absorbed water from the air prior to weighing, the mass of the product would have been too high, rather than too low. Some product must have been lost during the reaction. Since the mass of magnesium was found before the reaction started, the missing mass would incorrectly appear to be only oxygen (LO 3.6).

11. C Assign oxidation numbers to all atoms. In this reaction, manganese is reduced from +7 in MnO_4^- to +2 in Mn^{2+}, and carbon is oxidized from +3 in $H_2C_2O_4$ to +4 in CO_2 (LO 3.8).

12. C $0.0300\ L \times \frac{0.0500\ mol\ KMnO_4}{1\ L} \times \frac{5\ mol\ H_2C_2O_4}{2\ mol\ KMnO_4} = 0.00375\ mol\ H_2C_2O_4$

(LO 3.9)

13. A The balanced equation for the reaction is $Ba(NO_3)_2 + K_2S \rightarrow BaS + 2\ KNO_3$. According to the calculations, there are 0.060 mole $Ba(NO_3)_2$ and 0.020 mole K_2S; therefore, the potassium sulfide is the limiting reactant (they react in a 1 to 1 ratio). According to the equation, 1 mole of potassium sulfide produces 1 mole of barium sulfide. Therefore, 0.020 mole of potassium sulfide will allow for the formation of 0.020 mole of the solid barium sulfide (LO 3.4).

14. C The mass of 5.83 g of $Mg(OH)_2$ represents 0.100 mole of this base; the equation indicates that twice that amount (0.200 mole) of acid is needed for neutralization to be complete. Since both the concentration and number of moles needed are known for the acid, the volume of the acid can be calculated (0.200 mol/5.00 mol/L = 0.0400 L = 40.0 mL) (LO 3.4).

15. D The balanced equation for the reaction is $CH_4 + 2\ O_2 \rightarrow CO_2 + 2\ H_2O$. You are given 16 grams of methane = 1.0 mole, and 32 grams of oxygen = 1.0 mole. The balanced equation shows that 1 mole of methane requires 2 moles of oxygen. Therefore, the 1 mole of oxygen will be used up, and 0.5 mole of methane will remain unreacted. Oxygen is the limiting reactant.

Thus, if 1 mole of oxygen is used up, according to the equation, 0.5 mole CO_2 would be produced. The molar volume of any gas at STP is 22.4 L/mol, so the expected yield would be 11.2 L. The actual yield was 5.6 L so (5.6 L/11.2 L) × 100 = 50% (LO 3.3).

FREE-RESPONSE QUESTIONS

1. (a)

Cu^{2+} = □ NO_3^- = ○ K^+ = △ OH^- = ♡

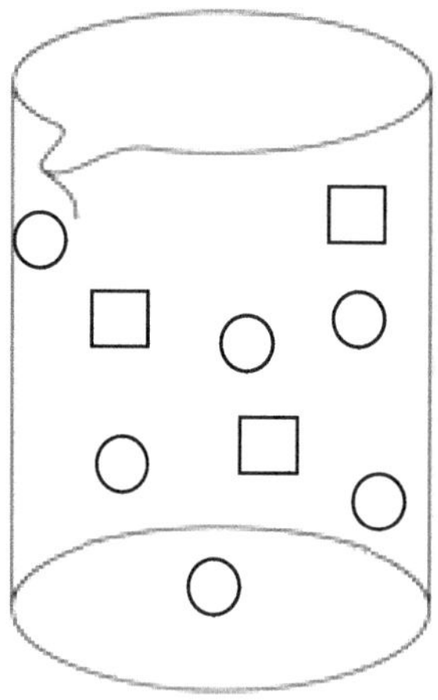

$Cu(NO_3)_2$

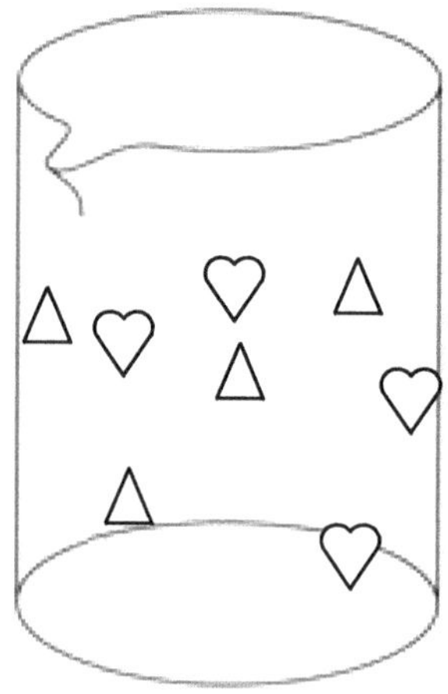

KOH

Both substances are soluble and will ionize completely (LO 3.1)

(b) (i)

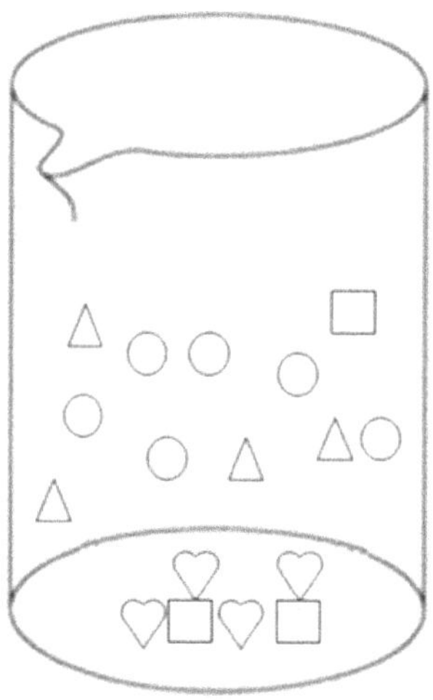

There are not enough OH^- ions to react with all the Cu^{2+} ions, so there will be 1 Cu^{2+} ion left in solution, along with the spectators K^+ and NO_3^- (LO 3.1).

(ii) $Cu^{2+} + 2\ OH^- \rightarrow Cu(OH)_2(s)$

(LO 3.2)

(c) (i)

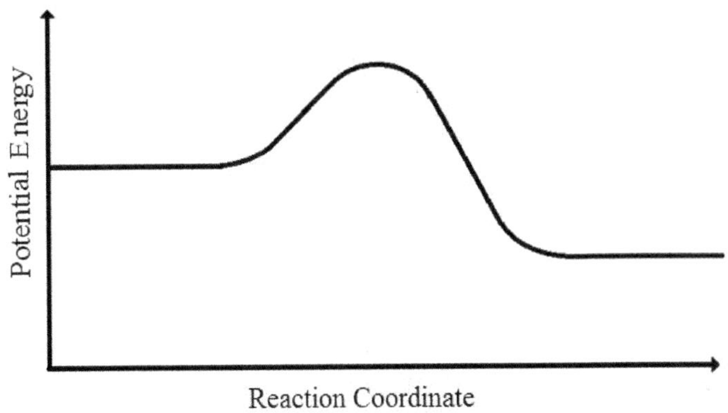

(LO 3.11)

(ii) Mass of dried copper oxide and mass of dried copper (LO 3.5)

(iii) Mass of copper/63.55 g = Moles of copper

Mass of copper oxide – Mass of copper = Mass of oxygen

Mass of oxygen/16.00 g = Moles of oxygen

Moles of copper: moles of oxygen to get mole ratio. To simplify the ratio, divide by the smaller number to obtain small whole numbers. Use the whole numbers as subscripts to obtain the Cu_xO_y formula (LO 3.6).

2. (a) Start by assigning oxidation numbers to determine which species are oxidized and which are reduced. Cr is reduced as its oxidation number goes from +6 to +3. C is oxidized as its oxidation number goes from –2 to +4 (LO 3.8).

(b) Use the half-reaction method.

STEP 1: *Write separate half-reactions:*

$$Cr_2O_7^{2-} \rightarrow Cr^{3+}$$

$$C_2H_5OH \rightarrow CO_2$$

STEP 2: *Balance all atoms except H and O:*

$$Cr_2O_7^{2-} \rightarrow 2\ Cr^{3+}$$

$$C_2H_5OH \rightarrow 2\ CO_2$$

STEP 3: *Balance oxygen using water:*

$$Cr_2O_7^{2-} \rightarrow 2\ Cr^{3+} + 7\ H_2O$$

$$3\ H_2O + C_2H_5OH \rightarrow 2\ CO_2$$

STEP 4: *Balance hydrogen with H^+:*

$$14\ H^+ + Cr_2O_7^{2-} \rightarrow 2\ Cr^{3+} + 7\ H_2O$$

$$3\ H_2O + C_2H_5OH \rightarrow 2\ CO_2 + 12\ H^+$$

STEP 5: *Balance charge using electrons:*

$$6\ e^- + 14\ H^+ + Cr_2O_7^{2-} \rightarrow 2\ Cr^{3+} + 7\ H_2O$$

$$3\ H_2O + C_2H_5OH \rightarrow 2\ CO_2 + 12\ H^+ + 12\ e^-$$

STEP 6: *Equalize electron transfer.* Multiply each reaction by numbers that will allow both reactions to have the same number of electrons exchanged:

$$12\ e^- + 28\ H^+ + 2\ Cr_2O_7^{2-} \rightarrow 4\ Cr^{3+} + 14\ H_2O$$

$$3\ H_2O + C_2H_5OH \rightarrow 2\ CO_2 + 12\ H^+ + 12\ e^-$$

Note: First reaction is multiplied by 2.

Add the two half-reactions, cancelling out all the electrons and the formulas that appear on both sides of the equation.

$$16\ H^+ + 2\ Cr_2O_7^{2-} + C_2H_5OH \rightarrow 4\ Cr^{3+} + 11\ H_2O + 2\ CO_2$$

Sum of Charges: $+16 - 4 + 0 = +12 \rightarrow +12 + 0 + 0 = +12$

STEP 7: *Double-check* that there is the same number of each kind of atom on both sides and that the sums of all charges are the same on both sides (LO 3.2).

(c)

$$0.00324\ L \times \frac{0.0432\ mol\ Cr_2O_7^{2-}}{1\ L} \times \frac{1\ mol\ C_2H_5OH}{2\ mol\ Cr_2O_7^{2-}} = 7.00 \times 10^{-5}\ mol\ C_2H_5OH$$

$$\frac{7.00 \times 10^{-5}\ mol\ C_2H_5OH}{0.00200\ L} = 0.0350\ M$$

(LO 1.20)

7

BIG IDEA 3: ELECTROCHEMISTRY

Big Idea 3

Changes in matter involve the rearrangement and/or reorganization of atoms and/or the transfer of electrons.

Electrochemistry is the study of the interchange of electrical and chemical energy through transfer of electrons. There are two types of electrochemical cells, galvanic and electrolytic. In galvanic (also known as voltaic) cells, thermodynamically favored (spontaneous) oxidation–reduction (redox) reactions generate electric current. In electrolytic cells, a non-thermodynamically favored (nonspontaneous) chemical reaction occurs with the application of an electric current.

You should be able to:

- Identify and compare the two types of electrochemical cells: galvanic and electrolytic.
- Draw and label a galvanic cell, including labeling the electrodes, the flow of electrons, and the flow of ions.
- Write balanced half-reactions and determine which reaction occurs at the anode and which reaction occurs at the cathode.
- Calculate the cell potential for a galvanic cell and an electrolytic cell.
- Determine if a reaction is thermodynamically favored (spontaneous) from its cell potential.
- Use reduction potentials to determine which substances are more likely to be reduced or oxidized.
- Draw and label an electrolytic cell.

- Determine the reactions occurring at the anode and the cathode during electrolysis.
- Perform stoichiometric calculations involving electrolysis.
- Use the Nernst equation to qualitatively determine the effects on voltage of a change in concentrations.

GALVANIC CELLS

COMPONENTS OF THE GALVANIC CELL

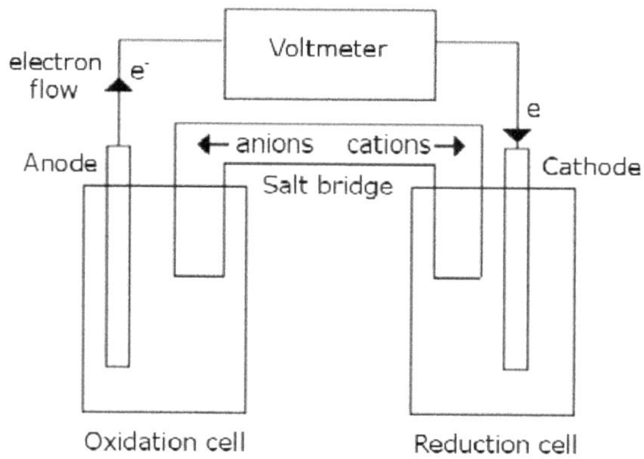

In a galvanic cell (also called a voltaic cell), a thermodynamically favored (or spontaneous) chemical reaction is used to produce electrical energy. The current produced by the cell is measured in volts by a voltmeter. One compartment of the galvanic cell contains the anode, and the other compartment contains the cathode. Oxidation, a loss of electrons, occurs at the anode. Reduction, a gain of electrons, occurs at the cathode. The electrodes (the anode and the cathode) are immersed in solutions containing metal cations. The two compartments of a galvanic cell are connected by a salt bridge or a porous disk that allows ions to flow between the compartments to maintain a net zero charge in each. This is shown in the figure above.

Since the anode is undergoing oxidation and is losing electrons, the half-reaction would be $X(s) \rightarrow X^{n+}(aq) + n\ e^-$. Since the solid electrode, X, is being "eaten away" and being transformed into the ion $X^{n+}(aq)$, the mass of the anode will decrease in a galvanic cell. Likewise, the cathode's mass will increase since the half-reaction that occurs at the cathode is $Q^{n+}(aq) + n\ e^- \rightarrow Q(s)$. The exam may provide you with a statement about the change in mass. This will allow you to determine the identity of the electrodes; hence, you are able to write the thermodynamically favored equation as the cell operates.

AP® Tip

Two common student misconceptions: Electrons ***never*** flow through the salt bridge, and the salt bridge ***does not*** complete the circuit.

In the diagram below, the anode is a strip of zinc metal immersed in a 1.0 *M* $ZnSO_4$ solution. The cathode is a special electrode known as the standard hydrogen electrode (SHE). The SHE has been given an arbitrary cell potential value of 0.00 volts, and all other cells are compared to it to determine their voltage. The SHE has a 1.0 *M* $H^+(aq)$ solution containing a square piece of platinum foil (connected by a platinum wire) where electrons can be exchanged inside a tube. Platinum is used because it is inert (unreactive). During the reaction, hydrogen gas is passed through the top of the tube and into the solution, causing the reaction $2\ H^+(aq) + 2\ e^- \rightarrow H_2(g)$.

In the example, Zn is oxidized to Zn^{2+}, and the electrons travel to the SHE, where H^+ in solution is reduced to $H_2(g)$. The salt bridge contains $KNO_3(aq)$, and the K^+ ions move toward the cathode to replace the reduced H^+. NO_3^- ions travel to the anode to balance the additional positive charge as Zn^{2+} is formed.

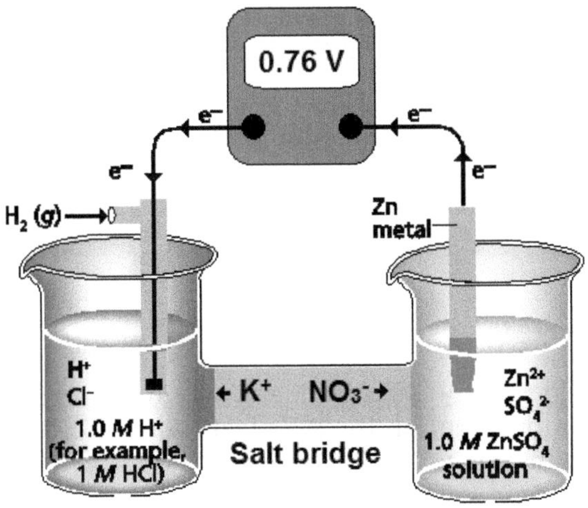

Inert electrodes may be necessary in other situations, as well. For example, Fe^{2+} may be oxidized to Fe^{3+} at the anode. Since these are both soluble ions, an electrode cannot be made of them. Instead, an inert solid electrode, usually platinum or graphite, is used. It does not experience any increase or decrease in mass throughout the reaction.

Although the terms "oxidizing agent" and "reducing agent" are no longer tested, students may still be asked to supply this information in a variety of ways. For example:

- Identify what is being reduced or oxidized.
- Which of the following can cause Zn to be oxidized or H^+ to be reduced?

Memory Tips

Here are some mnemonic devices to help you remember some facts about electrochemistry and redox reactions.

- "LEO" the lion goes "GER" (or "LEOpold was a great GERman") means Loss of Electrons is Oxidation and Gain of Electrons is Reduction.
- To recall what happens at the anode and the cathode: "The Red Cat chased An Ox" means reduction occurs at the cathode and oxidation occurs at the anode.
- To remember the direction of the migration of ions through the salt bridge or porous plate toward the electrodes for both types of cells, "CAT" ions move to the "CAT" hode and "AN" ions move to the "AN" ode. Another way to remember this is: Anions to the Anode, Cations to the Cathode! Charge!!
- To remember how electrons travel through the external wire, remember "Electrons travel FAT CAT" (from anode to cathode).

AP® Tip

- You will be provided any standard reduction potentials necessary to solve a problem on the AP® test, but will not be given a complete reduction potential table.
- You should be able to sketch a galvanic cell and label the electrodes, the flow of electrons, and the flow of ions.

STANDARD REDUCTION POTENTIALS

The cell potential, E_{cell}, is the potential of the cell to do work on its surroundings by driving an electric current through a wire. By definition, a potential of 1 volt is produced when 1 joule of energy moves 1 coulomb of electric charge across a potential. The magnitude of the cell potential is a measure of the driving force behind an electrochemical reaction. It may also be referred to as the electromotive force or *emf*. Tables of reduction potentials give $E°_{cell}$, standard voltages for reduction half-reactions measured at standard conditions of 1 atmosphere, 1 molar solution, and 25°C. The reduction cell potentials are always given in the form of $X^{m+} + m\,e^- \rightarrow X$. Thus, if you need the oxidation cell potential (i.e., $X \rightarrow X^{m+} + m\,e^-$), the voltage will be equal and opposite the value of the reduction potential.

Remember that a volt = $\frac{\text{joule}}{\text{coulomb}}$.

CALCULATING THE CELL POTENTIAL OF A GALVANIC CELL

The reaction occurring in a galvanic cell can be broken down into an oxidation half-reaction and a reduction half-reaction. Calculate the cell potential of the overall reaction.

Half-Reaction	$E°$ (V)
$Al^{3+}(aq) + 3\,e^- \rightarrow Al(s)$	–1.66
$Ni^{2+}(aq) + 2\,e^- \rightarrow Ni(s)$	–0.23

EXAMPLE: Consider a galvanic cell based on the unbalanced reaction:

$$Al(s) + Ni^{2+}(aq) \rightarrow Al^{3+}(aq) + Ni(s)$$

Balance the reaction above and calculate the cell potential, $E°_{cell}$, for the reaction.

SOLUTION:

STEP 1: Write the oxidation and reduction half-reactions.

Oxidation: $Al \rightarrow Al^{3+} + 3\,e^-$ **Reduction:** $Ni^{2+} + 2\,e^- \rightarrow Ni$

STEP 2: For the reduction half-reaction, look up the potential in the table.

$$Ni^{2+} + 2\,e^- \rightarrow Ni \qquad E°_{red} = -\,0.23\text{ V}$$

STEP 3: For the oxidation half-reaction, $E°_{ox} = -E°_{red}$.

$$\text{Oxidation: } Al \rightarrow Al^{3+} + 3\,e^-$$

$$E°_{ox} = -E°_{red} = -\,(-1.66\text{ V}) = +1.66\text{ V}.$$

STEP 4: The cell potential for the overall reaction is equal to the sum of the reduction potential, $E°_{red}$, and the oxidation potential, $E°_{ox}$.

$$E°_{cell} = E°_{ox} + E°_{red}$$

$$E°_{cell} = -\,0.23\text{ V} + 1.66\text{ V} = 1.43\text{ V}$$

To obtain the balanced cell reaction, you must make sure that the electrons lost equal the electrons gained and then add the two equations together to get the overall equation.

$3\ (Ni^{2+} + 2\ e^- \rightarrow Ni)$	$E°_{red} = -0.23$ V
$2\ (Al \rightarrow Al^{3+} + 3\ e^-)$	$E°_{ox} = +1.66$ V
$3\ Ni^{2+} + 2\ Al \rightarrow 3\ Ni + 2\ Al^{3+}$	$E°_{cell} = 1.43$ V

When multiplying the half-reactions through by a coefficient, do not change the value of $E°$, because potential is an intensive property. When the reaction $Ni^{2+} + 2\ e^- \rightarrow Ni$ is multiplied by 3 (to balance the electrons), the voltage does not change. Remember that a volt is equal to a joule/coulomb and when the equation is multiplied by 3, the energy from the reaction increases by 3 and the charge also increases by a factor of 3 and they cancel each other out.

$$\text{volt} = \frac{\cancel{3} \times \text{joule}}{\cancel{3} \times \text{coulomb}}$$

THERMODYNAMICALLY FAVORED (SPONTANEOUS) REACTIONS

Gibbs free energy, $\Delta G°$, is discussed in Chapter 10. One of the ways that it can be calculated is from the cell potential, $E°_{cell}$.

$$\Delta G° = -nFE°_{cell}$$

Faraday's constant, F, has a value of 96,485 coulombs/mol e^- (it is acceptable to use 96,500 C/mol e^-, especially in multiple-choice questions). The number of moles of electrons transferred in a redox reaction is represented by n.

A thermodynamically favored (spontaneous) reaction is one that has a negative value for $\Delta G°$ or a positive value for $E°_{cell}$.

Half-Reaction	$E°$ (V)
$2\ H^+(aq) + 2\ e^- \rightarrow H_2(g)$	0.00
$Ag^+(aq) + e^- \rightarrow Ag(s)$	0.80

EXAMPLE: Will 1 M HCl dissolve silver metal, forming Ag^+ in solution?

SOLUTION: Write the half-reactions and calculate $E°_{cell}$.

$2\ H^+ + 2\ e^- \rightarrow 2\ H_2$	$E°_{red} = 0.00$ V
$2\ (Ag \rightarrow Ag^+ + e^-)$	$E°_{ox} = -0.80$ V
$2\ H^+ + 2\ Ag \rightarrow H_2 + 2\ Ag^+$	$E°_{cell} = -0.80$ V

The negative value for $E°_{cell}$ indicates that the reaction will not occur.

If we calculate the value for $\Delta G°$, we will get a positive value using the equation $\Delta G° = -nFE°_{cell}$. In this case, the number of moles of electrons transferred comes from the overall balanced equation for the process, so $n = 2$.

$$\Delta G° = -(2 \text{ mol } e^-)\left(\frac{96{,}485 \text{ C}}{1 \text{ mol } e^-}\right)(-0.80 \text{ V}) = 154{,}376 \text{ CV} = 154{,}376 \text{ J} = 154 \text{ kJ}$$

Again, the positive value for $\Delta G°$ indicates that the reaction will not occur.

DETERMINING WHICH SUBSTANCES ARE MORE LIKELY TO BE OXIDIZED OR REDUCED

You may be asked if an element or ionic species is capable of reducing another element or ion. To determine if the reaction will occur, write the half-reactions and calculate the cell potential as in the previous example.

Half-Reaction	$E°$ (V)
$Br_2(l) + 2\,e^- \rightarrow 2\,Br^-(aq)$	1.09
$I_2(s) + 2\,e^- \rightarrow 2\,I^-(aq)$	0.54
$Cl_2(g) + 2\,e^- \rightarrow 2\,Cl^-(aq)$	1.36

EXAMPLE: Bromine, Br_2, can oxidize iodide ions, I^-, to iodine, I_2. However, Br_2 cannot oxidize chloride ions, Cl^-, to chlorine, Cl_2. Explain why the first reaction occurs, yet the second one does not.

SOLUTION: Begin by writing the appropriate half-reactions. Then calculate the cell potential for the overall reaction.

$$\begin{array}{ll} Br_2 + 2\,e^- \rightarrow 2\,Br^- & E°_{red} = 1.09 \text{ V} \\ 2\,I^- \rightarrow I_2 + 2\,e^- & E°_{ox} = -0.54 \text{ V} \\ \hline Br_2 + 2\,I^- \rightarrow 2\,Br^- + I_2 & E°_{cell} = 0.55 \text{ V} \end{array}$$

This reaction occurs; $E°_{cell}$ is positive.

$$\begin{array}{ll} Br_2 + 2\,e^- \rightarrow 2\,Br^- & E°_{red} = 1.09 \text{ V} \\ 2\,Cl^- \rightarrow Cl_2 + 2\,e^- & E°_{ox} = -1.36 \text{ V} \\ \hline Br_2 + 2\,Cl^- \rightarrow 2\,Br^- + Cl_2 & E°_{cell} = -0.27 \text{ V} \end{array}$$

This reaction does not occur; $E°_{cell}$ is negative.

For a substance to be oxidized, it must lose electrons and another substance must gain those electrons because oxidation and reduction always occur together. We balance the equations so that the number of electrons lost and the number of electrons gained are the same.

In any pair of half-reactions, there will be one of five cases: (1) both reduction potentials will be positive, (2) both will be negative, (3) one will be positive and one will be negative, (4) one will be zero and one will be negative, or (5) one will be zero and one will be positive. The substance with the more positive $E°_{red}$ is more likely to gain electrons. It will be the reactant in the reduction half-reaction. The substance with the more negative $E°_{red}$ is more likely to lose electrons. It will be the reactant in the oxidation half-reaction.

Half-Reaction	$E°$ (V)
$Cl_2(g) + 2\,e^- \rightarrow 2\,Cl^-(aq)$	1.36
$Br_2(l) + 2\,e^- \rightarrow 2\,Br^-(aq)$	1.09
$Fe^{3+}(aq) + e^- \rightarrow Fe^{2+}(aq)$	0.77
$I_2(s) + 2\,e^- \rightarrow 2\,I^-(aq)$	0.54
$Cu^{2+}(aq) + 2\,e^- \rightarrow Cu(s)$	0.34
$Fe^{2+}(aq) + 2\,e^- \rightarrow Fe(s)$	–0.44
$Mg^{2+}(aq) + 2\,e^- \rightarrow Mg(s)$	–2.37

EXAMPLE: Given the information above, rank the following substances from the most easily oxidized to the least easily oxidized.

$$Br^-, Cl^-, Cu, Fe^{2+}, I^-, Mg$$

SOLUTION: First, determine the oxidation potentials of each substance by reversing the sign on their reduction potentials. Then rank them from the highest oxidation potential (most easily oxidized) to the lowest oxidation potential (least easily oxidized). The correct order is:

$$Mg > Cu > I^- > Fe^{2+} > Br > Cl^-$$

PREDICTING CHANGES IN THE CELL POTENTIAL OF A GALVANIC CELL AT NONSTANDARD CONDITIONS

Often on the exam, students will be faced with determining the qualitative change to the voltage when the concentration of one of the species is changed. One of the easiest ways to predict or calculate this change is to use the Nernst equation. This equation is not on the equation sheet, but it is very helpful to use when explaining how changes in concentration affect cell potential.

$$E = E° - \frac{0.0591}{n}\log Q$$

The important part of this equation is the log Q term. Q is the reaction quotient. Q represents the concentrations of the products raised to the powers of their coefficients divided by the concentrations of the reactants raised to the powers of their coefficients. All solids and pure liquids are left out of the Q expression as their values are constant. Q will be discussed further in Chapter 11. Given a generic equation for a galvanic cell, shown below,

$$a\,R(s) + b\,T^{a+}(aq) \rightarrow a\,R^{b+}(aq) + b\,T(s)$$

$$Q \text{ will be } Q = \frac{[R^{b+}]^a}{[T^{a+}]^b}$$

There are three cases that one must consider when determining the effects of a change of concentration on the voltage; each case depends on the value of Q. We are dealing with the change in voltage on a battery as it operates, so the $E°$ must be a positive number. If the value of Q is less than 1, then the value of the log$Q$ term in the Nernst equation will be a negative number. Since $E_{cell} = E°$ minus the log$Q$ term, we have E_{cell} = *positive number* – *negative number*. Since subtracting a negative number is addition, then E_{cell} = *positive number* + *positive number*, and the E_{cell} voltage will increase. The three cases are shown below in the table.

Value of Q	Value of log$Q$	Signs used in the equation $E_{cell} = E° - \frac{0.0592}{n} \log Q$	Final effects on E_{cell}
Equal to 1	Zero	E_{cell} = (+ number) – (zero)	Stays the same
Less than 1	Negative number	E_{cell} = (+ number) – (– number)	Increases
More than 1	Positive number	E_{cell} = (+ number) – (+ number)	Decreases

Remember that you will ***not*** need to calculate the new voltage but will need to be able to explain it qualitatively. That can be done by looking at the log$Q$ value and then analyzing the effects on the value for E_{cell}.

A second way to qualitatively explain cell potential changes when concentrations of reactants and/or products are other than 1 M and 1 atm is to compare Q to K. K is an equilibrium constant and like Q, will be discussed further in Chapter 11. The only difference in K and Q is that K uses equilibrium concentrations while Q uses initial concentrations. Because standard conditions for an electrochemical cell specify that solutions are 1 M and gases have a pressure of 1 atm, the initial value of Q for an electrochemical cell under standard conditions is always 1. When an electrochemical cell reaches equilibrium, $Q = K$ and $E = 0$ V. For a galvanic cell, when Q gets closer to K (Q decreases), the voltage goes down. If Q gets further from K (Q increases), the voltage goes up.

AP® Tip

Although what happens to the voltage when the concentration of a species in a galvanic cell is changed may appear to be a Le Châtelier's principle question, you **cannot** use Le Châtelier's principle to explain this type of question on the free-response section. Le Châtelier is used for a system at equilibrium, and as a battery operates, it is not at equilibrium. When an electrochemical cell reaches equilibrium, it has a voltage of 0 (a dead battery). Using the same thought process as you would use with Le Châtelier's principle to determine the shift in the reaction with changes in concentration is an effective way to determine whether cell potential will increase or decrease when needed for a multiple-choice question, but it will receive **no credit** as an explanation on a free-response question.

ELECTROLYTIC CELLS

In an electrolytic cell, a non-thermodynamically favored (nonspontaneous) reaction is made to occur by forcing an electric current through the cell. In an earlier example, it was shown that the following reaction is thermodynamically favored (spontaneous):

$$3\ Ni^{2+} + 2\ Al \rightarrow 3\ Ni + 2\ Al^{3+}$$

The reverse of this reaction: $3\ Ni + 2\ Al^{3+} \rightarrow 3\ Ni^{2+} + 2\ Al$ is non-thermodynamically favored (nonspontaneous) and can be made to occur by the addition of an external power source. This electrolytic cell can be set up with two compartments just like in the galvanic cell, with the voltmeter being replaced with a power supply.

In the process of electroplating, the electrolytic cell can also be set up using only one compartment, as shown in the figure.

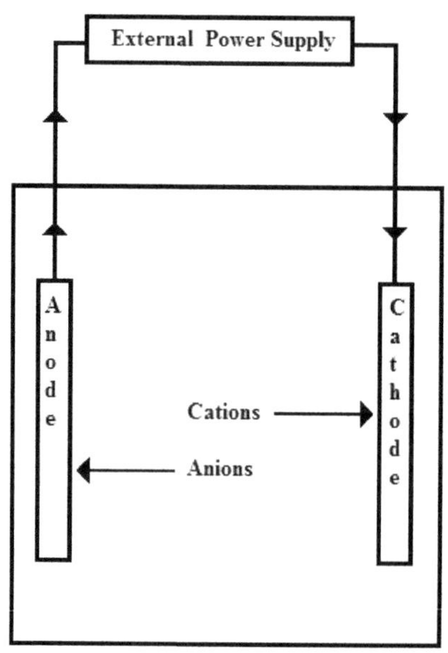

For example, an object to be plated with copper is the cathode and is immersed in a solution of copper(II) sulfate. The reduction reaction $Cu^{2+} + 2\ e^- \rightarrow Cu$ will deposit copper onto the object. The anode can also be made of copper, and the oxidation of the copper anode will replace the Cu^{2+} ions.

REACTIONS THAT OCCUR IN AN ELECTROLYTIC CELL

To determine which reactions occur at the anode and the cathode during electrolysis, you must consider all possible reactions and their reduction and oxidation potentials. If the reaction takes place in an aqueous solution, the oxidation and reduction of water must also be considered.

EXAMPLE: A solution of copper(II) sulfate is electrolyzed. Calculate the cell potential of the reaction, $E°_{cell}$.

SOLUTION:

	Possible reactions[1]	Cell potential, $E°$(V)
Cathode	$Cu^{2+} + 2e^- \rightarrow Cu$	0.34
	$SO_4^{2-} + 4H^+ + 2e^- \rightarrow H_2SO_3 + H_2O$	0.20
	$2H_2O + 2e^- \rightarrow H_2 + 2OH^-$	–0.83
Anode	$Cu \rightarrow Cu^{2+} + 2e^-$	–0.34
	$2H_2O \rightarrow O_2 + 4H^+ + 4e^-$	–1.23

For each electrode, the reaction with the more positive (or less negative) potential will occur. At the cathode, Cu^{2+} will be reduced. At the anode, Cu will be oxidized.

$$Cu^{2+} + 2e^- \rightarrow Cu \qquad E° = 0.34 \text{ V}$$

$$Cu \rightarrow Cu^{2+} + 2e^- \qquad E° = -0.34 \text{ V}$$

Overall cell potential is zero. This is not thermodynamically favored. As this is an electrolytic process, the current will be supplied by an outside power source to cause the reaction to occur.

This process is often more complicated in real life than the example shown here. There are other considerations to consider, such as the ease by which the species is oxidized or reduced and the overvoltage needed (overvoltage is the excess potential required for the reaction to occur). For these reasons, the problems likely to be found on the AP® exam will be simple and straightforward.

[1]There is no half-reaction for the oxidation of SO_4^{2-}. S in SO_4^{2-} is in its highest oxidation state, +6, and cannot be oxidized further.

AP® Tip

Frequently, the electrodes are inert for electrolysis. For example, during the electrolysis of KI(*aq*), K^+, I^-, and H_2O are the only species present. Only I^- and H_2O are present to be oxidized at the anode. *Note:* In aqueous KI, there is no K(*s*) to be oxidized and $K^+(aq)$ is already at its highest oxidation state. The overall reaction for the electrolysis of KI(*aq*) is $2\ I^-(aq) + 2\ H_2O(l) \rightarrow I_2(s) + H_2(g) + 2\ OH^-(aq)$.

STOICHIOMETRY OF ELECTROLYTIC PROCESSES

In this section, you will review how much chemical change occurs with the flow of a given current for a specified time. You might be asked how much metal was plated (or formed from the metal ion) or how long an electroplating process will take or how much current is required to produce a specified amount of metal over a period of time.

Three things to remember:

- A coulomb (C) is the fundamental unit of charge in the metric system.
- An ampere (amp) is the amount of charge that passes a given point in a second, so 1 amp = 1 C/s and 12 amps = 12 C/s.
- Faraday's constant (*F*, sometimes a much more fancy F is used, i.e., $\mathcal{F}$) is the amount of charge carried by a mole of electrons and has the value of 96,485 C = 1 mol e^-.

EXAMPLE: A current of 10.0 A is passed through a solution containing M^{2+} for 30.0 min. 5.94 g of metal, M, is produced. Determine the identity of metal, M.

SOLUTION: Since the metal forms a cation with a 2+ charge, 2 moles of electrons are required to reduce 1 mole of the metal.

Remember that 10.0 A is equal to 10.0 coulombs of charge that pass a given point in 1 second.

$$\frac{10.0\text{ C}}{\text{s}} \times \frac{60\text{ s}}{\text{min}} \times 30.0\text{ min} = 1.80 \times 10^4\text{ C}$$

$$1.80 \times 10^4\text{ C} \times \frac{1\text{ mol e}^-}{96{,}485\text{ C}} \times \frac{1\text{ mol M}}{2\text{ mol }e^-} = 9.33 \times 10^{-2}\text{ mol M}$$

$$5.94\text{ g M} \times \frac{1}{9.33 \times 10^{-2}\text{ mol M}} = 63.7\text{ g/mol}$$

The metal must be copper because the molar mass is very close to that of Cu.

This can also be shown in one step:

$$5.94\text{ g M} \times \frac{2\text{ mol }e^-}{1\text{ mol M}} \times \frac{96{,}485\text{ C}}{\text{mole }e^-} \times \frac{1\text{ s}}{10.0\text{ C}} \times \frac{1\text{ min}}{60\text{ s}} \times \frac{1}{30.0\text{ min}} = 63.7\text{ g/mol}$$

COMPARISON OF GALVANIC AND ELECTROLYTIC CELLS

Galvanic and electrolytic cells have a few features in common. For both types of cells, reduction always occurs at the cathode and oxidation at the anode. In an electrolytic cell, electrons travel from the battery to the cathode. In both cases, electrons travel in the wire moving from the anode to the cathode, but you wouldn't say the electrons spontaneously travel from the anode to the cathode in an electrolytic cell. Positive ions are always attracted to the cathode whether the cell is electrolytic or galvanic. The table below compares the galvanic cell to the electrolytic cell.

	Galvanic	Electrolytic
Process occurring at the anode	Oxidation	Oxidation
Process occurring at the cathode	Reduction	Reduction
Ions attracted to the cathode	Cations	Cations
Ions attracted to the anode	Anions	Anions
Direction of electron flow	From anode to cathode	From anode to cathode
Spontaneity	Thermodynamically favored (spontaneous)	Non-thermodynamically favored (nonspontaneous)

In both types of cells, the positive ions or cations move toward the cathode because there is an excess of negative ions at the cathode caused by the reduction of cations in solution. Likewise, oxidation at the anode produces cations, so negatively charged ions or anions in the salt bridge must move to the anode to maintain electrical neutrality. For many electrolytic cells that operate in a single cell, there is no salt bridge.

MULTIPLE-CHOICE QUESTIONS

No calculators are to be used on this part of the exam.

$MnO_4^- + 5\ Fe^{2+} + 8\ H^+ \rightarrow Mn^{2+} + 5\ Fe^{3+} + 4\ H_2O$

1. In the reaction shown above,
(A) Fe^{3+} is oxidized and Mn^{2+} is reduced.
(B) H^+ is oxidized and the Mn in MnO_4^- is reduced.
(C) Fe^{2+} is oxidized and the Mn in MnO_4^- is reduced.
(D) electrons are transferred from Mn^{2+} to Fe^{2+}.

2. Which statement correctly describes the function of a salt bridge (or porous barrier) in a galvanic cell?
(A) It allows electrons to pass through to complete the circuit.
(B) It encourages charge build-up in both cells.
(C) It keeps the net charge in each cell at zero.
(D) It measures the voltage produced by the galvanic cell.

3. What is the value of $\Delta G°$ for the reaction written below?

$2\ MnO_4^-(aq) + 16\ H^+(aq) + 5\ Zn(s) \rightarrow 2\ Mn^{2+}(aq) + 8\ H_2O(l) + 5\ Zn^{2+}(aq)$

You are given the following standard reduction potentials for the two ***unbalanced*** half-reactions:

$H^+(aq) + MnO_4^-(aq) + e^- \rightarrow Mn^{2+}(aq) + H_2O(l)$ $\quad E° = +1.51\ V$

$Zn^{2+}(aq) + e^- \rightarrow Zn(s)$ $\quad E° = -0.76\ V$

(A) -7.04×10^5 J
(B) -4.34×10^5 J
(C) -1.08×10^6 J
(D) -2.17×10^6 J

Questions 4–6 refer to the following information.

Half-Reaction	$E°$ (V)
$Mg^{2+}(aq) + 2\ e^- \rightarrow Mg(s)$	–2.37
$Fe^{3+}(aq) + e^- \rightarrow Fe^{2+}(aq)$	0.77
$Ag^+(aq) + e^- \rightarrow Ag(s)$	0.80

4. What is the $E°_{cell}$ value (cell potential) when the reaction $Mg(s) + 2\ Fe^{3+}(aq) \rightarrow Mg^{2+}(aq) + 2\ Fe^{2+}(aq)$ reaches equilibrium?
(A) 3.91 V
(B) 3.14 V
(C) 1.60 V
(D) 0.00 V

5. When only $Fe^{2+}(aq)$ and $Fe^{3+}(aq)$ are involved in a half-reaction with no Fe(s) present, an inert electrode such as Pt(s) or C(s) must be used. Which statement correctly describes the galvanic cell based on the reactions $Ag^+ + e^- \rightarrow Ag$ and $Fe^{3+} + e^- \rightarrow Fe^{2+}$?
(A) Electrons are lost by Ag^+ and gained by Fe^{2+}.
(B) The overall cell potential is +1.57 V.
(C) The mass of the silver electrode increases as the reaction proceeds.
(D) The mass of the inert electrode decreases as the reaction proceeds.

6. If a standard galvanic cell were made using solid silver and magnesium electrodes each immersed in a 1.0 *M* solution of their ions, what would be the reaction occurring at the cathode?
(A) $Ag^{+}(aq) + e^{-} \rightarrow Ag(s)$
(B) $Ag(s) \rightarrow Ag^{+}(aq) + e^{-}$
(C) $Mg^{2+}(aq) + 2\,e^{-} \rightarrow Mg(s)$
(D) $Mg(s) \rightarrow Mg^{2+}(aq) + 2\,e^{-}$

Half-Reaction	$E°$ (V)
$Cu^{2+}(aq) + 2\,e^{-} \rightarrow Cu(s)$	0.34
$Sn^{2+}(aq) + 2\,e^{-} \rightarrow Sn(s)$	–0.14
$Fe^{2+}(aq) + 2\,e^{-} \rightarrow Fe(s)$	–0.44
$Zn^{2+}(aq) + 2\,e^{-} \rightarrow Zn(s)$	–0.76
$Mg^{2+}(aq) + 2\,e^{-} \rightarrow Mg(s)$	–2.37

7. Cathodic protection is a method used to protect against the unwanted oxidation of an important reactive structural metal by either coating the structural metal with or connecting the structural metal to a "sacrificial" metal that is more easily oxidized. Cathodic protection is used in fuel pipelines, hulls of ships, bridges, and numerous other applications. Iron metal can be protected from rusting by putting it into contact with a substance that is more likely to be oxidized than iron. Given the reduction potentials in the table above, which metals could be used to protect iron?
(A) Zn or Mg
(B) Sn or Cu
(C) Zn or Sn
(D) Cu or Mg

8. Copper may be used for electroplating. If a current of 10.0 A is applied to a 1 *M* Cu^{2+} solution, which answer shows approximately how long it will take for 6.36 g of solid copper to be plated out?
(A) 1000 s
(B) 2000 s
(C) 10,000 s
(D) 20,000 s

Half-Reaction	$E°$ (V)
$Ag^{+}(aq) + e^{-} \rightarrow Ag(s)$	–0.80
$Cu^{2+}(aq) + 2\,e^{-} \rightarrow Cu(s)$	–0.34
$Sn^{2+}(aq) + 2\,e^{-} \rightarrow Sn(s)$	–0.14
$Mg^{2+}(aq) + 2\,e^{-} \rightarrow Mg(s)$	–2.37
$Ca^{2+}(aq) + 2\,e^{-} \rightarrow Ca(s)$	–2.87
$Sr^{2+}(aq) + 2\,e^{-} \rightarrow Sr(s)$	–2.89

9. A nonstandard voltaic cell is set up using a strip of metal X in an unknown concentration of nitrate solution of X ion for one half-cell and the other half-cell is set up using a strip of metal Y in an unknown concentration of nitrate solution of Y ion. If the measured voltage of the complete cell is 2.23 V, which metals are *least* likely to be X and Y?
(A) Sr and Ag
(B) Ca and Cu
(C) Mg and Sn
(D) Sn and Cu

10. In a certain galvanic cell, it is observed that the cell potential increases as the temperature increases. Which statement is correct?
(A) As the temperature increases, ΔG becomes more negative.
(B) As the temperature increases, ΔG becomes more positive.
(C) ΔG does not depend upon temperature in an electrochemical reaction.
(D) ΔG does not depend upon cell potential in an electrochemical reaction.

Questions 11 and 12 refer to the following information.

Half-Reaction	$E°$ (V)
$Al^{3+}(aq) + 3\,e^{-} \rightarrow Al(s)$	–1.66
$Ag^{+}(aq) + e^{-} \rightarrow Ag(s)$	0.80

A galvanic cell is created using solid silver and solid aluminum electrodes placed in beakers containing 100-mL portions of 1 *M* aqueous solutions of $AgNO_3$ and $Al(NO_3)_3$, respectively. The electrodes are connected via wires and a voltmeter, while the solutions are connected with a salt bridge saturated with 1 *M* $NaNO_3$. The *unbalanced* reaction is: $Al + Ag^{+} \rightarrow Al^{3+} + Ag$.

11. What would happen to the cell potential if 10 mL of 1.0 *M* $NH_3(aq)$ were added to the silver cell resulting in the formation of the complex ion $Ag(NH_3)_2^+(aq)$?
    (A) The voltage would increase.
    (B) The voltage would decrease.
    (C) The voltage would be zero.
    (D) There would be no change in voltage.

12. Which statement correctly identifies a way to increase the cell potential in this reaction?
    (A) Increase the size of the solid aluminum electrode.
    (B) Add solid $AgNO_3$ to the half-cell containing $AgNO_3(aq)$.
    (C) Add an additional 50 mL of 1 *M* $AgNO_3$ to the half-cell containing $AgNO_3(aq)$.
    (D) Use 2 *M* $NaNO_3$ in the salt bridge rather than 1 *M* $NaNO_3$.

Questions 13–15 refer to the following information.

Half-Reaction	$E°$ (V)
$Pb^{2+}(aq) + 2e^- \rightarrow Pb(s)$	–0.13
$Cu^{2+}(aq) + 2e^- \rightarrow Cu(s)$	0.34

An electrochemical cell is set up using two beakers. A strip of lead is placed in 100 mL of 1*M* $Pb(NO_3)_2(aq)$ in one beaker, while a strip of copper is placed in 100 mL of 1*M* $Cu(NO_3)_2(aq)$ in a second beaker. The two strips are connected to each other with a wire. The solutions in the two beakers are linked with a piece of filter paper saturated with 1*M* $KNO_3(aq)$.

13. If a voltmeter measures the voltage passing through the wire to be –0.21 V rather than +0.21 V, what would account for this?
    (A) Wires on the voltmeter need to be reversed.
    (B) Electrons are flowing through the salt bridge.
    (C) The two electrode metals need to be switched in the beakers.
    (D) Salt bridge ions are flowing in the wrong direction and need to be turned around.

14. In the galvanic cell described above, what travels through the filter paper?
    (A) Nitrate ions will travel through the paper to the cathode, while potassium ions travel through the paper to the anode.
    (B) Nitrate ions will travel through the paper to the anode, while potassium ions travel through the paper to the cathode.
    (C) Electrons will travel through the paper from anode to cathode.
    (D) Electrons will travel through the paper from cathode to anode.

15. Additional solid $Cu(NO_3)_2$ crystals are added to the beaker containing the $Cu(NO_3)_2(aq)$ and the solution is stirred. Which of the following best describes the resulting change in the voltage with the correct justification?
    (A) The voltage will increase because the increased molarity of the $Cu(NO_3)_2$ solution causes the value of Q to decrease.
    (B) The voltage will increase because the increased molarity of the $Cu(NO_3)_2$ solution causes the value of Q to increase.
    (C) The voltage will decrease because the increased molarity of the $Cu(NO_3)_2$ causes the value of Q to decrease.
    (D) The voltage will decrease because the increased molarity of the $Cu(NO_3)_2$ causes the value of Q to increase.

FREE-RESPONSE QUESTIONS

Calculators may be used on this section.

1. An electrochemical cell is constructed using as an electrode a strip of magnesium metal, placed into a 1.0 *M* solution of $Mg(NO_3)_2(aq)$ and a strip of aluminum metal as the other electrode, placed into a 1.0 *M* solution of $Al(NO_3)_3(aq)$. After the cell has operated for 30 minutes, the student makes the following diagram in her laboratory notebook. Note that the diagram is incomplete.

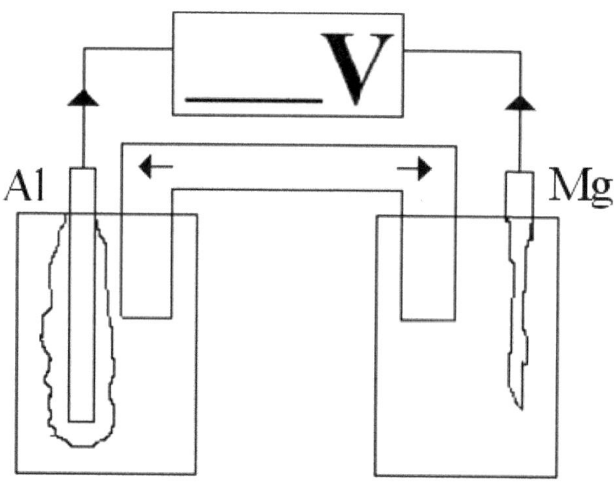

Half-Reaction	$E°$ (V)
$Al^{3+}(aq) + 3\,e^- \rightarrow Al(s)$	–1.66
$Mg^{2+}(aq) + 2\,e^- \rightarrow Mg(s)$	–2.37

(a) Identify the cathode. Justify your answer based on the student's observations, as shown in the partially completed diagram.
(b) Give the equation for the thermodynamically favored (spontaneous) reaction.
(c) In the appropriate place on the student's diagram, place the standard voltage, $E°_{cell}$ that would be indicated at the start of this experiment.

(d) Calculate the $\Delta G°$ for the reaction at 25°C.

(e) The student wishes to show the flow of electrons in the galvanic cell as it operates. Circle the arrowhead that correctly indicates the flow of electrons.

(f) The student uses a 1.0 M solution of $KNO_3(aq)$ as the salt in the salt bridge.

  (i) Indicate the correct movement of the ions in the salt bridge by completing that part of the diagram.

  (ii) If 50.0 mL of the KNO_3 solution is prepared for the experiment, how many total ions would be present in the 50.0 mL of solution?

(g) At the time that the diagram was made, would the voltage be greater than, less than, or equal to the standard voltage, $E°_{cell}$? Justify your answer.

2. In a laboratory experiment, a student filled a petri dish (shallow container) with a 1 M Na_2SO_4 solution. Two drops of phenolphthalein solution were added to the solution in the dish and gently stirred. Phenolphthalein is an indicator that is colorless at a pH less than 8 and turns bright pink at a pH greater than 8. Inert graphite electrodes were attached to a 9-volt battery and immersed in the solution. After 60 seconds, small bubbles were observed forming at both electrodes. At one electrode, the solution was turning pink.

(a)

Half-Reaction	$E°$ (V)
$O_2(aq) + 4\,H^+(aq) + 4\,e^- \rightarrow 2\,H_2O(l)$	1.23
$SO_4^{2-}(aq) + 4\,H^+(aq) + 2\,e^- \rightarrow H_2SO_3(aq) + H_2O(l)$	0.20
$2\,H_2O(l) + 2\,e^- \rightarrow H_2(g) + 2\,OH^-(aq)$	–0.83
$Na^+(aq) + e^- \rightarrow Na(s)$	–2.71

  (i) Write the half-reaction that occurs at the anode and justify your answer based on the student observations.

  (ii) Write the half-reaction that occurs at the cathode and justify your answer based on the student observations.

  (iii) If the gas produced at each electrode could be collected and measured, what would be the ratio by volume of the gas produced at the anode to the volume of the gas produced at the cathode? Assume that the solubilities of the gases in the solution do not affect the ratio. Justify your answer with the balanced overall equation.

(b) How, if at all, would the observations differ if the dish had been filled with pure water, rather than the sodium sulfate solution?

Answers

MULTIPLE-CHOICE QUESTIONS

1. C In changing from Fe^{2+} to Fe^{3+}, each iron(II) ion loses an electron, which is oxidation; MnO_4^- changing to Mn^{2+} is a gain of electrons, which is reduction. Also, Fe^{2+} going to Fe^{3+} is an increase in oxidation number from +2 to +3 (oxidation), while the oxidation number of manganese decreases from +7 in the permanganate ion to +2 in Mn^{2+} (reduction) (LO 3.8).

2. C While electrons flow through the wire from the anode to the cathode, the ions of the salt that are in the salt bridge flow in different direction to keep the net charge at zero. The anions of the salt will flow toward the compartment that contains the anode, and the cations of the salt will flow toward the compartment that contains the cathode (LO 3.13).

3. D Since you were given the following *unbalanced* half-reactions, you will need to balance them first. Remember to balance for mass and then for charge.

   $H^+(aq) + MnO_4^-(aq) + e^- \rightarrow Mn^{2+}(aq) + H_2O(l)$ $\quad E° = +1.51\ V$

   $Zn^{2+}(aq) + e^- \rightarrow Zn(s)$ $\quad E° = -0.76\ V$

   You will need to flip the more negative half-reaction before we balance them. Remember to change the sign of the voltage.

   $Zn(s) \rightarrow Zn^{2+}(aq) + e^-$ $\quad E° = +\ 0.76\ V$

   Balance for mass:

   $$8\ H^+(aq) + MnO_4^-(aq) + e^- \rightarrow Mn^{2+}(aq) + 4\ H_2O(l)$$

   $$Zn(s) \rightarrow Zn^{2+}(aq) + e^-$$

   Now balance for charge:

   $$8\ H^+(aq) + MnO_4^-(aq) + 5\ e^- \rightarrow Mn^{2+}(aq) + 4\ H_2O(l)$$

   $$Zn(s) \rightarrow Zn^{2+}(aq) + 2\ e^-$$

   Now balance for the flow of electrons:

   $$2 \times [8\ H^+(aq) + MnO_4^-(aq) + 5\ e^- \rightarrow Mn^{2+}(aq) + 4\ H_2O(l)] \quad E° = +1.49\ V$$

   $$5 \times [Zn(s) \rightarrow Zn^{2+}(aq) + 2\ e^-] \quad E° = +\ 0.76\ V$$

The overall equation is:

$2\ MnO_4^-(aq) + 16\ H^+(aq) + 5\ Zn(s) \rightarrow 2\ Mn^{2+}(aq) + 8\ H_2O(l) + 5\ Zn^{2+}(aq)$

and the

$$E° = +\ 0.76\ V + 1.49\ V = 2.25\ V$$

$$\Delta G° = -nFE°_{cell} = -\ (10\ mol\ e^-)(96{,}500\ C/mol\ e^-)(2.25\ V) = -\ 2{,}171{,}250\ V\ C$$
$$= -\ 2{,}171{,}250\ J$$

$$\Delta G° = -\ 2.17 \times 10^6\ J$$

Note: If you round the numbers and use estimation, the math can be done without a calculator:

$$\Delta G° = -nFE°_{cell} = -\ (10\ mol\ e^-)(96{,}500\ C/mol\ e^-)(2.25\ V) \approx -(10)(100{,}000)(2)$$
$$= -2{,}000{,}000$$

and the only answer that is close to this approximation is D (LO 3.11 and 5.15).

4. D When the cell reaction reaches equilibrium, no electrons flow and $E°_{cell} = 0$ (LO 6.25).

5. C To determine the overall thermodynamically favored reaction, flip the half-reaction that has the more negative (or less positive) $E°_{cell}$ potential. That would be the half-reaction that involves the iron.

$$Fe^{2+}(aq) \rightarrow Fe^{3+}(aq) + e^-$$

$$Ag^+(aq) + e^- \rightarrow Ag(s)$$

The overall equation will be: $Fe^{2+}(aq) + Ag^+(aq) \rightarrow Fe^{3+}(aq) + Ag(s)$. Since the $Ag^+(aq)$ is gaining electrons (undergoing reduction), the mass of the silver electrode will increase (LO 3.13).

6. A To determine the overall thermodynamically favored reaction, flip the half-reaction that has the more negative $E°_{cell}$ potential. That would be the half-reaction that involves the magnesium:

$$Mg(s) \rightarrow Mg^{2+}(aq) + 2\ e^-$$

$$Ag^+(aq) + e^- \rightarrow Ag(s)$$

The problem only wants the half-reaction that occurs at the cathode, which will be the one that undergoes reduction. Since a gain of electrons is reduction, find the half-reaction that has electrons as a reactant, which would be the $Ag^+(aq) + e^- \rightarrow Ag(s)$.

Note that only two answers show reduction; the other two choices show oxidation (LO 3.13).

7. A First, reverse all equations and change the sign of the reduction potentials to make them into oxidation reactions and oxidation potentials. The oxidation potential of Fe(*s*) is +0.44 V. The only two metals with higher oxidation potentials on this list are Mg(*s*) and Zn(*s*). Both are commonly used for cathodic protection of iron. *Note:* Since the costs of the two metals are very comparable, the choice of which metal to use would be determined by environmental considerations (soil conditions, amount of metals in the soil, location of ground water, etc.) (LO 3.12).

8. B $6.36 \text{ g Cu} \times \frac{1 \text{ mol Cu}}{63.6 \text{ g Cu}} \times \frac{2 \text{ mol } e^-}{1 \text{ mol Cu}} \times \frac{96{,}500 \text{ C}}{1 \text{ mol } e^-} \times \frac{1 \text{ s}}{10 \text{ C}} \approx 2000 \text{ s}$

   Note: In this set of calculations, 96,500 C was treated as 100,000 C to make the math easier to do (LO 3.3).

9. D A nonstandard cell indicates that the concentrations of one or more components will not be 1 *M*. The actual voltage will vary slightly from the standard cell voltage. When the standard cell potential is calculated for each pair, A = 2.09 V, B = 2.53 V, C = 2.23 V, and D = 0.20 V. D is far from the observed voltage of 2.23 V and is least likely to be the metals involved in the reaction (LO 3.13).

10. A The higher the cell potential value, the more thermodynamically favorable the reaction. If a reaction has an increased cell potential when the temperature is increased, the reaction is more thermodynamically favored and ΔG is more negative as the temperature is increased (LO 5.15).

11. B If you reduce the $[Ag^+]$ by formation of $Ag(NH_3)_2^+$, the Q ratio $[Al^{3+}]/[Ag^+]^3$ will be greater than 1 and closer to K. This causes a decrease in cell potential as the Q gets closer to K. See the answer to question 15 for an alternative approach to answering this question (LO 3.12).

12. B If you increase the $[Ag^+]$ by adding solid $AgNO_3$ (all nitrates are soluble), the Q ratio $[Al^{3+}]/[Ag^+]^3$ will decrease and become further from equilibrium (K). This causes an increase in cell potential as the Q value is less than 1. Increasing the size of an electrode will not change the cell potential. Adding more of the same concentration of a solution present will not change the cell potential. Increasing the concentration of salt in the salt bridge will not change the cell potential. See the answer to question 15 for an alternative approach to answering this question (LO 3.12).

13. A If the voltmeter reads a negative value for a galvanic cell (which involves a thermodynamically favored reaction), reversing the wires to the voltmeter should give the correct positive value. The correct value for this reaction is +0.21 V (LO 3.11).

14. B Ions travel through the salt bridge to maintain a charge balance. Without a salt bridge or other method of maintaining charge balance, electrons traveling from anode to cathode would quickly build up a negative charge and the reaction would stop. Anions, such as the nitrate ion, travel through the salt bridge to the anode. Cations travel through the salt bridge to the cathode (LO 3.11).

15. A An increase in the concentration of Cu^{2+} (a reactant) would decrease the value of Q ($[Pb^{2+}]/[Cu^{2+}]$); thus, $Q < 1$ and $\log Q$ would be a negative number. According to the Nernst equation, $E_{cell} = E° - (0.0592/n)\log Q$, so E_{cell} = positive number minus a negative number, so the cell potential increases. See the answer to question 11 for an alternative approach to answering this question (LO 3.12).

FREE-RESPONSE QUESTIONS

1. (a) Reduction occurs at the cathode, and as the reaction proceeds, the cathode will increase in mass. Oxidation occurs at the anode, and as the electrode loses electrons, its mass will decrease as the metal turns into the metal ion. The diagram that the student drew clearly shows an increase in the mass of the aluminum electrode and a decrease in the mass of the magnesium electrode, so the cathode is solid aluminum. *Note:* An answer based on the balanced equation would not receive credit since the question clearly says to base your answer on the student's diagram (LO 3.11).

   (b) Remember to balance the charge, $2\ Al^{3+} + 3\ Mg \rightarrow 2\ Al + 3\ Mg^{2+}$ (LO 3.12).

   (c) 2.37 V + (−1.66 V) = 0.71 V. This should be placed in the box representing the voltmeter. See diagram below (LO 3.12).

   (d) $\Delta G = -nFE$

   $\Delta G = -6(96{,}485)(0.71) = -410{,}000\ J/mol_{rxn}$ or $-410\ kJ/mol_{rxn}$

   (LO 5.14)

   (e) Electrons travel through the wire from the anode to the cathode. Since the aluminum electrode is the cathode, electrons in this reaction will travel right to left (anode to cathode). Circle the arrow pointing toward the aluminum cathode (right arrow). See diagram on the next page (LO 3.11).

   (f) (i) In the salt bridge, anions (NO_3^-) will travel to the anode (Mg), and cations (K^+) will travel to the cathode (Al). Place the symbols for the ions by the correct arrows (K^+ on the left and NO_3^- on the right). See diagram below (LO 3.12).

   (ii) $0.0500\ L \times \frac{1.0\ mol\ KNO_3}{1\ L} \times \frac{2\ mol\ ions}{1\ mol\ KNO_3} \times \frac{6.022 \times 10^{23}\ ions}{1\ mol\ ions} = 6.02 \times 10^{22}\ ions$

   (LO 2.9)

(g) As the reaction progresses, the concentration of reactants decreases, decreasing the number of collisions and electrons transferred. The cell potential decreases. Since the value of Q goes up, the cell potential decreases (LO 3.12).

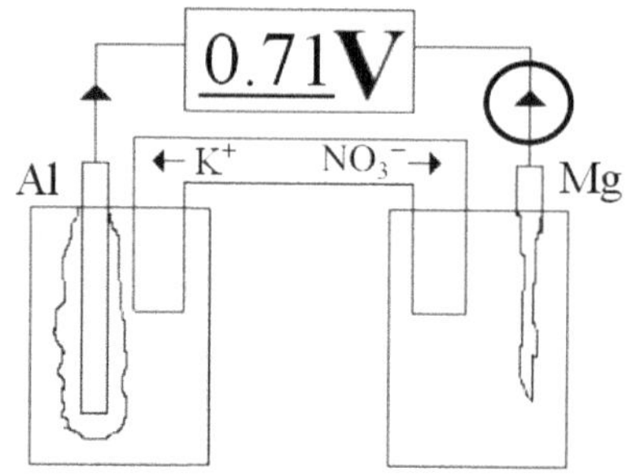

2. (a) (i) Oxidation occurs at the anode and there is only one possible oxidation: $2\ H_2O \rightarrow O_2 + 4\ H^+ + 4\ e^-$ $E° = -1.23$ V. Gas bubbles are produced at the anode. This must be oxygen gas. The oxidation of water is the only possibility as there is no solid sodium present, no H_2SO_3 present, and no H_2 gas and OH^- present to react (only H_2O, Na^+, and SO_4^{2-}) (LO 3.13).

(ii) Reduction occurs at the cathode. Possible reduction reactions:

$Na^+ + e^- \rightarrow Na$ $E° = -2.71$ V

$2\ H_2O + 2\ e^- \rightarrow H_2 + 2\ OH^-$ $E° = -0.83$ V

Gas is observed at the cathode, and the solution turns pink at the cathode. The gas is hydrogen. Production of hydroxide ions makes the solution basic and causes the phenolphthalein to turn pink. The observations are confirmed because we would expect the less negative reaction to be the one to occur. Water will be reduced (LO 3.12).

(iii) When the two reactions (oxidation of and reduction of water) are added together with charge and mass balance, the hydrogen ions combine with the hydroxide ions to form water. Four of the water molecules cancel out, leaving two: $2\ H_2O \rightarrow 2\ H_2 + O_2$. If the H_2 gas is formed twice as fast at the cathode as O_2 gas is formed at the anode, the volumes would be in a 2:1 ratio, with the gas at the cathode being twice the volume as the gas produced at the anode. This corresponds to the mole ratio (LO 3.6).

(b) Pure H_2O won't conduct electricity as it has essentially no ions present. The voltage will be zero (LO 3.13).

8

BIG IDEA 4: KINETICS

Big Idea 4

Rates of chemical reactions are determined by details of the molecular collisions.

Chemical reactions occur at different rates. A reaction may happen in the blink of an eye, or it may take millennia to reach equilibrium. There are many different factors that affect the rates of reactions and the rate can provide evidence for the mechanism of a reaction. Reaction rates can also be affected by certain added substances (catalysts or inhibitors).

You should be able to:

- Identify factors that affect reaction rates.
- Calculate the rate of production of a product or consumption of a reactant using mole ratios and the given rate.
- Determine the rate law for a reaction from given data, overall order, and value of the rate constant, including units.
- Determine the instantaneous rate of a reaction.
- Use integrated rate laws to determine concentrations at a certain time, *t*, and create or interpret graphs to determine the order of a reaction. Also, determine the half-life of a reaction.
- Explain how collision theory supports observations of reactions and reaction rates.
- Write the rate law from a given mechanism or determine if a mechanism is consistent with the observed rate law.

- Write the overall reaction for a mechanism and identify catalysts and intermediates present.
- Explain how different types of catalysts change reaction rates.

> **AP® Tip**
>
> Questions on kinetics may appear in both the multiple-choice and free-response sections. You may be asked to analyze data and construct an explanation based on evidence collected in an experiment.

REACTION RATES

The reaction rate is the change in the concentration of a reactant or product per unit time.

Consider the reaction, $A \rightarrow B$.

$$\text{Rate} = -\frac{\Delta[A]}{\Delta t}$$

The concentration of A in moles per liter is represented by [A]. The change in time is represented by Δt. The quantity, $\Delta [A]/ \Delta t$, is negative because the reactant concentrations are decreasing as products are formed.

EXAMPLE: Consider the reaction $4\ PH_3(g) \rightarrow P_4(g) + 6\ H_2(g)$.

If 0.0048 mole of PH_3 is consumed in a 2.0-L container during each second of the reaction, what are the rates of production for P_4 and H_2?

SOLUTION: The rate at which PH_3 is being consumed is

$$-\frac{\Delta[PH_3]}{\Delta t}.$$

0.0048 mol PH_3/(2.0 L × s) = – 0.0024 mol $L^{-1}\ s^{-1}$ PH_3

The rate at which P_4 and H_2 are being produced can be determined by using mole ratios:

$$\frac{0.0024 \text{ mol PH}_3}{\text{L} \times \text{s}} \times \frac{1 \text{ mol P}_4}{4 \text{ mol PH}_3} = 0.00060 \text{ mol L}^{-1}\text{ s}^{-1}\text{ P}_4$$

$$\frac{0.0024 \text{ mol PH}_3}{\text{L} \times \text{s}} \times \frac{6 \text{ mol H}_2}{4 \text{ mol PH}_3} = 0.0036 \text{ mol L}^{-1}\text{ s}^{-1}\text{ H}_2$$

The last two values are positive since they are products and are appearing as the reaction proceeds. The rate of change for any reactant will be negative, since the reactants disappear as the reaction proceeds.

AP® Tip

[ ] are not a unit. They mean molar concentration, and the units are $\frac{\text{mol}}{\text{L}}$. Students often use [ ] for the unit and do not receive credit on the test.

COLLISION THEORY

The collision theory model accounts for the observed characteristics of reaction rates. Molecules must collide in the correct orientation with sufficient energy, called the *activation energy* (E_a), for a reaction to occur. In most reactions, only a small fraction of collisions meet these criteria. The theory assumes that most reactions occur in a series of steps where one or more reactant particles collide. This sequence of collisions is called the reaction mechanism. In a unimolecular reaction, the reactant molecule may collide with a solvent particle or with other nonreactive particles that may be present. In a bimolecular reaction, two reactant particles collide. A termolecular reaction, where three particles collide simultaneously, is rare.

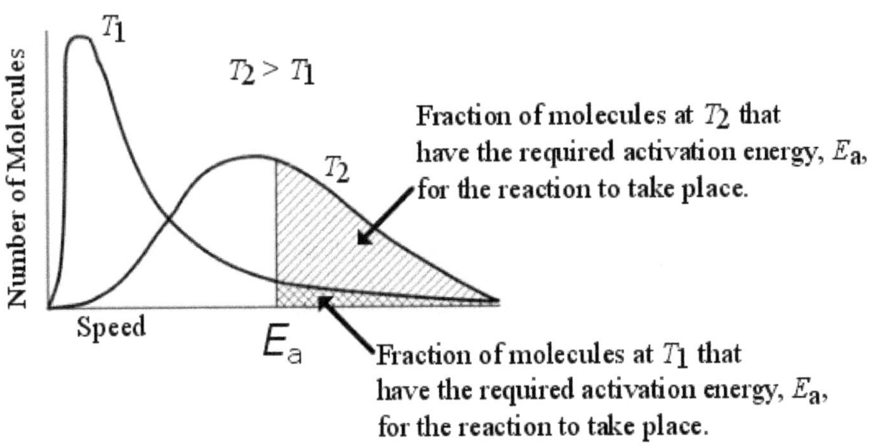

The Maxwell-Boltzmann energy diagram shows how temperature affects the number of particles that have sufficient energy to react. This graph can be presented in a number of ways: the *y*-axis can be labeled as probability, frequency of molecules, or number of molecules. The *x*-axis may measure kinetic energy or the speed of the molecules. The diagram shown reflects the most recent example from the exam. Temperature measures the average kinetic energy of the particles in a system, so the *x*-axis does reflect the temperature (either with kinetic energy or the speed of the molecules). At low temperatures, the graph is "squeezed together," but at higher temperatures, the graph's "peak" is shifted to the right and "squashed" down.

At a lower temperature (T_1), the fraction of particles with the required activation energy is very low. When temperature is higher (T_2), a higher fraction of particles has sufficient energy for the reaction to proceed; therefore, the reaction occurs faster.

Remember that in addition to having sufficient energy, the particles must collide with correct orientation. The diagram below illustrates how correct orientation is necessary for a reaction to occur. The reaction shown in the diagram below is between the gases hydrogen chloride and ethane, $HCl + C_2H_4 \rightarrow CH_3CH_2Cl$.

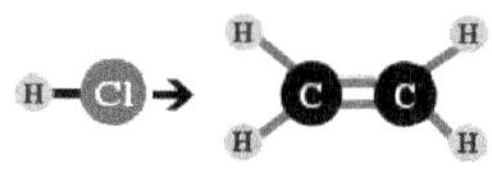

Collision 1

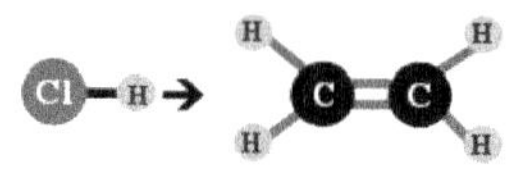

Collision 2

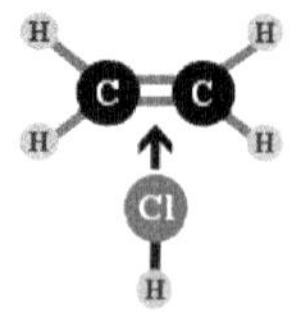

Collision 3

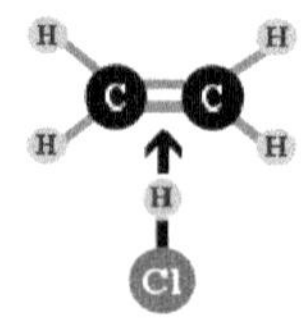

Collision 4

In order to add a chlorine atom to one of the carbon atoms in a C_2H_4 molecule and a hydrogen atom to the other carbon atom, both the H–Cl single bond and the C–C double bond must be broken. It may be helpful to draw out the Lewis structures of all reactants and products to see this. It is actually only the pi bond, rather than the entire C = C bond (sigma + pi), that must be broken. In Collisions 1, 2, and 3, the HCl and the C_2H_4 molecules will simply bounce off each other. The only reaction that produces a successful reaction is Collision 4.

You may wonder why Collision 3 does not also produce a successful collision. The double bond in ethene has a large concentration of electrons; hence, a large negative charge is associated with the bond. The chlorine atom has a large electronegativity value compared to hydrogen, resulting in a large negative dipole. The two negative charges cause repulsion, and the two molecules simply bounce off each other. The rate of this reaction would increase as you move down Group 17 (HF will be the slowest and HI will be the fastest). There are two reasons for this. First, as you go down Group 17, the electronegativity of the halogens decrease, resulting in a smaller negative dipole on the halogen. As the negative dipole decreases in strength, Collision 3 can become more effective in causing a reaction to take place. Second, as you go from H–F, to H–Cl, to H–Br, to H–I, the size of the halogen atom gets bigger and the bond energy drops. The smaller the bond energy, the easier it is to break the hydrogen–halogen bond and more collisions will result in bond breakage.

The potential energy diagram, also known as the reaction progress diagram, for the reaction

$$CO + NO_2 \rightarrow CO_2 + NO$$

is shown below. Energy must be added to break bonds of reactants in order to create a transition state at the top of the "hill," or energy barrier. The transition state (also known as activated complex) is the highest energy state of the reactants in the reaction. Once the energy in the transition state, the activation energy, is overcome, the reactants can form products.

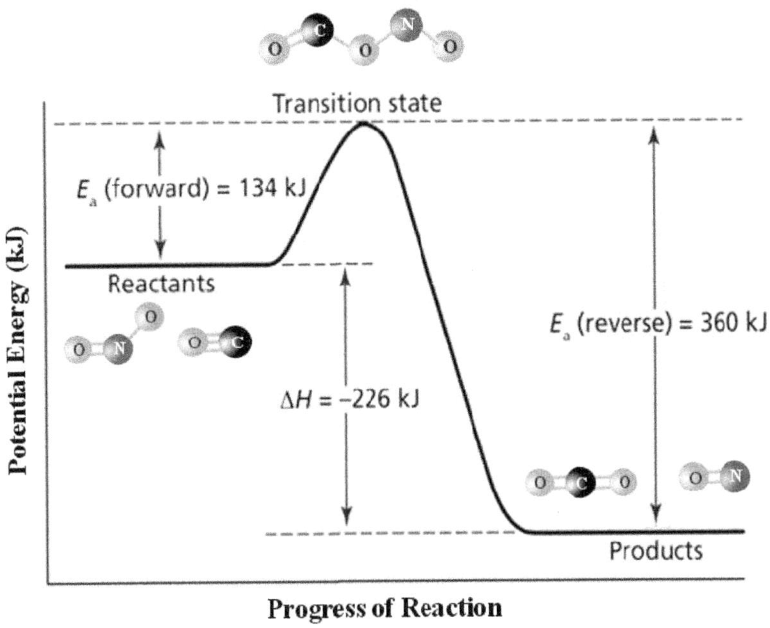

Because the products have a lower potential energy than the reactants, the reaction represented in the above diagram is exothermic. Recall that $\Delta H_{rxn} = \sum \Delta H^o_{f_{products}} - \sum \Delta H^o_{f_{reactants}}$. The difference between the potential energies of the products and the reactants is the enthalpy of the reaction.

If you are asked to draw the reaction progress for an endothermic reaction, the potential energy of the products will be higher than that of the reactants.

RATE LAWS

Rate laws show the relationship between the rate of a reaction and the reactant concentrations. Except for an elementary reaction (one which occurs in a single step, hence, the balanced equation represents the mechanism), the rate law cannot be determined from the balanced equation alone.

The rate law for the reaction

$$2\ NO_2(g) \rightarrow 2\ NO(g) + O_2(g)$$

can be written as

$$\text{Rate} = k[NO_2]^m$$

The proportionality constant, k, is called the rate constant and is determined by experiment. For a given reaction at a given temperature, this value is constant. Its units depend on the orders of the reactants (the overall order of the reaction). The rate constant is a calculated quantity that characterizes a chemical reaction. Rate constants vary over many orders of magnitude because reaction rates vary widely. The larger the value of k, the faster the reaction takes place. The rate constant will change if a catalyst or an inhibitor is present.

The order, m, of the reactant must also be determined by experiment. It is the power to which the reactant concentration must be raised in the rate law. Note that it is possible to have fractional or negative reactant orders.

For the equation A → B, different rate laws are described in the table below.

Rate Law	Order with Respect to A	Effects of Doubling [A]	Units of k
Rate = $k[A]^0$	zero	no effect on rate	$M \times \text{time}^{-1}$
Rate = $k[A]^1$	first	rate will double	time^{-1}
Rate = $k[A]^2$	second	rate will quadruple	$M^{-1} \times \text{time}^{-1}$

The overall order is the sum of all of the individual orders in a rate law. For example, in the rate law

$$\text{Rate} = \frac{k[A]^2[C]}{[B]}$$

which can be written as Rate = $k[A]^2[C][B]^{-1}$, the order with respect to A is second, the order with respect to B is –1, and the order with respect to C is first (when the order is 1, the exponent of 1 is generally not written in the rate law). The overall order is 2 + 1 + (–1) = 2. It is very important to be able to find the overall order since that will help you with the units on the rate constant, k. The units on k will always be $M^{-(\text{overall order } - 1)} \times \text{time}^{-1}$.

INSTANTANEOUS RATES

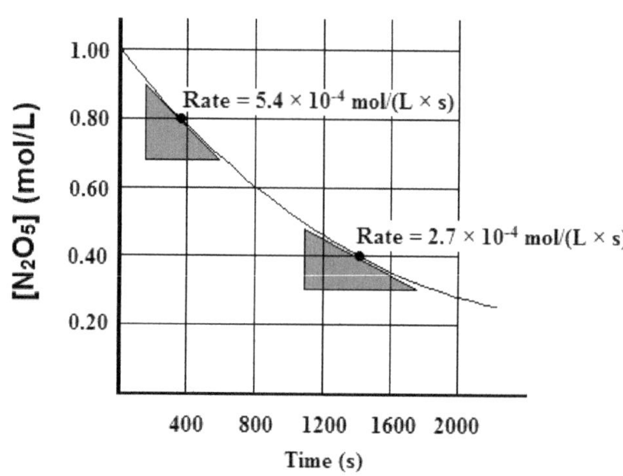

One way to determine the rate of a reaction at a particular time, the *instantaneous rate,* is to plot the reactant concentration versus time and take the slope of the tangent to the curve at time t. If you have had calculus

(calculus is not a pre-requisite for AP® Chemistry), you take the derivative of the curve at the given point.

If the slopes of tangents to the curve at two different concentrations are calculated, the rate law of a reaction can be determined by comparing the changes in rate to the changes in concentration.

In the graph on the previous page, when the concentration of the reactant, N_2O_5, is halved, the rate is also halved. The reaction is first order. The rate law for the reaction is Rate = $k[N_2O_5]$.

METHOD OF INITIAL RATES

The initial rate of a reaction is the instantaneous rate just after the reaction begins (just after $t = 0$ and before the initial concentrations of the reactants have changed).

The rate law for a reaction can be determined by performing a few trials with different reaction concentrations and measuring the initial rate for each trial. This type of rate law is called a differential rate law. To find the order of one reactant (or catalyst if it is present), change its concentration while holding the concentration of the other reactants constant.

EXAMPLE: Determine the rate law and the value of the rate constant for the following reaction:

$$2\,A(g) + B(g) \rightarrow 2\,C(g)$$

Trial	$[A]_o$ (mol/L)	$[B]_o$ (mol/L)	Initial Rate (mol $L^{-1}\,s^{-1}$)
1	0.10	0.10	0.18
2	0.10	0.20	0.72
3	0.20	0.20	1.45

SOLUTION: Write the general form of the rate law:

$$\text{Rate} = k[A]^x[B]^y$$

Your goal is to determine x and y.

There are two common methods to solve for this: (1) algebraic and (2) verbal (what many people refer to as "table logic"). We will solve the problem both ways here. Of the two, the verbal approach is much better on the multiple-choice section, since you don't need to justify your answers and both methods take about the same amount of time on the free-response section. If you are a little weaker with algebra, then stick with the "verbal" approach.

First, the algebraic method: To find x, the order with respect to A, pick two trials in which A changes and B remains the same.

Compare Trial 3 to Trial 2.

Write a ratio of the rate law of Trial 3 to the rate law for Trial 2. Substitute the values for the rates and known concentrations and solve for the order, *x*.

$$\frac{\text{Rate (Trial 3)}}{\text{Rate (Trial 2)}} = \frac{\cancel{k}[A]^x\cancel{[B]^y}}{\cancel{k}[A]^x\cancel{[B]^y}};\ \frac{1.45}{0.72} = \frac{(0.20)^x}{(0.10)^x};\ 2.0 = 2^x;\ x = 1$$

The order, *x*, with respect to A is 1. This means that when the concentration of A is doubled, the rate doubles. Note that the rate constant and the concentration of B cancel out because they are the same in Trials 2 and 3.

To find *y*, the order with respect to B, pick two trials in which B changes and A remains the same.

Compare Trial 2 to Trial 1.

$$\frac{\text{Rate (Trial 2)}}{\text{Rate (Trial 1)}} = \frac{\cancel{k[A]^x}[B]^y}{\cancel{k[A]^x}[B]^y};\ \frac{0.72}{0.18} = \frac{(0.20)^x}{(0.10)^x};\ 4.0 = 2^y;\ y = 2$$

The order, *y*, with respect to B, is 2. This means that when the concentration of B is doubled, the rate quadruples. Note that the rate constant and the concentration of A cancel out because they are the same in Trials 1 and 2.

The rate law for the reaction is Rate = $k[A][B]^2$.

For the verbal method: Again to find *x*, the order with respect to A, pick two trials in which A changes and B remains the same. However, you will need to write out the justification.

Trial	$[A]_o$ (mol/L)	$[B]_o$ (mol/L)	Initial Rate (mol L^{-1} s^{-1})
1	0.10	0.10	0.18
2	0.10	0.20	0.72
3	0.20	0.20	1.45
	× 2	× 1	× 2

In Trials 2 and 3, [B] is constant, the [A] goes from 0.10 to 0.20, up by a factor of 2; initial rate goes from 0.72 to 1.45, up by a factor of 2 and since $2^1 = 2$, the order with respect to A is first order.

To find *y*, the order with respect to B, pick two trials in which B changes and A remains the same.

Trial	$[A]_o$ (mol/L)	$[B]_o$ (mol/L)	Initial Rate (mol L^{-1} s^{-1})
1	0.10	0.10	0.18
2	0.10	0.20	0.72
	× 1	× 2	× 4
3	0.20	0.20	1.45

In Trials 1 and 2, [A] is constant, the [B] goes from 0.10 to 0.20, up by a factor of 2. The initial rate goes from 0.18 to 0.72, up by a factor of 4 and since $2^2 = 4$, the order for B is second order.

The overall order of a reaction is the sum of the reaction orders. For this example, the overall order is 2 + 1 = 3.

AP® Tip

Whenever you use the verbal/table logic method on the free-response section, you will need to justify your answer. Use the following template for your answer.

In experiments ____ and ____, the concentration of ____ remains constant, while the concentration of ____ goes from ____ to ____, up by a factor of ____. The initial rate goes from ____ to ____, up by a factor of ____, since ____ = ____, the order with respect to ____ is ____ order.

Note: The order with respect to each reactant may be the same as the coefficients in the balanced equation, but generally are not (as in this case) and the order may be a fraction such as ½, or a negative number. Fractional rate constants are rare but have appeared on recent exams. Negative orders indicate that the species is an inhibitor and would most likely be a product that is interfering with the collisions between the reactants.

To determine the value of the rate constant, including its units, use the rate law and experimental data from any given trial.

$$k = \text{Rate}/[A][B]^2$$

Using values from Trial 1:

$$= (0.18\ M\ s^{-1})/(0.10\ M)(0.10\ M)^2 = 180\ L^2\ mol^{-2}\ s^{-1}.$$

AP® Tip

If possible, use the trial that has equal concentrations for the reactants when calculating *k*. That way, if you substitute them incorrectly (substitute the value for A for B and vice versa), you will still get the correct answer.

It helps to work out the units separately.

$$\left(\frac{\text{mol}}{\text{L} \times \text{s}}\right)\left(\frac{\text{L}}{\text{mol}}\right)\left(\frac{\text{L}^2}{\text{mol}^2}\right) = \frac{\text{L}^2}{\text{mol}^2 \times \text{s}}$$

AP® Tip

The units on the rate constant, *k*, will always be

$$\frac{L^{(\text{overall order} - 1)}}{\text{mol}^{(\text{overall order} - 1)} \times (\text{unit for time})}.$$

If the rate constant is known and the order with respect to one of the reactants has been determined, the order with respect to the other can be calculated, using the algebraic method, even if its concentration is not held constant between any of two trials.

EXAMPLE: Determine the rate law for the reaction of ammonium and nitrite ions in aqueous solution at 25°C:

$$NH_4^+(aq) + NO_2^-(aq) \rightarrow N_2(g) + 2\,H_2O(l)$$

Trial	$[NH_4^+]_0$ (mol/L)	$[NO_2^-]_0$ (mol/L)	Initial Rate (mol L^{-1} min^{-1})
1	0.0100	0.200	5.37×10^{-7}
2	0.0400	0.200	2.15×10^{-6}
3	0.0800	0.400	1.72×10^{-5}

SOLUTION: Write the general form of the rate equation.

$$\text{Rate} = k\,[NH_4^+]^m[NO_2^-]^n$$

Using Trials 1 and 2, determine the order, *m*, of NH_4^+ using one of the two techniques described above. You should find that $m = 1$. Since $[NH_4^+]$ does not remain constant, you must use your newly found order with respect to NH_4^+ and the concentrations given to solve for *n*. You must use two trials where the $[NO_2^-]$ changes.

$$\frac{\text{Rate (Trial 3)}}{\text{Rate (Trail 2)}} = \frac{k\,[NH_4^+]^1\,[NO_2^-]^n}{k\,[NH_4^+]^1\,[NO_2^-]^n}$$

$$\frac{1.72 \times 10^{-5}}{2.15 \times 10^{-6}} = \frac{\cancel{k}\,[0.0800]^1\,[0.400]^n}{\cancel{k}\,[0.0400]^1\,[0.200]^n}$$

$$8 = 2(2)^n$$

$$4 = (2)^n$$

$$n = 2$$

Using the chart–logic (or verbal) method:

Since it was determined that $[NH_4^+]$ was first order, in Experiments 2 and 3, doubling the concentration of $[NH_4^+]$ alone would have caused the rate to double. Doubling the concentration of both $[NH_4^+]$ and $[NO_2^-]$ caused the rate to be 8 times greater. Negating the effect of doubling $[NH_4^+]$ alone, doubling $[NO_2^-]$ caused the rate to quadruple, indicating that the reaction is second order in NO_2^-.

The reaction is first order in NH_4^+ and second order for NO_2^-. The rate law is

$$\text{Rate} = k\,[NH_4^+][NO_2^-]^n$$

For practice, the value for the rate constant is $\frac{1.34 \times 10^{-3}\ \text{L}^2}{\text{mol}^2 \times \text{min}}$.

INTEGRATED RATE LAWS

An integrated rate law, derived from the differential rate law, expresses the reactant concentration as a function of time. The table below summarizes the integrated rate laws for the reaction A → Products.

	Order: Zero	First	Second
Rate Law:	Rate = k	Rate = k[A]	Rate = k[A]^2
Integrated Rate Law:	$[A] - [A]_0 = -kt$	$\ln[A] - \ln[A]_0 = -kt$	$\frac{1}{[A]} - \frac{1}{[A]_0} = kt$
Linear form of integrated rate law ($y = mx + b$)	$[A] = -kt + [A]_0$	$\ln[A] = -kt + \ln[A]_0$	$\frac{1}{[A]} = kt + \frac{1}{[A]_0}$
Plot Needed to Give a Straight Line:	[A] versus t	ln[A] versus t	$\frac{1}{[A]}$ versus t
Relationship of Rate Constant to the Slope of Straight Line:	Slope = $-k$	Slope = $-k$	Slope = k
Half-Life:	$t_{1/2} = \frac{[A]_0}{2k}$	$t_{1/2} = \frac{0.693}{k}$	$t_{1/2} = \frac{1}{[A]_0 k}$

You can determine the order of the reactant graphically if you know the concentration of A at various times, t, during the reaction. For example, if you plot ln[A] vs. t and obtain a straight line, the reaction is first order in A. If the graph is not linear, then the reaction is not first order. The table above summarizes what is graphed to test for the order of the reactant.

AP® Tip

Regardless of the order of the graph, |slope| = k

ZERO ORDER

Time (min)	[A]	ln [A]	1/[A]
0	1000.0	6.9078	0.0010000
100	922.3	6.8269	0.0010842
200	844.6	6.7389	0.0011840
300	766.9	6.6424	0.0013040
400	689.2	6.5355	0.0014510
500	611.5	6.4159	0.0016353
600	533.8	6.2800	0.0018734
700	456.1	6.1227	0.0021925
800	378.4	5.9360	0.0026427

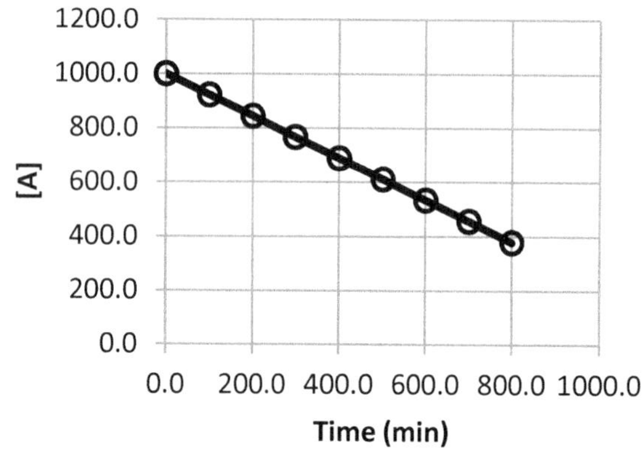

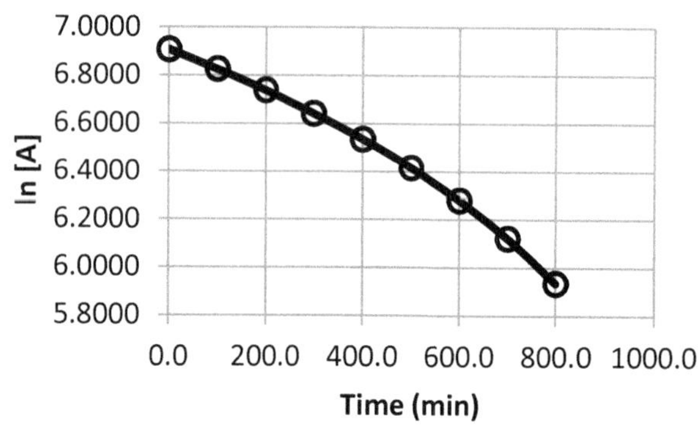

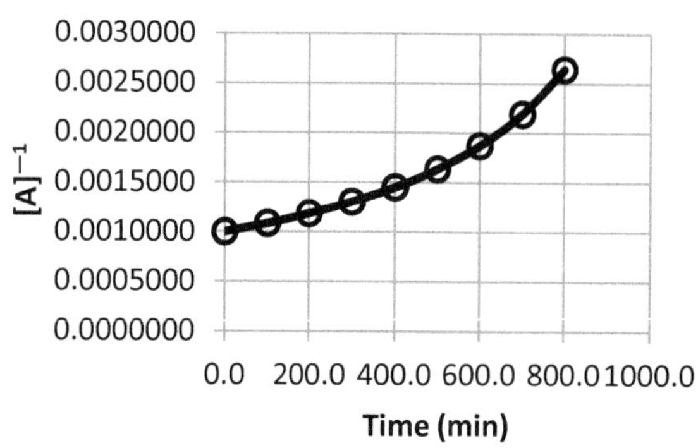

FIRST ORDER

Time (s)	[B]	ln [B]	1/[B]
0.0	1,000.0	6.90776	0.0010000
100.0	748.3	6.6178	0.001336
200.0	559.9	6.3278	0.001786
300.0	419.0	6.0378	0.002387
400.0	313.5	5.7478	0.003190
500.0	234.6	5.4578	0.004263
600.0	175.5	5.1678	0.005697
700.0	131.3	4.8778	0.007614
800.0	98.3	4.5878	0.010176

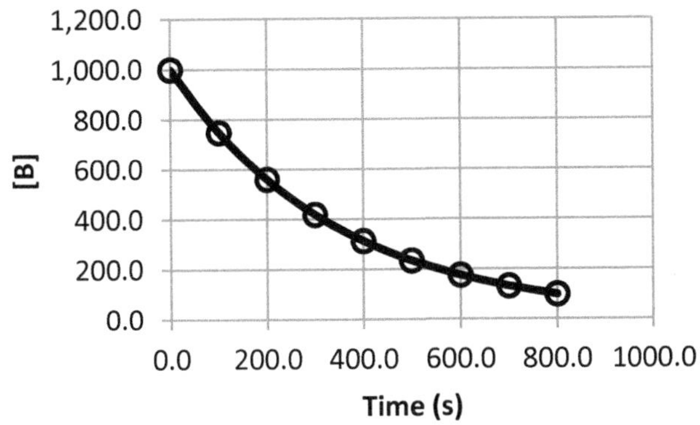

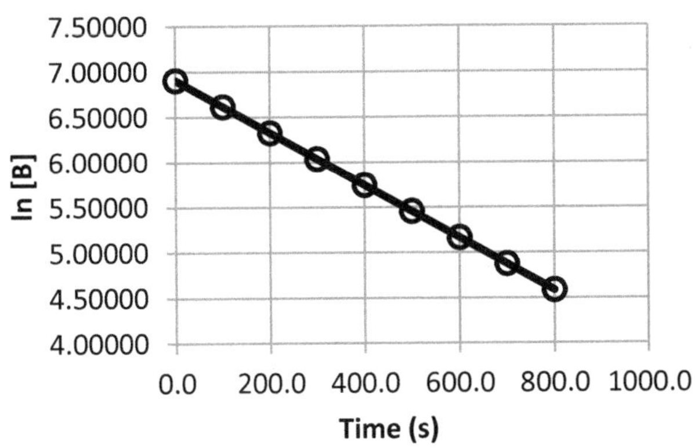

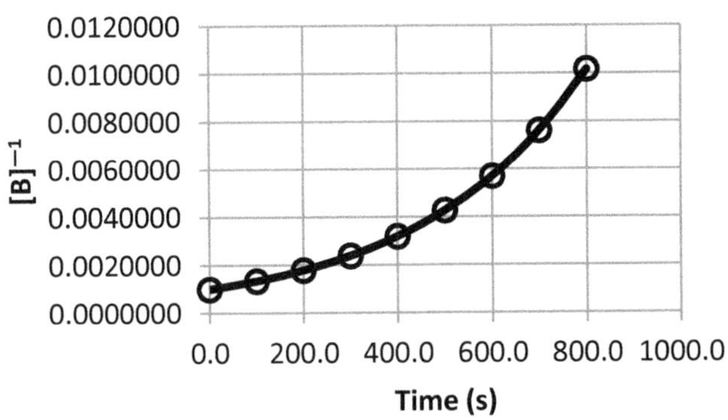

SECOND ORDER

Time (hr)	[C]	ln [C]	1/[C]
0.0	1000.0	6.90776	0.0010000
100.0	505.1	6.2247	0.001980
200.0	337.8	5.8226	0.002960
300.0	253.8	5.5366	0.003940
400.0	203.3	5.3144	0.004920
500.0	169.5	5.1328	0.005900
600.0	145.3	4.9791	0.006880
700.0	127.2	4.8460	0.007860
800.0	113.1	4.7285	0.008840

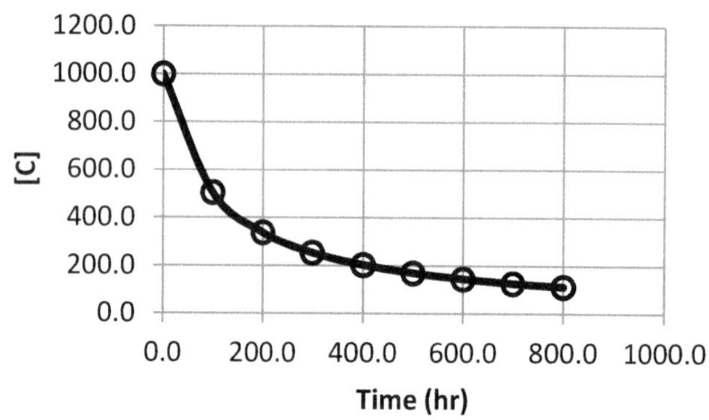

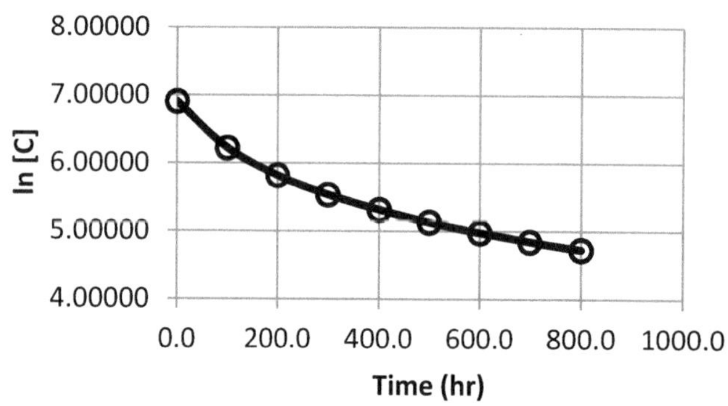

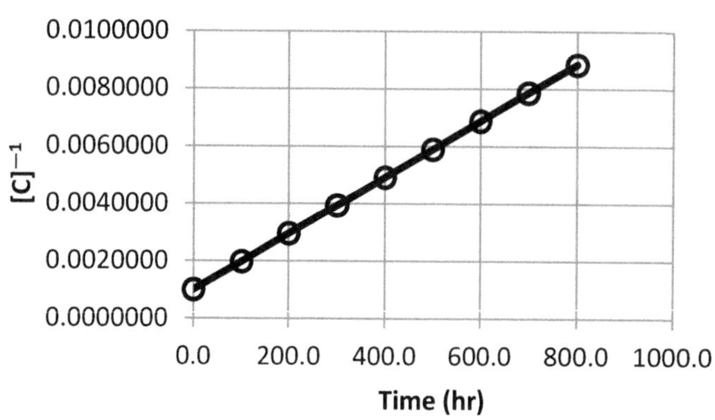

EXAMPLE: The rate of the reaction

$$NO_2(g) + CO(g) \rightarrow NO(g) + CO_2(g)$$

depends only on the concentration of nitrogen dioxide below 225°C. At a temperature below 225°C, the following data were collected. Determine the order of the reaction with respect to NO_2.

Time (s)	$[NO_2]$	$\ln[NO_2]$	$1/[NO_2]$
0	0.800	−0.223	1.25
1.00×10^3	0.597	−0.511	1.68
2.00×10^3	0.476	−0.734	2.10
3.00×10^3	0.396	−0.916	2.53
4.50×10^3	0.316	−1.14	3.16
6.00×10^3	0.263	−1.35	3.80
7.75×10^3	0.220	−1.51	4.54
1.20×10^4	0.157	−1.83	6.35

SOLUTION: Make three graphs: $[NO_2]$ vs. time, $\ln[NO_2]$ vs. time, and $1/[NO_2]$ vs. time. Determine which is a linear graph and based on the units on the y-axis, that will determine your order.

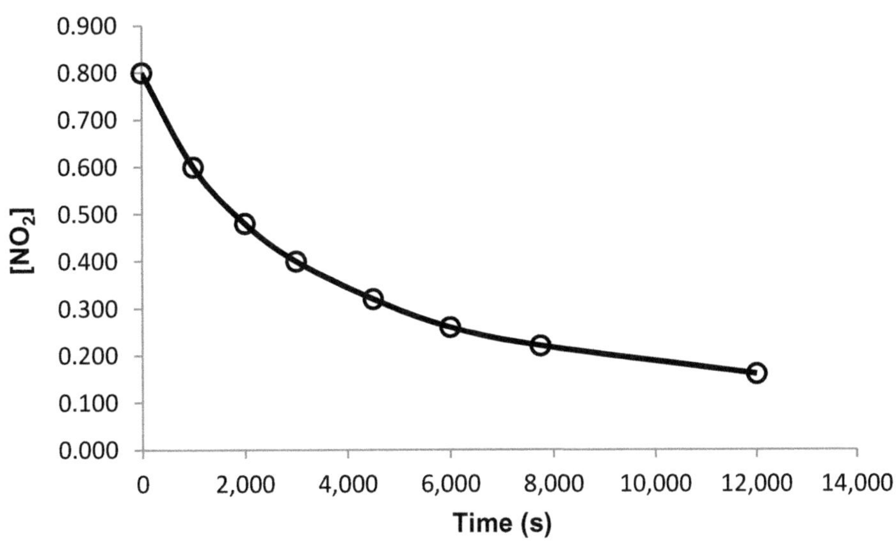

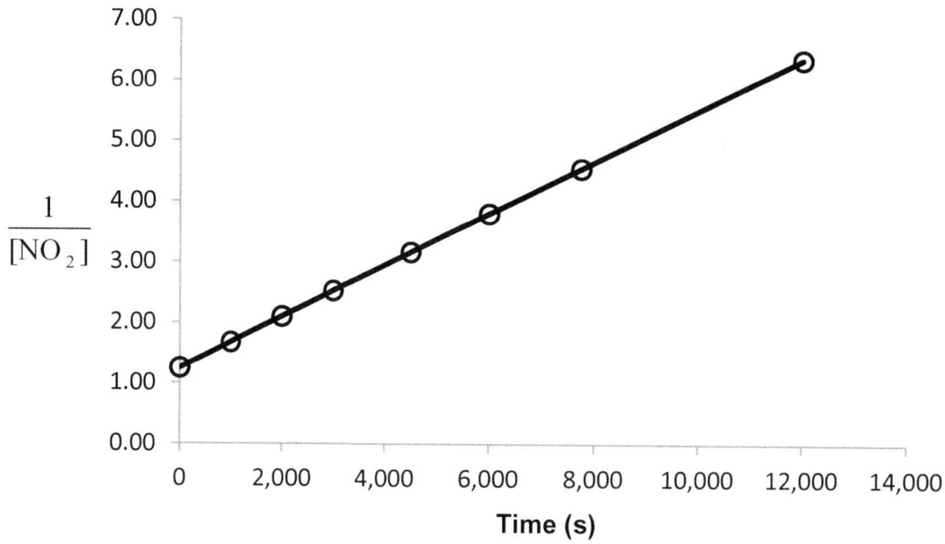

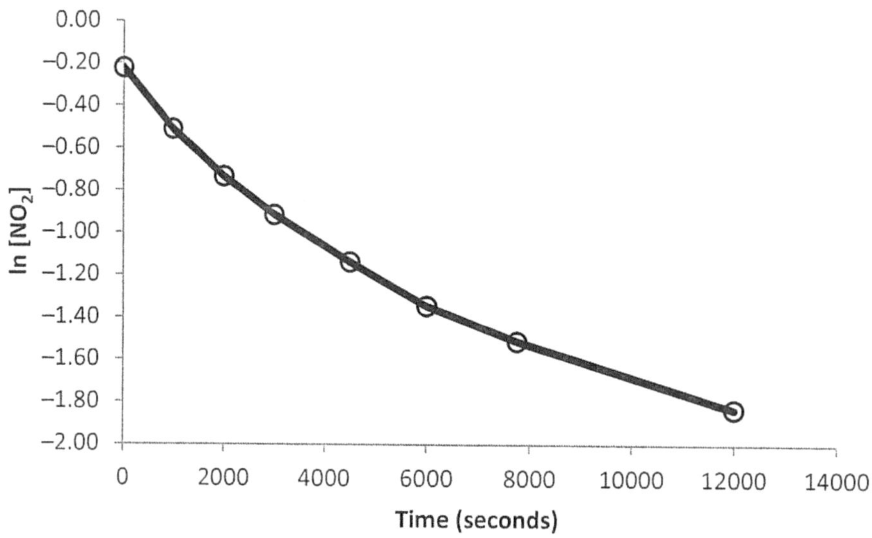

The plot of $1/[NO_2]$ vs. time is linear, which means the reaction is second order, so the slope of this line gives the value of k.

$$\text{Slope} = k = \frac{\Delta y}{\Delta x} = \frac{(5.75 - 2.00)\,M^{-1}}{(1.8 \times 10^4 - 0)\text{ s}} = 2.08 \times 10^{-4}\text{ L mol}^{-1}\text{ s}^{-1}$$

To determine $[NO_2]$ at 2.70×10^4 s, use the integrated rate law where $1/[NO_2]_0 = 1/0.500\ M = 2.00\ M^{-1}$.

$$1/[NO_2] = kt + 1/[NO_2]_0$$

$$\frac{1}{[NO_2]} = \frac{2.08 \times 10^{-4}\text{ L}}{\text{mol s}} \times 2.70 \times 10^4\text{s} + 2.00\text{ L mol}^{-1}$$

$$\frac{1}{[NO_2]} = 7.62 \text{ L mol}^{-1}$$

$$[NO_2] = 0.131\ M$$

The half-life of a reactant, $t_{1/2}$, is the time required for a reactant to reach half of its original concentration. The general equation for the half-life for a first-order reaction is $t_{1/2} = 0.693/k$.

THE KINETICS OF RADIOACTIVE DECAY

Radioactive decay is a common example of first-order kinetics in a real-life context. The decay of nuclides is represented by $\ln\left(\frac{A}{A_0}\right) = -kt$

Where A_0 = original amount of nuclides
A = amount remaining at time t
k = first-order rate constant
t = time

Note that the amount of nuclides can be measured in a variety of ways: grams or atoms. If you use a device that measures nuclear decay, such as a Geiger counter, then you might have clicks per time or disintegrations per minute per gram.

HALF-LIFE

The half-life, $t_{1/2}$, is the time required for the concentration of the reactants to reach half of their original value. It can be determined by the equation:

$$t_{1/2} = 0.693/k$$

EXAMPLE: Iodine-131, used in the diagnosis and treatment of thyroid diseases, has a half-life of 8.1 days. When a patient with thyroid disease consumes a sample containing 300. gray of iodine-131 (a gray is a unit that measures ionizing radiation), they are instructed to carry a letter of explanation with them for 3 months (90.0 days) to explain the elevated radiation levels. This is necessary as radiation detection devices used at airports and federal buildings may be sensitive to the elevated radiation levels present. How much of the I-131 remains before it can no longer be detected by the radiation detection devices?

SOLUTION: Use the equation: $\ln(A/A_0) = -kt$ and solve for A

Since $t_{1/2} = 0.693/k$, substitute 0.693/8.1 days for k.

$$\ln\left(\frac{A}{300.\text{ gray}}\right) = -\left(\frac{0.693}{8.1\text{ days}}\right) \times 90\text{ days} \Rightarrow A = 0.14\text{ gray}$$

If you refer to the table in the "Integrated Rate Laws" section, you will notice the half-life equations for zero-, first-, and second-order reactions. Notice that both zero- and second-order half-life equations have the $[A_0]$ term, the original amount of the reactant. In other words, the half-life of both zero- and second-order reactions depends upon the starting amounts. First-order half-lives are independent of the starting amount and will remain constant.

AP® Tip

The only half-life that will be tested is for first-order reactions, and the first-order half-life equation is provided on the formula chart. It is important to be able to use dependence of the half-life on the starting concentration as a means to identify the rate law.

FACTORS AFFECTING REACTION RATES

Ways to increase the rate of a reaction:

- Increase concentration of reactants
- Decrease volume (gaseous reaction)
- Increase surface area
- Increase temperature
- Add a catalyst

Increasing the reactant concentrations increases the chance for more molecular collisions that make products. Except for zero-order reactions, increasing reactant concentrations will usually increase the reaction rate. For gas phase reactions, decreasing the volume of the container increases reaction rate by increasing reactant concentration.

Increasing the reactant's surface area increases the frequency of particle collisions. For example, granular zinc reacts more quickly with hydrochloric acid than a strip of zinc.

Increasing the reaction temperature affects the rate of the reaction in two ways: energy factor and collision factor. Temperature is a measure of the average kinetic energy of the particles in the system; higher temperature means higher kinetic energy. More particles will have sufficient activation energy to react; the collisions are more energetic. The second way temperature affects the rate is by increasing the number of collisions. At higher temperatures, the particles are moving faster, thus increasing the rates of both the forward and reverse reactions. The number of particle collisions increases in any given period of time. The general rule of thumb states that for each 10°C increase in temperature, the reaction rate doubles. This works well for reactions performed close to room temperature with activation energies of around 50 kJ mol^{-1}. Of the two factors, the increase in the energy of the collisions is more important.

MEASURING REACTION RATES

The rate of a reaction is experimentally determined by measuring the amount of reactants converted to products in a given time. A common method of measuring the reaction rate is to use a spectrophotometer or colorimeter to monitor the absorption of light of a specific wavelength by a colored reactant or product. The absorbance of the chemical is directly related to its concentration via Beer's law. Beer's law was discussed in Chapter 1.

For instance, the rate of disappearance of color from the reaction of blue food dye with bleach (NaOCl) can be analyzed. According to Beer's law, absorbance is directly related to [dye].

$$\text{Rate} = k[\text{dye}]^x[\text{OCl}^-]^y$$

If blue food dye is the limiting reagent and the bleach concentration is so large that it remains constant throughout the reaction, then the rate expression simplifies to

$$\text{Rate} = k_{pseudo}\ [\text{dye}]^x$$

where k_{pseudo} is the pseudo rate constant for the reaction. The actual rate constant, k, can be determined by

$$k = k_{pseudo}/[\text{OCl}^-]$$

1.00 mL 0.0264 *M* sodium hypochlorite added to 25.00 mL of blue food dye.

Absorbance of blue food dye at 630 nm

Time (min)	Abs	ln Abs	1/Abs
0	0.591	–0.526	1.69
1	0.369	–0.997	2.71
2	0.262	–1.339	3.82
3	0.157	–1.852	6.37
4	0.113	–2.180	8.85
5	0.071	–2.645	14.1
6	0.045	–3.101	22.2
7	0.025	–3.689	40.0

In determining the order for this reaction, three graphs are produced, shown below. Please note that this is laboratory data and, as such, will not produce perfect results.

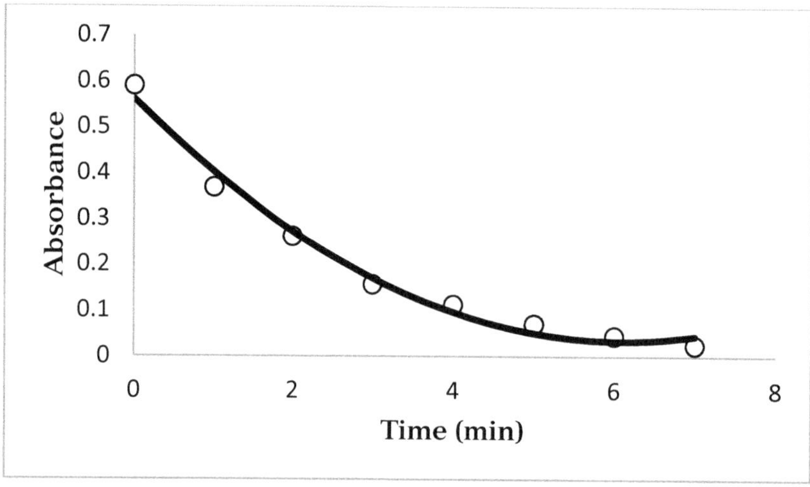

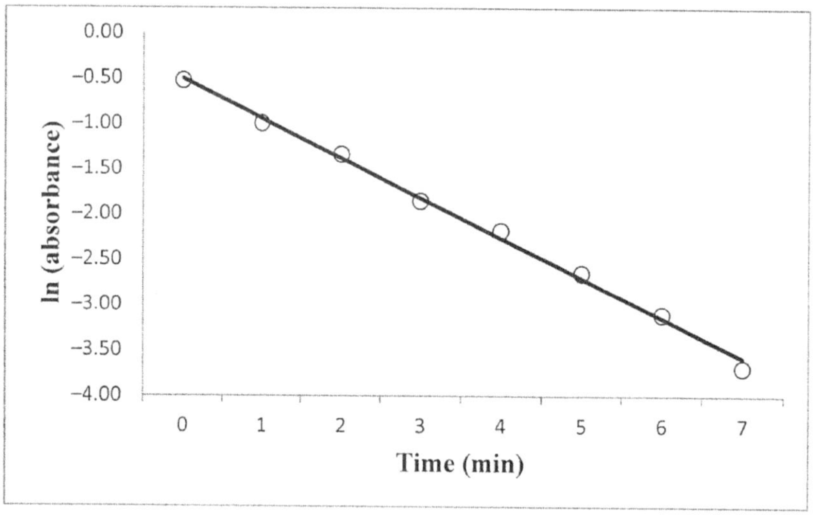

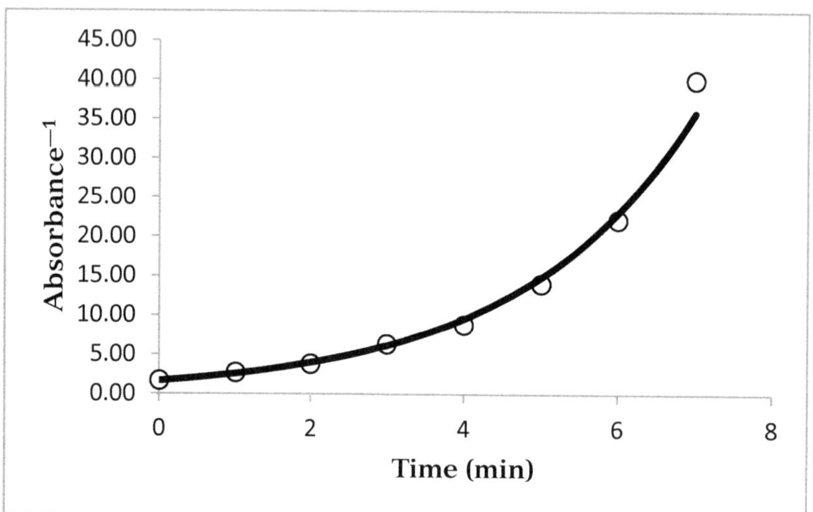

Since the graph of ln(absorbance) vs. time is linear, the reaction is first order with respect to the dye. Since the slope of the line is −0.439, the value of the rate constant will be the absolute value of the slope, so $k_{pseudo} = 0.439\ \text{min}^{-1}$.

A second experiment was carried out under the same conditions, except with 0.0132 *M* NaOCl. The slope was half that of the first experiment, meaning the reaction is also first order in OCl^-. The actual rate constant is

$$k = k_{pseudo}/[OCl^-]$$

$$k = 0.439/0.0264 = 16.6 \text{ L mol}^{-1}$$

and the rate law is

$$\text{Rate} = k[\text{dye}][OCl^-]$$

CATALYSIS

The diagram below shows the pathway for both catalyzed and uncatalyzed reactions. A catalyst speeds up a reaction by lowering the activation energy of both the forward and reverse reactions without being consumed itself. The activation energy is lowered because the mechanism of the reaction has changed and the transition state is stabilized. The pathway from reactants to products is different, which is depicted by the lower squiggly line—with four peaks representing a new 4-step mechanism for the reaction. This lower squiggly line (or multi-peak line) shows the catalyzed pathway, which represents a 4-step mechanism in this case. The important point is that the highest point in the new pathway (the second peak) has a lower activation energy than the original pathway; hence, the rate of the reaction increases.

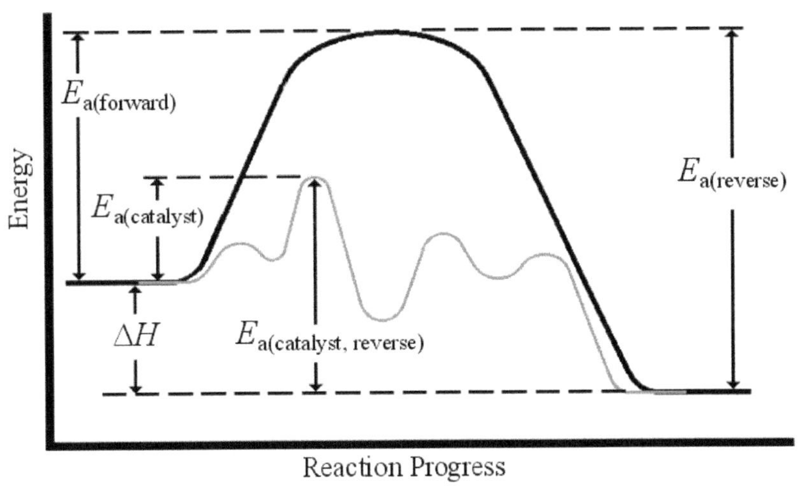

Three classes of catalysis reactions are important in AP® Chemistry: acid–base, surface, and enzyme.

Acid–base catalysis involves a reactant gaining or losing a proton which causes the reaction rate to increase. In acid catalysis, the catalyst donates a proton to the reactant and the reaction rate increases with decreasing pH. A common example is the acid-catalyzed esterification reaction between a carboxylic acid and an alcohol. The acid typically used is concentrated sulfuric acid since this acid contains almost no water which can hydrolyze esters. In base catalysis, the reactant donates a proton to the basic catalyst causing an increase in reaction rate with increasing pH. An industrial application of base catalysis is the production of polyurethane foams.

Surface catalysis involves the formation of new intermediates or an increase in the probability of successful collisions by improving the orientation of the reactants. The diagram below shows the de-composition of N_2O on rhodium clusters, which mimics the chemistry in catalytic converters in automobiles.

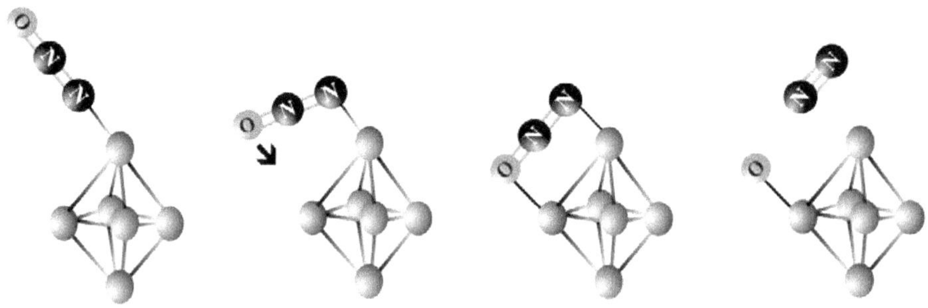

Another example of surface catalysis includes the use of finely divided iron in the synthesis of ammonia from nitrogen and hydrogen gas. The gases a<u>d</u>sorb onto the surface of the iron, and their bonds are weakened. Since the fragments of nitrogen and hydrogen are in close proximity, formation of new nitrogen–hydrogen bonds is favored.

Enzyme catalysis plays an important role in living organisms. Enzymes are homogenous catalysts that make products necessary for the organism. After a substrate (a substrate is the chemical that is being observed or reacted to form products) binds to the enzyme, the shape of the enzyme changes slightly. The substrate is converted into products which then leave the enzyme to be used elsewhere in the body. The enzyme is again available to bind with more substrate. Enzymes lower the activation energy by providing alternative pathways for biochemical reactions to occur.

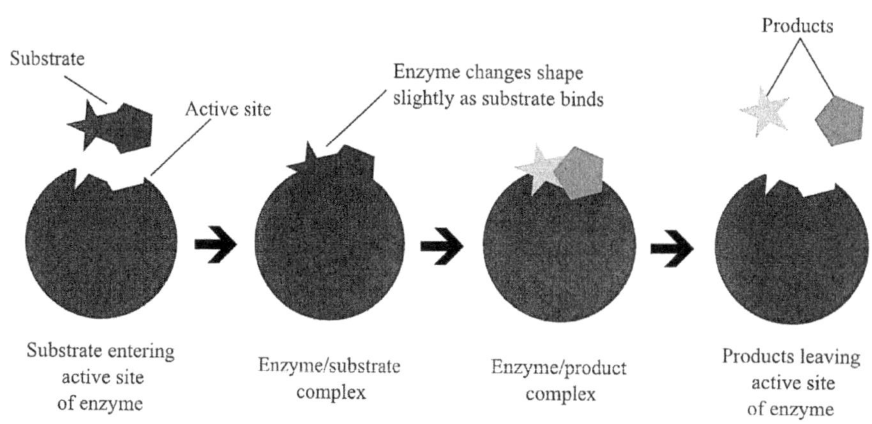

MECHANISMS

A mechanism is a series of steps by which a reaction occurs according to collision theory. The elementary steps, each step in the mechanism, must add up to give the overall balanced equation for the reaction. The slowest step in the mechanism must agree with the experimentally determined rate law.

Once the rate law has been determined experimentally, chemists propose several mechanisms consistent with that rate law and design experiments to attempt to determine which mechanism is consistent with the data. It is important to be able to identify catalysts and intermediates in a mechanism and write the overall equation for a mechanism. You will also need to be able to write a rate law from a given mechanism. You will not be asked to develop an experiment to verify a specific mechanism.

A catalyst and one or more reaction intermediate(s) may be present in a mechanism and they can easily be confused by students. Below is a simple chart that will help you identify and justify their identity.

	Where Does It First Appear	Where Does It Reappear
Catalyst	As a reactant in an early step in the mechanism.	As a product in a later step in the mechanism.
Intermediate	As a product in an early step in the mechanism.	As a reactant in a later step in the mechanism.

Notice that both catalysts and intermediates will cancel out and will not appear in the overall reaction. Recall that the catalyst is not consumed in the reaction. The detection and identification of intermediates provides evidence that scientists use to justify one mechanism over another.

EXAMPLE: Identify the catalyst and reaction intermediate in the decomposition of ozone by atoms of chlorine.

$$Cl(g) + O_3(g) \rightarrow ClO(g) + O_2(g)$$

$$O(g) + ClO(g) \rightarrow Cl(g) + O_2(g)$$

$$O(g) + O_3(g) \rightarrow 2\ O_2(g)$$

SOLUTION: $Cl(g)$ is the catalyst because it is a reactant in the first reaction and a product in the second reaction. $Cl(g)$ does not appear in the overall reaction. $ClO(g)$ is the reaction intermediate since it is produced in the first step of the mechanism and is used up in the second step. It also does not appear in the overall equation.

In order to determine if a given mechanism is consistent with the experimental rate law that was determined, two items must be checked: (1) the steps of the mechanism add up to the stoichiometry of the overall equation and (2) the rate-determining step (the slow step in the process) produces the experimental rate law found. The term ***molecularity*** is used to describe the number of species that must collide to produce the reaction. An ***elementary step*** is a reaction whose rate law is determined by its molecularity. The most common elementary steps are shown below.

$A \rightarrow$ Product	unimolecular	Rate = $k[A]$
$A + B \rightarrow$ Product	bimolecular	Rate = $k[A][B]$
$A + A \rightarrow$ Product	bimolecular	Rate = $k[A]^2$
$2A + B \rightarrow$ Product	termolecular	Rate = $k[A]^2[B]$
$A + B + C \rightarrow$ Product	termolecular	Rate = $k[A][B][C]$

Notice that unimolecular elementary steps produce a rate law that has an overall order of 1. Bimolecular elementary steps produce a rate law that has an overall order of 2, and termolecular elementary steps produce a rate law that has an overall order of 3.

Termolecular reactions are extremely rare, as the likelihood of 3 atoms colliding with the proper orientation and sufficient energy is very small. You will not see a termolecular elementary step on the test.

EXAMPLE: $NO_2(g) + CO(g) \rightarrow CO_2(g) + NO(g)$ follows the law: Rate = $k[NO_2]^2$

Show that the following mechanism is consistent with the given information.

(slow) $\quad NO_2 + NO_2 \rightarrow NO_3 + NO$

(fast) $\quad NO_3 + CO \rightarrow NO_2 + CO_2$

SOLUTION: First, the overall equation should be determined to see if it matches the overall equation above.

$$\cancel{NO_2} + NO_2 \rightarrow \cancel{NO_3} + NO$$

$$\cancel{NO_3} + CO \rightarrow \cancel{NO_2} + CO_2$$

Notice that one of the NO_2 molecules cancels and the NO_3 molecule also cancels, which leaves the overall equation of $NO_2(g) + CO(g) \rightarrow CO_2(g) + NO(g)$. Next, determine if the slow step produces the overall rate equation. Since the slow step is bimolecular, it will produce the rate law of Rate = $k[NO_2]^2$. This mechanism is consistent with the information presented and is a possible mechanism for this reaction.

EXAMPLE: Consider the balanced equation for the following hypothetical reaction

$$2\,W + 2\,X \rightarrow 2\,Y + Z_2$$

The experimentally determined rate law is Rate = $k[W]^2[X]$. Which of the following two mechanisms is consistent with the rate law?

I. $W + X \rightarrow E + Y$ (slow)

$E + W \rightarrow E_2$ (fast)

$E_2 + X \rightarrow Z_2 + Y$ (fast)

II. $W + X \rightleftarrows E$ (fast, equilibrium)

$E + W \rightarrow E_2 + Y$ (slow)

$E_2 + X \rightarrow Z_2 + Y$ (fast)

SOLUTION: For a proposed mechanism to be correct, the overall reaction and the rate law for the mechanism must agree with the given reaction and the experimentally determined rate law.

Write the overall reaction for each mechanism. Add up the elementary steps. Cancel out the catalysts and intermediates that appear. E and E_2 are intermediates because they are produced in one step and consumed in the subsequent step of the mechanism.

I. $W + X \rightarrow \cancel{E} + Y$

$\cancel{E} + W \rightarrow \cancel{E_2}$

$\cancel{E_2} + X \rightarrow Z_2 + Y$

$2\,W + 2\,X \rightarrow Z_2 + 2\,Y$

II. $W + X \rightleftarrows \cancel{E}$

$\cancel{E} + W \rightarrow \cancel{E_2} + Y$

$\cancel{E_2} + X \rightarrow Z_2 + Y$

$2\,W + 2\,X \rightarrow Z_2 + 2\,Y$

The overall reaction for each mechanism matches the given reaction so either mechanism could be correct.

The next step is to determine which mechanism has a rate law that matches the experimental data. Write the rate law for the slowest step in each mechanism.

Mechanism I

The slow step (the rate-determining step, sometimes abbreviated RDS) is the first step and is bimolecular in the form of A + B → Products, which will produce the rate law of

$$\text{Rate} = k[W][X]$$

This rate law is not consistent with the rate law determined by experiment; mechanism I is not a possible mechanism for this reaction.

Mechanism II

The slow step is bimolecular and is in the form of A + B → Products, which will produce the rate law of

$$\text{Rate} = k[E][W]$$

However, [E] is an intermediate and may not be included in the rate law. There are two common methods to remove [E] and find the correct rate law. The first method is to use the rate law of the reverse reaction. Because the first step in the mechanism is reversible and fast, the rate of the forward reaction equals the rate of the reverse reaction.

$$\text{Rate (forward)} = k_f[\text{W}][\text{X}]$$

$$\text{Rate (reverse)} = k_r[\text{E}]$$

Rate (forward) = Rate (reverse); therefore, $k_f[\text{W}][\text{X}] = k_r[\text{E}]$. Rearranging the terms gives:

$$[\text{E}] = (k_f/k_r)[\text{W}][\text{X}] = k''\,[\text{W}][\text{X}]$$

Substitute for [E] in the rate law for the slow step in mechanism II.

$$\text{Rate} = k\,[\text{E}][\text{W}]$$

$$\text{Rate} = k/k'[\text{W}][\text{X}][\text{W}]$$

Combining the rate constants gives k'' and [W] × [W] is $[\text{W}]^2$ to get

$$\text{Rate} = k''\,[\text{W}]^2[\text{X}]$$

The second method is using the equilibrium expression that will be discussed in more detail in Chapter 11. The K_c expression is written for the first step:

$$K_c = \frac{[\text{E}]}{[\text{W}][\text{X}]}$$

And then the equation is solved for [E],

$$[\text{E}] = K_c\,[\text{W}][\text{X}]$$

Substitute for [E] in the rate law for the slow step:

$$\text{Rate} = k\,[\text{E}][\text{W}]$$

$$\text{Rate} = k\,\{K_c[\text{W}][\text{X}]\}[\text{W}]$$

The rate constant, k, multiplied by the equilibrium constant, K_c, produces another constant (which you can write as k' or just leave as k if you want) and [W][W] is $[\text{W}]^2$ to get

$$\text{Rate} = k\,[\text{W}]^2[\text{X}]$$

Mechanism II is possible because its rate law agrees with the experimentally determined rate law and the stoichiometry also agrees with the rate-determining step.

Using the equilibrium expression to solve for a reaction intermediate and then substituting into a rate law can be done even if the step that produces the reaction intermediate is not written as an equilibrium step. You assume that it is and do the problem.

EXAMPLE: For the reaction $H_2 + 2\ ICl \rightarrow 2\ HCl + I_2$, determine the rate law for the following mechanism.

(fast) $H_2 + ICl \rightleftarrows HI + HCl$

(slow) $HI + ICl \rightarrow HCl + I_2$

SOLUTION: The slow step is bimolecular, which will produce the rate law of Rate = k[HI][ICl]. However, HI is a reaction intermediate and needs to be removed from the rate law. Assume that the first step is an equilibrium reaction and write the equilibrium expression.

$$K_c = \frac{[HI][HCl]}{[H_2][ICl]}$$

And then the equation is solved for [HI],

$$[HI] = \frac{K_c[H_2][ICl]}{[HCl]}$$

substituting into the rate law:

$$\text{Rate} = k[HI][ICl] \Rightarrow \text{Rate} = k\left(\frac{K_c[H_2][ICl]}{[HCl]}\right)[ICl]$$

$$\text{Rate} = \frac{k''[H_2][ICl]^2}{[HCl]}$$

The product HCl is in the denominator and that indicates that it is acting as an inhibitor to the reaction. As the reaction takes place, the [HCl] increases and interferes with the collisions of the two reactants (H_2 and ICl). This rate law can also be written as Rate = $k[H_2][ICl]^2[HCl]^{-1}$, so the overall order is 2.

DETERMINATION OF ACTIVATION ENERGY

Activation energy can be experimentally determined using the Arrhenius equation

$$\ln(k) = -\frac{E_a}{R}\left(\frac{1}{T}\right) + \ln(A)$$

The rate constant is represented by k

E_a is the activation energy

R is the gas constant, 8.314 J mol^{-1} K^{-1}

T is the Kelvin temperature

A is the frequency factor that is related to the collisions and is temperature dependent

The Arrhenius equation is not found on the formula chart and will not be required in a quantitative manner. The graphical method, as explained here, may be on the exam. One way to determine the activation energy, E_a, is to measure the rate constant, k, at several different temperatures and then graph $\ln(k)$ vs. $1/T$, which gives a straight line with the slope equal to $-E_a/R$.

A catalyst works by (1) changing the geometry of the collisions to make them more favorable, (2) changing the pathway by which the reaction takes place, or (3) a combination of both. Testing a reaction at different temperatures with and without a catalyst and then graphing the data and comparing the two graphs will allow you to determine how a catalyst functions. If the graph with the catalyst has a different slope, the pathway (mechanism) was changed; if it has a different y-intercept, then the geometry of collisions was affected.

MULTIPLE-CHOICE QUESTIONS

No calculators are to be used in this section.

Questions 1 and 2 refer to the following information.

Octane, the major component in gasoline, has the formula C_8H_{18}. There are 18 different isomers of octane. Four of the isomers, labeled as W, X, Y, and Z, convert via an elementary step to produce a product. Each graph shows a different reactant (W, X, Y, and Z) converting to a product.

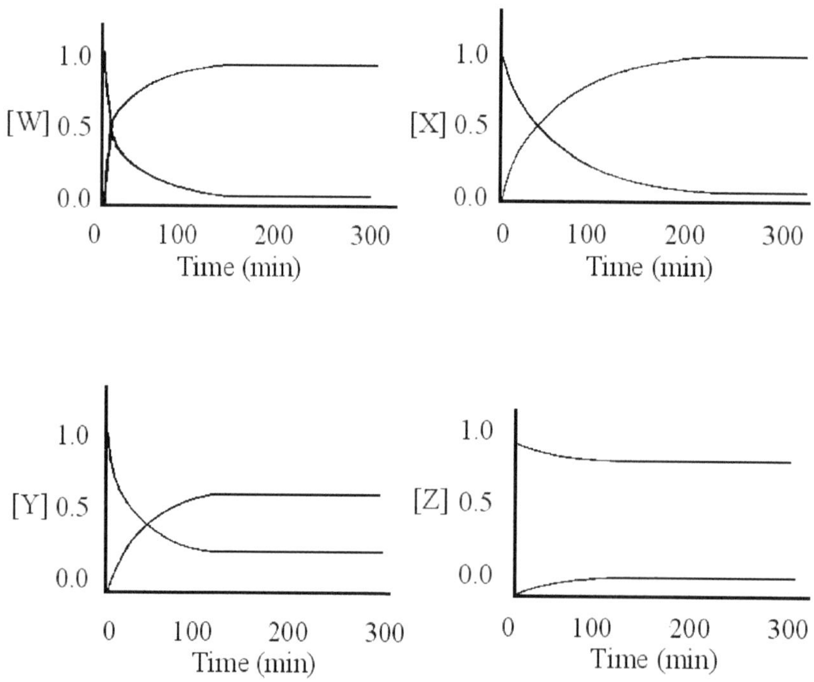

1. Which isomer produces the fastest initial rate for its conversion?
(A) W
(B) X
(C) Y
(D) Z

2. Which isomer's conversion proceeds via a first-order rate law?
(A) W
(B) X
(C) Y
(D) Z

3. The rate law for a certain reaction is determined to be

$$\text{Rate} = k[\text{A}][\text{B}]^2$$

Which statement would most likely describe this reaction?
(A) The reaction is catalyzed by an enzyme.
(B) The reaction involves a 1-step, termolecular mechanism.
(C) The stoichiometry of the overall reaction must be $\text{A} + 2\ \text{B} \rightarrow \text{Products}$.
(D) The reaction involves the formation of intermediates.

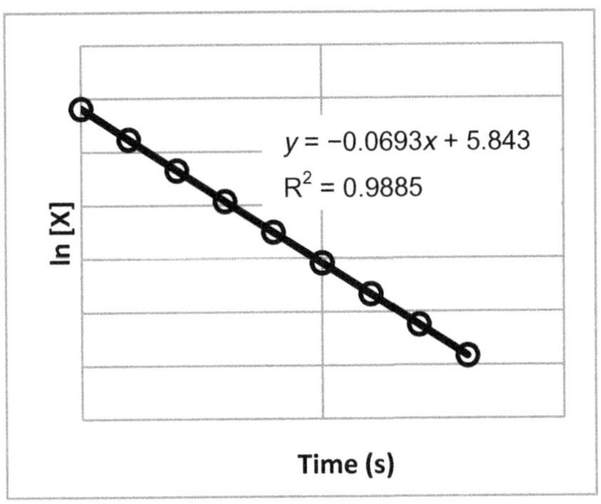

4. A student is investigating the reaction $\text{X} \rightarrow \text{Y}$ and produces the graph shown above. What are the rate law and the half-life of this reaction?
(A) Rate = $k[\text{X}]$; half-life = 0.0693 second
(B) Rate = $k[\text{X}]^2$; half-life = 0.0693 second
(C) Rate = $k[\text{X}]$; half-life = 10.0 seconds
(D) Rate = $k[\text{X}]^2$; half-life = 10.0 seconds

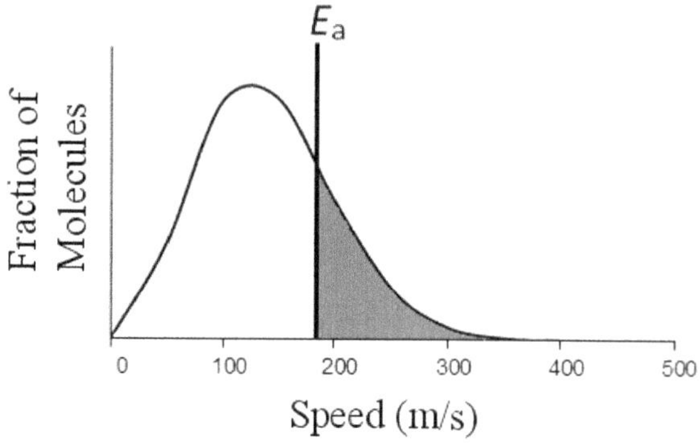

5. The graph above shows the number of molecules that have sufficient energy to overcome E_a at a certain temperature for a given reaction. How would adding an enzyme change the appearance of the graph?

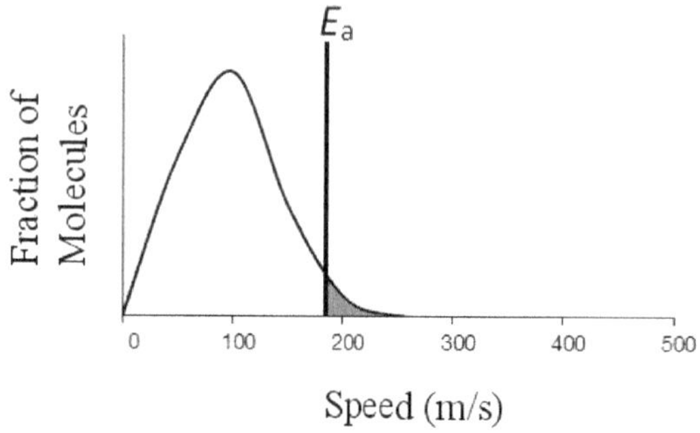

(A)

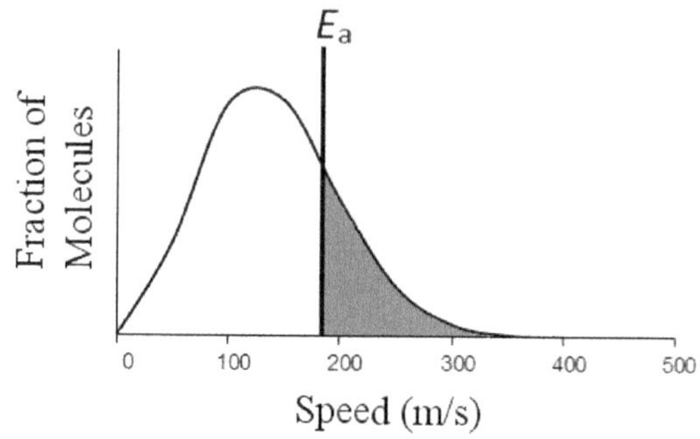

(B)

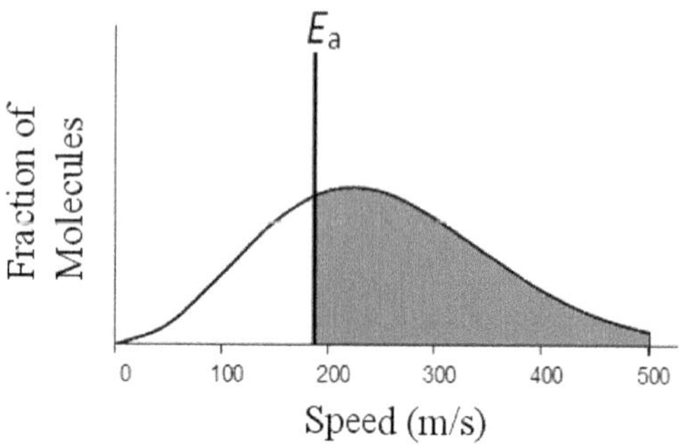

(C)

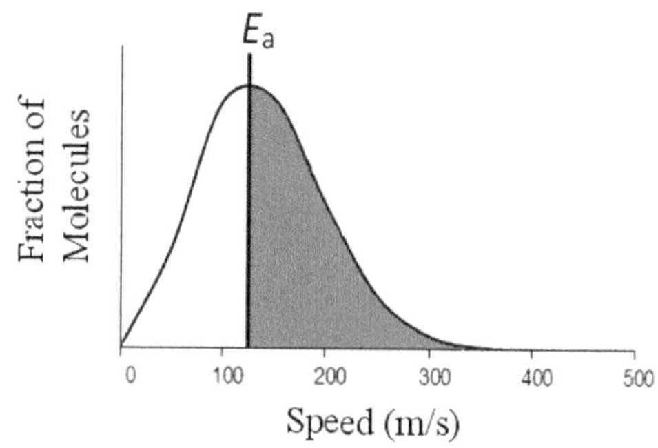

(D)

$$Zn(s) + 2\ HCl(aq) \rightarrow ZnCl_2(aq) + H_2(g)$$

6. The rate of the above reaction can be measured by collecting the hydrogen gas produced in the reaction by water displacement. If 5.0 g of zinc is combined with 100. mL of 1.0 *M* HCl and reacted for 3.0 minutes, which combination of zinc and temperature will produce the greatest volume of gas?
   (A) zinc pieces with average diameter of 0.5 cm and a temperature of 25°C
   (B) zinc pieces with average diameter of 0.5 cm and a temperature of 50°C
   (C) zinc pieces with average diameter of 0.05 cm and a temperature of 25°C
   (D) zinc pieces with average diameter of 0.05 cm and a temperature of 50°C

7. For two first-order chemical reactions of substances A and X,

   $A \rightarrow B \quad t_{1/2} = 30.0$ min

   $X \rightarrow Y \quad t_{1/2} = 60.0$ min

   Which statement is correct?
   (A) The rate constant for $A \rightarrow B$ is lower than the rate constant of $X \rightarrow Y$.
   (B) A certain number of moles of A will react twice as fast as the same number of moles of X.
   (C) In 60.0 minutes, half of X will have been converted to Y and all of A will have been converted to B.
   (D) When the graphs of the natural logs of concentration versus time are compared for A and X, the line representing ln[X] vs. time will have the greater slope.

Questions 8 and 9 refer to the following information.

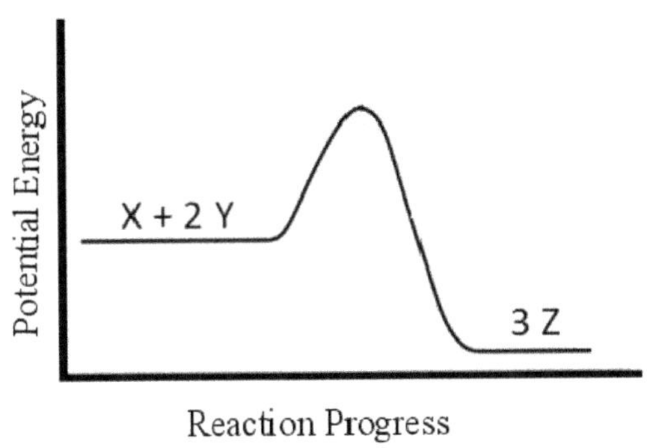

8. A reaction is first order with respect to [X] and second order with respect to [Y]. When [X] = 0.200 *M* and [Y] = 0.200 *M*, the rate is 8.00×10^{-3} *M* min^{-1}. The value of the rate constant, including correct units, is
   (A) 1.00 $M\ min^{-1}$.
   (B) 1.00 $M^{-2}\ min^{-1}$.
   (C) 2.00 $M^{-1}\ min^{-1}$.
   (D) 2.00 $M^{-2}\ min^{-1}$.

9. The activation energy for the reaction, $X + 2\ Y \rightarrow 3\ Z$, could be
   (A) increased by increasing [X] and [Y].
   (B) increased by increasing the temperature.
   (C) decreased by removing Z from the system as it forms.
   (D) decreased by adding a suitable catalyst.

10. Which of the following statements would be correct regarding the following reaction?

$$2\ H_2(g) + O_2(g) \rightarrow 2\ H_2O(g)$$

   (A) The rate of O_2 disappearance is twice the rate of H_2 disappearance.
   (B) The rate of H_2 disappearance is twice the rate of O_2 disappearance.
   (C) The rate of H_2O disappearance is twice the rate of O_2 disappearance.
   (D) The rate of H_2O formation is equal to the rate of O_2 disappearance.

Questions 11–13 refer to the following information.

Carbon monoxide reacts with oxygen according to the equation

$$CO(g) + \tfrac{1}{2}\ O_2(g) \rightarrow CO_2(g)$$

The rate law for the reaction is: Rate = $k[CO]^r[O_2]^s$

Experimental information for the reaction is given in the following table:

	[CO] mol L^{-1}	[O_2] mol L^{-1}	Initial rate, mol L^{-1} min^{-1}
Experiment 1	0.020	0.020	3.68×10^{-5}
Experiment 2	0.040	0.020	1.47×10^{-4}
Experiment 3	0.020	0.040	7.36×10^{-5}

11. Which values correspond to the reaction orders r and s for CO and O_2?
   (A) $r = 1$, $s = ½$
   (B) $r = 1$, $s = 1$
   (C) $r = 2$, $s = 1$
   (D) $r = 4$, $s = 2$

12. Which best describes why this reaction has a very high activation energy?
    (A) The first step in this process is to produce ½ of a mole of oxygen atoms, which requires a large amount of energy.
    (B) The geometric arrangement of the collisions needs to be aligned, and the molecules require sufficient energy to break bonds.
    (C) The reaction requires the formation of two double bonds, which requires a large amount of energy.
    (D) A triple bond and a double bond must both be broken for the reaction to take place.

13. If this reaction were to occur in a single, elementary process, which arrangement of collisions would most likely result in the formation of products for this reaction?

(A)

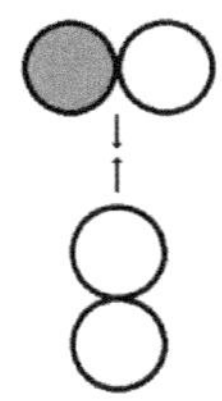

(B)

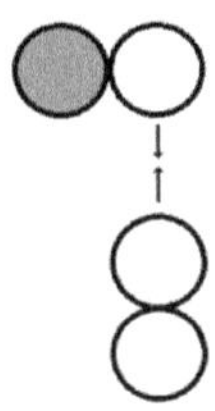

(C)

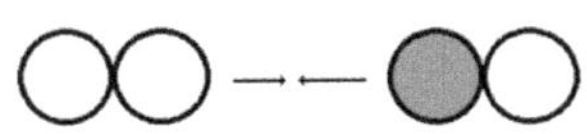

(D)

$Cl_2(g) \leftrightarrows 2\ Cl(g)$	fast equilibrium
$Cl(g) + CHCl_3(g) \rightarrow HCl(g) + CCl_3(g)$	slow
$CCl_3(g) + Cl(g) \rightarrow CCl_4(g)$	fast

14. Given the possible reaction mechanism shown above, what is the rate law for the corresponding reaction?
    (A) Rate = $k[Cl_2]$
    (B) Rate = $k[Cl][CHCl_3]$
    (C) Rate = $k[Cl_2]^{-1}[CHCl_3]$
    (D) Rate = $k[Cl_2]^{1/2}[CHCl_3]$

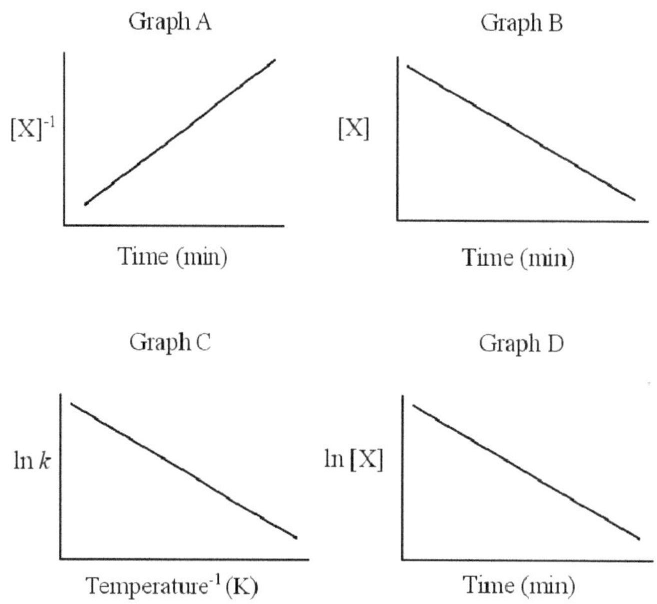

15. Students were asked to investigate the reaction X → Y to determine the activation energy. Each student made a graph, as shown above. Which student-produced graph is most appropriate for this investigation?
(A) Graph A
(B) Graph B
(C) Graph C
(D) Graph D

FREE-RESPONSE QUESTIONS

Calculators may be used for this section.

1. A common chemical demonstration is the "iodine clock" reaction. In one version of that demonstration, the following chemical reaction takes place:

$$S_2O_8^{2-} + 3\ I^- \rightarrow 2\ SO_4^{2-} + I_3^-$$

The kinetics of the reaction are well known, and the order with respect to each reactant is first order. A laboratory investigation was conducted to verify the order with respect to each reactant and to verify that a proposed mechanism is consistent with the data. Data for the kinetics experiment are presented on the next page.

Trial	$[S_2O_8^{2-}]_0$ (mol L^{-1})	$[I^-]_0$ (mol L^{-1})	Initial Rate of Formation of I_3^- (mol L^{-1} min^{-1})
1	0.040	0.060	0.332
2	0.040	0.120	0.662
3	0.080	0.060	0.663

(a) Select one of the two reactants and clearly justify why it is first order.
(b) Write the rate law expression.
(c) Calculate the value of the rate constant and include the units.
(d) Calculate the initial rate of formation of SO_4^{2-} in Experiment 3.
(e) A proposed mechanism is shown below.

Step 1 $I^- + S_2O_8^{2-} \rightarrow IS_2O_8^{3-}$

Step 2 $IS_2O_8^{3-} \rightarrow 2\ SO_4^{2-} + I^+$

Step 3 $I^+ + I^- \rightarrow I_2$

Step 4 $I_2 + I^- \rightarrow I_3^-$

It is known that step 1 is the rate-determining step.

(i) Circle one of the reaction intermediates.

(ii) Is the proposed mechanism consistent with the data presented above? Circle YES or NO and justify your answer.

YES NO

(f) Why is step 1 the rate-determining step? Justify your answer in terms of species present.
(g) Step 3 requires no activation energy for the reaction to take place. Account for this fact.

2. The Haber process is used to make ammonia, an important chemical feedstock for industry. Without the use of a catalyst, the reaction proceeds too slowly to be commercially viable.

The basic reaction is $N_2(g) + 3\ H_2(g) \rightleftarrows 2\ NH_3(g)$. However, the actual Haber process uses an iron catalyst, as shown in the diagram below. N_2 and H_2 adhere (*ad*) to the metal surface and both molecules dissociate. The first three steps of the mechanism are shown below.

Step 1. $N_2(g) \rightleftarrows N_2(ad)$ fast

Step 2. $N_2(ad) \rightleftarrows 2\ N(ad)$ slow

Step 3. $H_2(ad) \rightleftarrows 2\ H(ad)$ fast

Courtesy of the European Space Agency.

(a) Based on bonding principles, propose a reason that step 2 is the rate-determining step.

(b) Based on the diagram, propose another reason that step 2 is the rate-determining step.

Hydrogen atoms migrate across the surface in the following series of reactions.

$$N(ad) + H(ad) \rightleftarrows NH(ad)$$

$$NH(ad) + H(ad) \rightleftarrows NH_2(ad)$$

$$NH_2(ad) + H(ad) \rightleftarrows NH_3(ad)$$

The final step is for the NH_3 to desorb from the catalyst.

$$NH_3(ad) \rightleftarrows NH_3(g)$$

(c) Describe one physical characteristic of the catalyst that would enhance its usefulness to increase the reaction rate.

Reactions that take place on a heterogeneous catalyst generally display the following kinetic behavior:

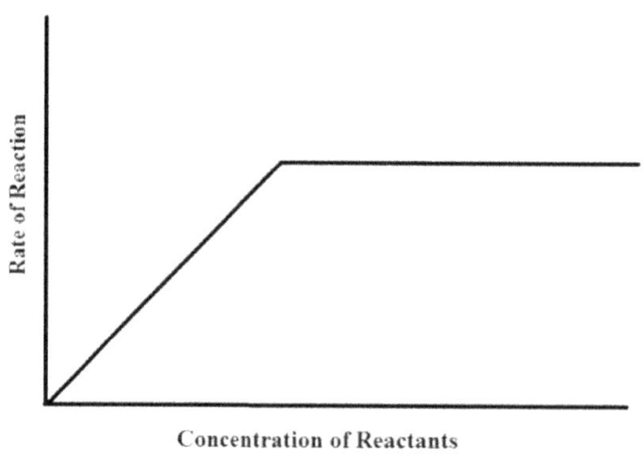

(d) Why does the graph change from a diagonal line to a flat line? In your answer, you may make reference to the catalyst sketch for the Haber process.

Answers

MULTIPLE-CHOICE QUESTIONS

1. A Look for the fastest decrease in reactant concentration or the fastest increase in product formation (LO 4.1).

2. B The half-life of a first-order reaction is constant. Look for the graph that has concentrations of reactants continually decreasing by ½ in a constant amount of time (LO 4.2).

3. D The chance that three particles will collide with correct energy and orientation is very rare, so this reaction most likely involves at least one intermediate (LO 4.7).

4. C Because the graph of ln[X] versus time is linear, the reaction is first order in X. The slope of the graph is $-k$. Half-life for a first-order reaction = $0.693/k$, so 0.693/0.0693 = 10.0 seconds (LO 4.2).

5. D The *average* kinetic energy of a particle is directly related to the Kelvin temperature; however, there will still be a variety of energies. The enzyme lowers E_a so more of the particles have sufficient energy to overcome the energy barrier at a given temperature (LO 4.9).

6. D The reaction will have the highest rate at a high temperature and small particle size. At a high temperature, the particles will have a higher kinetic energy and will collide with greater force and more frequently. A small particle size exposes more surface area to collisions to increase the rate of reaction (LO 4.1).

7. B It takes half as much time for A to form B as for X to form Y, as seen by the smaller half-life. The absolute value of the slope of the graph of the natural log of concentration is equal to k. A greater slope would mean a larger k and a faster reaction. In a chemical reaction, we must compare moles, not mass (LO 4.3).

8. B The rate law is Rate = $k[X][Y]^2$. Solving for the rate constant and substituting data for this reaction:
   $k = \text{Rate}/[X][Y]^2$
   $= 8.00 \times 10^{-3}\,M\,\text{min}^{-1}/(0.200\,M)(0.200\,M)^2$
   $= 0.00800\,M\,\text{min}^{-1}/0.00800\,M^3$
   $= 1.00\,M^{-2}\,\text{min}^{-1}$
   (LO 4.2)

9. D Adding a catalyst suitable for this reaction will increase the speed of the reaction in both directions by lowering the energy barrier (activation energy) by forming a different activated complex (due to a different pathway) which has a lower potential energy (LO 4.8).

10. B Look at the coefficients in the balanced reaction to determine the ratio in which they react (LO 3.4).

11. C Compare Experiments 1 and 2. Doubling the concentration of CO while holding O_2 constant has caused the initial rate to increase by a factor of 4 or 2^2. Therefore, $r = 2$. Compare Experiments 1 and 3. Doubling the concentration of O_2, keeping the [CO] the same, resulted in the initial rate also being doubled, or 2^1, hence $s = 1$. *Note:* When doing this problem on the exam, make sure that all exponents are the same before you start to solve the problem (LO 4.2).

12. D Breaking multiple bonds, such as the triple bond in CO and the double bond in O_2, requires a great amount of energy (LO 5.8).

13. D The C in CO must collide with an oxygen atom in O_2 for the collision to be effective. If 2 oxygen atoms collide, repulsion occurs. If oxygen (very electronegative) collides with the bond (very negative because of shared electrons), repulsion occurs (LO 4.5).

14. D Writing a rate law from the slow step gives Rate = $k[Cl][CHCl_3]$. However, Cl is an intermediate and is not in the overall equation. Something equivalent to Cl must be substituted. In the fast equilibrium step, $k_f\,[Cl_2] = k_r[Cl]^2$, so $k_f\,[Cl_2]^{1/2} = k_r\,[Cl]$. Substituting $[Cl_2]^{1/2}$ for [Cl], the rate law becomes Rate = $k\,[Cl_2]^{1/2}[CHCl_3]$. An alternative approach is to use the equilibrium expression to solve for the concentration of [Cl]. $K_c = \frac{[Cl]^2}{[Cl_2]} \Rightarrow [Cl] = \sqrt{K_c \times [Cl_2]}$, which will result in the order with respect to Cl_2 being ½ (LO 4.7).

15. C Graphs A, B, and D can all be used to determine the order of a reaction as well as its specific rate constant. To determine activation energy, a graph of ln k vs. the reciprocal of temperature in Kelvin is needed (LO 4.6).

FREE-RESPONSE QUESTIONS

1. (a) To determine how each of the concentrations of the reactants is related to the reaction rate, only the one reactant concentration may change.

   $S_2O_8^{2-}$ = Compare $[S_2O_8^{2-}]_0$ in Trials 1 and 3 where $[I^-]$ is held constant. Doubling $[S_2O_8^{2-}]$ increases the rate by a factor of about 2 which indicates the reaction is first order in $S_2O_8^{2-}$ because $2^1 = 2$.

   I^- = Compare $[I^-]_0$ in Trials 1 and 2 where $[S_2O_8^{2-}]$ is held constant. Doubling the concentration of $[S_2O_8^{2-}]$ doubles the rate which indicates that the reaction is first order in I^- because $2^1 = 2$ (LO 4.2).

   (b) Rate = $k\,[S_2O_8^{2-}][I^-]$ (LO 4.2)

(c) Solving the rate law for k yields:

$k = \text{Rate}/([S_2O_8^{2-}][I^-])$

$= 0.332 \text{ mol L}^{-1} \text{ min}^{-1}/(0.040 \text{ mol L}^{-1})(0.060 \text{ mol L}^{-1})$

$= 140 \text{ L mol}^{-1} \text{ min}^{-1}$ or $140\ M^{-1} \text{ min}^{-1}$

(LO 4.2)

(d) From the coefficients in the balanced equation, 2 moles of SO_4^{2-} ions are formed for every 1 mole of I_3^- ion formed. Since the rate of formation of I_3^- is 0.663 M min^{-1} in Trial 3, the rate of formation of SO_4^{2-} is equal to 2 × 0.663, or 1.33 M min^{-1} (LO 3.4).

(e) (i) There are three intermediates. Any one of them may be circled. They are $IS_2O_8^{3-}$, I^+, and I_2. An intermediate is the product of one step of a reaction mechanism that is used up in a later step. Intermediates don't appear in the overall equation (LO 4.7).

(ii) Yes. The rate law must match the rate-determining step. Step 1 is the RDS and the rate law written from the RDS would be Rate = $k[I^-][S_2O_8^{2-}]$ which does match the experimentally determined rate law (LO 4.4, LO 4.7).

(f) Two possible answers for this question.

First answer: The effective collision in step 1 would be much more difficult to achieve than in the other steps because of the complexity of the ion $S_2O_8^{2-}$. Few collisions would have the proper molecular orientation. The collisions in the other steps are much simpler, either involving a single molecule de-composing (step 2) or collisions of simple molecules or ions (steps 3 and 4).

Second answer: The first step involves two anions colliding together. They will likely repel from each other before colliding. This also explains why this reaction is faster at higher temperatures: the higher the temperature, the faster the particles are moving and the less likely the anions are to repel from each other (LO 4.5).

(g) The reaction in step 3 involves two oppositely charged ions that would have great attraction for each other (LO 2.1, LO 4.5).

2. (a) :N≡N: The reaction is thermodynamically unfeasible because a lot of energy would need to be added to break the very stable triple bond (LO 5.8).

(b) The nitrogen atoms, after being adsorbed on the metal catalyst, migrate to between the iron atoms. There is limited space between the adjoining atoms of iron. This limited space slows the reaction down, as a few atoms of nitrogen can fit between the atoms of iron (LO 4.9).

(c) One physical characteristic of the catalyst that will enhance its usefulness is high surface area so the reactants have lots of places to adsorb and react (LO 4.1).

(d) There are a limited number of active sites on the catalyst. Once all available active sites on the catalyst are filled, the reaction rate cannot increase any further (LO 4.9).

9

BIG IDEA 5: THERMOCHEMISTRY

Big Idea 5

The laws of thermodynamics describe the essential role of energy and explain and predict the direction of changes in matter.

Thermochemistry describes the heat flow of a chemical reaction or physical change. Changes in enthalpy are calculated through calorimetry, Hess's law, and standard heats of formation. You may be asked to interpret energy diagrams, or answer lab questions regarding the use of a calorimeter to determine the heat of a reaction.

You should be able to:

- Perform stoichiometric calculations with the enthalpy of the reaction.
- Perform calculations with specific heat.
- Discuss how a calorimeter is used and perform related calculations.
- Draw, label, and perform associated calculations for heating curves involving specific heat and changes in enthalpy for phase changes.
- Draw and label potential energy diagrams for chemical reactions.
- Use Hess's law, standard heats of formation, $\Delta H_f°$, and the bond energy method to determine the heat of a reaction and to identify a process as exothermic or endothermic.
- Write reactions representing standard heats of formation, $\Delta H_f°$.

TEMPERATURE VS. HEAT

Temperature is a measure of the average kinetic energy of particles in a system. *Heat* is the amount of energy transferred from one object to another due to a difference in temperature and always flows from a higher temperature to a lower temperature.

FIRST LAW OF THERMODYNAMICS

The first law of thermodynamics (also known as the law of conservation of energy) states that energy can be converted from one form to another, but can neither be created nor destroyed: The energy of the universe is constant.

ENTHALPY

At constant pressure, the change in enthalpy ($\Delta H°$) of a system is equal to the energy flow as heat. Enthalpy is a state function, so the change in H is independent of the pathway.

There are three terms that are used to discuss the nature of thermodynamics. A **system** is any object or group of objects that is under investigation. The **surroundings** are everything that is not the system, and the **universe** is the system plus the surroundings.

If a salt is placed into the container shown at the right and allowed to dissolve, then everything inside the container would be the system. Everything else (the container, the tabletop, the air, the clock on the wall, even your great Aunt Sally) would be the surroundings. Together, the system and the surroundings make up the universe.

EXOTHERMIC

A reaction that produces heat energy is said to be **exothermic**. Exothermic reactions have several characteristics:

- The heat energy flows from the system to the surroundings.
- The potential energy of the products is lower than the potential energy of the reactants.
- Exothermic reactions give off heat resulting in a change in enthalpy that is negative ($\Delta H° < 0$).
- The temperature of the system and the surroundings increases during an exothermic process.
- A typical graph showing an exothermic reaction is shown on the next page.

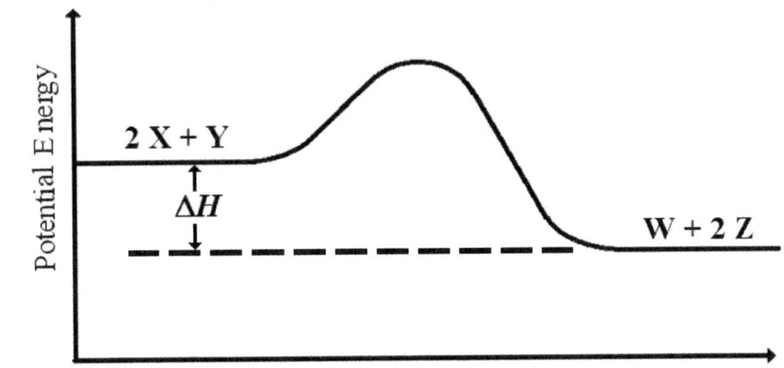

ENDOTHERMIC

A reaction that absorbs heat energy is said to be **endothermic**. Endothermic reactions have several characteristics:

- The heat energy flows from the surroundings to the system.
- The potential energy of the products is higher than the potential energy of the reactants.
- Endothermic reactions absorb heat. The change in enthalpy is positive ($\Delta H° > 0$).
- The temperature of the system and surroundings decreases in an endothermic process.
- A typical graph showing an endothermic reaction is shown below.

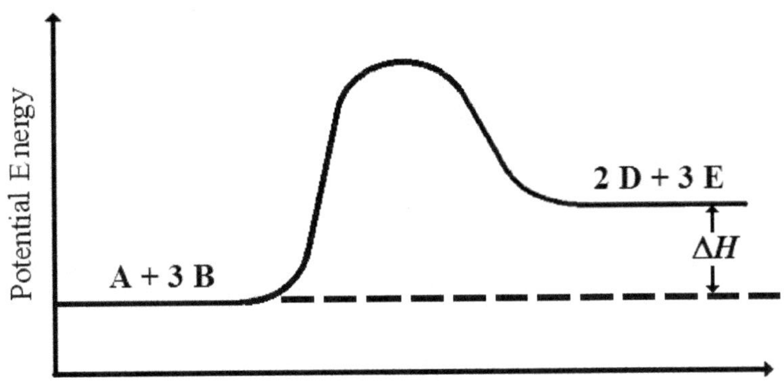

Reaction Coordinate

EXAMPLE: How much heat is released when 4.04 g of hydrogen gas is reacted with excess oxygen gas?

SOLUTION:

$$2\ H_2(g) + O_2(g) \rightarrow 2\ H_2O(l) \qquad \Delta H°_{rxn} = -572\ kJ/mol_{rxn}$$

First, place the heat term into the equation to make a thermochemical equation. Since $\Delta H°_{rxn}$ is negative, the reaction is exothermic; heat energy is being produced and flowing from the system to the surroundings. The heat term will be a product.

$$2\ H_2(g) + O_2(g) \rightarrow 2\ H_2O(l) + 572\ kJ$$

$$4.04g\ H_2 \times \frac{1\ mol\ H_2}{2.02g\ H_2} \times \frac{572kJ}{2\ mol\ H_2} = 572kJ$$

Note: Since the problem asked "How much heat energy is released?", the answer can be 572 kJ. However, if the question asked "How much heat energy is associated with ...?", the answer needs to indicate that the energy is released. This can be done one of two ways: writing "572 kJ of heat energy are released" or using a negative sign on the 572 kJ to indicate that it is exothermic.

Note that conversion factors can be written using the relationship between the enthalpy, $\Delta H°$, and the moles of reactants and products. 572 kJ of heat is produced when 1 mole of O_2 reacts with 2 moles of H_2 to produce 2 moles of H_2O.

Chemical reactions are not the only processes that can be exothermic or endothermic. The dissolving of salts in water can, in part, be considered a physical change. Some salts, like NH_4NO_3 and NaCl, dissolve, and the temperature of the solution decreases, indicating an endothermic process. Energy is being drawn out of the water—the kinetic energy of the water particles is being used to help supply the energy for the solvation of the solute. Since the kinetic energy of the water is decreasing, the temperature of the solution decreases. Some other solutes, such as LiCl, NaOH, and KOH, dissolve in an exothermic process. Heat energy is released as the solute dissolves, flowing into the water and increasing the temperature of the solution. Phase changes are also endothermic or exothermic. During a phase change, the temperature doesn't change; however, energy will flow from the system to the surroundings (in the case of freezing or condensing) or from the surroundings to the system (boiling or melting).

SPECIFIC HEAT

The specific heat capacity, *c* (some textbooks use "*s*", but the formula chart has "*c*" so that is what we will use here), is the amount of energy required to change the temperature of 1.0 g of a substance by 1°C and typically has units of J/(g × °C). Molar heat capacity is the amount of energy required to change the temperature of 1 mole of a substance by 1°C [J/(mol × °C)].

EXAMPLE: The specific heats of solid aluminum, solid iron, mercury liquid, and carbon graphite are:

Element	Specific Heat J/(g × °C)
Al(*s*)	0.89
Fe(*s*)	0.45
Hg(*l*)	0.14
C(*s*)	0.71

When the same amount of heat energy is applied to 1.0 g of each of these substances, which one will reach the highest temperature?

SOLUTION: The answer is Hg(*l*). Mercury has the lowest specific heat of the four mentioned substances, which means it takes the least amount of heat to raise 1 g of mercury 1°C.

CALORIMETRY

A calorimeter is a device used to experimentally determine (1) the heat energy associated with a chemical reaction or (2) the specific heat of an object. The "system" is either the chemical reaction or the object of which the specific heat is being determined. The "surroundings" are usually water or another liquid with a high boiling point, along with the calorimeter itself and the surrounding air. Most calorimetry is carried out under constant pressure (calorimeter is open to the atmosphere). By observing the temperature change in a well-insulated calorimeter and using the relationship that the change in the heat energy, q, of the surroundings is equal and opposite in sign to the heat released or absorbed by the reaction (the system), the following equations can be used.

$$q = m \times c \times \Delta T$$

$$\Delta H = \frac{-q}{\text{moles of limiting reagent}}$$

Where:

m = mass of solution
c = the specific heat capacity of the solution
ΔT = the temperature change in the solution during the reaction

AP® Tip

When working a problem with ΔT, it is not necessary to change °C to K as the difference between units is the same in °C and K. All other mathematical problems in chemistry involving temperature must use Kelvin units.

EXAMPLE: A 1.00-g sample of magnesium metal is added to a calorimeter containing 100. mL of hydrochloric acid (specific heat capacity 4.184 J/(g × °C), initially at 25.0°C. The solution temperature increases to 29.5°C. What is the heat of reaction, in kJ/mol$_{rxn}$? Assume that the hydrochloric acid is in excess and that the density of the solution is 1.00 g/mL.

SOLUTION: First, determine the total mass of the solution: 100. g HCl + 1.00 g Mg = 101 g. The heat gained by the solution is equal to:

$$101 \text{ g solution} \times \frac{4.184 \text{ J}}{\text{g °C}} \times (29.5\text{°C} - 25.0\text{°C}) = 1902 \text{ J} = 1.9 \text{ kJ}$$

Note: This answer will have only two significant figures since the determination of ΔT will produce an answer (4.5°C) with only two significant figures.

Since the solution's temperature increased, the reaction must be exothermic. The amount of heat released by the reaction is equal and opposite to the heat gained by the solution:

– (heat gained by solution) = (heat of reaction) = –1.9 kJ

This is the heat of reaction per gram of Mg reacted. The heat of reaction per mole of Mg reacted is:

$$\frac{-1.9 \text{ kJ}}{1 \text{ g}} \times \frac{24.31 \text{ g}}{1 \text{ mol}} = -46 \text{ kJ/mol}$$

AP® Tip

Students sometimes have trouble deciding what numbers to use when calculating ΔH from calorimetry data. Remember that the thermometer is in the surroundings, so the "mass" should be the mass of water or other heat-transfer fluid. The masses of reactants and products are part of the system and must be included in this mass. By convention, the change in temperature, ΔT, is calculated $T_{final} - T_{initial}$. According to the first law of thermodynamics, the heat lost by one substance equals the heat gained by the other substance, so ΔH will be equal in magnitude but opposite in sign to q.

Memory Trick

An easy way to remember the calculations for ΔT is to use alphabetical order: f comes before i, so it will be $T_{final} - T_{initial}$.

HESS'S LAW

Enthalpy is a state function, which means that it only depends on the starting position (reactants) and ending position (products). It is independent of the path by which you go from products to reactants. Hess's law states that the enthalpy change for an overall process is the sum of the changes in enthalpy for each step in the process. When doing a Hess's law problem, the idea is to manipulate the equations given in the problem statement so that they add up to the overall reaction. There are two ways that the equations can be manipulated.

1. Reverse the reaction. The sign of ΔH is also reversed.
2. Multiply the coefficients in a balanced reaction by an integer. The value of ΔH is multiplied by the same integer.

AP® Tip

Steps for Hess's law:

1. Put the "target" equation at the top of the problem.
2. Select one species that is in the target equation and appears in only one of the given equations.
3. Start with that equation on your list, flipping it if needed to get the species in the correct location.
4. Repeat steps 2 and 3 until all equations are placed.
5. If you cannot repeat steps 2 and 3 to get all of the equations placed, then use the equation(s) that you have already placed and look for species that need to be cancelled out, and place other equations to cancel those species.

EXAMPLE: Given the following reactions and $\Delta H°$ values,

	Equation	$\Delta H°_{rxn}$
(1)	$Si(s) + 2\ H_2(g) \rightarrow SiH_4(g)$	+34 kJ/mol
(2)	$Si(s) + O_2(g) \rightarrow SiO_2(s)$	–911 kJ/mol
(3)	$H_2(g) + \frac{1}{2}\ O_2(g) \rightarrow H_2O(g)$	–242 kJ/mol

Calculate ΔH for

$SiH_4(g) + 2\ O_2(g) \rightarrow SiO_2(s) + 2\ H_2O(g)$ $\qquad \Delta H° = ?$

SOLUTION:

Target: $SiH_4(g) + 2\ O_2(g) \rightarrow SiO_2(s) + 2\ H_2O(g)$

1. Start by finding a substance that appears only once in all of the reactions; in this case, SiO_2, SiH_4, or H_2O would all be a good starting point. We are going to start with H_2O.

$H_2(g) + \frac{1}{2}\ O_2(g) \rightarrow H_2O(g)$ $\qquad$ –242 kJ/mol

Look to see where that substance (H_2O) is in the target equation (product side) and in what amount (need 2 of them). Modify the reaction so that the substance appears where it should be and in the correct amount. Do not use ( ) around the entire equation ... it is better if you multiply by 2 and change all of the coefficients—this will avoid a common error on these types of problems and maximize your points on this problem.

$$2\ H_2(g) + O_2(g) \rightarrow 2\ H_2O(g) \qquad 2(-242\text{ kJ/mol})$$

2. SiH_4 will be the next species to be placed. Look to see where that substance (SiH_4) is in the target equation (reactant side) and in what amount (need 1 of them). Modify the reaction so that the substance appears where it should be and in the correct amount. SiH_4 needs to be a reactant, so Reaction 1 is reversed—remember to change the sign on the ΔH term.

$$SiH_4(g) \rightarrow Si(s) + 2\ H_2(g) \qquad -(34\text{ kJ/mol})$$

3. SiO_2 is the last substance to be placed. Look to see where that substance (SiO_2) is in the target equation (product side) and in what amount (need 1 of them). Reaction 2 is not changed.

$$Si(s) + O_2(g) \rightarrow SiO_2(s) \qquad -911\text{ kJ/mol}$$

4. The last step would be to make sure that the equations add up to give you the target equation and then to add the $\Delta H°$ values to give you the overall $\Delta H°_{rxn}$.

Putting the whole problem together, it would like the following:

Target: $SiH_4(g) + 2\ O_2(g) \rightarrow SiO_2(s) + 2\ H_2O(g)$	$\Delta H°$
$2\ H_2(g) + O_2(g) \rightarrow 2\ H_2O(g)$	2(–242 kJ/mol)
$SiH_4(g) \rightarrow Si(s) + 2\ H_2(g)$	–(34 kJ/mol)
$Si(s) + O_2(g) \rightarrow SiO_2(s)$	–911 kJ/mol

Overall:

$$2\ H_2(g) + O_2(g) + SiH_4(g) + Si(s) + O_2(g) \rightarrow$$
$$SiO_2(s) + Si(s) + 2\ H_2(g) + 2\ H_2O(g)$$

Cancel the $Si(s)$:

$$2\ H_2(g) + O_2(g) + SiH_4(g) + \cancel{Si(s)} + O_2(g) \rightarrow$$
$$SiO_2(s) + \cancel{Si(s)} + 2\ H_2(g) + 2\ H_2O(g)$$

Cancel the $H_2(g)$:

$$\cancel{2\ H_2(g)} + O_2(g) + SiH_4(g) + O_2(g) \rightarrow SiO_2(s) + \cancel{2\ H_2(g)} + 2\ H_2O(g)$$

Group the $O_2(g)$:

$$SiH_4(g) + 2\ O_2(g) \rightarrow SiO_2(s) + 2\ H_2O(g)$$

Since the overall equation is the same as the target equation, add up the $\Delta H°$ values to get the overall $\Delta H°_{rxn}$. $\Delta H°_{rxn} = (2 \times -242) + -34 + -911 = -1429$ kJ/mol$_{rxn}$. This reaction is exothermic.

SPECIFIC TYPES OF ENTHALPIES

ENTHALPY OF FORMATION: The standard enthalpy of formation of a substance is the change in enthalpy that accompanies the formation of 1 mole of a substance from its elements with all in their standard states.

The naught symbol (looks like a degree symbol) on $\Delta H°$ indicates that the process occurred under standard conditions of 25°C, 1 atm, and 1 M solutions.

By definition, the standard heat of formation for elements in their standard states equals zero.

EXAMPLE: Write the balanced molecular equation representing the standard heat of formation reaction of ethanol, $\Delta H_f°$, $C_2H_5OH(l)$.

SOLUTION:

$$2\ C(s) + 3\ H_2(g) + \tfrac{1}{2}\ O_2(g) \rightarrow C_2H_5OH(l)$$

Note that 1 mole of product is produced according to the definition of the standard heat of formation. All of the reactants are elements in their standard states.

Standard heats of formation can be used to calculate the enthalpy change of a reaction by subtracting the enthalpies of formation of the reactants from the enthalpies of formation of the products.

$$\Delta H^{\circ}_{rxn} = \sum \Delta H^{\circ}_{f_{products}} - \sum \Delta H^{\circ}_{f_{reactants}}$$

EXAMPLE: Using the standard heats of formation, $\Delta H_f°$, calculate the change in enthalpy, $\Delta H°$, for the following reaction.

$$2\ H_2O_2(l) \rightarrow 2\ H_2O(l) + O_2(g)$$

The standard heats of formation values for compounds are generally given on the exam in one of two manners, a table of values or by providing the reactions.

Table:

Substance	Standard Heat of Formation, $\Delta H_f°$ (kJ/mol)
$H_2O_2(l)$	–187
$H_2O(l)$	–285

Equations:

$H_2(g) + O_2(g) \rightarrow H_2O_2(l)$ $\Delta H°_{rxn} = -187$ kJ/mol

$H_2(g) + \frac{1}{2} O_2(g) \rightarrow H_2O(l)$ $\Delta H°_{rxn} = -285$ kJ/mol

$$2\ H_2O_2(l) \rightarrow 2\ H_2O(l) + O_2(g)$$

SOLUTION:

$$\Delta H°_{reaction} = [2(-285\ kJ) + 0] - [2(-187\ kJ)] = -196\ kJ$$

EXAMPLE: Given the information below, calculate the heat of formation of gaseous carbon monoxide.

$$2\ CO(g) + C(s) \rightarrow C_3O_2(g) \rightarrow \Delta H° = 127.3\ kJ$$

$\Delta H°_f$ for carbon suboxide, $C_3O_2(g)$ is –93.7 kJ/mol.

SOLUTION:

$$\Delta H^{\circ}_{rxn} = \sum \Delta H^{\circ}_{f_{products}} - \sum \Delta H^{\circ}_{f_{reactants}}$$

$$\Delta H°_{rxn} = [\Delta H_f° \text{ for } C_3O_2(g)] - \{[2 \times \Delta H_f° \text{ for } CO(g)] - \Delta H_f° \text{ for } C(s)\}$$

Rearrange the equation and solve for the missing variable; note $\Delta H_f°$ for C(s) is zero.

$$\Delta H_f° \text{ for } CO(g) = [\Delta H_f° \text{ for } C_3O_2(g) - \Delta H°_{rxn}]/2$$

$$= (-93.7\ kJ - 127.3\ kJ)/2 = -110.5\ kJ/mol$$

ENTHALPY OF COMBUSTION: The enthalpy of combustion, $\Delta H°_{comb}$, of a compound is the change in enthalpy that accompanies the combustion of 1 mole of a compound.

EXAMPLE: Write the balanced molecular equation representing the $\Delta H°_{comb}$, for ethanol, $C_2H_5OH(l)$.

SOLUTION:

$$C_2H_5OH(l) + 3\ O_2(g) \rightarrow 3\ H_2O(g) + 2\ CO_2(g)$$

EXAMPLE: Write the balanced molecular equation representing the $\Delta H°_{comb}$, for (i) carbon, (ii) hydrogen gas, and (iii) magnesium.

SOLUTION:

(i) $C_{graphite}(s) + O_2(g) \rightarrow CO_2(g)$

(ii) $H_2(g) + ½\ O_2(g) \rightarrow H_2O(g)$

(iii) $Mg(s) + ½\ O_2(g) \rightarrow MgO(s)$

Note that in each case, there is only 1 mole of each product produced from its elements in their standard states. Therefore,

(i) $\Delta H°_{comb}[C_{graphite}] = \Delta H_f°[CO_2(g)]$

(ii) $\Delta H°_{comb}[H_2(g)] = \Delta H_f°[H_2O(g)]$

(iii) $\Delta H°_{comb}[Mg(s)] = \Delta H_f°[MgO(s)]$

If the problem provides a table of enthalpy of combustion values, this is another method of providing you with the enthalpy of formation values.

BOND ENERGIES AND ENTHALPY

THE ENERGETICS OF BOND FORMATION

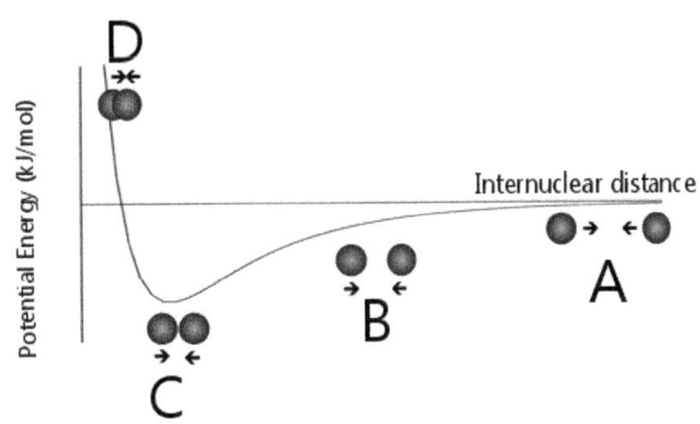

A chemical bond forms when a system of bonded atoms is lower in potential energy than that of the independent atoms. The optimum distance between atoms is the bond length, which represents this lowest energy state. The bond length is a balance between the attractive electrostatic forces between the nucleus of 1 atom and the electrons of another, and the repulsive forces between the positively charged nuclei of the 2 atoms and negatively charged electrons of the 2 atoms.

In the diagram above, at point A, the nucleus of each atom is attracting the valence electrons of the other atom. As they move closer together (point B), the electrostatic attraction increases between the nuclei and the valence electrons of the other atom. At point D, the valence electrons do not effectively screen the 2 nuclei and repulsive forces between the 2 nuclei are the predominate force. At point C, the repulsive forces between the 2 nuclei and the attractive forces between the valence electrons and the nuclei are in balance. Point C provides you with the bond length and the bond energy. All bond formation is exothermic; hence, breaking bonds (the reverse process) must be endothermic. Note also that the graph asymptotes at $y = 0$ due to Coulomb's law and the distance approaching infinite.

Breaking bonds requires an input of energy to overcome the attractive forces. When new bonds are formed, energy is released. The difference between input and output determines whether a process is endothermic or exothermic.

CALCULATING ENTHALPY FROM BOND ENERGIES

The change of enthalpy for a reaction can also be calculated from the bond energies of the reactants and the products. This method requires a bond energy table, similar to the one shown below. The $\Delta H°_{rxn}$ calculated from bond energies and the $\Delta H°_{rxn}$ calculated from a table of enthalpy of formations will not produce the same value—but their values will be close to each other. This is due to the experimental methods used to determine bond energies.

Bond Energies (kJ/mol)

Single Bonds						Multiple Bonds	
H—H	432	N—H	391	I—I	149	C=C	614
H—F	565	N—N	160	I—Cl	208	C≡C	839
H—Cl	427	N—F	272	I—Br	175	O=O	495
H—Br	363	N—Cl	200			C=O*	745
H—I	295	N—Br	243	S—H	347	C≡O	1072
		N—O	201	S—F	327	N=O	607
C—H	413	O—H	467	S—Cl	253	N=N	418
C—C	347	O—O	146	S—Br	218	N≡N	941
C—N	305	O—F	190	S—S	266	C≡N	891
C—O	358	O—Cl	203			C=N	615
C—F	485	O—I	234	Si—Si	340		
C—Cl	339			Si—H	393		
C—Br	276	F—F	154	Si—C	360		
C—I	240	F—Cl	253	Si—O	452		
C—S	259	F—Br	237				
		Cl—Cl	239				
		Cl—Br	218				
		Br—Br	193				

*C=O in CO_2 has a bond energy of 799.

To use this method, draw Lewis structures of the reactants and products, if they are not already provided in the problem. You will use the equation

$$\Delta H^{\circ}_{rxn} = \sum \textit{Bond Energies}_{\text{bonds broken}} - \sum \textit{Bond Energies}_{\text{bonds formed}}$$

Memory Trick

Alphabetical order: Bonds **B**roken then Bonds **F**ormed.

It is best to make a table of the bonds on each side of the arrow or to cross them out as you do the calculation. This will help to avoid a common mistake of missing a bond in the calculations.

EXAMPLE: Estimate the change in enthalpy, ΔH, for the following reaction using the table of bond energies.

$$N_2 + 3\ H_2 \rightarrow 2\ NH_3$$

SOLUTION: First, draw the Lewis structures for the reactants and products.

$$N_2 + 3\ H_2 \rightarrow 2\ NH_3$$

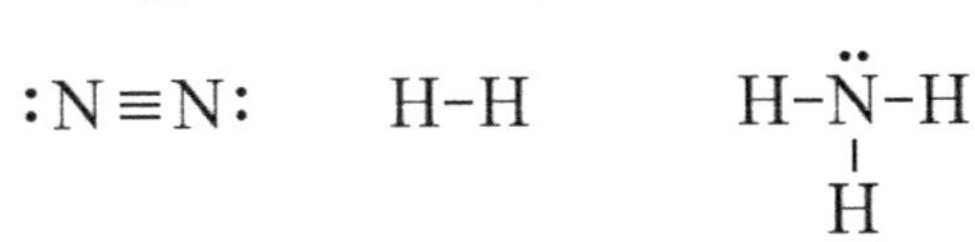

Second, total the number of bonds that need to be broken and formed, remembering to take into account the coefficients from the balanced equation. In the reactants, one triple bond between the atoms of nitrogen and one single bond between atoms of hydrogen are broken. Bond breaking is an endothermic process that has a positive value for enthalpy. Energy must be added to break the bonds. Three single bonds between hydrogen and nitrogen are formed in the products. Bond formation is exothermic having a negative value for enthalpy. Make a chart similar to this:

Bonds Broken	Bond Energy	Bonds Formed	Bond Energy
N≡N	1 × 941	N–H	6 × 391
H–H	3 × 432		
Total	2237 kJ		2346 kJ

$$\Delta H = 2237\text{ kJ} - 2346\text{ kJ} = -109\text{ kJ/mol}_{rxn}$$

An alternative way to do this type of problem is to simply use positive (+) values for bonds broken and negative (–) values for bonds formed and add them together.

INTERMOLECULAR FORCES

Intermolecular forces are attractions between molecules that arise when molecules are near each other. Breaking these forces requires the input of energy, as illustrated with the heating curve on the next page. One way to determine the relative strengths of intermolecular forces is to measure how much energy is required to cause a phase change. However, the bonds between the atoms within a molecule (*intra*molecular forces) are not broken, so the molecules themselves remain intact. Intermolecular forces are discussed in more detail in Chapter 2, Big Idea 2 in this book.

HEATING CURVES

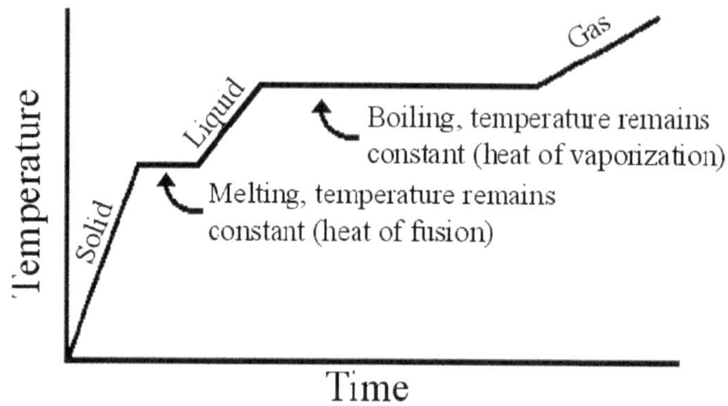

When a system is heated, energy is transferred into it. In response to the energy it receives, the system changes, for example, by increasing its temperature. A plot of the temperature versus time is called a **heating curve**. One such heating curve is shown above. The physical states of the substance and the phase transitions are identified along the curve. If the graph went from a high temperature to a low temperature, it would be a cooling curve, as shown below.

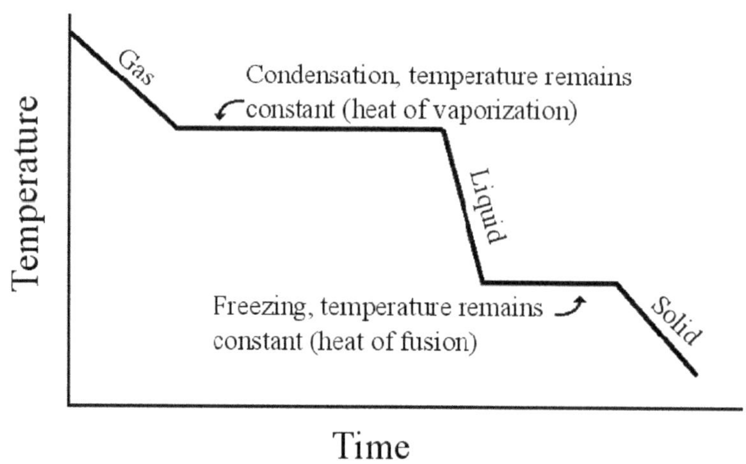

The temperature of the system usually increases when energy is applied. However, when the energy absorbed is used for phase transition, a change in the physical state, the temperature (average kinetic energy) remains constant because the potential energy of the system is being increased as the molecules are rearranged in the phase change. Phase changes only affect the intermolecular forces of attraction; no intramolecular bonds are broken.

The heat of fusion (also called enthalpy of fusion, $\Delta H°_{fus}$) is the change in enthalpy of a substance resulting from the amount of energy needed for the change of state from a solid to a liquid. The heat of vaporization (also called enthalpy of vaporization, $\Delta H°_{vap}$) is the change in enthalpy of a substance resulting from the amount of energy needed for the change of state from a liquid to a gas. The units are generally J/g or kJ/mol.

The calculation for the amount of heat energy is similar to a specific heat problem. For specific heat, $q = mc\Delta T$. However, for heat of fusion or vaporization problems, since there is no change in temperature, effectively, the ΔT is removed from the equation. We are left with $q = m(\Delta H°_{fus})$ and $q = m(\Delta H°_{vap})$. If the values of the heats of fusion and vaporization are given in kJ/mol, the mass must be converted to moles and the equations would be $q = n(\Delta H°_{fus})$ and $q = n(\Delta H°_{vap})$.

HEAT OF FUSION

The heat of fusion, ΔH_{fus}, is the enthalpy change that occurs in melting a solid at its melting point. Stronger intermolecular forces result in higher heats of fusion.

$$\uparrow \text{IMF} = \uparrow \Delta H_{fus}$$

EXAMPLE: What quantity of heat is required to melt 1.00 kg of ice at its melting point? For ice, $\Delta H_{fus} = 6.02$ kJ/mol.

SOLUTION:

$$1.00 \text{ kg} \times \frac{1000 \text{ g}}{1 \text{ kg}} \times \frac{1 \text{ mol}}{18.0 \text{ g}} \times \frac{6.02 \text{ kJ}}{1 \text{ mol}} = 334 \text{ kJ}$$

HEAT OF VAPORIZATION

The heat of vaporization, ΔH_{vap}, is the energy required to vaporize 1 mole of a liquid at a pressure of 1 atm. Stronger intermolecular forces result in higher heats of vaporization.

$$\uparrow \text{IMF} = \uparrow \Delta H_{vap}$$

EXAMPLE: What quantity of heat is required to vaporize 130. g of water? For liquid water, $\Delta H_{vap} = 43.9$ kJ/mol.

SOLUTION:

$$130 \text{ kg} \times \frac{1 \text{ mol}}{18.0 \text{ g}} \times \frac{43.9 \text{ kJ}}{1 \text{ mol}} = 317 \text{ kJ}$$

EXAMPLE: The substance, X, has the following properties:

	Specific Heat
ΔH_{vap} = 20.0 kJ/mol	Solid = 3.0 J/(g °C)
ΔH_{fus} = 5.0 kJ/mol	Liquid = 2.5 J/(g °C)
Boiling point = 75°C	Gas = 1.0 J/(g °C)
Melting point = –15°C	

Calculate the energy required to convert 250.0 g of substance X from a solid at –50.0°C to a gas at 100°C. Assume that X has a molar mass of 75.00 g/mol.

SOLUTION:

There are five steps involved.

1. Heating the solid from –50.0°C to –15°C.
2. Melting the solid at –15°C.
3. Heating the liquid to its boiling point from –15°C to 75°C.
4. Boiling the liquid at 75°C.
5. Heating the gas from 75°C to 100°C.

The energy used is the sum of the energies from the individual steps.

There are 3.333 moles of X in 250.0 g of X.

Step 1: $q = m \times c \times \Delta T$; 250.0 g × [3.0 J/(g °C)] × 35°C = 26 kJ

Step 2: mol × ΔH_{fus} = 3.333 mol × 5.0 kJ/mol = 17 kJ

Step 3: $q = m \times c \times \Delta T$; 250.0 g × [2.5 J/(g °C)] × 90°C = 56 kJ

Step 4: mol × ΔH_{vap} = 3.333 mol × 20.0 kJ/mol = 66.7 kJ

Step 5: $q = m \times c \times \Delta T$; 250.0 g × [1.0 J/(g °C)] × 25°C = 6.3kJ

The total energy required is the sum of the energies of the individual steps:

$$(26 + 17 + 56 + 66.7 + 6.3)\text{ kJ} = 172\text{ kJ}$$

MULTIPLE-CHOICE QUESTIONS

No calculators may be used in this part of the exam.

1. The enthalpy change of which reaction represents the standard enthalpy of formation for nitrogen dioxide?
   (A) ½ $N_2(g) + O_2(g) \rightarrow NO_2(g)$
   (B) $N_2(g) + 2\ O_2(g) \rightarrow N_2O_4(g)$
   (C) $N_2(g) + 2\ O_2(g) \rightarrow 2\ NO_2(g)$
   (D) $NO(g)$ + ½ $O_2(g) \rightarrow NO_2(g)$

2. The enthalpies of vaporization for water and ammonia are given below. What accounts for the much higher value for water?

Substance	Enthalpy of Vaporization (kJ/mol)
H_2O	40.7
NH_3	23.4

(A) Water's bent shape and 2 lone pairs of electrons allow more hydrogen bonding than ammonia's trigonal pyramidal shape and 1 lone pair of electrons; hence, it has a higher dipole moment.
(B) Water has a higher molar mass than ammonia so it requires more energy for water molecules to reach the same kinetic energy as the ammonia molecules.
(C) Water has more polarizable electrons due to its bent shape being more spread out than ammonia's more compact trigonal pyramidal shape; hence, it has a higher dipole moment.
(D) Water is a liquid at room temperature and ammonia is a gas. A gas requires less energy to vaporize at a lower temperature than a liquid does, so water will require more energy to vaporize.

3. For endothermic reactions at constant pressure
(A) heat energy flows from the system to the surroundings and $\Delta H < 0$.
(B) heat energy flows from the system to the surroundings and $\Delta H > 0$.
(C) heat energy flows from the surroundings to the system and $\Delta H < 0$.
(D) heat energy flows from the surroundings to the system and $\Delta H > 0$.

4. Why is the heat of vaporization always higher than the heat of fusion for the same substance?
(A) When a substance boils, it requires more energy to break the intramolecular bonds.
(B) When a substance melts, it takes less energy than when a substance boils due to the particles vibrating more.
(C) When a substance melts, only some of the intermolecular forces are overcome, but when a substance vaporizes, all of the intermolecular forces are overcome.
(D) When a substance melts, only the intermolecular forces have to be overcome, but when a substance boils, both the intermolecular and intramolecular forces have to be overcome.

5. The heats of vaporization for carbon tetrachloride and ammonia are shown below. What accounts for the much higher value for carbon tetrachloride?

Substance	Heat of Vaporization (kJ/mol)
CCl_4	30.0
NH_3	23.4

(A) Carbon tetrachloride's C–Cl bonds are more polar than the N–H bonds in ammonia.
(B) Carbon tetrachloride has a higher molar mass than ammonia so it requires more energy to get the carbon tetrachloride molecules to the same kinetic energy as the ammonia molecules.
(C) The dispersion forces in carbon tetrachloride are stronger than the hydrogen bonding in ammonia due to carbon tetrachloride having more polarizable electrons.
(D) Carbon tetrachloride does not have a lone pair of electrons on the central atom and exhibits sp^3 hybridization, whereas ammonia has one lone pair of electrons on the central atom and exhibits sp^2 hybridization. The greater the degree of hybridization, the higher the heat of vaporization due to molecular orbital interactions.

Questions 6–8 use the following information.

A student is asked to determine the enthalpy of the reaction between Zn(*s*) and HCl(*aq*). She wrote in her laboratory notebook the following observations and made the drawing shown on the next page.

- The zinc metal reacted with HCl and disappeared in approximately 1 minute.
- There was a production of gas. I could see the bubbles coming from the surface of the zinc metal.
- When I put my finger over the second hole in the rubber stopper gas would build up, and when I removed it I could hear a puff or hiss sound.
- The bubbles occurred very rapidly at first and then slowed down as the reaction proceeded.
- The outside of the container became hot to the touch.
- After holding my finger over the second hole in the rubber stopper for 30 seconds, I removed my finger and brought a lighted wooden splint near the top of the flask. A loud "popping" noise could be heard.

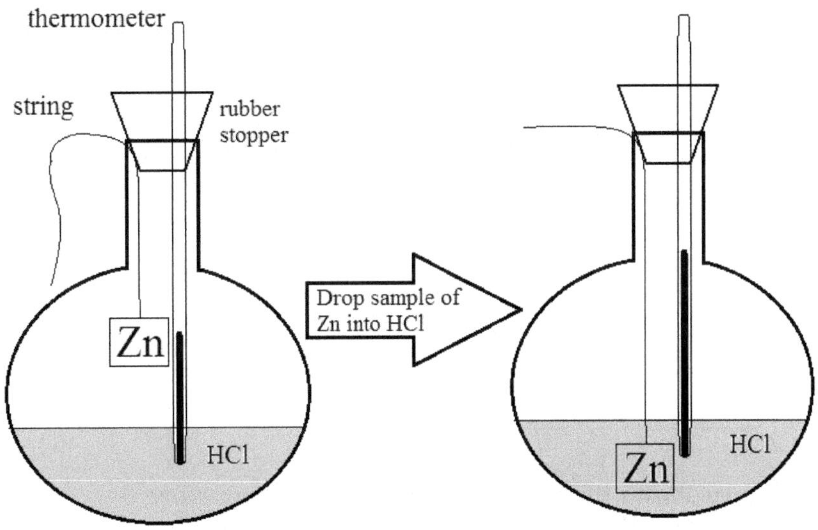

6. What is the sign of the $\Delta H°_{rxn}$, and what provided the experimental evidence to support the student's conclusion?
   (A) $\Delta H°_{rxn}$ is positive; the disappearance of the Zn provides the experimental evidence.
   (B) $\Delta H°_{rxn}$ is positive; the production of bubbles provides the experimental evidence.
   (C) $\Delta H°_{rxn}$ is negative; the loud "popping" noise provides the experimental evidence.
   (D) $\Delta H°_{rxn}$ is negative; the outside of the container becoming hot provides the experimental evidence.

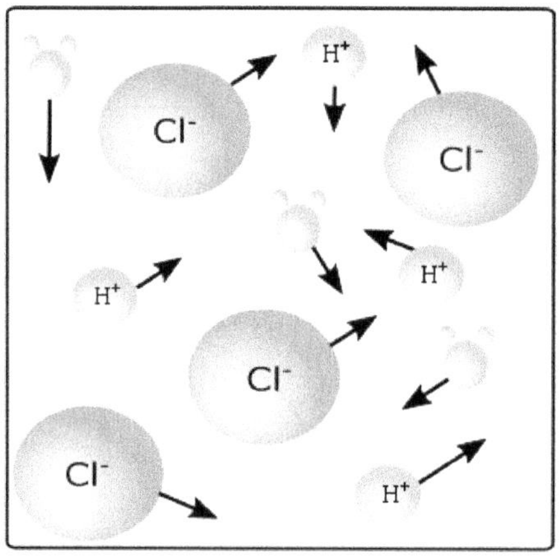

7. The diagram above shows the hydrochloric acid solution before the reaction takes place. Which diagram below most correctly shows the particles in solution in the flask after the reaction has taken place?

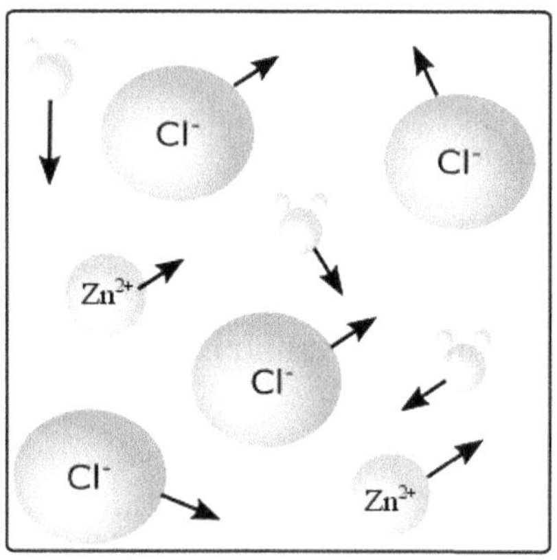

(A) Diagram A

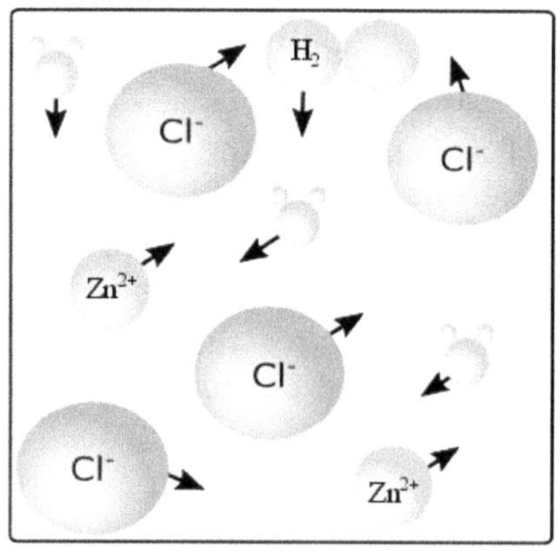

(B) Diagram B

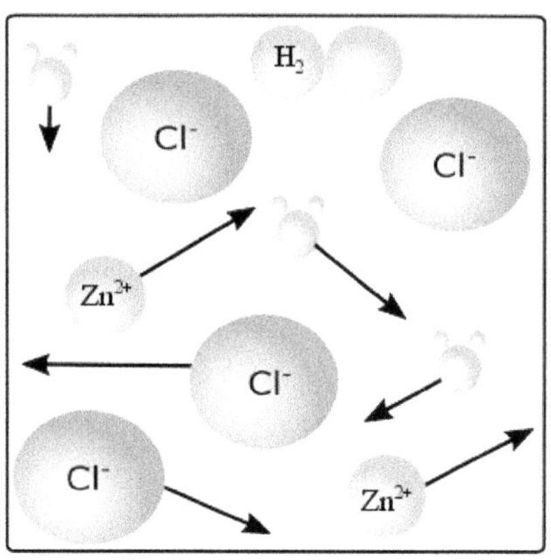

(C) Diagram C

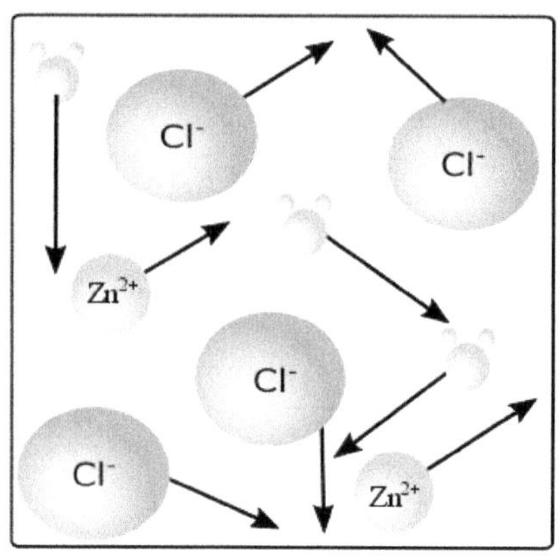

(D) Diagram D

8. A student's data table is shown below.

Mass of 2.0 *M* HCl(*aq*) used	76.0 g
Initial temperature of HCl(*aq*)	21.5°C
Final temperature of HCl(*aq*)	38.6°C
Mass of zinc metal used	5.0 g

The student's calculations are shown below.

$$q = mc\Delta T$$

$$q = (76.0\ \text{g})\left(\frac{4.2\ \text{J}}{\text{g}\ °\text{C}}\right)(38.6 - 21.5°\text{C}) = 5.46\ \text{kJ}$$

$$5.00\ \text{g Zn} \times \frac{1\ \text{mol Zn}}{65.38\ \text{g Zn}} = 0.0765\ \text{mol Zn}$$

$$\Delta H = \frac{q}{\text{mol Zn}} = \frac{5.46\ \text{kJ}}{0.0765\ \text{mol Zn}} = \frac{71.4\ \text{kJ}}{\text{mol}}$$

Since the reaction is exothermic

$$\Delta H = \frac{-71.4\ \text{kJ}}{\text{mol}}$$

The accepted value is –80.7 kJ/mol_{rxn}. Based on the information presented, what possible error would explain this result?

(A) The mass of the solution used in the calculations should have included the mass of the reacted zinc.
(B) The change of temperature used in the calculations didn't take into account the heat energy produced from the reaction of the calorimeter with the hydrochloric acid.
(C) The mass of the solution used in the calculations should include the mass of the hydrogen gas produced.
(D) The moles used in the calculations should be the moles of the limiting reagent, in this case, the moles of HCl.

9. A student drops a block of Zn into a container of water, as shown below, and notices a change in temperature. Which statement best describes this situation?

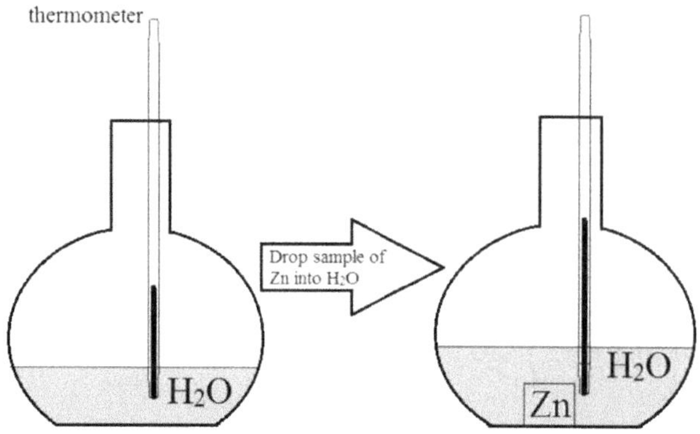

(A) $q_{zinc} = q_{water}$ and heat energy flowed from the water to the zinc.
(B) $q_{zinc} = -q_{water}$ and heat energy flowed from the water to the zinc.
(C) $q_{zinc} = q_{water}$ and heat energy flowed from the zinc to the water.
(D) $q_{zinc} = -q_{water}$ and heat energy flowed from the zinc to the water.

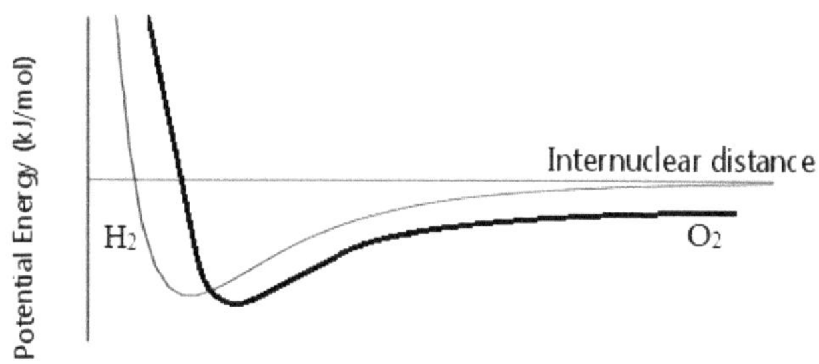

10. The internuclear distance versus potential energy diagram for oxygen and hydrogen is shown above, along with their Lewis dot diagrams. What explains why oxygen and hydrogen have approximately the same bond energy?
(A) The lone pair of electrons in oxygen form pi bonds, which increase the bond energy of oxygen.
(B) Hydrogen's sigma bond is using the electrons from the first energy level, while oxygen's bonds are using electrons from the second energy level. However, the increase in nuclear charge in oxygen almost balances the increase in distance between the nuclei in oxygen.
(C) The double bond in oxygen has more potential energy than the single bond in hydrogen; however, hydrogen's bond is shorter, so there is more attraction between the nuclei of hydrogen and the valence electrons.
(D) The lone pair of electrons in oxygen causes increased electron–electron repulsion, pushing the 2 atoms apart and reducing the bond energy in the oxygen bond to approximately the same strength as hydrogen's bond.

11. What is the enthalpy change for the following reaction under standard conditions?

$$CS_2(l) + 3\ O_2(g) \rightarrow CO_2(g) + 2\ SO_2(g)$$

ΔH_f° for $CS_2(l)$ = +88 kJ/mol

ΔH_f° for $CO_2(g)$ = –394 kJ/mol

ΔH_f° for $SO_2(g)$ = –297 kJ/mol

(A) –603 kJ/mol_{rxn}
(B) –779 kJ/mol_{rxn}
(C) –900. kJ/mol_{rxn}
(D) –1076 kJ/mol_{rxn}

12. A 57-g block of metal at 92°C is dropped into an insulated flask containing approximately 45.0 g of ice and 30.0 g of water at 0°C. After the system reaches equilibrium, it is determined that 9.5 g of the ice have melted. What is the specific heat of the metal? (Heat of fusion of ice = 333 J/g.)
(A) 0.22 J/g °C
(B) 0.32 J/g °C
(C) 0.60 J/g °C
(D) 0.92 J/g °C

13. Human hair is made up mostly of bundles of proteins called keratins. For people who have naturally curly hair, bonds between sulfur atoms (disulfide bonds) cause the keratin strands to fold back on themselves to produce waves and curls. To eliminate these waves and curls, a permanent chemical straightening process can be used to break these bonds. Even people with straight hair can experience waves and curls (often unwanted, the so-called "bad hair day") when the humidity (water content) in the air is high. Temporary bonds that involve water molecules form between keratin strands. Why does it require so much less energy to break these bonds (which cause the temporary "frizzy" hair) in comparison to the more permanent disulfide bonds?
(A) Disulfide bonds are covalent bonds, while the bonds caused by high humidity are hydrogen bonds.
(B) Disulfide bonds are ionic bonds, while the bonds caused by high humidity are covalent bonds.
(C) Disulfide bonds are covalent bonds, while the bonds caused by high humidity are ionic bonds.
(D) Disulfide bonds are ionic bonds, while the bonds caused by high humidity are hydrogen bonds.

14. A sample of dry ice [$CO_2(s)$] is initially at a temperature of –80°C. As it is gradually warmed to room temperature, the solid changes to a gas without the formation of any liquid. What is the best explanation for the observation?
    (A) The CO_2 undergoes a chemical change as the addition of thermal energy breaks covalent bonds between carbon and oxygen atoms.
    (B) The CO_2 undergoes a chemical change as the addition of thermal energy increases the kinetic energy of the particles, breaking all dispersion forces.
    (C) The CO_2 undergoes a physical change as the addition of thermal energy breaks the covalent bonds between carbon and oxygen atoms.
    (D) The CO_2 undergoes a physical change as the addition of thermal energy increases the kinetic energy of the particles, breaking all dispersion forces.

15. A 40.0-g sample of an unknown metal was heated to 69.0°C and dropped into an insulated foam cup containing 100.0 g of water initially at 20.0°C. The metal and water were stirred, and the temperature of the water plus the metal increased to 24.0°C. Assume that the specific heat capacity of water is 4.0 J g^{-1}°C^{-1}. What is the identity of the unknown metal?

	Metal	Specific Heat Capacity, J/(g °C)
(A)	Sn	0.21
(B)	Fe	0.45
(C)	Al	0.89
(D)	Mg	1.02

FREE-RESPONSE QUESTIONS

1. Chromium has been used in electroplating to increase the wear resistance of softer metals and to prevent oxidation of the underlying metal. Two forms of chromium are used: hexavalent chromium (oxidation state is +6) and trivalent chromium (oxidation state is +3). In both types of electroplating, highly concentrated sulfuric acid is used. The process of hexavalent chromium electroplating is highly regulated due to the toxicity of the chromium.
    (a) The ground state electron configuration of chromium is expected to be $1s^22s^22p^63s^23p^64s^23d^4$, but experimental evidence shows that it is $1s^22s^22p^63s^23p^64s^13d^5$. What is the reason for chromium's exception to the expected electron configuration?
    (b) What spectroscopy technique is used to provide evidence for chromium's exception to the expected electron configuration?
    (c) Write the electron configuration for the Cr^{3+} ion.
    (d) The most stable radioactive isotope of chromium is Cr-51, with a half-life of 27.7 days. If a 25.6-g sample of Cr-51 is allowed to sit for 79.2 days, how much of the isotope remains?

In the process of electroplating, a large amount of heat energy is released during the neutralization of the electrolytic solutions and the removal of the impurities produced in the process. Because of the increase in temperature and its environmental impact on air and water, thermal pollution is of growing concern to the environmental chemist.

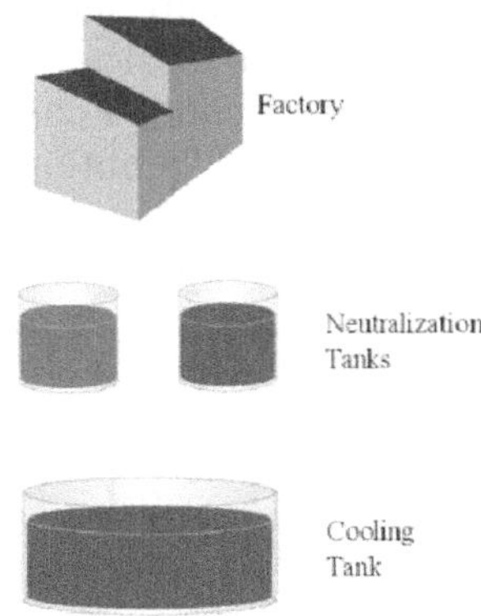

An industrial electroplating firm has several primary "neutralization tanks" and dumps 4000. L per week of 11.8 *M* sulfuric acid into each of two neutralization ponds where sufficient lime slurry, $Ca(OH)_2$, is added to neutralize the acid. After the resulting calcium sulfate has settled, the water from the neutralization tanks is discharged into a secondary "cooling" tank to dissipate the heat before the resulting water is sent to a waste water treatment plant.

(e) Write a balanced equation for this reaction.

(f) Identify the type of reaction by circling the correct term(s). Justify your answer.

gas forming reaction | precipitation reaction

neutralization reaction | redox reaction

(g) How much water, in liters, is produced in each of the neutralization tanks per week? Assume that the density of water produced is 1.00 g/mL.

The discharging of water from the neutralization tanks into the cooling tank is done at 2-week intervals, which provides enough time for the calcium sulfate to settle out. The cooling tank contains 8.000×10^4 L of water whose normal temperature is 21.25°C.

(h) Given the information in the table below, calculate the ΔH_{rxn} for the reaction in part (e).

Substance	Standard Heats of Formation (kJ/mol)
$H_2SO_4(aq)$	–814
$Ca(OH)_2(aq)$	–926
$CaSO_4(s)$	–1432
$H_2O(l)$	–286

(i) Assuming that the waste water drainage from the two neutralization tanks adds heat which is equivalent to 35% of the standard heat of the neutralization reaction for the sulfuric acid waste, what is the peak temperature that the cooling tank will reach? Assume that the volume of waste water released is 4000. L from each tank, the density of the combined water in the "heat regulation" pond is 1.00 g/mL, and that the specific heat of the solution is 4.22 J g^{-1} K^{-1}.

2. Butene, C_4H_8, may be obtained by catalytic decomposition of long chain hydrocarbons during the refining process for petroleum. It has many applications in the chemical industry as a precursor in the production of other organic compounds.
   (a) Write a balanced chemical equation for the complete combustion of butene gas.
   (b) Complete the Lewis structure for butene gas by showing all bonded and lone-pair electrons.

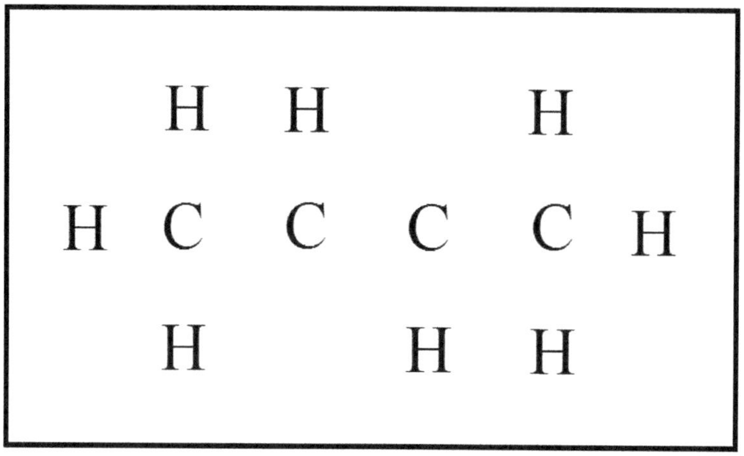

   (c) Given the following table of bond energies, calculate the enthalpy change, ΔH, for the reaction noted in part (a).

Average Bond Energies (kJ/mol)			
C–H	413	C=O	799
C–C	347	H–O	467
C=C	614	H–H	432
C≡C	839	O=O	495
C–O	358		

   (d) There are four isomers of butene, all of which are alkenes. Would these isomers have the same or different values for the ΔH_{rxn} calculated in part (c)? Explain your answer.

Answers

MULTIPLE-CHOICE QUESTIONS

1. A This question addresses your understanding of the term "standard enthalpy of formation." It is the energy involved in forming 1 mole of a compound from elements in standard state at 25°C and 1 atm pressure. Write the equation showing the formation of $NO_2(g)$ as the product from the elements nitrogen and oxygen (both are diatomic elements under these conditions). Next, balance it so that only 1 mole of the product compound is formed (LO 5.8).

2. A The stronger the intermolecular forces, the higher the enthalpy of vaporization. Both ammonia and water exhibit hydrogen bonding, but water has 2 lone pairs of electrons and is bent, which allows more opportunity for hydrogen bonding than ammonia (LO 5.9).

3. D In an endothermic reaction, heat energy flows from the surroundings to the system. Since endothermic reactions absorb heat, the change in enthalpy is positive ($\Delta H° > 0$) (LO 5.4).

4. C When a substance melts (enthalpy of fusion), the particles in the liquid are held together by intermolecular forces. When a substance vaporizes (enthalpy of vaporization), the particles are in the gas phase; effectively, the intermolecular forces are not a significant factor (LO 5.9).

5. C The stronger the intermolecular forces, the higher the enthalpy of vaporization. Since the data supports that CCl_4 has a higher enthalpy of vaporization than NH_3, this indicates that the dispersion forces caused by the large number of polarizable electrons in CCl_4 are stronger than the hydrogen bonding in NH_3 (LO 5.9).

6. D In exothermic reactions, heat energy flows from the system to the surroundings. Since the heat energy is flowing into the surroundings, the surroundings will increase in temperature. This is supported by the observation that the container gets hot (LO 5.3).

7. D Since the reaction is exothermic, the temperature of the particles will increase. Since temperature measures the average kinetic energy of all of the particles in the system, the particles would have greater kinetic energy. The kinetic energy is shown by the length of the arrows, so the higher kinetic energy would be reflected in the greater length of the arrows (LO 5.2).

8. A When solving for q, the mass of the solution needs to be used. The solution includes the hydrochloric acid and the reacted zinc. If the mass used in the calculations does not include the mass of the zinc, the mass is too low, resulting in the calculation of the heat energy being too low. Therefore, the enthalpy of the reaction will be too low (LO 5.6).

9. D Since the thermometer indicates that the temperature of the water increases, this indicates that the zinc was at a higher temperature than the water. Since the zinc was higher in temperature, the heat energy flows from the zinc to the water. The amount of heat energy lost by the zinc was all gained by the water, so they will be equal in magnitude but opposite in sign (LO 5.7).

10. B Bond energy is related to the distance from the nucleus to the electrons in the bond and the amount of attraction between the nucleus and electrons in the bond. According to Coulomb's law, the amount of attraction is directly proportional to the size of the charges and inversely proportional to the distance between them.

The similar bond energy of the two substances indicates that the increase distance is offset by the increase in the nuclear charge (LO 5.1).

11. D $\Delta H^\circ_{rxn} = \sum \Delta H^\circ_{f_{products}} - \sum \Delta H^\circ_{f_{reactants}}$

$$\Delta H^\circ_{rxn} = [(1)(-394) + (2)(-297)] - [(3)(0) + (1)(88)] = -1076 \text{ kJ/mol}$$

Note: This can be estimated as [–400 + (2)(–300)] – 100 = –1100, which will give you only one realistic answer (LO 5.8).

12. C Heat energy needed to melt ice

$$q = (\text{mass})(\text{heat of fusion})$$

$$q = (9.5 \text{ g})\,(333 \text{ J/g}) \approx (10 \text{ g})(300 \text{ J/g}) = 3000 \text{ J}$$

Since only some of the ice melted, the final temperature will be 0°C.

$$c = \frac{q}{m\Delta T} \approx \frac{3000 \text{ J}}{(57 \text{ g})(92°\text{C})} \approx \frac{3000 \text{ J}}{(60 \text{ g})(90°\text{C})} \approx \frac{3000 \text{ J}}{5400 \text{ g} \times °\text{C}} \approx \frac{30 \text{ J}}{54 \text{ g} \times °\text{C}}$$

$$\approx \frac{3 \text{ J}}{5 \text{ g} \times °\text{C}} \approx 0.60 \frac{\text{J}}{\text{g} \times °\text{C}}$$

(LO 5.7)

13. A The disulfide bonds are covalent bonds between 2 atoms, while the bonds caused by the high humidity are hydrogen bonds between 2 adjacent water molecules (LO 5.11).

14. D When a substance undergoes a phase change (melting, vaporization, boiling, sublimation, deposition, condensation), the intramolecular bonds are not involved. The force of attraction between particles accounts for the phase change (LO 5.10).

15. C

$$q_{water} = mc\Delta T \Rightarrow q_{water} = (100.0 \text{ g})(4.0°\text{C})(4.0 \frac{\text{J}}{\text{g °C}}) = 1600 \text{ J}$$

$$q_{water} = -q_{metal}$$

$$c = \frac{q_{metal}}{m\Delta T} \approx \frac{-1600 \text{ J}}{(40.0 \text{ g})(-45.0°\text{C})} = \frac{1600}{1800} \frac{\text{J}}{\text{g} \times °\text{C}} = \frac{16}{18} \frac{\text{J}}{\text{g} \times °\text{C}} = \frac{8}{9} \frac{\text{J}}{\text{g} \times °\text{C}}$$

$$c = 0.89 \frac{\text{J}}{\text{g} \times °\text{C}}$$

Note: For 40 × 45, think of it as 40 × (40 + 5) = 1600 + 200 = 1800 (LO 5.5).

FREE-RESPONSE QUESTIONS

1. (a) In order to decrease the electron to electron repulsion that occurs in doubly occupied orbitals, chromium promotes the electron from 4*s* to 3*d* (LO 1.8).
   (b) Photoelectronic spectroscopy (PES) (LO 1.15).
   (c) $1s^22s^22p^63s^23p^63d^3$ or $[Ar]3d^3$ (LO 1.7)
   (d)

$$k = \frac{0.693}{t_{1/2}} \Rightarrow k = \frac{0.693}{27.7 \text{ days}} \qquad k = 0.0250 \text{ days}^{-1}$$

$$\ln\left(\frac{A}{A_O}\right) = -kt \Rightarrow \ln\left(\frac{A}{25.6 \text{ g}}\right) = -\frac{0.0250}{\text{days}} \times 79.2 \text{ days}$$

$$A = 3.53 \text{ g}$$

   (LO 4.3)
   (e) $Ca(OH)_2 + H_2SO_4 \rightarrow 2\ H_2O + CaSO_4$ (LO 3.2)
   (f) Neutralization reaction and precipitation reaction (LO 3.10)
   (g)

$$\frac{11.8 \text{ mol}}{\text{L}} \times 4000 \text{ L} \times \frac{2 \text{ mol } H_2O}{1 \text{ mol } H_2SO_4} \times \frac{18.02 \text{ g } H_2O}{1 \text{ mol } H_2O} \times \frac{1 \text{ mL}}{1 \text{ g}} \times \frac{1 \text{ L}}{1000 \text{ mL}}$$

$$= 1.70 \times 10^3 \text{ L}$$

   (LO 3.4)
   (h)

$$\Delta H^\circ_{rxn} = \sum \Delta H^\circ_{f\,\text{products}} - \sum \Delta H^\circ_{f\,\text{reactants}}$$

$$\Delta H^\circ_{rxn} = [(2)(-286) + (1)(-1432)] - [(1)(-926) + (1)(-814)]$$

$$\Delta H^\circ_{rxn} = -264 \text{ kJ/mol}$$

   (LO 5.8)
   (i)

$$\frac{11.8 \text{ mol } H_2SO_4}{\text{L}} \times 4000 \text{ L} \times \frac{264 \text{ kJ}}{1 \text{ mol } H_2SO_4} = 12{,}500{,}000 \text{ kJ}$$

$$2 \text{ tanks} \times 12{,}500{,}000 \text{ kJ} = 2.50 \times 10^7 \text{ kJ}$$

$$35\% \text{ of } 2.50 \times 10^7 \text{ kJ} = 8{,}750{,}000 \text{ kJ}$$

$$\Delta T = \frac{q}{mc} \Rightarrow \Delta T = \frac{8{,}750{,}000{,}000 \text{ J}}{\left[(80{,}000 \text{ L} + 8000 \text{ L}) \times \frac{1000 \text{ g}}{1 \text{ L}}\right] \times 4.22 \frac{\text{J}}{\text{g °C}}} = 23.6\text{°C}$$

   Final temperature is 21.25 + 23.6 = 44.9°C
   (LO 5.6)

2. (a) $C_4H_8 + 6\ O_2 \rightarrow 4\ CO_2 + 4\ H_2O$ (LO 3.2)

(b)

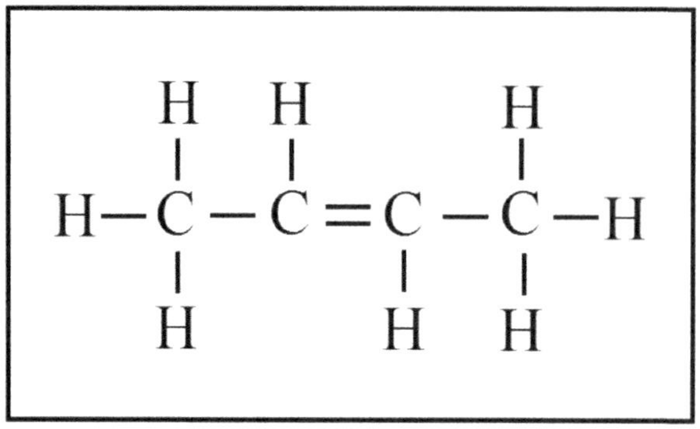

(LO 2.21)

(c) The Lewis structures for all reactants and products for this reaction are either shown below or in part (b).

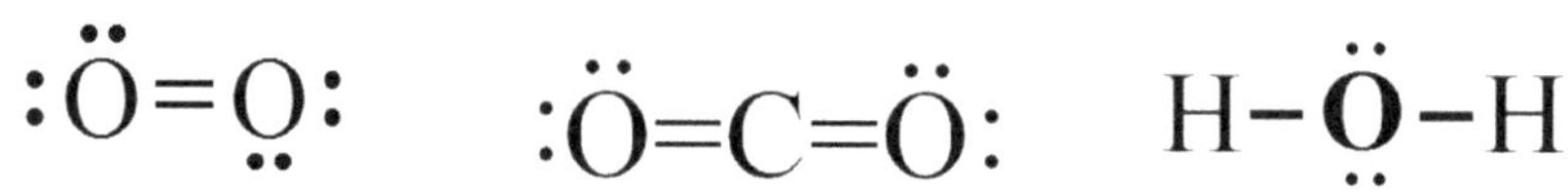

Bonds broken:
C–C: 2 × 347 = 694
C=C: 1 × 614 = 614
C–H: 8 × 413 = 3304
O=O: 6 × 495 = 2970
Total bond energy of bonds broken: 7582 kJ
Bonds formed:
C=O: 8 × 799 = 6392
O–H: 8 × 467 = 3736
Total bond energy of bonds formed: 10,128 kJ

$$\Delta H^\circ_{rxn} = \sum Bond\ Energies_{\text{bonds broken}} - \sum Bond\ Energies_{\text{bonds formed}}$$

$$\Delta H^\circ_{rxn} = 7582 - 10{,}128 = -2546 \text{ kJ/mol}$$

(LO 5.8)

(d) Since they are isomers, they have the same number of atoms and the same molecular formula—they just have different arrangements. Since the problem states that they are all alkenes, they will have 1 double bond between 2 carbon atoms; therefore, they should have the same number and type of bonds broken (LO 5.8).

10

BIG IDEA 5: THERMODYNAMICS

Big Idea 5

The laws of thermodynamics describe the essential role of energy and explain and predict the direction of changes in matter.

This is a continuation of the thermochemistry covered in the previous chapter. Two additional topics will be covered, entropy and Gibbs free energy. Entropy looks at the disorder of the system and its effects on the thermodynamic favorability of a reaction. Determining the value of Gibbs free energy for a process allows one to predict whether or not a process is thermodynamically favored (spontaneous). The dependence of Gibbs free energy on temperature, entropy, and enthalpy will be identified. The connections between Gibbs free energy and electrochemistry as well as the magnitude of free energy and its relationship to the equilibrium constant will also be addressed. The laws of thermodynamics will also be explored.

The free-response portion of the exam may include calculations and predictions involving entropy, enthalpy, and free energy.

You should be able to:

- Draw, label, and perform associated calculations for heating (cooling) curves to determine the change of energy.
- Use heating and cooling curves to determine the temperature at which a phase change process happens.
- Relate $\Delta G°_{rxn}$ to the voltage of a galvanic cell.
- Relate $\Delta G°_{rxn}$ to the equilibrium constant.
- Compare the absolute entropies, $S°$, of compounds.

- Predict relative entropy changes in chemical reactions and phase changes.
- Perform calculations with $\Delta S°$.
- Perform calculations with free energy, $\Delta G°_f$.
- Couple non-thermodynamically favored reactions to thermodynamically favored reactions to drive the reactions to increase production of a desired product.

THREE LAWS OF THERMODYNAMICS

FIRST LAW OF THERMODYNAMICS

The first law of thermodynamics states that energy can be converted from one form to another, but can neither be created nor destroyed: The energy of the universe is constant. This is the law of conservation of energy.

SECOND LAW OF THERMODYNAMICS

The second law of thermodynamics states that in any spontaneous process there is always an increase in the entropy of the universe. The change in entropy of the universe is equal to the change in the entropy of the system plus the change in entropy of the surroundings.

$$\Delta S_{univ} = \Delta S_{sys} + \Delta S_{surr}$$

$$\Delta S_{surr} = -\Delta H / T$$

THIRD LAW OF THERMODYNAMICS

The third law of thermodynamics states that the entropy of a pure, perfectly formed, crystalline solid at absolute zero is zero, that is, $S° = 0$. Since absolute zero cannot be reached, effectively, (almost) every species has a positive value for entropy. The ion H^+ is a single proton (there are no neutrons or electrons) and, as such, it has a defined entropy, $S° = 0$.

ENTROPY

An increase in the entropy of the universe, ΔS_{univ}, is the driving force for a thermodynamically favored process. If a reaction is thermodynamically favored, the process occurs without intervention. Entropy measures disorder of a system or the number of "microstates" in a system. A microstate is a different arrangement of energy in the system. Many microstates are indistinguishable from each other and the more microstates that a system has, the higher the entropy. Let's say that we have 3 molecules and 3 units of energy that are spread over those 3 molecules. One possible arrangement would be as shown in the following table:

Units of Energy	Molecule "Larry"	Molecule "Curly"	Molecule "Moe"
3			
2			
1	X	X	X
0			

If Larry, Curly, and Moe are different, then you would be able to identify that there are three different cases (i.e., three different microstates). However, if Larry, Curly, and Moe are three identical triplets, then we would have this arrangement:

Units of Energy	Molecule A	Molecule B	Molecule C
3			
2			
1	X	X	X
0			

A, B, and C would be indistinguishable from each other, so this would be one possible arrangement (one microstate). There are other ways to distribute 3 units of energy over the 3 molecules A, B, and C, as shown charts below.

Units of Energy	A	B	C
3	X		
2			
1			
0		X	X

Units of Energy	A	B	C
3		X	
2			
1			
0	X		X

Units of Energy	A	B	C
3			X
2			
1			
0	X	X	

Since there are three arrangements to accomplish this, there are three microstates. However, each of these three microstates is identical or equivalent (since A, B, and C are identical).

There is one other possible way to distribute 3 units of energy over the 3 molecules: 1 molecule receives 2 units of energy, another molecule receives 1 unit of energy, and the last molecule receives 0 units of energy. It turns out that there are six arrangements for this case, as shown in the charts below.

Units of Energy	A	B	C
3			
2	X		
1		X	
0			X

Units of Energy	A	B	C
3			
2		X	
1	X		
0			X

Units of Energy	A	B	C
3			
2	X		
1			X
0		X	

Units of Energy	A	B	C
3			
2			X
1	X		
0		X	

Units of Energy	A	B	C
3			
2		X	
1			X
0	X		

Units of Energy	A	B	C
3			
2			X
1		X	
0	X		

This would provide a total of 10 different microstates. It is important to note that as the number of molecules increases, so does the number of microstates. As the units of energy increase, there will be more microstates (if there were 4 units of energy to spread over the 3 molecules, then there would be 15 microstates). As temperature increases, so does the entropy.

An increase in the number of possible arrangements or positions available to a system in a given state increases the entropy. This positional entropy increases when going from a solid to liquid to gas. In the solid state, molecules are much closer together than in the gaseous state, with very few positions available for them.

Entropy change during a process is said to be increasing when the value of $\Delta S > 0$; the value of ΔS is positive. Entropy change during a process is said to be decreasing when the value of $\Delta S < 0$; the value of ΔS is negative. This is because of the second law of thermodynamics; since the natural tendency is to an increase in disorder, there is no energy expended to increase disorder, just in making something more ordered. Hence, the sign on decreasing entropy is negative (i.e., making it more ordered), since that would require energy to be expended.

EXAMPLE: Which of the following pairs is likely to have the higher positional entropy per mole at a given temperature?

1. Solid CO_2 or gaseous CO_2?
2. N_2 gas at 1.0 atm or N_2 gas at 0.001 atm?

SOLUTION:

1. Gaseous CO_2 has more positional entropy than solid CO_2 since there are more positions for the molecules in the gaseous state to move to than in the solid state.
2. N_2 gas at 0.001 atm has more positional entropy than N_2 gas at 1.0 atm, because at a lower pressure, there is more volume for the molecules to move about than at a higher pressure.

EXAMPLE: Describe the change in entropy when a solid salt is added to water.

SOLUTION: The entropy increases when salt in its solid state is dissolved in water. The ions fixed in a crystal lattice in the solid are free to move about in the water due to solute–solvent interactions. *Note:* There are a very few exceptions where the hydration spheres that form when ions are solvated cause the overall entropy change to be negative.

ENTROPY CHANGES IN A CHEMICAL REACTION

The change in entropy, ΔS, for a chemical reaction can be predicted without calculation. Entropy generally increases (ΔS is positive) when:

- the number of moles of gas increases from reactants to products in a chemical reaction.
- the phase changes from solid to liquid or from liquid to gas.
- a reaction has more moles of products than of reactants.
- the complexity of compounds increases; as you move from methane to ethane to propane to butane, etc., the entropy is greater.
- temperature increases (since temperature is the average kinetic energy of the particles in the system, an increase in entropy results from the particles moving faster).
- solutions are formed (there are exceptions).
- the volume of a gas is expanded.

The determination of entropy changes when dissolving an ionic compound can be tricky. As the ionic solid dissolves, the ions are free to float around in the solution, increasing the number of microstates that exist for the ions. When the reference is the water molecules (the solvent), there is a slight decrease in the disorder as the water molecules are held around the ions by ion–dipole interactions. However, the overall process does increase the entropy of the system in most cases.

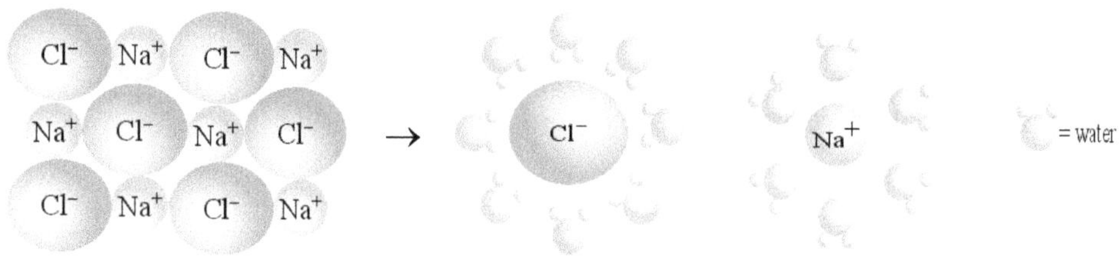

EXAMPLE: Predict the sign of $\Delta S°$ for the following reaction:

$$N_2(g) + 3\ H_2(g) \rightarrow 2\ NH_3(g)$$

SOLUTION: The entropy decreases as the reaction proceeds from reactants to products because the total number of moles of gas decreases from 4 moles to 2 moles. The sign of $\Delta S°$ is negative.

The change in entropy for a reaction, $\Delta S°$, can also be calculated using tabulated thermodynamic values provided on the exam.

$$\Delta S°_{rxn} = \Sigma\Delta S°_{products} - \Sigma\Delta S°_{reactants}$$

EXAMPLE: Given the following information, calculate $\Delta S°$ for the following reaction.

Substance	$S°$ (J/mol × K)
$N_2(g)$	192
$H_2(g)$	131
$NH_3(g)$	193

$N_2(g) + 3\ H_2(g) \rightarrow 2\ NH_3(g)$

SOLUTION: $\Delta S°_{rxn} = \Sigma\Delta S°_{products} - \Sigma\Delta S°_{reactants}$

$$\Delta S° = [2(193)] - [1(192) + 3(131)] = -199\ J/(K \times mol_{rxn})$$

The sign of $\Delta S°$ is negative, confirming the prediction in the previous example.

AP® Tip

Pay very close attention to the phases of matter when doing thermodynamic calculations.
The $S°$ for $NH_3(g)$ is 193 J/(mol × K), but the $S°$ for $NH_3(aq)$ is 111 J/(mol × K). Phases matter—especially in thermodynamics problems.

FREE ENERGY

Free energy, G, is a thermodynamic function whose value describes whether or not a process is thermodynamically favored in the forward direction (reactions are usually written such that the forward direction is thermodynamically favored).

A term that is very common but sometimes misunderstood by students is "spontaneous." *Spontaneous* and *thermodynamically favored* have the same meaning. On the new exam, the terms will be *thermodynamically favored* (spontaneous) and *non-thermodynamically favored* (nonspontaneous). The switching of the terms on the test avoids a common misconception that students have—namely, that if a process is spontaneous, students often assume that the reaction will take place very quickly (if not instantaneously). However, ΔG provides no information about the rate of reaction. Spontaneous reactions can be very slow; rusting of iron is one example of a slow spontaneous reaction. Another example is the natural decay of uranium-238 into lead-206 which has a half-life of 4.47 billion years—a thermodynamically favored reaction (spontaneous process) that is extremely slow!

The preferred terms on the exam are *thermodynamically favored* and *non-thermodynamically favored* and those will be the terms used here. However, if a student uses the terms "spontaneous" and "nonspontaneous" to refer to the sign of ΔG, then they should receive credit. Gibbs free energy is dependent on the change in enthalpy, change in entropy, and temperature of the system. ΔG is negative for all thermodynamically favored processes.

$$\Delta G = \Delta H - T\Delta S; \; T \text{ is the Kelvin temperature}$$

Using the chart on the next page, you can predict if a reaction will occur without the exact values for ΔH and ΔS. There are four possible cases involving the calculations for ΔG. Included in the chart is the term "driven by" meaning which term or terms will make ΔG negative and what temperature will make ΔG negative?

For instance, if ΔH is positive (+) and ΔS is also positive (+), then to determine the value for ΔG, we would use the equation $\Delta G = \Delta H - T\Delta S$.

$\Delta G = \Delta H - T\Delta S$; so ΔG = (+ number) – (+ number). The sign of ΔG would depend on the magnitude (value) of ΔH and $T\Delta S$. If ΔH is bigger than $T\Delta S$, then ΔG is positive. Since we are looking for the conditions where ΔG is negative, we would want $T\Delta S$ to be bigger than the ΔH term, which means that the reaction is driven by entropy (i.e., it is the entropy that will make ΔG negative). Since $T\Delta S$ needs to be bigger than the ΔH term, it will be more favored at higher temperatures.

Using Gibbs Free Energy to Determine if a Reaction Is Thermodynamically Favored or Non-Thermodynamically Favored							
ΔG		ΔH		ΔS		Reaction Driven by	Favorable Temperature
Sign	**Meaning**	**Sign**	**Meaning**	**Sign**	**Meaning**		
+ or –	????????	+	Endothermic	+	Increase disorder	Entropy	High temperatures
+	Non-Thermodynamically Favored	+	Endothermic	–	Decrease disorder	Nothing	None
–	Thermodynamically Favored	–	Exothermic	+	Increase disorder	Enthalpy and entropy	All temperatures
+ or –	????????	–	Exothermic	–	Decrease disorder	Enthalpy	Low temperatures

Notice the pattern here: Reactions are driven by entropy, enthalpy, both entropy and enthalpy, and finally, by neither entropy nor enthalpy. Also the temperatures follow the similar pattern: favored at high temperatures, low temperatures, all temperatures, and finally no temperatures.

FREE ENERGY AND CHEMICAL REACTIONS

The standard free energy change, $\Delta G°$, is the change in free energy that will occur if the reactants in their standard states are converted to the products in their standard states. This value cannot be measured directly, but it can be calculated from other measured quantities such as the equilibrium constant and the standard cell potential.

$\Delta G°$ can also be calculated from the changes in enthalpy and entropy.

$$\Delta G° = \Delta H° - T\Delta S°$$

Note that the units of $\Delta G°$ and $\Delta H°$ are typically provided in kJ/mol, while $\Delta S°$ is provided in J/(mol × K). Make sure your units are consistent before proceeding with calculations!

EXAMPLE: Consider the reaction 2 $POCl_3(g) \rightarrow$ 2 $PCl_3(g) + O_2(g)$.

The value of $\Delta S°$ is 179 J/K. The value of $\Delta H°$ is 542 kJ. At what temperature does this reaction become thermodynamically favored? Assume that $\Delta H°$ and $\Delta S°$ do not depend on temperature.

SOLUTION: The temperature at which $\Delta G° = 0$ is where the process shifts from thermodynamically favored to non-thermodynamically favored. Set $\Delta G° = 0$ and rearrange $\Delta G° = \Delta H° - T\Delta S°$ to solve for T.

$$T = \Delta H°/\Delta S°$$

$$T = 542 \text{ kJ}/(0.179 \text{ kJ/K}) = 3030 \text{ K}$$

The standard free energy change, $\Delta G°$, is a state function and can be calculated by

$$\Delta G°_{rxn} = \Sigma\Delta G°_{f_{products}} - \Sigma\Delta G°_{f_{reactants}}$$

Notice that this is the same basic equation that is used to find $\Delta H°_{rxn}$ and $\Delta S°$.

EXAMPLE: Consider the reaction 2 $POCl_3(g) \rightarrow$ 2 $PCl_3(g)$ + $O_2(g)$.

Determine the value for $\Delta G°_{rxn}$ given the following table.

Substance	$\Delta G°_f$ (kJ/mol)
$O_2(g)$	0
$POCl_3(g)$	–502.50
$PCl_3(g)$	–267.78

SOLUTION:

$$\Delta G°_{rxn} = \Sigma\Delta G°_{f_{products}} - \Sigma\Delta G°_{f_{reactants}}$$

$$\Delta G°_{rxn} = [(2)(-267.78) + (1)(0)] - [(2)(-502.50)] = 469.44 \text{ kJ/mol}_{rxn}$$

Since $\Delta G°_{rxn}$ is positive, the reaction is non-thermodynamically favored.

FREE ENERGY AND EQUILIBRIUM

$$\Delta G° = -RT \ln K;\ K = \text{equilibrium constant}$$

This is discussed in Chapter 11 (Big Idea 6) of this book.

FREE ENERGY AND CELL POTENTIAL

$$\Delta G° = -nFE°;\ E° \text{ is the standard cell potential}$$

This is discussed in Chapter 7 (Big Idea 3) of this book.

HEATING AND COOLING CURVES

When a system is heated, energy is transferred into it. In response to the energy it receives, the system changes, for example, by increasing its temperature. A plot of temperature versus time is called the heating curve. One such heating curve is shown on the next page.

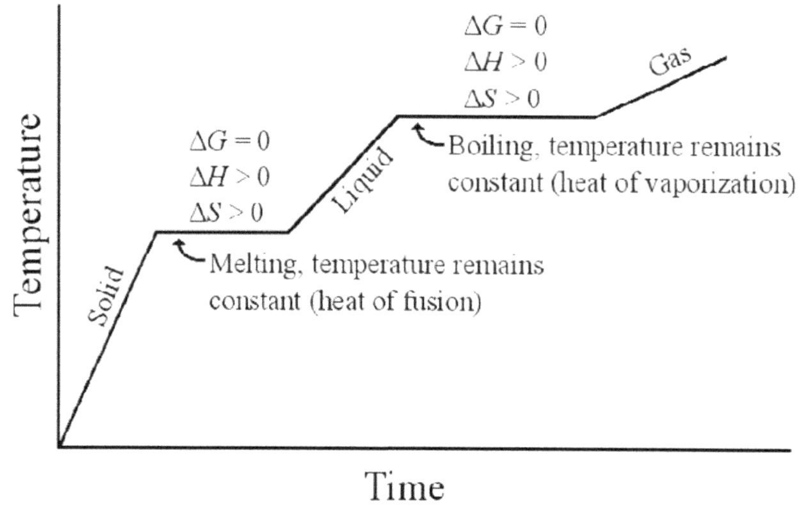

The physical states of the substance and the phase transitions are identified along the curve. If the graph went from high temperature to low temperature, it would be a cooling curve, as shown below.

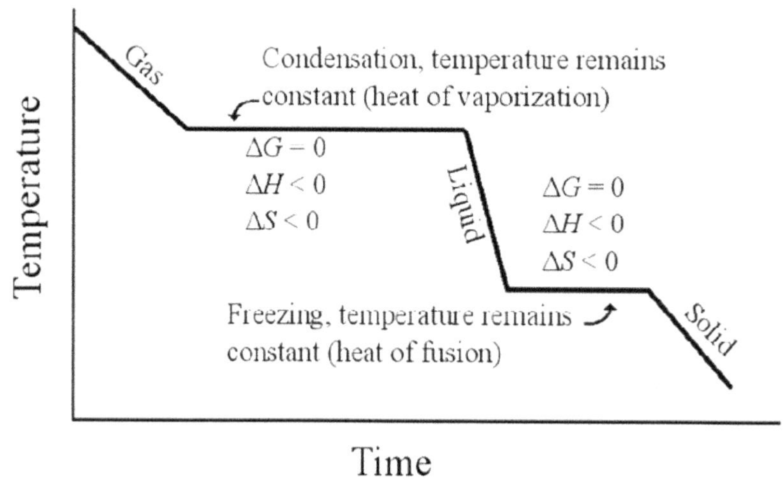

The temperature of the system usually increases when energy is applied. However, when the energy absorbed is used for phase transition, a change in the physical state, the temperature (average kinetic energy) remains constant because the potential energy of the system is being increased as the molecules are rearranged in the phase change. Phase changes only affect the intermolecular forces of attraction; no intramolecular bonds are broken. During a phase change, the system is at equilibrium, so $\Delta G° = 0$. The values for $\Delta H°_{rxn}$ and $\Delta S°$ can then be used to determine the temperature at which the phase change occurs.

EXAMPLE: Using the following thermodynamic values, determine the boiling point of water.

	ΔH°_f (kJ/mol)	S° (J/mol K)
H_2O (liquid)	–285.83	69.95
H_2O (gas)	–241.83	188.84

SOLUTION: The reaction is $H_2O(l) \rightarrow H_2O(g)$. We need to find both ΔH°_{rxn} and ΔS° and then use those values to find T.

$$\Delta S^\circ_{rxn} = \Sigma\Delta S^\circ_{products} - \Sigma\Delta S^\circ_{reactants}$$

$$\Delta S^\circ = 188.84 \text{ J/(mol K)} - 69.95 \text{ J/(mol K)} = 118.89 \text{ J/(mol K)}$$

$$\Delta H^\circ_{rxn} = \Sigma\Delta H^\circ_{f_{products}} - \Sigma\Delta H^\circ_{f_{reactants}}$$

$$\Delta H^\circ_{rxn} = -241.83 \text{ kJ/mol} - (-285.83 \text{ kJ/mol}) = 44.00 \text{ kJ/mol}$$

$$T = \Delta H^\circ/\Delta S^\circ$$

$$T = \frac{44.00 \text{ kJ/mol}}{0.11889 \text{ kJ/(mol} \times \text{K)}} = 370.1 \text{ K} = 96.94^\circ \text{ C}$$

The calculated temperature (96.94°C) is less than the actual boiling point (100.°C) because the values for ΔH°_f and S° are calculated at standard conditions (25°C, 1 atm of pressure, and 1 M solutions). We are using values calculated at 25°C to extrapolate behavior at temperatures that are not 25°C. Thus, the temperature calculated is only an approximate temperature. This is the reason why problems of this type will say "Assume that ΔH° and ΔS° do not depend on temperature." In reality, both of them do depend slightly on temperature.

COUPLED REACTIONS

Coupling of reactions is very common in chemistry and biological systems. When two or more reactions are combined to produce a desired outcome, the reactions are said to be "coupled." Coupled reactions are usually in the form of taking a non-thermodynamically favored process and combining it with a thermodynamically favored process so that the combined reactions produce a thermodynamically favored overall process.

One of the most common and useful coupled reactions is the refining of metal from ore. The general form of the process is (MO represents a metallic oxide):

Reaction 1: $MO \rightarrow M + \frac{1}{2} O_2$ $\quad \Delta G^\circ > 0$

Reaction 2: $C + \frac{1}{2} O_2 \rightarrow CO$ $\quad \Delta G^\circ = -137 \text{ kJ/mol}_{rxn}$

Reaction 3: $CO + ½ O_2 \rightarrow CO_2$ $\Delta G° = -257$ kJ/mol$_{rxn}$

Overall process is: $MO + C + ½ O_2 \rightarrow M + CO_2$ $\Delta G_{rxn} < 0$

Reaction 1 must be thermodynamically favored in the reverse direction ($M + ½ O_2 \rightarrow MO$) since the ores have formed in a spontaneous process (that process may have taken millions of years, but it did happen on its own). Reactions 2 and 3 have their $\Delta G°$ at standard conditions shown, and they are not sufficiently negative to drive the overall process into being thermodynamically favored. However, since $\Delta G = \Delta H - T\Delta S$, both Reactions 2 and 3 are temperature dependent and are driven in part by the entropy. As the temperature increases, the $T\Delta S$ term becomes larger in value, thus increasing the magnitude of the ΔG of the overall reaction and making it more negative. Removal of the oxygen from the ore always increases the entropy, so the ΔS is always positive. As the temperature increases, the $T\Delta S$ drives the overall process to have a negative ΔG. That is why this type of process is carried out in a blast furnace!

Other ways that make non-thermodynamically favored reactions "work" are the addition of electricity, as in electrolysis, or exposure to UV radiation, as in photosynthesis.

WORK

The sign of the change in Gibb's free energy ($\Delta G°$) allows you to determine if a reaction is thermodynamically favored or not; however, since it is energy it also measures the capacity of the system or surroundings to do work or transfer heat. "Work" on the exam is limited to what is referred to as "$P\Delta V$."

The AP® exam is not a test where equations or relationships are derived. However, several equations that relate to work need to be discussed. Work is the force by which an object is moved, multiplied by the distance that it is moved; $w = f \times d$. Pressure is force divided by area, and the international standard unit of pressure is the pascal, which is a newton per meter squared (N/m^2). For these problems, volume is usually measured in cubic meters (m^3).

Notice in the diagram below that we really only need to measure the change in the volume of the piston to calculate work that was done to move the piston; the change in volume is directly related to the distance that the piston moves.

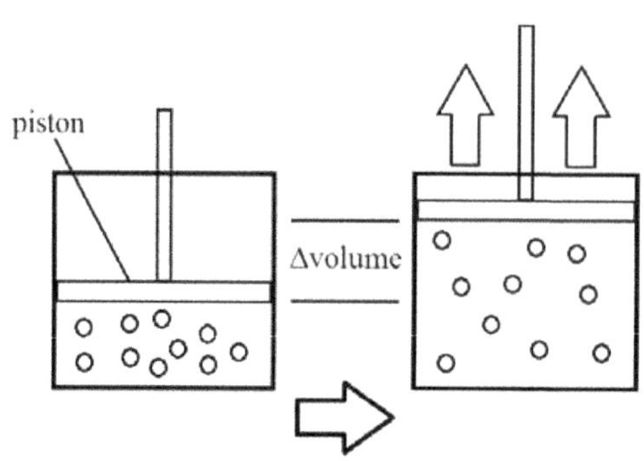

The magnitude (the size of the number) of work is equal to the product of pressure and the change in volume, so work has the units of $(N/m^2) \times m^3 = N \times m = J$. More commonly, pressure is measured in atmospheres (atm) and ΔV is measured in liters, so we use the conversion $\frac{101.3\text{ J}}{\text{L} \times \text{atm}}$. It is not likely that calculations of work will be required on the exam, but changes in work, the sign of work, and the general concept may be tested.

However, what about the sign of the work? The gas is expanding, pushing the cylinder up. The external pressure is trying to push the piston down, so the gas inside the piston is doing work on the surroundings. The signs for heat and work are always from the system's point of view (if a system can have a point of view).

- If energy flows out of the system, the sign of *q* is negative.
- If energy flows into the system, the sign of *q* is positive.
- If work is done on the system (the surroundings are doing work on the system), the sign of *w* is positive.
- If work is done on the surroundings (the system is doing work on the surroundings), the sign of *w* is negative.

In chemistry, *q* and *w* always reflect what happens with the system. In other fields, *q* and *w* may reflect what happens with the surroundings. In engineering, machines do work on the surroundings so *q* and *w* are defined the opposite as they are in chemistry (same magnitude, opposite signs). For chemistry:

$$w = -P\Delta V$$

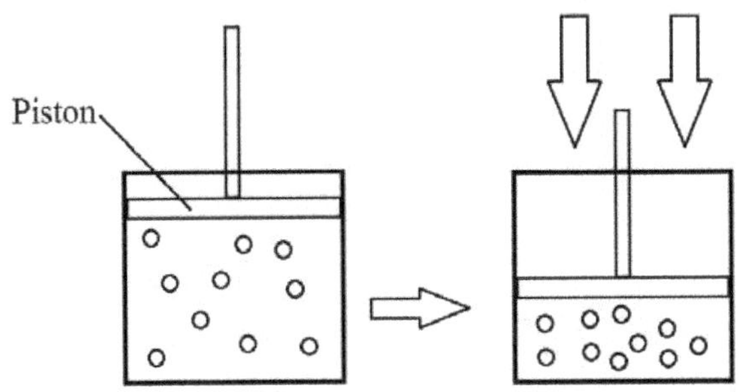

In this diagram, work is being done on the piston from the surroundings, causing the piston to be moved in, compressing the gas on the inside of the cylinder (the system). The sign of *w* will be positive.

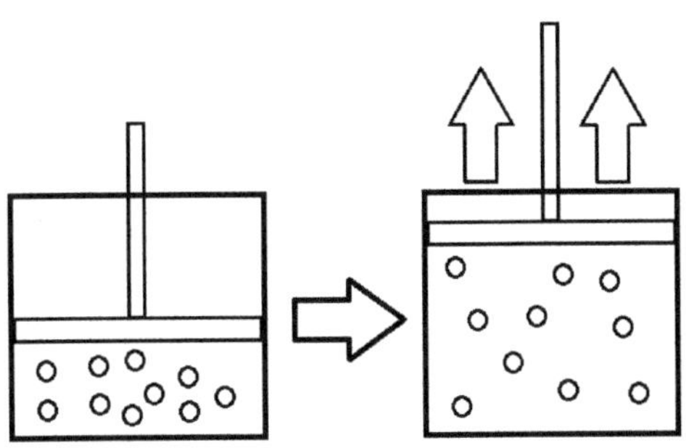

In this diagram, the gas in the piston is expanding (the system), causing work to be done on the surroundings (the piston is pushing up, against the air pressure); the sign of *w* will be negative.

THE RELATIONSHIP BETWEEN ΔG, ΔH, ΔS, WORK, AND KINETICS

A sample of aluminum is suspended on a string. A thermometer is placed in 3.0 *M* hydrochloric acid solution. The rubber stopper is lifted briefly to allow the aluminum to fall into the hydrochloric acid, and then the rubber stopper is quickly replaced. The rubber stopper at the top prevents any gas from escaping. The reaction is monitored every 2 minutes, as shown in the series of diagrams shown below.

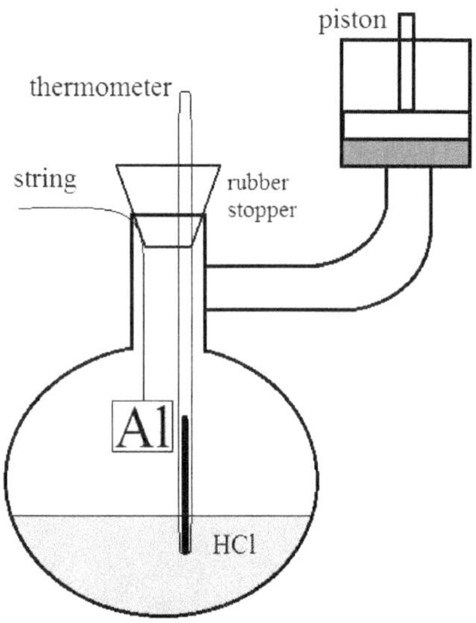

Time = 0 minutes

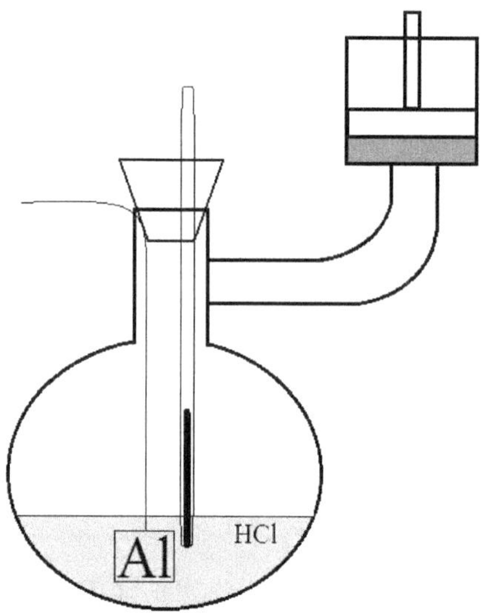

Time = 2 minutes

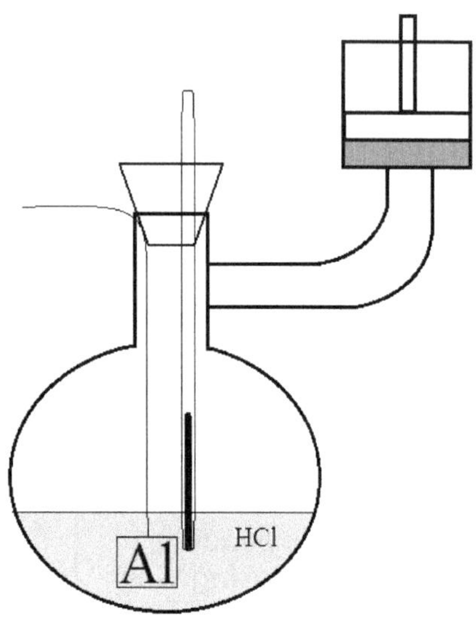

Time = 4 minutes

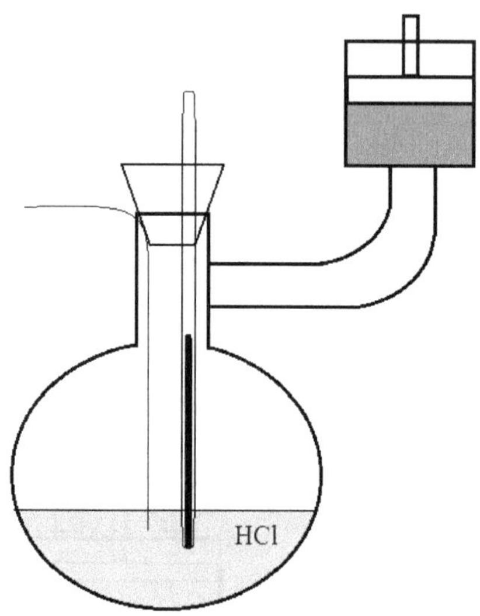

Time = 6 minutes

GIBB'S FREE ENERGY: Since the reaction took place, the reaction must be thermodynamically favored, so ΔG is negative (–).

ENTHALPY: Since the temperature increased (as shown by the thermometer), heat energy is flowing from the system to the surroundings, so ΔH is negative (–).

ENTROPY: The reaction is 2 Al(*s*) + 6 HCl(*aq*) → 3 H_2(*g*) + 2 $AlCl_3$(*aq*), which shows that the reaction converts a solid to a gas, so the entropy increases; thus, ΔS is positive (+). This is also indicated by the piston's movement.

WORK: The hydrogen gas produced in the reaction causes the piston to move up. Work is being done by the system on the surroundings, so *w* is negative (–).

KINETICS: The reaction appears to start out very slowly, since there seems to be no change in the aluminum for the first 4 minutes. It appears to be very rapid between 4 and 6 minutes since the block of aluminum has disappeared. The reason for the slow start is that the hydrochloric acid has to first react with a coating of aluminum oxide on the surface of the block. Once that coating is removed, the reaction proceeds extremely quickly.

Sign Conventions		
	Positive sign (+)	Negative sign (–)
q	System ***gains*** heat	System ***loses*** heat
w	Work done ***on*** system	Work done ***by*** system
ΔE	Net ***gain*** of energy by system	Net ***loss*** of energy by system
ΔH	Endothermic reaction, heat energy flows into the system	Exothermic reaction, heat energy flows out of system into the surroundings
ΔS	Entropy increases (greater disorder or an increase in the number of microstates)	Entropy decreases (greater order, or a decrease in the number of microstates)
ΔG	Non-thermodynamically favored (not spontaneous)	Thermodynamically favored (spontaneous)

MULTIPLE-CHOICE QUESTIONS

No calculators may be used in this part of the exam.

$C(_{diamond}) + O_2(g) \rightarrow CO_2(g) \quad \Delta G° = -396 \text{ kJ/mol}_{rxn}$

1. The reaction above represents the oxidation of diamond to form carbon dioxide. Diamonds are very valuable, and they have not been found to change mass appreciably over time. Explain this observation in terms of the thermodynamic favorability of the reaction.
   (A) The reaction is not thermodynamically favored at any temperature so it does not occur.
   (B) The reaction is thermodynamically favored at all temperatures. This reaction occurs but is not noticed since $CO_2(g)$ is colorless.
   (C) The reaction is thermodynamically favored but is under kinetic control. The reaction rate is extremely slow.
   (D) The reaction is only thermodynamically favored at very high temperatures and cannot occur at room temperature.

2. A student is instructed to place 0.50 g of crystals of ammonium chloride into 20.0 mL of water at 25°C. The student made the following observations:
   - The crystals disappeared.
   - The temperature of water decreased.

   Based on the student's observations, which graph shows the correct relationship between ΔG and temperature for this process?

(A)

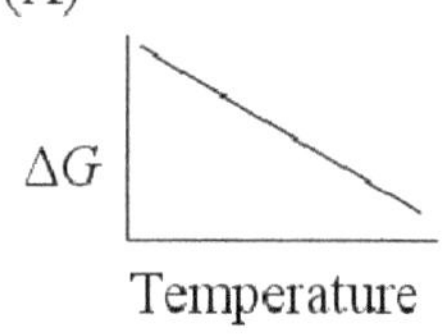

(B)

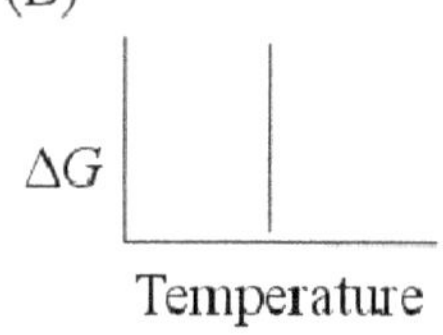

(C)

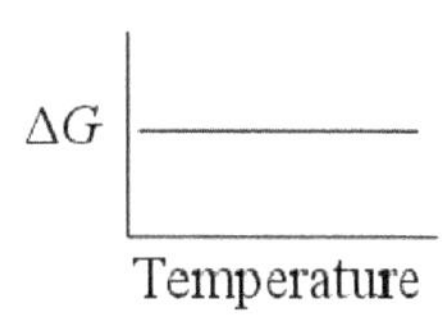

(D)

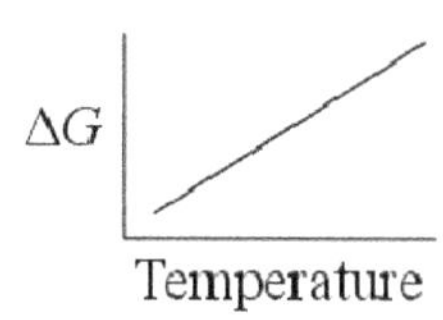

3. A student is asked to determine the enthalpy change of the reaction between Zn(*s*) and HCl(*aq*). She wrote in her laboratory notebook the following observations and made the following drawing.
   - The zinc metal reacted with HCl and disappeared in approximately 1 minute.
   - There was a production of gas. I could see the bubbles coming from the surface of the zinc metal.
   - When I put my finger over the second hole in the rubber stopper, gas would build up and when I removed it, you can hear a puff or hiss sound.
   - The bubbles occurred very rapidly at first and then slowed down as the reaction proceeded.
   - The solution remained clear and colorless.
   - The outside of the container became hot to the touch.
   - After holding my finger over the second hole in the rubber stopper for 30 seconds, I removed my finger and brought a lighted wooden splint near the top of the flask. A loud "popping" noise could be heard.

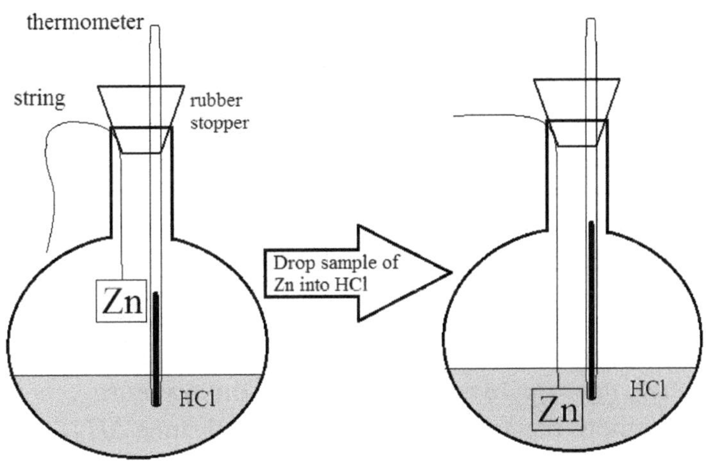

   Which of the student's observations is unrelated to an increase in entropy?
   (A) "you can hear a puff or hiss sound"
   (B) "solution remained clear and colorless"
   (C) "a loud "popping" noise could be heard"
   (D) "the bubbles occurred very rapidly"

4. Which of these four processes is not thermodynamically favorable at a low temperature but becomes more favorable as the temperature rises?
   (A) ΔH_{rxn} = + and ΔS_{rxn} = +
   (B) ΔH_{rxn} = + and ΔS_{rxn} = −
   (C) ΔH_{rxn} = − and ΔS_{rxn} = −
   (D) ΔH_{rxn} = − and ΔS_{rxn} = +

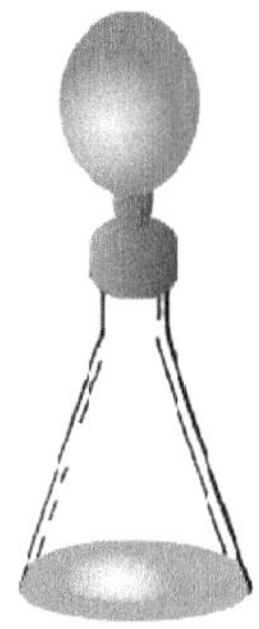

5. A student reacts magnesium metal with hydrochloric acid in a flask. A balloon is attached to the top of the flask and captures the resulting gas, as shown in the diagram above. Which statement below is correct?
   (A) The sign for work is positive (+) since the system did work on the surroundings.
   (B) The sign for work is positive (+) since the surroundings did work on the system.
   (C) The sign for work is negative (–) since the system did work on the surroundings.
   (D) The sign for work is negative (–) since the surroundings did work on the system.

6. In the production of nitric acid, the first step (shown below) is usually carried out at temperatures between 500 K and 1000 K and between 4 and 10 atm of pressure.

$$4\ NH_3(g) + 5\ O_2(g) \rightarrow 4\ NO(g) + 6\ H_2O(g)$$

   This reaction is highly exothermic. Is the use of high temperature and high pressure necessary due to thermodynamic conditions?
   (A) Yes, the higher temperatures result in ΔG_{rxn} becoming more negative (hence, more thermodynamically favored) since ΔH_{rxn} is negative and ΔS_{rxn} is negative.
   (B) Yes, the higher temperatures result in ΔG_{rxn} becoming more negative (hence, more thermodynamically favored) since ΔH_{rxn} is negative and ΔS_{rxn} is positive.
   (C) No, since ΔH_{rxn} is negative and ΔS_{rxn} is negative, ΔG_{rxn} will always be a negative value and the increase in temperature is not needed to make the reaction thermodynamically favored.
   (D) No, since ΔH_{rxn} is negative and ΔS_{rxn} is positive, ΔG_{rxn} will always be a negative value and the increase in temperature is not needed to make the reaction thermodynamically favored.

7. While most substances will have standard entropy values that are positive, some common ions do not, as shown in the table below.

Substance	$S°$ (J/mol × K)
$OH^-(aq)$	–11
$Al^{3+}(aq)$	–321.75
$Be^{2+}(aq)$	–129.70
$F^-(aq)$	–9.6

Account for the fact that these ions do not have a positive value.
(A) Since the salts containing these ions dissolve in water, ΔG_{rxn} is negative; in order for this to happen, ΔS_{rxn} needs to be negative.
(B) The hydrating water molecules are arranged around the ions in a very ordered state, resulting in a decrease in the disorder of the system.
(C) The relatively small size of the ions minimizes interaction with the surroundings and creates a more ordered system.
(D) To remove an electron from an atom requires energy to overcome the electrostatic attraction between the nucleus and the valence electrons. The addition of an electron requires energy to overcome the electron–electron repulsion.

8. In which of the following four processes is there an increase in entropy?
(A) $2\ SO_2(g) + O_2(g) \rightarrow 2\ SO_3(g)$
(B) $H_2O(g) \rightarrow H_2O(s)$
(C) $Hg(g) \rightarrow Hg(l)$
(D) $H_2O_2(l) \rightarrow H_2O(l) + \frac{1}{2}\ O_2(g)$

9. $H_2O(l) \rightarrow H_2O(g)$
What are the signs of $\Delta H°$, $\Delta S°$, and $\Delta G°$ for this process at 25°C?

	$\Delta H°$	$\Delta S°$	$\Delta G°$
(A)	–	–	+
(B)	–	+	–
(C)	+	+	+
(D)	+	+	–

10. In the reaction $PCl_3(g) + Cl_2(g) \rightarrow PCl_5(g)$, $\Delta G_{rxn} = -92.5$ kJ/mol$_{rxn}$. Which is correct concerning the changes in ΔG as the temperature of the reaction is increased?
    (A) ΔG_{rxn} becomes more negative. $\Delta G_{rxn} = \Delta H - T\Delta S$. Since ΔH is positive and ΔS is positive, a higher temperature will result in a more negative ΔG_{rxn}.
    (B) ΔG_{rxn} becomes less negative (more positive). $\Delta G_{rxn} = \Delta H - T\Delta S$. Since ΔH is positive and ΔS is negative, a higher temperature will result in a less negative ΔG_{rxn}.
    (C) ΔG_{rxn} becomes more negative. $\Delta G_{rxn} = \Delta H - T\Delta S$. Since ΔH is negative and ΔS is positive, a higher temperature will result in a more negative ΔG_{rxn}.
    (D) ΔG_{rxn} becomes less negative (more positive). $\Delta G_{rxn} = \Delta H - T\Delta S$. Since ΔH is negative and ΔS is negative, a higher temperature will result in a less negative ΔG_{rxn}.

11. A reaction takes place within a system. As a result, the entropy of the system decreases. Which of the following statements *must* be true?
    (A) The reaction is endothermic.
    (B) The entropy of the universe decreases.
    (C) The amount of work that the system can do will be dependent on the decrease in entropy.
    (D) The entropy of the surroundings increases.

12. When propane burns in air, heat is released:

    $C_3H_8(g) + 5\ O_2(g) \rightarrow 3\ CO_2(g) + 4\ H_2O(g)$

    What are the signs of ΔH, ΔS, and ΔG for this process as illustrated by the above equation?

	ΔH	ΔS	ΔG
(A)	–	+	+
(B)	–	+	–
(C)	–	–	+
(D)	+	+	–

$2\ Al(s) + 6\ HCl(aq) \rightarrow 2\ AlCl_3(aq) + 3\ H_2(g)$

13. A student reacted aluminum metal with excess $HCl(aq)$ and collected two different volumes of hydrogen gas as shown below. Sample 1 produced 1.1 L of hydrogen gas, and Sample 2 produced 3.3 L of hydrogen gas at STP.

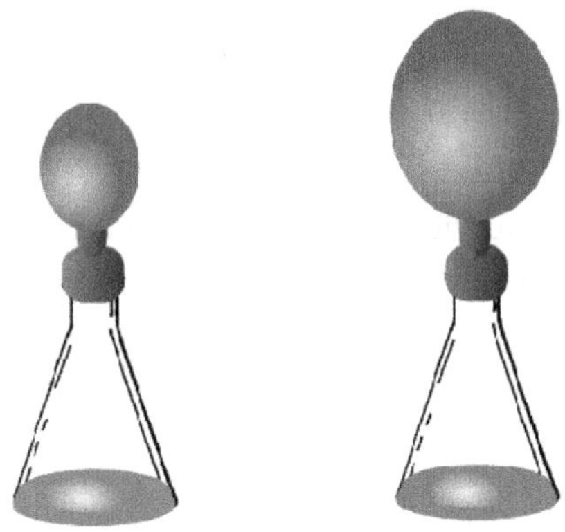

Sample 1 **Sample 2**

Which statement is correct?

(A) The sign of work, *w*, in both cases is positive as the system did work on the surroundings. The average kinetic energy of the particles in Sample 2 is greater than the average kinetic energy of the particles in Sample 1.
(B) The sign of work, *w*, in both cases is negative as the system did work on the surroundings. The average kinetic energy of the particles in Sample 2 is the same as the average kinetic energy of the particles in Sample 1.
(C) The sign of work, *w*, in both cases is negative as the surroundings did work on the system. The average kinetic energy of the particles in Sample 2 is greater than the average kinetic energy of the particles in Sample 1.
(D) The sign of work, *w*, in both cases is positive as the surroundings did work on the system. The average kinetic energy of the particles in Sample 2 is the same as the average kinetic energy of the particles in Sample 1.

14. Determine ΔG_{rxn} for the following reaction

$2\ Fe_2O_3(s) + 6\ CO(g) \rightarrow 4\ Fe(s) + 6\ CO_2(g)$

given the following information.

$2\ Fe_2O_3(s) \rightarrow 4\ Fe(s) + 3\ O_2(g)$ $\Delta G_{rxn} = +1487\ kJ/mol_{rxn}$

$2\ CO(g) + O_2(g) \rightarrow 2CO_2(g)$ $\Delta G_{rxn} = -514\ kJ/mol_{rxn}$

(A) $-2001\ kJ/mol_{rxn}$
(B) $-55\ kJ/mol_{rxn}$
(C) $+973\ kJ/mol_{rxn}$
(D) $+3947\ kJ/mol_{rxn}$

15. One of the most important ores of copper is chalcocite, which is copper(I) sulfide (Cu_2S). It has been mined and processed for centuries. It is one of the most profitable copper ores since it is easy to reduce the copper and it has a very high copper percent by mass (80%).

Reaction	ΔG°_f(kJ/mol)
$Cu_2S(s) \rightarrow 2\ Cu(s) + S(s)$	+86.2
$SO_2(g) \rightarrow S(s) + O_2(g)$	+300.1

The overall process for the production of copper from chalcocite is $Cu_2S(s) + O_2(g) \rightarrow 2\ Cu(s) + SO_2(g)$. What is ΔG_{rxn} for the production of copper?
(A) –386.3 kJ/mol_{rxn}
(B) –213.9 kJ/mol_{rxn}
(C) +213.9 kJ/mol_{rxn}
(D) +386.3 kJ/mol_{rxn}

FREE-RESPONSE QUESTIONS

1. Hydrazine, N_2H_4, is used as a rocket fuel. When used along with dinitrogen tetroxide, N_2O_4, the reaction is hypergolic; the two fuels instantly ignite when they come into contact with each other. Another use for hydrazine is as a monopropellant for maneuvering thrusters on space craft. The Voyager spacecraft uses hydrazine thrusters to align its transmitter antennae to send data to Earth.
   The Lewis diagrams for hydrazine and dinitrogen tetroxide are shown below.

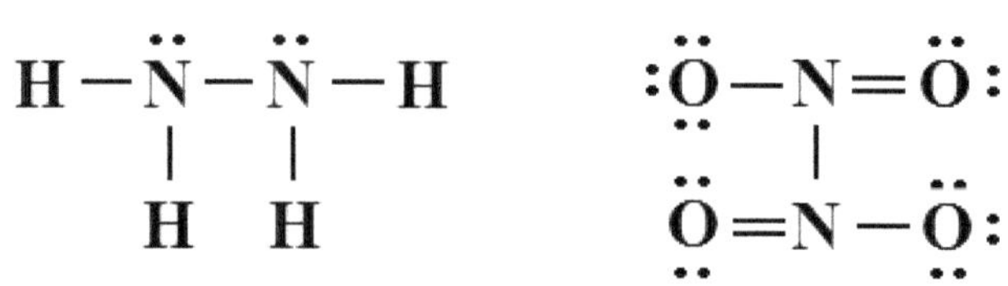

(a) Is the bond angle of the H–N–N bond in hydrazine greater than, equal to, or less than the O–N–N bond in dinitrogen tetroxide? Justify your answer.

When hydrazine is used in thrusters, the following three reactions occur and the overall equation is shown.

Step 1	$3\ N_2H_4(l) \rightarrow 4\ NH_3(g) + N_2(g)$	$\Delta H_{rxn} = -235\ kJ/mol_{rxn}$
Step 2	$N_2H_4(l) \rightarrow N_2(g) + 2\ H_2(g)$	$\Delta H_{rxn} = ?$
Step 3	$4\ NH_3(g) + N_2H_4(l) \rightarrow 3\ N_2(g) + 8\ H_2(g)$	$\Delta H_{rxn} = +133\ kJ/mol_{rxn}$
Overall reaction	$5\ N_2H_4(l) \rightarrow 5\ N_2(g) + 10\ H_2(g)$	$\Delta H_{rxn} = -153\ kJ/mol_{rxn}$

(b) Calculate ΔH_{rxn} for step 2.
(c) What is the sign of the change in entropy, ΔS_{rxn}, for step 3? Explain your reasoning.

(d) Is the reaction for step 3 driven by entropy, enthalpy, or both? Explain your reasoning.

(e) ΔH_{rxn} can be calculated three ways: combining reactions (which was done to get ΔH_{rxn} for the overall reaction), from a table of $\Delta H°_f$ values (which was done to get ΔH_{rxn} values for steps 1, 2, and 3 above) or from a table of bond energies, shown below.

Bonds	Bond Energy (kJ/mol)
N≡N	941
H–H	432
N–H	391
N–N	160

(i) Calculate ΔH_{rxn} for step 3 using the table of bond energies provided above.

(ii) Account for the reason why the two values do not agree.

The reaction between hydrazine and dinitrogen tetroxide is shown below.

$$2\ N_2H_4(l) + N_2O_4(l) \rightarrow 3\ N_2(g) + 4\ H_2O(g)$$

The value for ΔH_{rxn} is –1049.0 kJ/mol$_{rxn}$, and the value for ΔG_{rxn} is –1310.7 kJ/mol$_{rxn}$.

(f) Given the table below, calculate the absolute entropy, $S°$, of nitrogen gas, N_2 at 25°C.

Substance	$\Delta H°_f$ (kJ/mol)	$\Delta G°_f$ (kJ/mol)	$S°$ (J/mol × K)
$N_2H_4(l)$	51	149	121
$N_2O_4(l)$	–20	97	209
$N_2(g)$	0	0	?
$H_2O(g)$	–242	–229	189

(g) The normal boiling point of $N_2H_4(l)$ is 114°C and that of $N_2O_4(l)$ is 21.1°C. Account for the difference in boiling points in terms of the intermolecular forces present in **each** liquid.

2. The Fischer–Tropsch process was first developed in 1925 and consists of a series of chemical reactions that ultimately convert a mixture of carbon monoxide gas and hydrogen gas into liquid hydrocarbons which can be used for the production of synthetic lubrication oil and synthetic fuel. Typically, the starting point for the conversion is coal, natural gas, or biomass.

The Fischer–Tropsch process was widely used by Germany in World War II to produce fuel.

There are several different starting points for the Fischer–Tropsch process, depending on the starting reactants used as the feedstock. If using methane as the source of carbon monoxide and hydrogen gases, the "steam reforming" reaction, shown below, is used.

$$H_2O(g) + CH_4(g) \rightarrow CO(g) + 3\ H_2(g)$$

Useful thermodynamic data are shown in the table below.

	$\Delta H°_f$ (kJ/mol)	$S°$ (J/mol × K)	$\Delta G°_f$ (kJ/mol)
$H_2O(g)$	–241.82	188.83	–228.57
$CH_4(g)$	–74.81	186.26	–50.72
$CO(g)$	–110.53	197.67	–137.17
$H_2(g)$	0	130.68	0

(a) Calculate ΔH_{rxn}.

(b) Which has higher bond energy—the reactants or products?

(c) Calculate ΔS_{rxn}.

(d) Is this reaction favored at low temperatures, high temperatures, all temperatures, or no temperatures? Circle your choice and justify your answer.

low high all no

Answers

MULTIPLE-CHOICE QUESTIONS

1. C The reaction is thermodynamically favored as indicated by the negative ΔG value. However, since the reaction does not seem to take place over time, this reaction must be under kinetic control. Diamonds are a covalent network solid; therefore, the bonds between the carbon atoms are very strong and require a large amount of energy to break (LO 5.18).

2. A The temperature of the system decreased so the reaction is endothermic, and ΔH is positive. Since the crystals dissolved, ΔG must be negative, which means that ΔS must be positive. Both ΔH and ΔS are positive; therefore, the increase in temperature will result in a more negative ΔG ($\Delta G = \Delta H - T\Delta S$) (LO 5.13).

3. B The increase in entropy resulting from the reaction described would be the production of a gas and the dissolving of the metal block. Choices A, C, and D all are associated with the production of a gas (LO 5.3).

4. A From $\Delta G = \Delta H - T\Delta S$, if the temperature is high, the $T\Delta S$ factor is large and "overcomes" the influence of the ΔH term (+). Therefore, ΔG is negative, and the reaction becomes thermodynamically favorable at higher temperatures (LO 5.13).

5. C When the system does work on the surroundings, the sign is negative (–). As shown by the balloon, the system did work on the surroundings by pushing the air molecules out of the way when the balloon expanded (LO 5.4).

6. D ΔH is negative (exothermic), and ΔS is positive (the reaction shown goes from 9 moles of gas reactants to 10 moles of gas product, an increase in entropy). $\Delta G = \Delta H - T\Delta S$ results in a positive number – a positive number, so ΔG will always be a negative number (i.e., thermodynamically favored). The high heat and pressure must be for other reasons (kinetic—breaking the double bond in the oxygen molecule requires the molecules to be moving faster and to increase the yield) (LO 5.13).

7. B The first thing to notice is that all of these ions are dissolved in water. When the particles dissolve, there will be strong ion–dipole interactions between the ions and the water molecules. This strong attraction makes the formation of the aqueous ions more ordered than the individual ions (LO 5.10 and 5.13).

8. D If there is an increase in entropy, then the products must be more disordered than the reactants. In choice (A), the system goes from 3 moles of gas to 1 mole of gas. In choice (B), the system goes from high entropy gas to low entropy solid. In choice (C), the system goes from high entropy gas to lower entropy liquid. The last choice goes from 1 mole of liquid to 1 mole of liquid and ½ mole of gas, an increase in entropy (LO 5.12).

9. C This reaction is known to happen; however, it does not become thermodynamically favored until the temperature reaches 100°C. ΔG is positive, but some evaporation will happen at temperatures below 100°C. It requires energy to move from the liquid phase to the gas phase so ΔH is positive (for humans sweating, heat is drawn from the body, cooling it in the process). Since a liquid is converted into a gas, entropy increases, so ΔS is positive (LO 5.13).

10. D Since the reaction goes from 2 moles of gas reactants to 1 mole of gas product, the entropy decreases, so ΔS is negative. In order for ΔG to be negative, ΔH must be negative. $\Delta G = \Delta H - T\Delta S$ and if both ΔH and ΔS are negative, as the temperature increases, the value for ΔG becomes less negative (more positive) (LO 5.13).

11. D The entropy of the universe must increase and the free energy of the system must decrease, as these are expressions of the second law. If the entropy of the system decreases, the entropy of the surroundings must increase. If the entropy of the surroundings

is to increase, the reaction must be exothermic. Therefore, only the last statement is true (LO 5.13).

12. B Since the reaction occurs, we know it is thermodynamically favored, which means ΔG must be negative. The reaction is exothermic, meaning that ΔH is negative. Since there is an increase in the number of molecules of gas, there is an increase in the randomness, meaning that ΔS is positive (LO 5.13).

13. B When the system does work on the surroundings, the sign is negative (–). As shown by the balloon, the system did work on the surroundings by pushing the air molecules out of the way when the balloon expanded. Since the temperature is the same in the two balloons, the average kinetic energy of the particles is the same (LO 5.2 and 5.6).

14. B In order to get the overall equation of

$$2\ Fe_2O_3(s) + 6\ CO(g) \rightarrow 4\ Fe(s) + 6\ CO_2(g)$$

the equations

$2\ Fe_2O_3(s) \rightarrow 4\ Fe(s) + 3\ O_2(g)$ $\qquad \Delta G_{rxn} = +1487\ kJ/mol_{rxn}$

$2\ CO(g) + O_2(g) \rightarrow 2\ CO_2(g)$ $\qquad \Delta G_{rxn} = -514\ kJ/mol_{rxn}$

need to be arranged as shown above, and the last equation needs to be multiplied by 3.

$3 \times [2\ CO(g) + O_2(g) \rightarrow 2\ CO_2(g)]$ $\qquad \Delta G_{rxn} = 3 \times -514\ kJ/mol_{rxn}$

Next, add up the equations:

$2\ Fe_2O_3(s) \rightarrow 4\ Fe(s) + 3\ O_2(g)$ $\qquad \Delta G_{rxn} = +1487\ kJ/mol_{rxn}$

$6\ CO(g) + 3\ O_2(g) \rightarrow 6\ CO_2(g)$ $\qquad \Delta G_{rxn} = -1542\ kJ/mol_{rxn}$

Overall equation is

$2\ Fe_2O_3(s) + 6\ CO(g) + \cancel{3\ O_2(g)} \rightarrow 6\ CO_2(g) + 4\ Fe(s) + \cancel{3\ O_2(g)}$

and $\Delta G_{rxn} = (-1542 + 1487) = -55\ kJ/mol_{rxn}$ (LO 5.15).

15. B In order to get the overall equation of $Cu_2S(s) + O_2(g) \rightarrow 2\ Cu(s) + SO_2(g)$, the equations need to be arranged:

	$\Delta G°_f$ (kJ/mol)
$Cu_2S(s) \rightarrow 2\ Cu(s) + S(s)$	+86.2
$S(s) + O_2(g) \rightarrow SO_2(g)$	–300.1

$\Delta G_{rxn} = (-300.1 + 86.2) = -213.9\ kJ/mol_{rxn}$

FREE-RESPONSE QUESTIONS

1. (a) Hydrazine is sp^3 hybridized and has bond angles of < 109.5°, while dinitrogen tetroxide is sp^2 hybridized and has bond angles of 120°. Therefore, the bond angle in hydrazine is less than the bond angle in dinitrogen tetroxide (LO 2.21).

   (b) –51 kJ/mol$_{rxn}$. The overall reaction is the sum of steps 1, 2, and 3.

   $$\Delta G_{overall} = \Delta G_{step\ 1} + \Delta G_{step\ 2} + \Delta G_{step\ 3}$$

   $$-153 = -235 + ? + 133 \qquad \text{so } ? = -51 \text{ kJ/mol}_{rxn}$$

   (LO 5.8)

   (c) Positive (+). The reaction has 4 moles of gas and 1 mole of liquid reactants and produces 11 moles of gas products. This is an increase in entropy (LO 5.12)

   (d) Entropy. Since ΔH is positive and ΔS is positive in order for $\Delta G < 0$, the magnitude of $T\Delta S$ must be greater than ΔH (LO 5.13)

   (e) (i) $\Delta H°_{rxn} = \sum \textit{Bond Energies}_{\text{bonds broken}} - \sum \textit{Bond Energies}_{\text{bonds formed}}$

Bonds Broken	Bond Energy (kJ/mol)		Bonds Formed	Bond Energy (kJ/mol)
16 N–H	16 × 391 = 6256		3 N≡N	3 × 941 = 2823
1 N–N	1 × 160 = 160		8 H–H	8 × 432 = 3456
Total	6416 kJ		Total	6279 kJ

   $\Delta H_{rxn} = 6416 - 6279 = 137$ kJ/mol$_{rxn}$ (LO 5.8)

   (ii) The ΔH_{rxn} given in step 3 is 133 kJ/mol. This value was calculated using ΔH_f values. Bond energy found in tables is the average of that type of bond and may not represent the energy needed to break a specific bond of that type, just the average of breaking all of the bonds of that type (LO 5.8).

   (f) In order to find the absolute entropy of nitrogen, find ΔS of the reaction using $\Delta G = \Delta H - T\Delta S$.

   $$\Delta S = \frac{\Delta H - \Delta G}{T} \Rightarrow \Delta S = \frac{-1{,}049{,}000 - 1{,}310{,}700 \text{ J/mol}}{(25 + 273)\text{ K}} = \frac{878 \text{ J}}{\text{mol} \times \text{K}}$$

   Then to find $S°$ of nitrogen gas, use the equation

   $$\Delta S°_{rxn} = \Sigma\Delta S°_{products} - \Sigma\Delta S°_{reactants} \Rightarrow$$

   $$\frac{878 \text{ J}}{\text{mol} \times \text{K}} = [[(3 \times X) + (4 \times 189)] - [(1 \times 209) + (2 \times 121)]]$$

   $$X = \frac{191 \text{ J}}{\text{mol} \times \text{K}} = S° \text{ for nitrogen gas}$$

   (LO 5.13)

(g) $N_2H_4(l)$ exhibits London dispersion forces, dipole–dipole interactions, and hydrogen bonding. $N_2O_4(l)$ exhibits London dispersion forces and dipole–dipole interactions. The intermolecular forces in $N_2H_4(l)$ are stronger than the intermolecular forces in $N_2O_4(l)$ which accounts for the higher boiling point in $N_2H_4(l)$ (LO 2.16).

2. (a)

$$\Delta H^\circ_{rxn} = \Sigma\Delta H^\circ_{f_{products}} - \Sigma\Delta H^\circ_{f_{reactants}}$$

$$\Delta H^\circ_{rxn} = [(3 \times 0) + (1 \times -110.53)] - [(1 \times -241.82) + (1 \times -74.81)]$$

$$\Delta H^\circ_{rxn} = \frac{206.10 \text{ kJ}}{\text{mol}}$$

(LO 5.6)

(b) Since the change in enthalpy is positive, more energy was needed to break the bonds of the reactants than was released in the formation of the bonds of the products (LO 5.8).

(c)

$$\Delta S^\circ_{rxn} = \Sigma\Delta S^\circ_{products} - \Sigma\Delta S^\circ_{reactants} \Rightarrow$$

$$\Delta S^\circ_{rxn} = [[(3 \times 130.68) + (1 \times 197.67)] - [(1 \times 188.83) + (1 \times 186.26)]] \Rightarrow$$

$$\Delta S^\circ_{rxn} = \frac{214.62 \text{ J}}{\text{mol} \times \text{K}}$$

(LO 5.13)

(d) High should be circled. The justification is that $\Delta G_{rxn} = \Delta H_{rxn} - T\Delta S_{rxn}$ and since the enthalpy is positive and the entropy is positive, ΔG_{rxn} = + number – + number and the sign of ΔG_{rxn} will be determined by the magnitude of $T\Delta S$ (LO 5.13).

11

BIG IDEA 6: GENERAL EQUILIBRIUM

Big Idea 6

Any bond or intermolecular attraction that can be formed can be broken. These two processes are in a dynamic competition, sensitive to initial conditions and external perturbations.

In this chapter, you will review the characteristics of equilibrium and calculations involving the concentrations of reactants and products for a given system at equilibrium. Your knowledge of the basic equilibrium problems in this chapter will enable you to qualitatively and quantitatively explain pH of acids, bases, salts, and buffers as well as the solubility of ionic compounds. Understanding general equilibrium is fundamental to understanding the next three chapters (all involving types of equilibrium) as well.

AP® Tip

Equilibrium is one of the big ideas in AP® Chemistry that underpins much of what you study. Make sure you are comfortable with particle and graphical representations as well as with the calculations.

You should be able to:

- Write equilibrium expressions for a given reaction.
- Calculate Q and compare it to K to determine if a reaction is at equilibrium.
- Manipulate K if a reaction is reversed or multiplied by a coefficient.

- Determine the value of K for an overall equation given the K values of the individual reactions that make up the overall equation.
- Calculate K from given equilibrium concentrations, or if given K and all except one equilibrium concentration, solve for the missing value.
- Calculate equilibrium concentrations (or one of the missing variables) if given any two of the following values: K, the initial concentrations, or one equilibrium concentration.
- Do calculations involving gaseous equilibria, partial pressures, and K_p.
- Use Le Chatelier's principle to determine in what direction the position of equilibrium will shift when a change is imposed.
- Calculate the value of K from thermodynamic values such as $\Delta G°$.
- Calculate the value of K from the cell potential, $E°$.

EQUILIBRIUM CONDITION

Many chemical and physical changes are reversible. When equilibrium is reached, the *rate* of the forward reaction equals the *rate* of the reverse reaction.

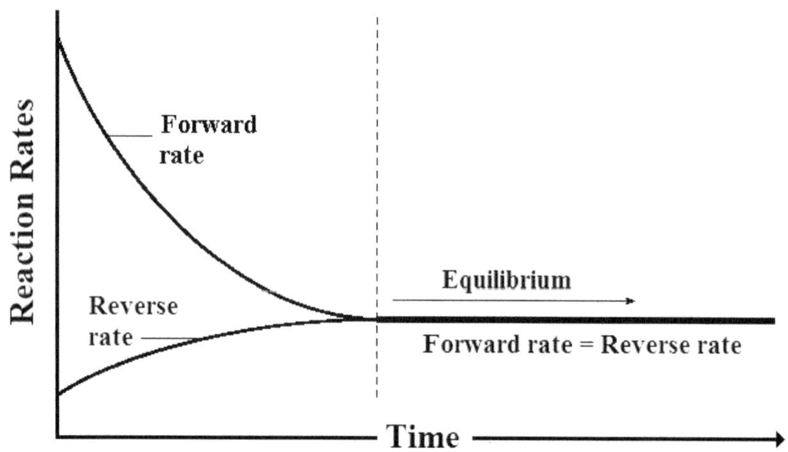

The ratio of the product concentrations to the reactant concentrations is constant at equilibrium. At a given set of conditions, macroscopic variables such as concentrations, partial pressures, and temperature do not change over time. Most often, the concentrations of products and reactants are not equal at equilibrium. Only the rates of forward and reverse reactions are equal.

For the reaction, $N_2(g) + 3\ H_2(g) \leftrightarrows 2\ NH_3(g)$, the change in concentration of reactants and products over time is shown below. Note the constant concentrations when equilibrium is reached.

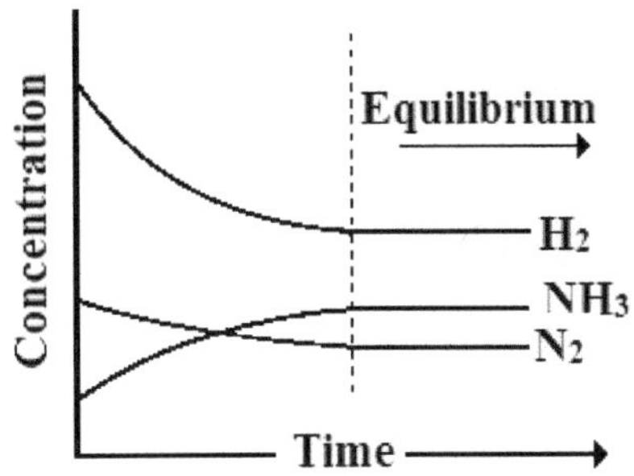

Some examples of reversible processes include biological processes such as oxygen binding to hemoglobin in red blood cells and the sense of smell due to molecules binding to receptor sites in the nose. Environmental examples include nitrogen and carbon biogeochemical cycles. Chemistry examples include dissolution of solids as well as proton transfer in acid–base reactions and electron transfer in redox reactions.

THE EQUILIBRIUM EXPRESSION

The equilibrium constant, *K*, measures the ratio of product concentrations (in molarity) or pressures (in atmospheres) to reactant concentrations or pressures.

For a hypothetical reaction $a\,\mathrm{A} + b\,\mathrm{B} \rightleftarrows c\,\mathrm{C} + d\,\mathrm{D}$

$$K = \frac{[\mathrm{C}]^c\,[\mathrm{D}]^d}{[\mathrm{A}]^a\,[\mathrm{B}]^b}$$

K is the equilibrium constant, usually given without units. Brackets, [ ], represent the concentration of the reactants and products at equilibrium in moles/liter. The concentrations of products and reactants are raised to the powers of their respective coefficients in the balanced chemical equation. Note that this is different from kinetics where the orders of each reactant must be determined experimentally. If pressures are used, *P* or (*P*) (for partial pressure) is used instead of brackets. Often, the *K* will then have the subscript *p* added to indicate that it is the equilibrium constant for pressure.

$$K_P = \frac{(P_\mathrm{C})^c(P_\mathrm{D})^d}{(P_\mathrm{A})^a(P_\mathrm{B})^b}$$

You will be expected to write the equilibrium expression for a given reaction. The concentrations of pure solids and pure liquids are not included because their concentrations are constant and are built into the *K* value.

TYPES OF EQUILIBRIUM EXPRESSIONS

K and K_{eq} are generic terms for an equilibrium constant. More specific terms specify the type of reaction.

K_c = equilibrium constant for general reactions using molarity

K_p = equilibrium constant for general gas phase reactions using pressure in atm

K_a = equilibrium constant for the dissociation of a weak acid (uses *M*)

K_b = equilibrium constant for the dissociation of a weak base (uses *M*)

K_w = equilibrium constant for the dissociation of water (uses *M*)

K_{sp} = equilibrium constant for the dissociation of a sparingly soluble salt (uses *M*)

AP® Tip

Always set up the equilibrium expression without any numbers in it before adding concentrations (or pressures). This will greatly lower the chance of error. Sometimes, writing the equilibrium expression will be a separate step on the AP® exam.

EXAMPLE: Write the equilibrium expression for the reaction

$$P_4(s) + 5\ O_2(g) \rightleftarrows P_4O_{10}(s)$$

SOLUTION: The equilibrium expression for the reaction is

$$K = \frac{1}{[O_2]^5} \quad \text{or} \quad K_p = \frac{1}{(P_{O_2})^5}$$

The concentrations of the solids are not included in the expression because the concentration of a solid or liquid is a constant.

The value of the equilibrium constant measures the extent to which a reaction occurs.

$K >> 1$: The concentrations of the products are much greater than the concentrations of the reactants so the reaction is product-favored.

$K << 1$: The concentrations of the reactants are much greater than the concentrations of the products. In this case, the reaction does not proceed in the direction written to any great extent and is called a reactant-favored reaction.

$K \approx 1$: Reactants and products are both present in significant concentrations at equilibrium when *K* is between 0.001 and 1000.

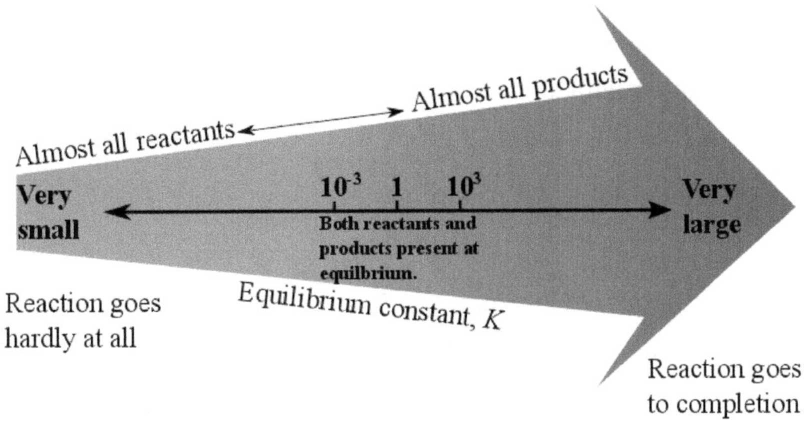

REACTION QUOTIENT

When reactants and products are mixed together, they may not initially be at equilibrium. The reaction quotient, Q, is set up like K, but uses initial concentrations instead of equilibrium concentrations. Comparing Q to K will determine which way a system will shift to reach equilibrium. If a system is not at equilibrium, it will move in a direction to reach equilibrium.

If a reactant or product in a reaction is not present and is at zero concentration, the reaction will move in the direction that produces the missing component.

If all reactants and products are present and have an initial concentration, you must determine the value of Q, the reaction quotient.

To determine if the reaction is at equilibrium or the direction it will shift to attain equilibrium, plug all of the initial concentrations into the reaction quotient, which is the same as the equilibrium expression (also called the law of mass action), and compare the value of Q to K.

AP® Tip

If $Q = K$, the reaction is at equilibrium and the rate of the forward and reverse reactions are the same.

If $Q > K$, the reaction will shift to the left. A shift toward the reactants will consume products because the rate of the reverse reaction is greater than the rate of the forward reaction until equilibrium is established.

If $Q < K$, the reaction will shift to the right. A shift to produce more products will consume reactants because the rate of the forward reaction is greater than the rate of the reverse reaction until equilibrium is established.

EXAMPLE: For the reaction $2\ NO(g) \rightleftarrows N_2(g) + O_2(g)$, $K = 2.4 \times 10^3$ at a particular temperature.

(a) If the initial concentrations are 0.024 M NO, 2.0 M N_2, and 2.6 M O_2, is the system at equilibrium?

(b) If it is not at equilibrium, in which direction will the reaction shift?

SOLUTION:

$$Q = \frac{[N_2][O_2]}{[NO]^2} = \frac{(2.0)(2.6)}{(0.024)^2} = 9.0 \times 10^3$$

The reaction is not at equilibrium, $Q > K$, so it will shift to the left.

Note: If the problem asked for a justification, your answer can be written: "Since $Q > K$ the numerator is too large and the denominator is too small, so the reaction will shift to reduce the amount of product and increase the amount of reactant."

Memory Trick

One method of remembering which direction the equilibrim shifts is to place Q and K on a number line. Then draw an arrow from Q to K (since you will always move toward equilibrium).

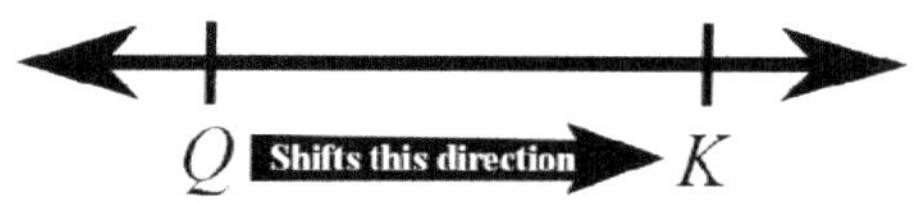

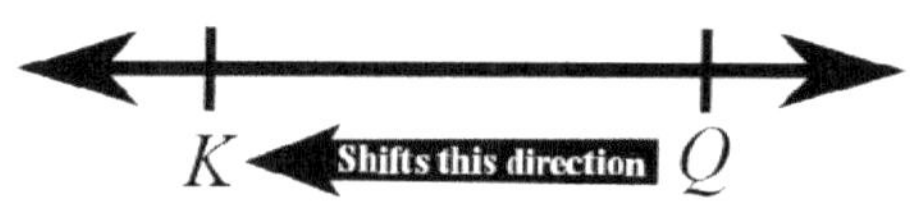

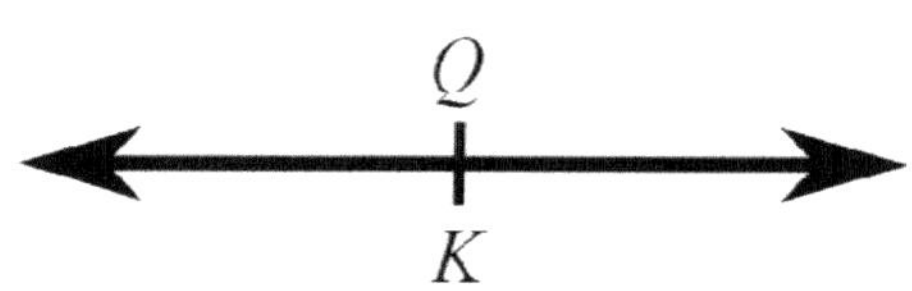

GASEOUS EQUILIBRIUM

For equilibrium in the gas phase, the equilibrium expression can be written in terms of the partial pressures of the gases.

For the reaction $2\ AsH_3(g) \rightleftarrows 2\ As(s) + 3\ H_2(g)$,

$$K_p = \frac{(P_{H_2})^3}{(P_{AsH_3})^2}$$

K_p is the equilibrium constant in terms of the partial pressures of the gases.

P represents the partial pressure of the gases in atmospheres. They are raised to their coefficients in the balanced chemical equation. Notice that parentheses, ( ), may be used for the partial pressure. The use of brackets, [ ], in this instance could result in loss of points on the exam since brackets represent concentration in mol/L.

TYPES OF EQUILIBRIUM PROBLEMS

The types of equilibrium problems you can expect to solve on the AP® Chemistry exam include:

1. **Manipulation of the equilibrium constant, *K*.**
   a. If the reaction is reversed, the equilibrium expression is the reciprocal of the expression for the forward reaction.

   EXAMPLE:

$$\mathrm{A} \rightleftarrows \mathrm{B} \quad K = \frac{[\mathrm{B}]}{[\mathrm{A}]}$$

$$\mathrm{B} \rightleftarrows \mathrm{A} \quad K' = \frac{1}{\mathrm{K}} = \frac{[\mathrm{A}]}{[\mathrm{B}]}$$

   b. If the coefficients in a balanced equation are multiplied by a number, *n*, the equilibrium constant is raised to the power *n*.

   EXAMPLE:

$$\mathrm{A} \rightleftarrows \mathrm{B} \quad K = \frac{[\mathrm{B}]}{[\mathrm{A}]}$$

$$2\,\mathrm{A} \rightleftarrows 2\mathrm{B} \qquad K'' = K^2 = \frac{[\mathrm{B}]^2}{[\mathrm{A}]^2}$$

   c. If two or more reactions are added together through the presence of one or more common intermediates, the equilibrium constant of the resulting reaction is the product of the values of *K* for the original reactions.

   EXAMPLE:

$$\text{eqn 1. } \mathrm{A} \rightleftarrows \mathrm{B} : K_1 = \frac{[\mathrm{B}]}{[\mathrm{A}]}$$

$$\text{eqn 2. } 3\,\mathrm{B} \rightleftarrows \mathrm{C} : K_2 = \frac{[\mathrm{C}]}{[\mathrm{B}]^3}$$

$$\text{overall eqn. } \mathrm{A} + 2\,\mathrm{B} \rightleftarrows \mathrm{C} \qquad K = K_1 \times K_2 = \frac{[\mathrm{C}]}{[\mathrm{A}][\mathrm{B}]^2}$$

EXAMPLE: (Type 1) For the reaction

$N_2(g) + 3\ H_2(g) \rightleftarrows 2\ NH_3(g)$, $K = 1.3 \times 10^{-2}$ at a certain temperature.

Calculate the value of K, called K', for the reaction

$NH_3(g) \rightleftarrows \frac{1}{2}\ N_2(g) + \frac{3}{2}\ H_2(g)$

SOLUTION: The reaction is the reverse and one-half of the one that is given.

$$K' = \frac{([N_2]^{\frac{1}{2}}\,[H_2]^{\frac{3}{2}})}{[NH_3]} = \left(\frac{1}{K}\right)^{\frac{1}{2}} = \left(\frac{1}{1.3 \times 10^{-2}}\right)^{\frac{1}{2}} = 8.8$$

2. **Given all equilibrium concentrations, calculate the value of the constant, *K*.**

   EXAMPLE: (Type 2) Consider the reaction $2\ NOCl(g) \leftrightarrows 2\ NO(g) + Cl_2(g)$ at 35°C. Varying amounts of $NOCl(g)$, $NO(g)$, and $Cl_2(g)$ are mixed in a 10.0-L flask. After the system has reached equilibrium, the concentrations are observed to be:

   $$[Cl_2] = 0.150\ M,\ [NO] = 0.00350\ M,\ [NOCl] = 0.400\ M$$

   Calculate the value of K for this system at 35°C.

   SOLUTION:

   $$K = \frac{[NO]^2[Cl_2]}{[NOCl]^2} = \frac{(0.00350)^2(0.150)}{(0.400)^2} = 1.15 \times 10^{-5}$$

3. **Given the value of the equilibrium constant, *K*, and all but one of the equilibrium concentrations, solve for the missing concentration.**

   EXAMPLE: (Type 3) Given the following reaction:

   $$2\ NO(g) + Br_2(g) \rightleftarrows 2\ NOBr(g),\ K_p = 109 \text{ at } 25°C$$

   The equilibrium partial pressure of Br_2 = 0.0159 atm and NOBr = 0.0768 atm. Calculate the equilibrium partial pressure of NO.

   $$K_P = \frac{(P_{NOBr})^2}{(P_{NO})^2(P_{Br_2})}$$

   $$109 = \frac{(0.0768\ \text{atm})^2}{(P_{NO})^2(0.0159\ \text{atm})}$$

   $$P_{NO} = 0.0583\ \text{atm}$$

4. **Given the value of the initial concentrations of the reactants and one of the equilibrium concentrations of the reactants or**

products (or total pressure, if gaseous), solve for all equilibrium concentrations and the value of *K*.

The fourth and fifth types of problems can be solved by using an ICE chart. ICE is an acronym for Initial, Change, Equilibrium and is a convenient method for organizing your work. The use of an ICE chart is illustrated below. An ICE (or RICE, with the R representing the Reaction) chart is extremely helpful for showing work, both for the person working the problem and the person grading the problem.

EXAMPLE: Given the following reaction:

$$2\ NO(g) + O_2(g) \rightleftarrows 2\ NO_2(g)$$

NO and O_2 gases were added to a rigid, evacuated container at pressures of 0.620 atm and 0.540 atm, respectively, with a temperature of 200. K. Once the gases reacted and reached equilibrium, the total pressure in the container was found to be 0.950 atm. Determine the value of K_p for this reaction at 200. K.

SOLUTION:

	$2\ NO(g)$	+	$O_2(g)$	$\rightleftarrows$	$2\ NO_2(g)$	P_{total}
I	0.620 atm		0.540 atm		0	1.160 atm
C	$-2x$		$-x$		$+2x$	$-x$
E	$0.620 - 2x$		$0.540 - x$		$2x$	0.950 atm

The addition of a P_{total} column makes this easier to solve with an ICE diagram. The change in total pressure is $(-x)$ and is equal to 1.160 – 0.950 or 0.210 atm.

$P_{NO} = 0.620 - 2(0.210) = 0.200$ atm

$P_{O_2} = 0.540 - 0.210 = 0.330$ atm

$P_{NO_2} = 2(0.210) = 0.420$ atm

$$K_p = \frac{(P_{NO_2})^2}{(P_{NO})^2(P_{O_2})} = \frac{(0.420)^2}{(0.200)^2(0.330)} = 13.4$$

Two alternative methods in which the same information may be presented in a question are shown below.

ALTERNATIVE PROBLEM: Given the following reaction:

$$2\ NO(g) + O_2(g) \rightleftarrows 2\ NO_2(g)$$

NO and O_2 gases were added to a rigid, evacuated container at pressures of 0.620 atm and 0.540 atm, respectively, with a temperature of 200. K. As the reaction proceeded, the total pressure in the container was monitored, and the results are shown in the table below.

Time (hours)	Total Pressure of Gases in Container at 200. K (atm)
0.0	1.160
1.0	1.094
2.0	1.002
3.0	0.986
4.0	0.950
5.0	0.950

The student will have to interpret the data in the table to determine the equilibrium pressure. In this case, look to see where the total pressure remains unchanged. There will be at least two measurements where the total pressure remains unchanged to ensure that the reaction has reached equilibrium.

ALTERNATIVE PROBLEM: Given the following reaction:

$$2\ NO(g) + O_2(g) \rightleftarrows 2\ NO_2(g)$$

NO and O_2 gases were added to a rigid, evacuated container at pressures of 0.620 atm and 0.540 atm, respectively, with a temperature of 200. K. As the reaction proceeded, the total pressure in the container was monitored, and the results are shown in the graph below.

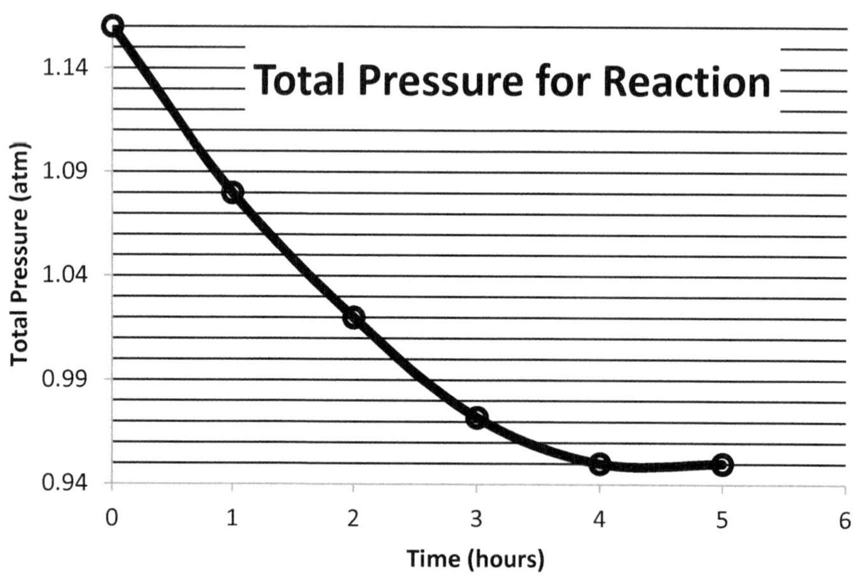

The student will have to interpret the graph to determine the equilibrium pressure. In this case, look to see where the graph goes flat and remains unchanged over two successive data points.

5. **Given the initial concentrations and the value of *K*, solve by approximation for the equilibrium concentrations.**

EXAMPLE: (Type 5) At a particular temperature, $K = 1.00 \times 10^2$ for the reaction

$$H_2(g) + I_2(g) \rightleftarrows 2\ HI(g)$$

In an experiment, 1.00 moles H_2, 1.00 moles I_2, and 1.00 moles HI are introduced into a 1.00-L container. Calculate the equilibrium concentrations of all reactions and products.

SOLUTION: To begin, write the balanced equation for the reaction and the equilibrium expression, omitting pure solids and liquids.

	$H_2(g)$	+	$I_2(g)$	$\rightleftarrows$	$2\ HI(g)$	$K = \frac{[HI]^2}{[H_2][I_2]}$
I	1.00		1.00		1.00	
C	$-x$		$-x$		$+2x$	
E	$1.00 - x$		$1.00 - x$		$1.00 + 2x$	

Make an ICE chart under the balanced chemical equation.

I = **Initial concentration** in mol/L. Most often, the reaction starts with only reactants and no products. In this case, the concentrations of each reactant and product are equal. To determine which direction the reaction will go to reach equilibrium, *Q* could be calculated. However, since *K* is much greater than 1, the reaction favors the products and will go in the forward direction to increase the concentration of products.

C = **Change to reach equilibrium** represented by $+x$ or $-x$. A minus sign indicates a decrease in concentration; a plus sign indicates an increase. The coefficient in front of the reactant in the balanced equation is placed in front of the x in the change line.

E = **Equilibrium concentrations** which are obtained by adding the I and C lines together.

Plug the equilibrium values into the expression and solve for x:

$$K = 100. = \frac{(1.00 + 2x)^2}{(1.00 - x)^2}$$

This is an example of a "perfect square" in algebra. Taking the square root of both sides:

$$10.0 = \frac{1.00 + 2x}{1.00 - x}$$

$$10.0 - 10.0x = 1.00 + 2x$$

$$12x = 9.0$$

$$x = 0.75\ M$$

Use this value of x to solve for the equilibrium concentrations of all reactants and products.

$$[H_2] = [I_2] = 1.00 - 0.75 = 0.25\ M$$

$$[HI] = 1.00 + 2(0.75) = 2.50\ M$$

Finally, check your equilibrium concentrations by making sure that they equal the correct value of K.

EXAMPLE: (Type 5) For the reaction

$N_2O_4(g) \rightleftarrows 2\ NO_2(g)$, $K = 4.0 \times 10^{-7}$ at a specific temperature.

In an experiment, 1.0 mole of N_2O_4 is placed in a 10.0-L vessel.

Calculate the equilibrium concentrations of NO_2 and N_2O_4.

SOLUTION: To solve this problem, proceed as in the previous example.

	$N_2O_4(g)$	$\rightleftarrows$	$2\ NO_2(g)$
I	0.10		0
C	$-x$		$+2x$
E	$0.10 - x$		$2x$

$$K = \frac{[NO_2]^2}{[N_2O_4]} = \frac{(2x)^2}{0.10 - x} = 4.0 \times 10^{-7}$$

Since the value of K is much smaller than 1, you can assume that the change from the initial concentration, x in $0.10 - x$, is so small that it is negligible, that is, $0.10 - x$ is about equal to 0.10.

This greatly simplifies the math to

$$K = \frac{(2x)^2}{0.10} = 4.0 \times 10^{-7}$$

$$4x^2 = 4.0 \times 10^{-8},\ x = 1.0 \times 10^{-4}\ M$$

It is okay to make this assumption if the change from the initial concentration, in this case, x, is less than 5% of the initial concentration.

$$\frac{x}{0.10} \times 100 = \frac{1.0 \times 10^{-4}}{0.10} \times 100\% = 0.10\%$$

This is less than 5%, so it is okay to make the assumption.

The equilibrium concentrations are as follows:

$$[N_2O_4] = 0.10 - x = (0.10 - 1.0 \times 10^{-4}) = 0.10\ M$$

$$[NO_2] = 2x = 2(1.0 \times 10^{-4}) = 2.0 \times 10^{-4}\ M$$

AP® Tip

When solving equilibrium problems, you will never have to solve a quadratic equation on the exam. All of the problems will be set up so that you can find the answer without using the quadratic formula. This is accomplished in three ways:

- by having the change in concentration less than 5% (by providing a very small K value that you can assume that the reaction does not effectively produce products)
- by providing a very large K value that you can assume that the reaction goes to completion
- by setting up a perfect square situation where you can take the square root

LE CHATELIER'S PRINCIPLE

The value of K is constant at a particular temperature. The only factor that changes the value of K for a particular reaction is temperature. Pressure, a catalyst, and changes in concentration will not affect the value of K, but will affect the speed of the reaction and how long it takes to reach equilibrium.

Le Chatelier's principle states that if a change or stress is imposed on a system at equilibrium, the position of the equilibrium will shift in a direction that will counteract the change.

EXAMPLE 1: Consider the following changes on the system below at equilibrium.

$$2\ SO_3(g) \rightleftarrows 2\ SO_2(g) + O_2(g),\ \Delta H = 197\text{ kJ}$$

(a) Addition of O_2

SOLUTION: The equilibrium will shift in such a way as to use up the O_2; hence, the reaction will shift to the left to form more reactants. The addition of oxygen increases the rate of the reverse reaction so more reactants will form, until the rate of the forward reaction again equals the rate of the reverse reaction. On the particulate level, more O_2 is available to collide with the SO_2 so more reactant will be made. Another way to explain the shift is that with the added O_2, Q is now greater than K. The reverse reaction is favored so the product concentrations decrease and reactant concentrations increase until $Q = K$ again.

(b) Removal of SO_2

SOLUTION: Removal of an equilibrium component such as SO_2 at constant pressure and temperature will cause the equilibrium to shift toward the removed component to increase its concentration. The reaction in this example will shift to products. When SO_2 is removed, Q is less than K. The forward reaction is favored until enough product has been made for $Q = K$.

(c) Increase in temperature

SOLUTION: The direction of the shift can be predicted in the same way as the addition or removal of a reactant or product.

The decomposition of SO_3 is endothermic. Treating heat as a reactant, an increase in temperature will cause the reaction to shift right, producing more products, increasing the value of K. More SO_3 molecules will have sufficient energy to break the bonds and make products. A decrease in temperature will cause a shift to the left, increasing the reactant concentrations and lowering the value of K.

(d) Increase in pressure

SOLUTION: If the pressure is increased in an equilibrium system, the reaction will shift toward the side with fewer moles of gas. Decreasing the volume has the same effect because the only way the pressure can be increased without changing the temperature or number of moles is to decrease the volume. In this example, the reaction will shift to the left, toward SO_3, when the pressure is increased or the volume is decreased.

Another way to understand this is to look at the collisions of the particles. If the pressure increases, the volume of the container decreases and the particles are closer together. This will increase the number of collisions and the reaction will shift to try to reduce the number of collisions and hence will shift toward the side with the fewer moles of gas. If the number of moles of gas is the same on both sides of the reaction, no shift will occur.

(e) Addition of a solid or an inert gas, such as Ne

SOLUTION: If a solid or an inert gas (with no change in volume) is added to the reaction, there will be no shift in equilibrium. Neither the solid nor the inert gas is part of the equilibrium expression. The concentrations of all the components of the equilibrium expression remain unchanged. An inert gas will increase the total pressure but will not affect the partial pressures of the gases in the reaction.

(f) An inert gas, such as Ne, is added at constant pressure.

SOLUTION: There will be a shift in the equilibrium to the side of the equation with more moles of gas. To add Ne at constant pressure, the volume of the container must increase so the concentrations or partial pressures of all gases have decreased. If the moles of gas are greater in the reactants, then $Q < K$, so equilibrium can be reestablished only by increasing the products.

EXAMPLE 2: Dilution of a reaction system with a solvent may have an effect on the equilibrium concentrations. Consider how the following will affect equilibrium.

Water is added to a dilute $Fe(SCN)^{2+}$ solution.

$$Fe^{3+} + SCN^{-} \rightleftarrows Fe(SCN)^{2+}$$

SOLUTION: When water is added to this equilibrium, the reaction will shift to the reactants because the more dilute the solution, the fewer chances the individual ions have of colliding to make the complex ion. Since there are two aqueous reactants and only one aqueous product, dilution has twice the effect on the reactants as products. The denominator will decrease in value more than the numerator, making Q larger. The reaction will shift to the reactant side to reach K.

EXAMPLE 3: The addition of a chemical species not already present in the reaction that will react with one of the reactants or products to remove it (neutralization, formation of a precipitate, formation of a complex ion) will cause a shift in the reaction.

$$HgO(s) + 4\ I^-(aq) + H_2O(l) \rightleftarrows HgI_4^{2-}(aq) + 2\ OH^-(aq)$$

Consider how the following additions will affect the above reaction at equilibrium.

(a) a 0.10 M solution of $HC_2H_3O_2$

SOLUTION: The addition of acetic acid will neutralize OH^-, producing water. The removal of the hydroxide ions shifts the reaction toward products to replace some of the OH^-.

(b) a 0.10 M solution of $AgNO_3$

SOLUTION: The $AgNO_3$ will react with the I^-, producing a precipitate of AgI. This removes I^-, shifting the reaction back toward the reactants to replace some of the I^- removed.

EXAMPLE 4: Create three particulate models to illustrate the system described below. Solve for K or Q in each situation.

Hydrogen gas can react with several different metal oxides (XO) according to the general equation shown below.

$$H_2(g) + XO(s) \rightleftarrows H_2O(g) + X(s)$$

6 molecules of H_2 and 2 molecules of H_2O are at equilibrium. 4 molecules of H_2 are added, disturbing equilibrium.Equilibrium is reestablished with temperature and pressure remaining constant.

	$H_2(g)$ +	$XO(s)$	$\rightleftarrows$	$H_2O(g)$ +	$X(s)$	
Initial Equilibrium	6	----		2	----	$K = 0.33$
Disturb Equilibrium	+4 = 10	----		2	----	$Q = 0.20$
Change to Equilibrum	–1	----		+1	----	
New Equilibrium	9	----		3	----	$K = 0.33$

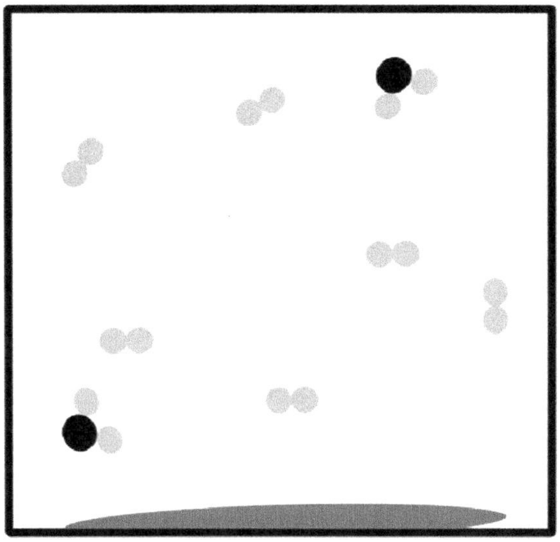

Initial Equilibrium

6 H_2 molecules
2 H_2O molecules

$$K = \frac{[H_2O]}{[H_2]}$$

$$K = \frac{2}{6} = \frac{1}{3}$$

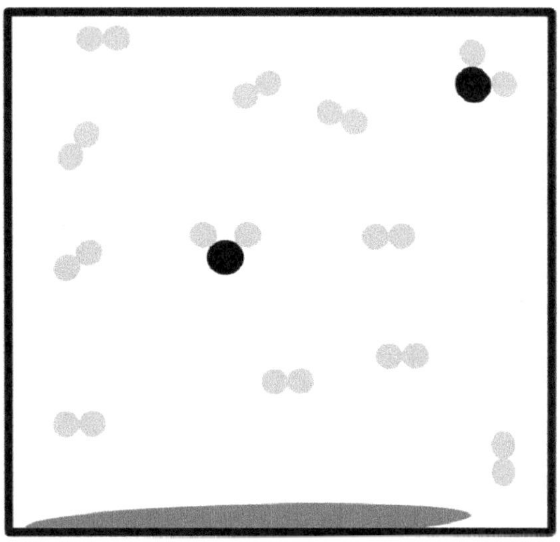

Disturb Equilibrium

Add 4 H_2 molecules

$$Q = \frac{2}{10} = \frac{1}{5}$$

Since $Q < K$,
reaction shifts right.

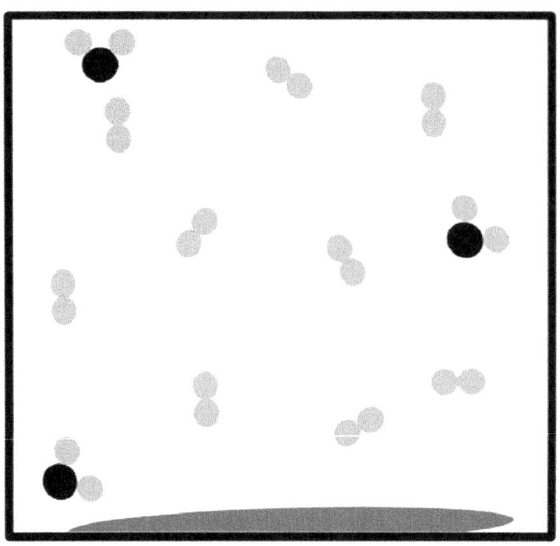

Restore Equilibrium

9 H_2 molecules
3 H_2O molecules

$$K = \frac{3}{9} = \frac{1}{3}$$

RELATIONSHIP OF K TO FREE ENERGY, $\Delta G°$

The magnitude of the equilibrium constant, K, is directly related to the change in Gibbs free energy ($\Delta G°$) associated with the reaction.

The equilibrium constant at standard temperature can be determined from the thermodynamic values of a reaction using the equation:

$$\Delta G° = -RT\ln K$$

$$R = \frac{8.314\text{ J}}{(\text{K} \times \text{mol})}$$

Alternatively, the equation can be expressed in terms of the equilibrium constant:

$$K = e^{-\Delta G°/RT}$$

The species that have lower free energy will have larger relative concentrations at equilibrium. When both reactants and products have significant concentrations at equilibrium, then the value of K will be approximately 1 and the value of $\ln K$ will result in the magnitude of $\Delta G°$ to be similar to the thermal energy (RT).

For example, for the reaction A ⇄ B as the value of K gets larger, the $\ln K$ term will have a greater effect on the value of ΔG, as shown in the table below.

[A] at Equilibrium	[B] at Equilibrium	K value	$\ln K$	Value of ΔG	How far away from RT is ΔG?
1.00	1.00	1.00	0	0	N-A
1.50	1.00	1.50	0.405	–1.00	0.405
0.500	1.00	2.00	0.693	–1.72	0.693
1.00	10000.	10000.	9.21	–22.8	9.21
0.100	10000.	100,000	11.5	–28.5	11.5
10000.	0.100	0.00001	–11.5	+28.5	11.5

If the value of K is the reciprocal of any of the K values listed (the final two entries in the table), then $\ln K$ and ΔG will be opposite in sign, but the magnitude will be the same.

The equilibrium constant can be determined quantitatively through calculation or qualitatively through estimation. The thermal energy (RT) at room temperature is 2.4 kJ/mol. If $\Delta G°$ is large compared to RT, then K will be significantly larger than 1.

EXAMPLE: Calculate the value of K at 25°C for the reaction $2\ NO_2(g) \rightleftarrows N_2O_4(g)$

The values of $\Delta H°$ and $\Delta S°$ are –58.03 kJ/mol and –176.6 J/K × mol, respectively.

SOLUTION:

Notice that the units for energy must both be in J.

At 25°C, $\Delta G° = \Delta H° - T\Delta S°$

$$= -5.803 \times 10^4 \text{ J/mol} - (298 \text{ K})(-176.6 \text{ J K}^{-1} \text{mol}^{-1})$$

$$= -5.40 \times 10^3 \text{ J mol}^{-1}$$

$\Delta G° = -RT \ln K$

$$\ln K = -\frac{\Delta G°}{RT} = -\frac{-5.40 \times 10^3 \text{ J mol}^{-1}}{8.314 \text{ J K}^{-1} \text{ mol}^{-1} \times 298 \text{ K}} = 2.18$$

$K = 8.85$

This reaction is product-favored since $K > 1$.

AP® Tip

If $\Delta G° < 0$, then $K > 1$ and the reaction is product-favored.

If $\Delta G° > 0$, then $K < 1$ and the reaction is reactant-favored.

Exergonic reactions ($\Delta G° < 0$) and endergonic reactions ($\Delta G° > 0$) are important in many biological applications where the magnitude of the equilibrium constant is important.

RELATIONSHIP OF K TO CELL POTENTIAL, $\Delta E°$

Equilibrium constants can also be calculated using cell potentials (and vice versa). At equilibrium $E = 0$ and $Q = K$.

$$E° = (0.0592/n) \log K$$

or you can use $\Delta G° = -RT \ln K$ and $\Delta G° = -nFE°$ and set them equal to each other to get

$$nFE° = RT \ln K$$

EXAMPLE: Determine whether the following reaction is thermodynamically favorable and calculate its equilibrium constant at 25°C.

$$\text{Sn}(s) + \text{Ni}^{2+}(aq) \rightarrow \text{Sn}^{2+}(aq) + \text{Ni}(s)$$

SOLUTION: The half-reactions would be as follows:

$$Sn(s) \rightarrow Sn^{2+} + 2\,e^- \quad E° = 0.14 \text{ V}$$

$$Ni^{2+} + 2\,e^- \rightarrow Ni(s) \quad E° = -0.23 \text{ V}$$

$$E°_{cell} = 0.14 + (-0.23) = -0.09 \text{ V}$$

Because cell potential is negative, it is not thermodynamically favored.

$$E° = (0.0592/n)\log K$$

$$-0.09 = (0.0592/2)\log K$$

$$K = 9 \times 10^{-4}$$

Notice that the K is much less than 1, indicating that the reactants are favored—just as the negative $E°_{cell}$ also indicates that the reaction is non-thermodynamically favored (i.e., favors the reactants).

MULTIPLE-CHOICE QUESTIONS

No calculators are to be used in this section.

1. In the reaction $3\text{ W} + \text{X} \rightleftarrows 2\text{ Y} + \text{Z}$, all substances are gases and K = 10. The reaction is initiated by adding an equal number of moles of W and of X. When equilibrium is reached,
(A) [Y] = [Z].
(B) [X] = [Y].
(C) [W] = 0.
(D) [X] > [W].

2. An esterification reaction occurs when an alcohol reacts with a carboxylic acid as shown below. Note the acid catalyst and the arrows that indicate where the reaction takes place.

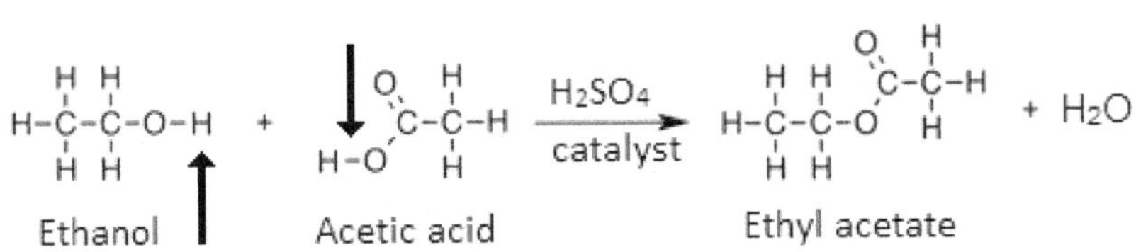

The equilibrium constant is approximately 5 at room temperature, but the rate of reaction without a catalyst is very slow. How do the rates of the forward and reverse reactions compare at equilibrium?
(A) The forward rate is 5 times faster than the reverse rate.
(B) The reverse rate is the same as the forward rate.
(C) The reverse rate is 5 times faster than the forward rate.
(D) The rates can only be determined if concentrations are known.

Questions 3–5 refer to the following information:

$3\ H_2(g) + N_2(g) \rightleftarrows 2\ NH_3(g)$ $\Delta H = -92$ kJ/mol$_{rxn}$.

The synthesis of ammonia (NH_3) is represented by the equation above.

3. Increasing the temperature of this equilibrium system, at constant pressure, causes
(A) an increase in $[NH_3]$ and an increase in the value of K.
(B) an increase in $[N_2]$ and the value of K stays constant.
(C) a decrease in $[N_2]$ and a decrease in the value of K.
(D) an increase in $[H_2]$ and a decrease in the value of K.

4. At a certain temperature, the synthesis of ammonia gas has a K of 3.0×10^{-2}. If $[H_2] = [N_2] = 0.10\ M$ and $[NH_3] = 0.20\ M$,
(A) the reaction would shift toward the ammonia.
(B) the reaction would shift toward N_2 and H_2.
(C) the system is at equilibrium, therefore, no shifting will occur.
(D) the reaction is too slow to shift and must be carried out at higher temperatures (500. K) and increased pressure (200. atm) for a shift to occur.

5. According to the graph below, what happened to disturb the equilibrium, and what is true about the value of Q at the point at which equilibrium was disturbed?

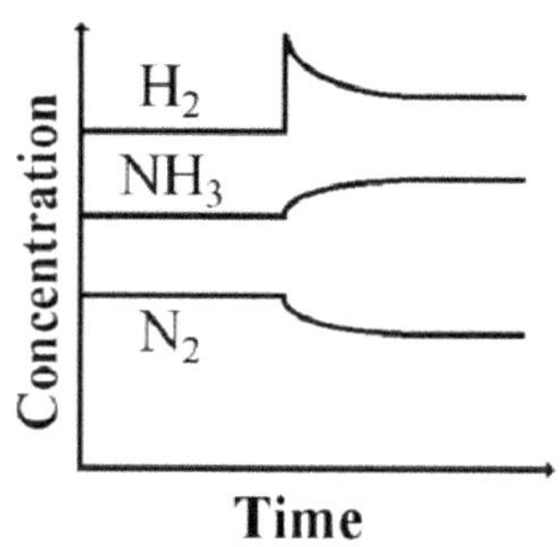

(A) Ammonia was added and the value of $Q < K$.
(B) Ammonia was added and the value of $Q > K$.
(C) Hydrogen was added and the value of $Q < K$.
(D) Hydrogen was added and the value of $Q > K$.

6. $N_2(g) + O_2(g) \rightleftarrows 2\ NO(g)$

One product of the oxidation of nitrogen is NO. The reaction was initiated by adding 15.0 moles of NO to a 1.0-L flask. At equilibrium, 3.0 moles of oxygen are present in the 1.0-L flask. The value of K_c must be
(A) 0.33.
(B) 3.0.
(C) 5.0.
(D) 9.0.

7. One way to produce nitric acid is through the Ostwald process. In the first reaction of this process, ammonia is oxidized to nitrogen monoxide.

   $4\ NH_3(g) + 5\ O_2(g) \rightleftarrows 4\ NO(g) + 6\ H_2O(g) \quad \Delta H = -905.2\text{ kJ}$

   Which of the following sets of stresses would shift equilibrium toward products?
   (A) Raise the temperature of the reaction and remove nitrogen monoxide as it is made.
   (B) Lower the temperature of the reaction and remove nitrogen monoxide as it is made.
   (C) Raise the temperature of the reaction and add a catalyst.
   (D) Lower the temperature of the reaction and add a catalyst.

8. At a certain temperature, the K_c for the reaction $2\ X \rightleftarrows 2\ Y + Z$ is equal to 8.0. Which of the diagrams below represents this reaction at equilibrium?

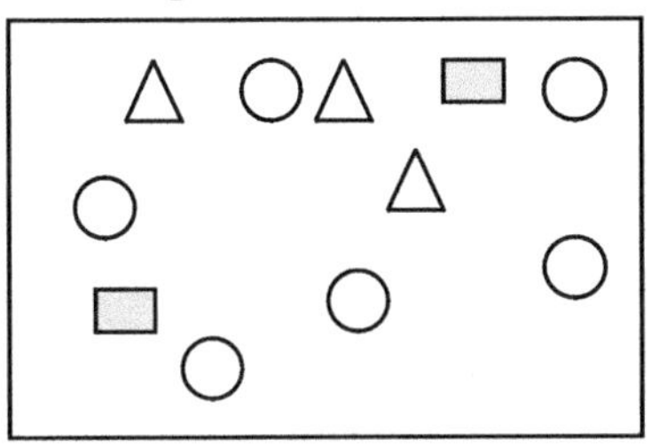

Diagram 1

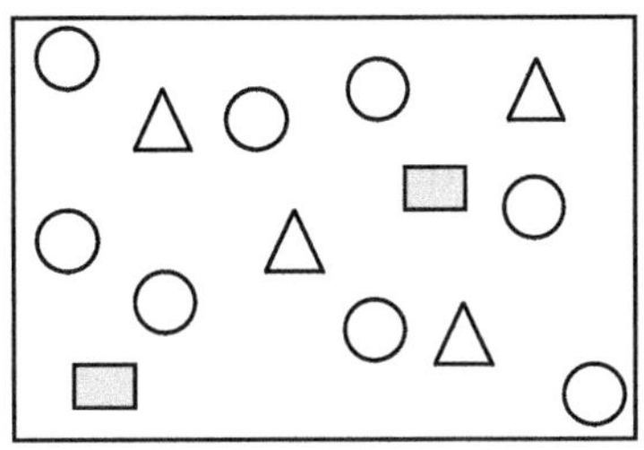

Diagram 2

Substance	Symbol
X	△
Y	○
Z	□

   (A) Diagram 1
   (B) Diagram 2
   (C) Both Diagrams 1 and 2
   (D) Neither Diagram 1 or 2

9. In which system will pressures of all gaseous reactants and products increase by the same factor when the volume of the system is decreased at constant temperature?
   (A) $H_2(g) + Cl_2(g) \rightleftarrows 2\ HCl(g)$
   (B) $H_2(g) + I_2(s) \rightleftarrows 2\ HI(g)$
   (C) $N_2(g) + 3\ H_2(g) \rightleftarrows 2\ NH_3(g)$
   (D) $2\ NaHCO_3(s) \rightleftarrows Na_2CO_3(s) + CO_2(g) + H_2O(g)$

$P_4(g) + 6\ Cl_2(g) \rightleftarrows 4\ PCl_3(l)$

10. Phosphorus trichloride is an important industrial chemical as it is a precursor in the production of PCl_5 and $POCl_3$, along with other compounds. Industrially, PCl_3 is produced by the reaction of "white" phosphorus with chlorine gas. As the PCl_3 is produced, it is continually removed to drive the reaction forward and prevent the formation of PCl_5. Choose the correct equilibrium expression for K_p for this reaction.
    (A) $K_p = \frac{(P_{PCl_3})}{(P_{P_4})(P_{Cl_2})}$
    (B) $K_p = \frac{(P_{PCl_3})^4}{(P_{P_4})(P_{Cl_2})^6}$
    (C) $K_p = \frac{(P_{P_4})(P_{Cl_2})^6}{(P_{PCl_3})^4}$
    (D) $K_p = \frac{1}{(P_{P_4})(P_{Cl_2})^6}$

$NH_4HS(s) \rightleftarrows NH_3(g) + H_2S(g)$

11. Ammonium hydrogen sulfide will decompose into ammonia gas and hydrogen sulfide gas when heated. Given the reaction shown above, 1.000 mole of NH_4HS is placed in a 10.0-L evacuated container and carefully heated. When equilibrium is reached, the total pressure is found to be 4.00 atm. What is the value of K_p?
    (A) 1.00
    (B) 2.00
    (C) 4.00
    (D) 16.0

$U_3O_8(s) + 3\ Cl_2(g) \rightleftarrows 3\ UO_2Cl_2(g) + O_2(g)$

12. Triuranium octoxide (U_3O_8) is the most stable form of uranium and is the most common form found in nature. When it is exposed to chlorine gas, highly toxic uranyl chloride (UO_2Cl_2) is formed. At a certain high temperature, the K_p value for this reaction is 3×10^{-13}. At this temperature, what is the correct relationship among the pressures of the reactants and products?
    (A) Pressure of Cl_2 > Pressure of UO_2Cl_2 > Pressure of O_2
    (B) Pressure of Cl_2 = Pressure of UO_2Cl_2 > Pressure of O_2
    (C) Pressure of UO_2Cl_2 > Pressure of O_2 > Pressure of Cl_2
    (D) Pressure of UO_2Cl_2 = Pressure of Cl_2 = Pressure of O_2

13. $K = 0.25$ for $2\ NOBr(g) \rightleftarrows 2\ NO(g) + Br_2(g)$ at a certain temperature. At the same T and P, what is the K for the reaction $NO + \frac{1}{2}\ Br_2 \rightleftarrows NOBr$?
    (A) 0.50
    (B) 0.63
    (C) 2.0
    (D) 4.0

Questions 14–15 refer to the following information.

Reaction 1: $2\ NOCl(g) \rightleftarrows 2\ NO(g) + Cl_2(g)$ $K = 1.6 \times 10^{-5}$

Reaction 2: $2\ NO(g) \rightleftarrows N_2(g) + O_2(g)$ $K = 1.0 \times 10^{31}$

Both reactions are at the same temperature.

14. Which is true concerning the set of reactions shown above?
(A) Reaction 1 occurs at a higher rate than Reaction 2.
(B) Reaction 2 occurs at a higher rate than Reaction 1.
(C) Reactions 1 and 2 occur at equal rates.
(D) Relative reaction rates cannot be determined with the information provided.

15. What is the K value for the following reaction at the same temperature as the reactions above?
$2\ NOCl(g) \rightleftarrows N_2(g) + Cl_2(g) + O_2(g)$
(A) 1.6×10^{-155}
(B) 1.6×10^{26}
(C) 2.6×10^{26}
(D) 1.0×10^{31}

FREE-RESPONSE QUESTIONS

1. The Haber process for the synthesis of ammonia is kinetically and thermodynamically controlled.

$$N_2(g) + 3\ H_2(g) \rightleftarrows 2\ NH_3(g)$$

(a) For the reaction above:

$\Delta H° = -92\ \text{kJ/mol}_{rxn}$ $\Delta S° = -199\ \text{J/(K} \times \text{mol}_{rxn})$

(i) Calculate the value of $\Delta G°$ for the reaction.
(ii) Is the reaction thermodynamically favorable at 25°C? Justify your answer.
(iii) At what temperature does $\Delta G°$ become zero? What does this mean in regard to the temperature at which this reaction can be successfully carried out at?

(b) Calculate the equilibrium constant (K_c) at 25°C.
(c) If additional hydrogen gas is added to the reaction at equilibrium, how will the value of Q compare to K?
(d) The rate of this reaction at room temperature is very low. Explain.
(e) What one thing can be done to the reaction process that would both increase the rate of the reaction and also shift equilibrium toward the products? Explain.

2. $PCl_3(g) + Cl_2(g) \rightleftarrows PCl_5(g)$

To an empty 15.0-L cylinder, 0.500 mole of gaseous PCl_5 is added and allowed to reach equilibrium where the pressure of PCl_3 is found to be 0.677 atm. Assume a temperature of 375 K.

(a) What is the initial pressure of PCl_5, before the reaction occurs?

(b) Write the equilibrium constant expression (K_p) for the above reaction.
(c) Determine the value of K_p for this system at 375 K.
(d) How would the value of K_p be affected by increasing the temperature of the system at equilibrium for this exothermic reaction?

Answers

MULTIPLE-CHOICE QUESTIONS

1. D Since every time 1 mole of X reacts, 3 moles of W must react, the amount of W remaining must be less than the amount of X remaining (recall that you started with an equal number of moles of W and X). The concentration of W cannot equal zero, as the reaction reaches equilibrium. With a K of 10., there will still be a substantial amount of all products and reactants (LO 6.1).

2. B Equilibrium is established when the forward rate is equal to the reverse rate (LO 6.3).

3. D Increasing the temperature causes the equilibrium to shift to the left. For an exothermic reaction, increasing the temperature increases the rate of both the forward and reverse reactions, but proportionally makes a greater increase in the reverse reaction since it has the higher activation energy, favoring the formation of more hydrogen gas and more nitrogen gas, and lowering the concentration of the ammonia (LO 6.8, 6.10).

4. B

$$Q = \frac{[NH_3]^2}{[H_2]^3[N_2]} = \frac{(0.20)^2}{(0.10)^3(0.10)} = 400.$$

   Since Q is greater than K, the reaction will shift toward N_2 and H_2 (LO 6.4).

5. C The graph shows a spike where H_2 was added. At the point at which equilibrium was disturbed, the increase in $[H_2]$ causes the value of the denominator of Q to increase, decreasing the value of Q (LO 6.10).

6. D If 3.0 moles of oxygen are formed, 6.0 moles of NO must have reacted, leaving 9.0 moles of NO at equilibrium (15.0 – 6.0 = 9.0 mol/L for [NO]). Each time 3.0 moles of oxygen form, the same number of moles of nitrogen is produced. Since the reaction is written showing NO as a product, the equilibrium constant expression is

$$K = \frac{[NO]^2}{[N_2][O_2]} = \frac{(9.0)^2}{(3.0)(3.0)} = \frac{81.0}{9.0} = 9.0$$

   (LO 6.5)

7. B The reaction is exothermic, so lowering the temperature and removing the nitrogen monoxide will shift the reaction to products according to Le Chatelier's principle (LO 6.9).

8. C The equilibrium expression for the reaction 2 X ⇄ 2 Y + Z would be $K = \frac{[Y]^2[Z]}{[X]^2}$. The value of K for Diagram 1 would be $K = \frac{(6)^2(2)}{(3)^2} = 8$. The value of K for Diagram 2 would be $K = \frac{(8)^2(2)}{(4)^2} = 8$ (LO 6.6).

9. A When gaseous systems have an equal number of moles of gas on each side, a change in pressure (results from decreased volume) will not cause the reaction to shift and all pressures will increase proportionately (LO 6.8).

10. D The equilibrium constant is a ratio of the partial pressures of products divided by the partial pressures of reactants, each taken to a power represented by their coefficients. Pure liquids and solids are not shown in the equilibrium constant expression (LO 6.5).

11. C The two gases are formed in equal molar amounts (1:1); therefore, half of the pressure is due to each gas (4.00 atm/2 = 2.00 atm).

$$K_P = (P_{NH_3})(P_{H_2S}) = (2.00)^2 = 4.00$$

(LO 6.5)

12. A The K_p value of 3×10^{-13} is very small and indicates that very few products have formed and the pressure due to the products will be very small (LO 6.7).

13. C First, recognize that the reaction requested is the reverse of the one for which the K is given. The value of K for the reversed equilibrium reaction is the reciprocal of K or $1/K$, which is 1/0.25 = 4. The new reaction as written is then multiplied by a factor of ½; therefore, the equilibrium expression for the new reaction is the original K raised to the ½ power, or in this case, the square root of 4, which is 2.0 (LO 6.2).

14. D The magnitude of the K value indicates the extent to which the reactants progress to products and has nothing to do with the rate of the reaction (LO 6.3).

15. B The reaction given is the sum of Reactions 1 and 2. To find the new K, multiply the K values of the two reactions, remembering that exponents are added when numbers in scientific notation are multiplied.

(1.6×10^{-5}) $(1.0 \times 10^{31}) = 1.6 \times 10^{26}$ (LO 6.2)

FREE-RESPONSE QUESTIONS

1. (a) (i) $\Delta G° = \Delta H° - T\Delta S°$

$\Delta G° = -92 \text{ kJ} - (298 \text{ K})(0.199 \text{ kJ/K})$

$\Delta G° = -33 \text{ kJ/mol}_{rxn}$

(LO 5.13)

(ii) Yes, $\Delta G°$ is negative, which indicates that the reaction is thermodynamically favored at that temperature (LO 5.14).

(iii) $\Delta G° = \Delta H° - T\Delta S°$

$0 \text{ kJ mol}^{-1} = -92 \text{ kJ mol}^{-1} - T(-0.199 \text{ kJ K}^{-1}\text{mol}^{-1})$

$T = 462$ K; Since both $\Delta H°$ and $\Delta S°$ have negative values, the reaction is thermodynamically favored at relatively low temperatures. To be thermodynamically favored, the reaction must be carried out at temperatures below 462 K (LO 6.6).

(b) $K = e^{-\Delta G°/RT}$ or $\Delta G = -RT \ln K$

$K = e^{33000\text{J}/(8.314\text{J/K mol})(298 \text{ K})}$ or $-33{,}000 \text{ J} = -8.314(298)\ln K$

$K = 6.1 \times 10^5$

(LO 6.25).

(c) If additional hydrogen gas is added, the value of Q will be lower (since hydrogen is a reactant, the denominator increases) than K. The reaction will shift to the product side until $Q = K$ (LO 6.6).

(d) The rate of reaction is probably very low due to a high E_a that must be overcome. The very stable triple bond in the nitrogen molecule requires a lot of energy to break; that is the activation energy for the reaction (LO 5.18).

(e) Increasing the pressure would cause the reaction to shift to the side with fewer moles of gas (the product side has 2 moles of gas while the reactant side has 4). Also, increasing pressure would cause the particles to be closer together, increasing the chance for collisions between particles. This will cause an increase in the rate of both the forward and reverse reactions (LO 4.1, LO 6.9).

2. (a) $PV = nRT$ $\quad P(15.0) = 0.500(0.08206)(375)$ $\quad P = 1.03$ atm

(LO 2.6)

(b)

$$K_p = \frac{(P_{PCl_5})}{(P_{PCl_3})(P_{Cl_2})}$$

(LO 6.5)

(c)

	$PCl_3(g)$	+	$Cl_2(g)$	⇄	$PCl_5(g)$
I	0		0		1.03 atm
C	$+x$		$+x$		$-x$
E	x		x		$1.03 - x$
	$x = 0.677$ atm		0.677 atm		0.35 atm

$$K_p = \frac{(0.35)}{(0.677)(0.677)} = 0.76$$

(LO 6.5)

(d) Since the reaction is exothermic, increasing the temperature will shift the reaction to the reactant side, increasing the reactant concentration and decreasing the product concentration. This will cause a decrease in the value of K_p (LO 6.10).

12

Big Idea 6: Applications of Equilibria: Solubility Equilibria

Big Idea 6

Any bond or intermolecular attraction that can be formed can be broken. These two processes are in a dynamic competition, sensitive to initial conditions and external perturbations.

Concepts from the last chapter will be applied in this section covering the equilibria of sparingly soluble salts. You will apply equilibrium concepts involving solids dissolving to form aqueous solutions.

You should be able to:

- Write balanced equations for the dissolution of a salt and its corresponding solubility product expression.
- Predict the relative solubilities of salts which dissolve to give the same number of ions from their K_{sp} values.
- Calculate the K_{sp} value from the solubility of a salt and also calculate the solubility of the salt in units of mol/L or g/L from the given K_{sp} value.
- Predict the effect of a common ion on the solubility of a salt and perform calculations.

- Perform calculations to predict if a precipitate will form when two solutions are mixed.

AP® Tip

Be systematic. Follow the same series of steps for all equilibrium problems. The use of ICE tables will prevent many careless errors with K_{sp} problems. Pay close attention to coefficients in balanced equations as they must be used both in the ICE table and in the equilibrium expression.

SOLUBILITY PRODUCT

You must memorize the fact that all sodium, potassium, ammonium, and nitrate salts are soluble in water. For all other salts, you will either be given solubility information or will need to calculate solubility based on equilibrium data and/or constants provided to you.

Even salts that are considered to be "insoluble" will dissolve to a very small extent and are sometimes called "sparingly soluble salts." For slightly soluble or insoluble salts, equilibrium exists between the solid and its aqueous ions. For example, lead(II) chloride dissolves in water as follows:

$$PbCl_2(s) \rightleftarrows Pb^{2+}(aq) + 2\ Cl^-(aq)$$

At first, when the salt is added to the water, there are no ions present. As the solid dissolves, the concentration of the ions increases. A simultaneous competing process is the reverse of the dissolution, that is, the reforming of the solid called crystallization. At some point, the maximum amount of dissolution is achieved, which is called the saturation point. However, remember that on a molecular level, a dynamic equilibrium exists between dissolved solute and undissolved solid solute. (The rate of dissolution equals the rate of crystallization.) The solution is saturated when the amount of solid is constant and equilibrium is reached.

The equilibrium expression for the dissolution of lead(II) chloride is

$$K_{sp} = [Pb^{2+}][Cl^-]^2$$

The constant, K_{sp}, is the solubility product constant. For salts producing the same number of ions, the K_{sp} value can be used to measure the extent to which the solid dissolves. The larger the K_{sp} value, the more soluble the salt.

EXAMPLE: Given the following salts and their K_{sp} values, which salt is the most soluble? Which salt is the least soluble?

Formula	K_{sp}
$NiCO_3$	1.4×10^{-7}
MnS	2.3×10^{-13}
$CaSO_4$	6.1×10^{-5}

SOLUTION: Since all three compounds have the same number of ions (in this case, one cation and one anion), a direct comparison to determine the relative solubilities of the salts can be made of their K_{sp} values. The most soluble salt is the salt with the largest K_{sp} value, $CaSO_4$.

The least soluble salt is the salt with the lowest value of K_{sp}, MnS.

CALCULATIONS INVOLVING SOLUBILITY

The solubility of a salt is the amount of salt that will dissolve in 1 L of water. The solubility of a salt can be given in units of mol/L or g/L. The solubility of a salt can be used to determine the K_{sp} value for the salt.

CALCULATING K_{SP} FROM SOLUBILITY

EXAMPLE: The solubility of $Pb_3(PO_4)_2$ is 6.2×10^{-12} *M*. Calculate the K_{sp} value for the solid.

Step 1: Write the reaction for the dissolution of the solid.

$$Pb_3(PO_4)_2(s) \rightleftarrows 3\ Pb^{2+}(aq) + 2\ PO_4^{3-}(aq)$$

Step 2: Underneath the reaction, make an ICE chart.

	$Pb_3(PO_4)_2(s)$	$\rightleftarrows$	$3\ Pb^{2+}(aq)$	+	$2\ PO_4^{3-}(aq)$
I	-----		0		0
C	$-x$		$+3x$		$+2x$
E	------		$3x$		$2x$

In the ICE chart above, $-x$ represents x mol/L of $Pb_3(PO_4)_2(s)$ dissolving to reach equilibrium. In this example, $x = 6.2 \times 10^{-12}$ *M*. Remember that pure solids are not included in equilibrium expressions.

For every 1 mole per liter of $Pb_3(PO_4)_2$ which dissolves, 3 moles per liter of Pb^{2+} and 2 moles per liter of PO_4^{3-} form.

$3x$ is the mol/L of $Pb^{2+}(aq)$ produced when the solid, $Pb_3(PO_4)_2$, dissolves.

$2x$ is the mol/L of $PO_4^{3-}(aq)$ produced when the solid, $Pb_3(PO_4)_2$, dissolves.

Step 3: Write the equilibrium expression for the reaction and plug in the values from the equilibrium line of the ICE chart.

$$K_{sp} = [Pb^{2+}]^3[PO_4^{3-}]^2 = (3x)^3(2x)^2 = 108x^5$$

The value of x is the solubility of $Pb_3(PO_4)_2$, which equals 6.2×10^{-12} *M*.

$$K_{sp} = 108(6.2 \times 10^{-12})^5 = 9.9 \times 10^{-55}$$

Alternatively, the molar solubility (x) could have been multiplied by 3 (for Pb^{2+}) and by 2 (for PO_4^{3-}) and placed into the ICE table, resulting in the following K_{sp} expression.

$$K_{sp} = [Pb^{2+}]^3[PO_4^{3-}]^2 = (1.9 \times 10^{-11})^3(1.2 \times 10^{-11})^2 = 9.9 \times 10^{-55}$$

CALCULATING SOLUBILITY FROM K_{SP}

If you are given a group of salts which do not all have the same cation-to-anion ratio, along with their K_{sp} values, and asked which is more soluble, you must perform a calculation to determine the solubility of each salt.

EXAMPLE: Given the two salts in the table below, which is more soluble? Show calculations to support your answer.

	K_{sp}
FeC_2O_4	2.1×10^{-7}
$Cu(IO_4)_2$	1.4×10^{-7}

SOLUTION: You cannot directly compare the K_{sp} values to predict which is more soluble because the salts dissolve to produce a different number of ions. FeC_2O_4 dissolves to produce two ions and $Cu(IO_4)_2$ produces three ions. You must calculate the solubility for each salt from its K_{sp} value.

Step 1: Write the reaction for the dissolution of the solid.

$$FeC_2O_4(s) \rightleftarrows Fe^{2+}(aq) + C_2O_4^{2-}(aq)$$

Step 2: Underneath the reaction, make an ICE chart.

	$FeC_2O_4(s)$	$\rightleftarrows$	$Fe^{2+}(aq)$	+	$C_2O_4^{2-}(aq)$
I	----		0		0
C	$-x$		$+x$		$+x$
E	----		x		x

Step 3: Write the equilibrium expression for K_{sp}, plug in the equilibrium line, and solve for x.

$$K_{sp} = [Fe^{2+}][C_2O_4^{2-}]$$

$$2.1 \times 10^{-7} = x^2$$

$$x = 4.6 \times 10^{-4} \text{ mol } Fe^{2+}/L$$

Repeat the same calculations for the next salt.

Step 1: Write the reaction for the dissolution of the solid.

$$Cu(IO_4)_2(s) \rightleftarrows Cu^{2+}(aq) + 2\ IO_4^-(aq)$$

Step 2: Underneath the reaction, make an ICE chart.

	$Cu(IO_4)_2(s)$	$\rightleftarrows$	$Cu^{2+}(aq)$	+	$2\ IO_4^-(aq)$
I	----		0		0
C	$-x$		$+x$		$+2x$
E	----		x		$2x$

Step 3: Write the equilibrium expression for K_{sp}, plug in the equilibrium line, and solve for x.

$$K_{sp} = [Cu^{2+}][IO_4^-]^2$$

$$1.4 \times 10^{-7} = x(2x)^2$$

$$1.4 \times 10^{-7} = 4x^3$$

$$x = 3.3 \times 10^{-3} \text{ mol/L}$$

3.3×10^{-3} mol $Cu(IO_4)_2$/L is greater than 4.6×10^{-4} mol FeC_2O_4/L.

Therefore, $Cu(IO_4)_2$ is more soluble, even though the K_{sp} for FeC_2O_4 is slightly larger than the K_{sp} for $Cu(IO_4)_2$.

Math Trick

Don't forget that you must raise the coefficient of the "x" to the power and not just use the "x." Remember $(3x)^3$ is $27x^3$ and not $3x^3$ or $9x^3$.

COMMON ION EFFECT

When a salt is dissolved in water containing a common ion, its solubility is decreased.

EXAMPLE: The solubility product constant of silver sulfate is 1.20×10^{-5}. What is the solubility in g/L of silver sulfate in (i) pure water and (ii) in 0.100 M $AgNO_3$?

SOLUTION:

Part (i)

Step 1: Write the reaction for the dissolution of the solid.

$$Ag_2SO_4(s) \rightleftarrows 2\ Ag^+(aq) + SO_4^{2-}(aq)$$

Step 2: Underneath the reaction, make an ICE chart.

	$Ag_2SO_4(s)$	$\rightleftarrows$	$2\ Ag^+(aq)$	+	$SO_4^{2-}(aq)$
I	----		0		0
C	$-x$		$+2x$		$+x$
E	----		$2x$		x

Step 3: Write the equilibrium expression for K_{sp}, plug in the equilibrium line, and solve for x.

$$K_{sp} = [Ag^+]^2[SO_4^{2-}]$$

$$1.20 \times 10^{-5} = (2x)^2(x) \quad \text{so } 1.20 \times 10^{-5} = 4x^3$$

$$x = 1.44 \times 10^{-2} \text{ mol } Ag_2SO_4(s)/L$$

$$\frac{1.44 \times 10^{-2} \text{ mol } Ag_2SO_4}{L} \times \frac{311.80 \text{ g } Ag_2SO_4}{1 \text{ mol } Ag_2SO_4} = \frac{4.49 \text{ g } Ag_2SO_4}{L}$$

Part (ii)

Step 1: Repeat the same calculations for the salt dissolved in the 1.00 M $AgNO_3$, making sure to include the concentration of the silver ion in the ICE table.

	$Ag_2SO_4(s)$	$\rightleftarrows$	$2\ Ag^+(aq)$	+	$SO_4^{2-}(aq)$
I	----		0.100		0
C	$-x$		$+2x$		$+x$
E	----		$0.100 + 2x$		x

We can ignore the $2x$ term in the $0.100 + 2x$ as long as the $2x$ term does not affect the concentration by more than 5%. Given that the concentration of the silver ion is greater than $100 \times K_{sp}$ value, we should be able to ignore it. Remember that we will not have to solve a quadratic equation on the exam.

Step 2: Write the equilibrium expression for K_{sp}, plug in the equilibrium line, and solve for x.

$$K_{sp} = [Ag^+]^2[SO_4^{2-}]$$

$$1.20 \times 10^{-5} = (0.100)^2(x) \quad \text{so } 1.20 \times 10^{-3} = x$$

$$x = 1.20 \times 10^{-3} \text{ mol } Ag_2SO_4(s)/L$$

As a check to see if we can safely remove the $2x$ term, we have:

$$\frac{1.20 \times 10^{-3}}{0.100} \times 100 = 1.20\% \text{ so we can safely remove the term}$$

$$\frac{1.20 \times 10^{-2} \text{ mol } Ag_2SO_4}{L} \times \frac{311.80 \text{ g } Ag_2SO_4}{1 \text{ mol } Ag_2SO_4} = \frac{0.374 \text{ g } Ag_2SO_4}{L}$$

Note the greatly decreased solubility of the silver sulfate when dissolved in a solution of silver ions.

pH AND SOLUBILITY

Chromium(III) hydroxide dissolves according to the equilibrium

$$Cr(OH)_3(s) \rightleftarrows Cr^{3+}(aq) + 3\ OH^-(aq)$$

An increase in pH, caused by the addition of OH^- ions, will shift the equilibrium to the left, decreasing the solubility of $Cr(OH)_3$.

A decrease in pH, caused by the addition of H^+ ions, will shift the equilibrium to the right, increasing the solubility of $Cr(OH)_3$. The H^+ ions remove the OH^- ions from the solution by forming H_2O.

A salt with the general formula, MX, will show increased solubility in acidic solution if the anion, X^-, is an effective base (if HX is a weak acid). Common anions that make effective bases include F^-, S^{2-}, OH^-, and CO_3^{2-}.

PRECIPITATE FORMATION

A precipitate may or may not form when two solutions are mixed, depending on the concentrations of the ions involved in the formation of the solid.

ION PRODUCT

The ion product, Q, is written in the same way as the K_{sp} expression. For lead(II) chloride, $Q = [Pb^{2+}][Cl^-]^2$.

Calculation of the value, Q, involves the use of the initial concentrations of the solutions mixed, $[Pb^{2+}]_0$ and $[Cl^-]_0$, instead of the equilibrium concentrations. This is the same Q that was used in general equilibrium problems and the calculations are the same and so is the "shifting." However, since the solubility equations are always written in the form of

$$A_xB_y(s) \rightleftarrows x\ A^{y+}(aq) + y\ B^{x-}(aq)$$

the "shifting" has additional meaning. A comparison of the value of Q to K_{sp} determines if a precipitate is formed.

$Q > K_{sp}$: precipitation occurs

$Q < K_{sp}$: no precipitation occurs

$Q = K_{sp}$: the solution is saturated

EXAMPLE: Will a precipitate form when 100.0 mL of 4.0×10^{-4} M $Mg(NO_3)_2$ is added to 100.0 mL of 2.0×10^{-4} M NaOH?

SOLUTION: Step 1: Determine the identity of the precipitate formed. $Mg(NO_3)_2 + 2\ NaOH \rightarrow Mg(OH)_2 + 2\ NaNO_3$

$Mg(OH)_2$ must be the precipitate as nitrates are always soluble.

Step 2: Determine the concentration of the ions after they are mixed and before any reaction occurs.

Determine the molar concentration of each solute present. Be sure to divide by the total volume of the two solutions mixed.

$$0.1000\ \text{L} \times \frac{4.0 \times 10^{-4}\ \text{mol}}{1\ \text{L}} = 4.0 \times 10^{-5}\ \text{mol Mg}^{2+}$$

$$\frac{4.0 \times 10^{-5}\ \text{mol}}{0.2000\ \text{L}} = 2.0 \times 10^{-4}\ M\ \text{Mg}^{2+} = [\text{Mg}^{2+}]_0$$

$$0.1000\ \text{L} \times \frac{2.0 \times 10^{-4}\ \text{mol}}{1\ \text{L}} = 2.0 \times 10^{-5}\ \text{mol OH}^-$$

$$\frac{2.0 \times 10^{-5}\ \text{mol}}{0.2000\ \text{L}} = 1.0 \times 10^{-4}\ M\ \text{OH}^- = [\text{OH}^-]_0$$

Step 3: Write the ion product expression, calculate its value, and compare it to K_{sp}, which equals 8.8×10^{-12}.

$$Mg(OH)_2(s) \rightarrow Mg^{2+}(aq) + 2\ OH^-(aq)$$

$Q = [Mg^{2+}][OH^-]^2$

$Q = (2.0 \times 10^{-4})(1.0 \times 10^{-4})^2 = 2.0 \times 10^{-12}$

Since $Q < K_{sp}$, no precipitate will form.

MULTIPLE-CHOICE QUESTIONS

No calculators are to be used in this section.

Questions 1-3 refer to the following information.

	K_{sp}	Solubility (mol/L)
$AgC_2H_3O_2$	1.9×10^{-3}	4.4×10^{-2}
AgCl	1.6×10^{-10}	1.3×10^{-5}
AgI	1.5×10^{-16}	1.2×10^{-8}
$PbCl_2$	1.6×10^{-5}	2.5×10^{-2}
PbI_2	1.4×10^{-8}	2.4×10^{-3}

1. In a laboratory experiment, 500 mL of 0.010 *M* $Pb(NO_3)_2$ and 500 mL of 0.010 *M* $AgC_2H_3O_2$ are mixed. The resulting solution is clear and colorless. 0.10 *M* KI is added dropwise to the solution until a precipitate formed. After several drops are added, a second and visually different precipitate forms. What is the identity of the second precipitate?
   (A) $AgC_2H_3O_2$
   (B) AgI
   (C) $Pb(C_2H_3O_2)_2$
   (D) PbI_2

2. What is the molar solubility of silver chloride in a solution of 0.10 *M* sodium chloride?
   (A) 2.6×10^{-20}
   (B) 1.7×10^{-10}
   (C) 1.6×10^{-9}
   (D) 1.3×10^{-5}

3. The dissolving of $PbCl_2$ in 10.0 mL of water at 25.0°C causes the temperature of the water to decrease slightly. Will $PbCl_2$ be more soluble or less soluble in water at a temperature of 30°C?
   (A) The solubility of $PbCl_2$ will increase because ΔH is negative.
   (B) The solubility of $PbCl_2$ will increase because ΔH is positive.
   (C) The solubility of $PbCl_2$ will decrease because ΔH is negative.
   (D) The solubility of $PbCl_2$ will decrease because ΔH is positive.

4. Each of the following salts has a solubility of 1.0×10^{-5} *M*. Rank the salts in order of lowest to highest K_{sp}.
   AX, B_2X, C_2X_3, D_3X
   (A) $AX < B_2X < C_2X_3 < D_3X$
   (B) $B_2X < AX < D_3X < C_2X_3$
   (C) $C_2X_3 < D_3X < B_2X < AX$
   (D) $D_3X < C_2X_3 < B_2X < AX$

Questions 5 and 6 refer to the following information.

Native Americans would take field corn (also known as maize) and dry the grain and then treat it by soaking and cooking the kernels (grain) in a dilute solution of calcium hydroxide, which served several purposes. The treated corn has significant increases in the amount of niacin and the grain is easier to digest. Unlike untreated cornmeal, cornmeal made from the treated corn kernels has the ability to form dough, which is needed to make tortillas.

$$Ca(OH)_2(s) \rightleftarrows Ca^{2+}(aq) + 2\ OH^-(aq)$$

5. The K_{sp} of $Ca(OH)_2 = 1.3 \times 10^{-6}$. Which statement is correct concerning the reaction shown above?
   (A) The reaction is thermodynamically favored because the K_{sp} value is less than 1.0.
   (B) The reaction is not thermodynamically favored because the K_{sp} value is less than 1.0.
   (C) The reaction is thermodynamically favored because the reaction increases entropy.
   (D) The reaction is not thermodynamically favored because the reaction increases entropy.

6. What would increase the molar solubility of calcium hydroxide?
   (A) Add $Ca(NO_3)_2$
   (B) Add $HC_2H_3O_2$
   (C) Add H_2O
   (D) Add NaOH

Questions 7 and 8 refer to the following information.

	K_{sp}
Ag_2CO_3	8.1×10^{-12}
$CaCO_3$	8.7×10^{-9}

7. Which has a greater molar solubility, calcium carbonate or silver carbonate?
   (A) $CaCO_3$, because it has a higher K_{sp} and ionizes into fewer ions.
   (B) $CaCO_3$, because compounds containing Group 2 cations are quite soluble.
   (C) Ag_2CO_3, because it has a lower K_{sp} and ionizes into more ions.
   (D) Ag_2CO_3, because silver compounds tend to have low solubility.

8. Equal volumes of a 0.010 *M* solution of calcium chloride and a 0.010 *M* solution of sodium carbonate are mixed. Will a precipitate of calcium carbonate form?
   (A) A precipitate will form because $Q > K_{sp}$.
   (B) A precipitate will form because $Q < K_{sp}$.
   (C) A precipitate will not form because $Q > K_{sp}$.
   (D) A precipitate will not form because $Q < K_{sp}$.

9. Magnesium arsenate [$Mg_3(AsO_4)_2$] is an extremely toxic substance, sometimes used as an insecticide. If *s* represents the molar solubility of magnesium arsenate, which is the correct expression for its K_{sp}?
(A) $K_{sp} = 36s^5$
(B) $K_{sp} = 108s^5$
(C) $K_{sp} = 36s^6$
(D) $K_{sp} = 108s^6$

10. If the molar solubility of compound X_2Y is found to be 0.01 *M*, what is the K_{sp} of X_2Y?
(A) 1×10^{-6}
(B) 4×10^{-6}
(C) 1×10^{-4}
(D) 2×10^{-4}

11. All of these salts have similar K_{sp} values. Which salt would have the greatest molar solubility? All values are at 25°C.
(A) $Fe(OH)_2$ $K_{sp} = 1.6 \times 10^{-14}$
(B) MnS $K_{sp} = 3.0 \times 10^{-14}$
(C) $PbCrO_4$ $K_{sp} = 2.0 \times 10^{-14}$
(D) $Zn(OH)_2$ $K_{sp} = 1.8 \times 10^{-14}$

12. Which of the following describes what happens to the solubility of a slightly soluble ionic compound when a common ion is added to the solution?
(A) The solubility of the ionic compound is reduced.
(B) The solubility of the ionic compound is increased.
(C) There is no effect on the solubility of the ionic compound.
(D) The ionic compound dissolves more rapidly.

13. Lead sulfide (PbS, molar mass = 239.3 g) has a K_{sp} of 2.5×10^{-27} at 25°C. What mass of PbS would dissolve in 1.00 L of water at 25°C?
(A) 5.0×10^{-14}
(B) 5.0×10^{-13}
(C) 1.2×10^{-11}
(D) 1.2×10^{-10}

14. In which solution would silver chloride (AgCl) be the *least* soluble?
(A) pure water
(B) seawater ($[Cl^-] = 0.54\ M$)
(C) 0.10 *M* $HC_2H_3O_2$
(D) 0.010 *M* HCl

15. Two salts, AX and BX_2, have identical K_{sp} values at a given temperature. Which statement is correct?
(A) The salts are more soluble in 0.1 *M* NaX than in water.
(B) The molar solubility of AX is identical to BX_2.
(C) The addition of NaX will not affect the solubilities of the salts.
(D) The molar solubility of AX is greater than that of BX_2.

FREE-RESPONSE QUESTIONS

1. Zinc sulfide is an ionic compound with very unique properties. It can be prepared by the high-temperature reaction of solid zinc powder with sulfur powder.
   (a) Zinc sulfide ($K_{sp} = 2.5 \times 10^{-22}$) has a very low solubility in water.
      (i) Write the equation representing the dissolving of ZnS.
      (ii) Write the solubility equilibrium expression for the dissolving of ZnS.
      (iii) Calculate the molar solubility of ZnS.
      (iv) Will ZnS be more soluble, be less soluble, or have the same solubility in a solution of 0.1 *M* HCl than in pure water? Write a chemical equation to justify your answer.
   (b) 0.10 *M* Na_2S is added slowly to 100.0 mL of a solution containing 0.0500 mole of silver nitrate and 0.0500 mole of zinc nitrate. The K_{sp} for Ag_2S is 1.6×10^{-40}. Will ZnS or Ag_2S precipitate first? Circle your answer. Provide a mathematical explanation for your answer.

      ZnS Ag_2S

   (c) Zinc sulfide is a semiconductor and exhibits phosphorescence when small amounts of certain elements are added. A phosphorescent substance will absorb energy and then emit it back as light over a period of minutes or hours. Depending upon the element added, zinc sulfide can glow various colors such as yellow-green, blue, or red. When copper is added, the phosphorescent glow is the typical "glow-in-the-dark" color (yellow-green) that we are used to seeing. The maximum wavelength emitted by this substance is 520 nm.
      (i) What is the frequency of this light?
      (ii) Calculate the energy of one photon of this light.
   (d) Zinc sulfide used to be used extensively in paint for signs and other safety equipment because of its glow-in-the dark properties. However, it has been found that strontium aluminate ($SrAl_2O_4$) glows much more brightly and for a longer time period than zinc sulfide. While zinc sulfide is used in many novelty items (glow-in-the-dark stars, etc.), strontium aluminate is used in most safety applications, even though its cost is about 10 times that of zinc sulfide. Like ZnS, $SrAl_2O_4$ requires a small amount of another element to activate its phosphorescent properties.
      (i) There are several different structures of strontium aluminate, all of which consist of different arrangements of the basic aluminate ion shown below. Complete the structure with any unshared electrons.

$$\left[\mathrm{O=Al-O}\right]^{-}$$

      (ii) What is the formal charge of aluminum in this structure?

2. Human bodies maintain their balance (homeostasis) through a series of equilibria. One important equilibrium system involves a calcium ion because Ca^{2+} is important for muscle contractions in the heart as well as for strong bones. Bone is a complex tissue consisting of living cells and nonliving material. Calcium phosphate is part of the nonliving material and is how the body stores calcium.
   (a) Write the solubility product expression for the dissolution of calcium phosphate.
   (b) What is the specific interparticle attraction between the dissolved calcium phosphate and the water?
   (c) When the calcium ion concentration in blood decreases, a hormone is released that causes calcium ions from the bones to be released into the bloodstream. What effect does this have on the solubility of the calcium phosphate in the bones?

   $H_3PO_4 + H_2O \leftrightharpoons H_3O^+ + H_2PO_4^-$ $K_{a_1} = 7.5 \times 10^{-3}$

   $H_2PO_4^- + H_2O \leftrightharpoons H_3O^+ + HPO_4^{2-}$ $K_{a_2} = 6.2 \times 10^{-8}$

   $HPO_4^{2-} + H_2O \leftrightharpoons H_3O^+ + PO_4^{3-}$ $K_{a_3} = 4.8 \times 10^{-13}$

   (d) One way skeletons can be cleaned for study is to boil the bones in a solution to remove all material and leave the bone. Based on the equilibrium equations above, would a highly acidic or highly basic solution be best to clean the bones? Explain your answer.

Answers

MULTIPLE-CHOICE QUESTIONS

1. D Since there is no solid present when the lead(II) nitrate and silver acetate are mixed, the answer cannot be $AgC_2H_3O_2$ or $Pb(C_2H_3O_2)_2$ (which would be a product of the reaction, if it occurred). Since volume and molarity of the two solutions are equal, the concentration of Ag^+ and Pb^{2+} ions will be the same. When KI is added, the possible precipitates are AgI and PbI_2. AgI has a lower molar solubility, so pale yellow AgI(s) would precipitate with a smaller volume of Pb^{2+} added. The molar solubility of PbI_2 is higher. A bright yellow precipitate of PbI_2 would form as additional Pb^{2+} is added (LO 6.21).

2. C This is a common ion effect problem.

   $K_{sp} = [Ag^+][Cl^-] = 1.6 \times 10^{-10}$

   $1.6 \times 10^{-10} = [Ag^+](0.10\ M)$

   $[Ag^+] = 1.6 \times 10^{-9}$

   (LO 6.23)

3. B $PbCl_2(s) \leftrightarrows Pb^{2+}(aq) + 2\ Cl^-(aq)$ Since the dissolving of $PbCl_2$ lowers the temperature of the solution, the process is endothermic and ΔH is positive. The stress of increasing temperature on the system shifts the reaction to the dissolved ions, increasing the solubility of $PbCl_2$ (LO 5.13).

4. C C_2X_3 forms the most ions so it will have the lowest calculated K_{sp}. AX forms the fewest ions so it will have the highest K_{sp}.

Compound	K_{sp}
AX	1.0×10^{-10}
B_2X	4.0×10^{-15}
C_2X_3	1.1×10^{-23}
D_3X	2.7×10^{-19}

(LO 6.22)

5. B When K values of any type are less than 1, ΔG values are positive (not thermodynamically favored) (LO 6.25).

6. B $Ca(OH)_2(s) \leftrightarrows Ca^{2+}(aq) + 2\ OH^-(aq)$ Since there are hydroxide ions present as products, the addition of H^+ ions from the acetic acid ($HC_2H_3O_2$) would react with the OH^-, removing them from the reaction. The reaction would shift toward the dissolved ions, increasing the solubility (LO 6.23).

7. A If s is the molar solubility, the formula $4s^3 = K_{sp}$ would be used to determine the molar solubility for silver carbonate (which breaks into 3 ions), while s^2 would be used to determine the molar solubility for calcium carbonate (which only breaks into 2 ions). Since the K_{sp} of calcium carbonate is already larger than that of the silver carbonate, the value of s will be larger for calcium carbonate (LO 6.21).

8. A $Q = [Ca^{2+}][CO_3^{2-}] = (0.010\ M)(0.010\ M) = 1.0 \times 10^{-4}$

   Q of $1.0 \times 10^{-4} > K$ of 8.7×10^{-9}, there are too many ions present to stay dissolved and a precipitate of $CaCO_3$ forms (LO 6.21).

9. B $Mg_3(AsO_4)_2$ will ionize into 3 Mg^{2+} and 2 AsO_4^{3-}. The K_{sp} expression will be $K_{sp} = [Mg^{2+}]^3[AsO_4^{3-}]^2$. Using s for molar solubility, $K_{sp} = (3s)^3(2s)^2 = 108s^5$ (LO 6.21).

10. B $X_2Y \leftrightarrows 2\ X^+ + Y^{2-}$ $K_{sp} = [X^+]^2[Y^{2-}]$ $K_{sp} = (0.02)^2(0.01) = 4 \times 10^{-6}$ (LO 6.22)

11. B MnS and $PbCrO_4$ each ionize into 2 ions, while $Fe(OH)_2$ and $Zn(OH)_2$ each ionize into 3 ions. To determine the molar solubility of MnS and $PbCrO_4$, you would simply take the square root of K_{sp}. To find the molar solubility of the others, you would solve for x in $K_{sp} = 4x^3$. Since all of the compounds had similar solubilities, no calculations are really needed. The two larger molar solubilities will be those of MnS and $PbCrO_4$. Since the K_{sp} of MnS is a little

larger than that of $PbCrO_4$, MnS will also have the greater molar solubility (LO 6.21).

12. A The solubility of a slightly soluble ionic compound is always lowered whenever the solution already contains ions common in the solid or ions common to the solid that are added after a solution is prepared. This is due to the reaction shifting to the undissolved solid when the common dissolved ion is added. The common ion effect is an example of Le Chatelier's principle (LO 6.23).

13. C $PbS(s) \leftrightarrows Pb^{2+}(aq) + S^{2-}(aq)$ $\quad K_{sp} = [Pb^{2+}][S^{2-}] = 2.5 \times 10^{-27} = x^2$

$25 \times 10^{-28} = x^2$ so $x = 5.0 \times 10^{-14}$ mol/L

To solve (5.0×10^{-14} mol/L) × 239.3 g/mol, we need to estimate the numbers to make the problem easier to do, so (5.0×10^{-14} mol/L) × 239.3 g/mol becomes (5.0×10^{-14} mol/L) × (2.4×10^2 g/mol) = $(5 \times 2.4) \times 10^{(-14+2)} = 12 \times 10^{-12}$ which can be solved easily by writing it as $(1.2 \times 10^1) \times 10^{-12} = 1.2 \times 10^{(-12+1)} = 1.2 \times 10^{-11}$ g/L.

Estimate. Notice before you start that all answers have different powers of 10. (LO 6.23)

14. B Seawater and HCl each contain a common ion (Cl^-). This will shift the reaction to the undissolved solid, lowering the solubility. Acetic acid will not affect the solubility as it does not react with either ion. Since the seawater has a higher molar concentration of Cl^-, it will decrease the solubility more (LO 6.23).

15. D For each salt, the respective solubility equilibrium expressions are:

$$AX(s) \leftrightarrows A^+(aq) + X^-(aq) \quad \text{and} \quad BX_2(s) \leftrightarrows B^{2+}(aq) + 2\,X^-(aq)$$

Since they have the same K_{sp} values, let us call that value N. Let "a" represent the solubility of AX. Thus, (a)(a) = N or a = $\sqrt{N}$.

Let "b" represent the solubility of BX_2. Thus, $(b)(2b)^2 = N$ or b = $\sqrt[3]{N/4}$.

Since a > b, the molar solubility of AX is greater than that of BX_2 (LO 6.21).

FREE-RESPONSE QUESTIONS

1. (a) (i) $ZnS \leftrightarrows Zn^{2+} + S^{2-}$ (LO 3.2)
   (ii) $K_{sp} = [Zn^{2+}][S^{2-}]$ (LO 6.5)
   (iii) molar solubility = x, $K_{sp} = x^2 = 2.5 \times 10^{-22}$, $x = 1.6 \times 10^{-11}\ M$ (LO 6.21)
   (iv) ZnS will be more soluble in a solution of 0.1 M HCl. $H^+ + S^{2-} \rightarrow H_2S$ or $H^+ + S^{2-} \rightarrow HS^-$ (LO 6.23)

   (b) (0.0500 mol/0.100 L) = 0.500 M The concentrations of the silver ion and the zinc ion in the solution will each be 0.500 M. A precipitate will form when the Q value for the compound is greater than the K_{sp}.

   $Ag_2S \leftrightarrows 2\ Ag^+ + S^{2-}$ $K_{sp} = [Ag^+]^2[S^{2-}] = 1.6 \times 10^{-40}$

   $K = (0.500)^2(x) = 1.6 \times 10^{-40}$ $x = 6.4 \times 10^{-40}\ M$ (concentration of S^{2-} when ppt begins)

   For ZnS, $Q = (0.500)(x) = 2.5 \times 10^{-22}$ $x = 5.0 \times 10^{-22}\ M$ (concentration of S^{2-} when ppt begins)

   Since the $[S^{2-}]$ will reach $6.4 \times 10^{-40}\ M$ long before it reaches $5.0 \times 10^{-22}\ M$, Ag_2S will precipitate first (LO 6.21).

   (c) (i) $c/\lambda = \upsilon$ $(3.0 \times 10^8\ m/s)/5.2 \times 10^{-7}\ m = 5.8 \times 10^{14}\ s^{-1}$ (LO 1.15)
   (ii) $E = h\upsilon$ $E = 6.626 \times 10^{-34}\ Js(5.8 \times 10^{14}\ s^{-1}) = 3.8 \times 10^{-19}\ J$ (LO 1.15)

   (d) (i) 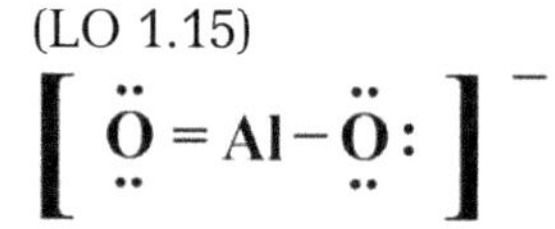
   (LO 2.21)
   (ii) The formal charge of aluminum is 0. A neutral aluminum atom has 3 valence electrons. In this ion structure, it has three bonds and no unshared electrons. Since it gets 1 electron from each bond (3), the formal charge is 3 – 3 = 0 (LO 2.21).

2. (a) $Ca_3(PO_4)_2(s) \leftrightarrows 3\ Ca^{2+}(aq) + 2\ PO_4^{3-}(aq)$

   $K_{sp} = [Ca^{2+}]^3[PO_4^{3-}]^2$ (LO 6.5)

   (b) The interparticle force (sometimes called intermolecular, even though ions are involved) between the calcium ions and water molecules and between the phosphate ions and water molecules is termed "ion–dipole attraction" (LO 2.14).

   (c) The solubility of the calcium phosphate would temporarily increase to replace the dissolved ions that were taken up by the blood and reestablish equilibrium (LO 6.8).

   (d) A highly basic solution will prevent the bones from being damaged in the cleaning. At high pH, acid–base equilibrium favors PO_4^{3-}. Higher PO_4^{3-} concentrations shift the solubility equilibrium back to solid $Ca_3(PO_4)_2$ (LO 6.8).

13

Big Idea 6: Applications of Equilibria: Acids, Bases, and Salts

Big Idea 6

Any bond or intermolecular attraction that can be formed can be broken. These two processes are in a dynamic competition, sensitive to initial conditions and external perturbations.

Concepts of equilibria will be applied in this section covering the equilibria of weak acids, weak bases, and salts. In addition, acid–base theories and properties of acids and bases will be reviewed.

You should be able to:

- Understand the Brønsted–Lowry acid–base theory.
- Identify strong acids and bases and calculate their pH values.
- Calculate the pH of a weak acid or base.
- Calculate the concentration of a strong or weak acid or base from its pH.
- Predict the relative pH of a salt from its formula and then calculate the pH of the salt.

AP® Tip

Be systematic. Follow the same series of steps for all equilibrium problems. The use of ICE tables will prevent many careless errors. To make the math easier for the ionization of weak acids and bases on the AP® Chemistry exam, the problem will be designed so that you will never have to solve a quadratic equation.

ACID–BASE THEORY

BRØNSTED–LOWRY

The Brønsted–Lowry theory states that an acid is a proton (H^+) donor and a base is a proton acceptor.

In the reaction below, HNO_3 transfers a proton to H_2O forming H_3O^+, the hydronium ion. H_3O^+ is the conjugate acid of H_2O, and NO_3^- is the conjugate base of HNO_3.

$$\underset{\text{Acid}}{HNO_3} + \underset{\text{Base}}{H_2O} \rightarrow \underset{\text{Conjugate acid}}{H_3O^+} + \underset{\text{Conjugate base}}{NO_3^-}$$

Note: Since HNO_3 is a strong acid, it ionizes 100%. This reaction goes to completion so we use a one-way arrow ($\rightarrow$) rather than the double equilibrium arrow ($\rightleftarrows$).

EXAMPLE: Give the formulas for the conjugate base of H_2SO_4 and the conjugate acid of CH_3NH_2.

SOLUTION: HSO_4^- is the conjugate base of H_2SO_4.

$CH_3NH_3^+$ is the conjugate acid of CH_3NH_2.

Note that each conjugate acid–base pair differs by 1 H^+.

H_2SO_4 donates a H^+ (a proton) to become HSO_4^-, while CH_3NH_2 accepts a H^+ (a proton) to become $CH_3NH_3^+$.

ACID AND BASE STRENGTH

COMMON STRONG ACIDS	
Formula	**Name**
$HCl(aq)$	Hydrochloric acid
$HBr(aq)$	Hydrobromic acid
$HI(aq)$	Hydroiodic acid
$H_2SO_4(aq)$	Sulfuric acid
$HNO_3(aq)$	Nitric acid
$HClO_4(aq)$	Perchloric acid

The names and formulas of six strong acids must be memorized. These strong acids are shown in the table on the previous page.

If an acid is not one of the six in the list, then for purposes of the AP® exam, you can assume it is a weak acid unless told otherwise. Common weak acids include carboxylic acids and most oxyacids. Several factors affect the magnitude of K_a of weak acids: bond strength, solvation, and electronegativity of the atom bonded to the proton that will ionize.

A particulate model is especially useful for understanding the differences between strong and weak acids.

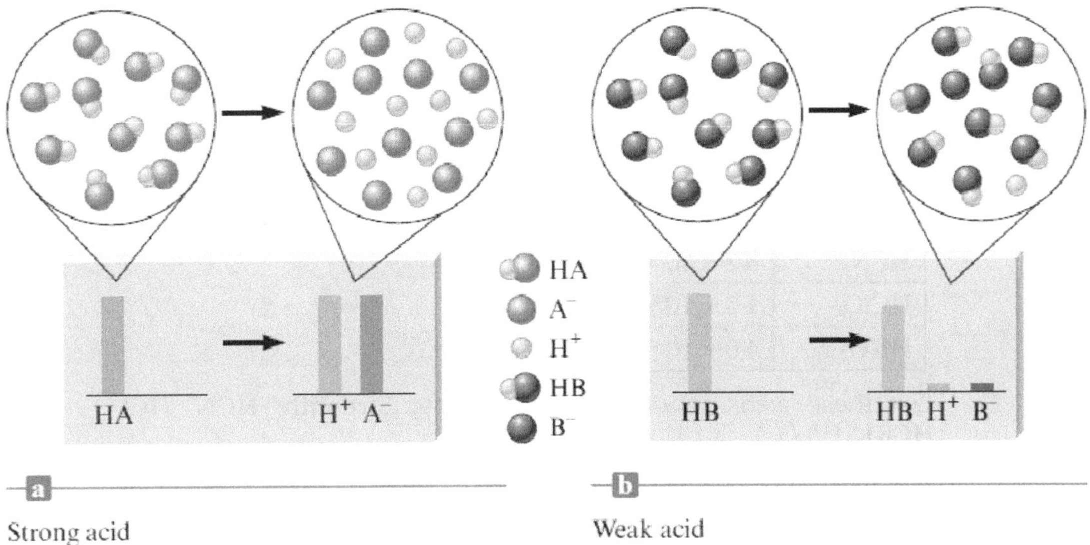

Weak acids exist in equilibrium with their ions in aqueous solution. The equilibrium constant, K_a, measures the extent to which the acid dissociates in water:

$$HA(aq) + H_2O(l) \rightleftarrows H_3O^+(aq) + A^-(aq)$$

Or simply

$$HA(aq) \rightleftarrows H^+(aq) + A^-(aq)$$

The equilibrium expression for the reaction representing the dissociation of a weak acid is:

$$K_a = \frac{[H_3O^+][A^-]}{[HA]} \quad \text{or} \quad K_a = \frac{[H^+][A^-]}{[HA]}$$

A table of K_a values for monoprotic acids (contain one acidic hydrogen) appears in Chapter 14 of the text and also in the Appendix and is worth studying. The larger the K_a value, the stronger the acid. The stronger the acid, the weaker its conjugate base. Conjugate bases of strong acids are weaker bases than water.

Comparison of Strong and Weak Acids				
Type of acid, HA	Reversibility of reaction	K_a value	Relative strength of conjugate base	Ions existing when acid, HA, dissociates in H_2O
Strong	Not reversible $\rightarrow$	Too large to measure >>1	Weaker base than water	H_3O^+ and A^-, only. No HA present.
Weak	Reversible $\rightleftarrows$	Small <1	The greater the K_a value of the weak acid, the weaker its conjugate base	H_3O^+, A^-, and HA (mostly HA)

EXAMPLE:

Acid	K_a
HCN	6.2×10^{-10}
$HClO_2$	1.2×10^{-2}
HNO_2	4.0×10^{-4}

List these acids in order of increasing strength: HCN, HCl, $HClO_2$, HNO_2.

SOLUTION: HCN < HNO_2 < $HClO_2$ < HCl. The first three weak acids are listed in order of increasing K_a values. HCl is stronger than all of the weak acids given. HCl is a strong acid.

EXAMPLE: Arrange the following species in order of increasing base strength: NO_2^-, ClO_2^-, CN^-, Cl^-.

SOLUTION: $Cl^- < ClO_2^- < NO_2^- < CN^-$. The bases are listed in reverse order of their conjugate acids in the previous example because the stronger the acid, the weaker its conjugate base. The conjugate base of a strong acid is a weaker base than water.

Strong bases include Group 1 (1A) and 2 (2A) hydroxides such as NaOH and $Ca(OH)_2$. Weak bases include ammonia, amines, and other nitrogenous bases. The equilibrium constant for the dissociation of a weak base, K_b, measures the extent to which a base reacts with water. The reaction of a weak base, B, with water and its corresponding equilibrium expression is

$$B(aq) + H_2O(l) \rightleftarrows BH^+(aq) + OH^-(aq)$$

$$K_b = \frac{[BH^+][OH^-]}{[B]}$$

AP® Tip

Writing the reaction for base ionization can be tricky. Always remember to react the base with water. The base reacts with water to produce hydroxide ions and accepts a hydrogen ion to become protonated. Remember to check the charges of the reactants and products and be sure that the charges on both sides of the reaction are equal.

The acid and conjugate base equilibrium constants, K_a and K_b, are related to each other through the equilibrium constant of water, K_w.

$$K_w = K_a \times K_b$$

K_w is the equilibrium constant for the self-ionization of water.

$$2\ H_2O \rightleftarrows H_3O^+ + OH^-$$

$K_w = 1.0 \times 10^{-14}$ at 25°C. Note the small value of the equilibrium constant. Although water ionizes to some extent, not enough ions form to allow pure water to conduct electricity.

This reaction is endothermic, which causes a greater ionization of water as temperature increases. Thus, K_w will have a higher value at higher temperatures.

THE EFFECT OF STRUCTURE ON ACID–BASE PROPERTIES

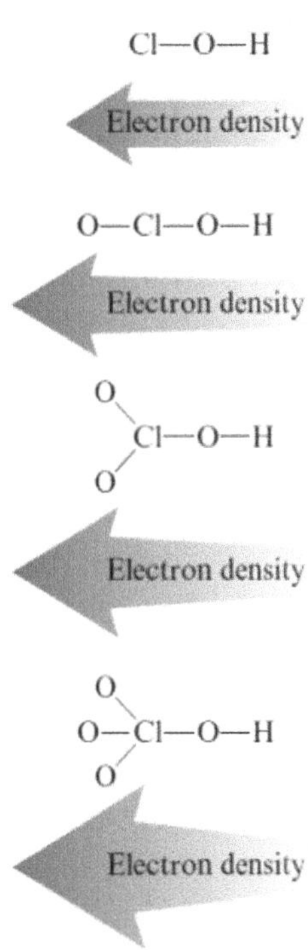

For a binary acid, HX, the strength of the H–X bond and the polarity of the bond will determine the behavior of the acid. The polarity of the bonds in hydrogen halides becomes less polar going down a group. The very strong H–F bond is what makes it a weak acid: F^- has a very high attraction for H^+ so a great deal of energy is needed to break the bond. The rest of the hydrogen halides are strong acids: HI > HBr > HCl. When these three acids are dissolved in water, their strength is equal because they all ionize 100%.

For a given series of oxyacids such as $HClO_4$, $HClO_3$, $HClO_2$, and HClO, the acid strength increases with an increasing number of oxygen atoms attached to the central atom. $HClO_4$ is a strong acid, as is $HClO_3$ (it is just not on the list of strong acids that we need to know). The remaining oxyacids are listed in order of decreasing strength (decreasing number of oxygen atoms). The O–H bond becomes more polarized and weakened due to the electron density drawn toward the highly electronegative oxygen atoms.

CALCULATING THE pH OF STRONG ACIDS AND BASES

The pH of a strong acid can be calculated directly from the hydronium (or hydrogen) ion concentration ($[H^+]$ and $[H_3O^+]$ are used interchangeably).

$$pH = -\log [H_3O^+] \quad \text{or} \quad pH = -\log [H^+]$$

$[H_3O^+]$, the molar concentration of the hydronium ion, is obtained from the molarity of the acid.

EXAMPLE: Calculate the pH of 0.025 *M* HCl.

SOLUTION: HCl is a strong, monoprotic acid so its $[H_3O^+]$ will be equal to its molarity. $pH = -\log [H_3O^+]$, so $pH = -\log (0.025) = 1.60$.

The pH of sulfuric acid (H_2SO_4) cannot usually be calculated directly from its molarity because the second proton does not dissociate completely. This will be discussed more thoroughly in the section on polyprotic acids.

The pH of a strong base can be calculated from its hydroxide ion concentration.

$$pOH = -\log[OH^-]$$

$$pH + pOH = 14.00$$

The concentration of a strong acid or strong base can be determined from the solution's pH.

EXAMPLE: The pH of a certain $Sr(OH)_2$ solution is 13.50. Calculate the concentration of $Sr(OH)_2$.

SOLUTION: $pOH = 14.00 - pH = 14.00 - 13.50 = 0.50$

$[OH^-] = 10^{(-pOH)}$; $10^{(-0.50)} = 0.32\ M\ OH^-$

$$\frac{0.32\ M\ OH^-}{1\ L} \times \frac{1\ \text{mol Sr(OH)}_2}{2\ \text{mol OH}^-} = 0.16\ M\ Sr(OH)_2$$

AP® Tip

The number of significant figures in a pH or other logarithmic value can be a little tricky.

If the pH = 3.45, the "3" is called the characteristic and the ".45" is the mantissa. The characteristic is **ALWAYS** determined by the exponent in the number. The mantissa is always determined by the numbers in front of the "×" in scientific notation. Say you have $\log_{10} 5.16 \times 10^{-9}$; the –9 will determine the characteristic and the 5.16 will provide you the mantissa. Since there are three numbers before the "×" sign, then your answer will need to indicate three significant figures, which would be equal to the number of decimal places in the pH.

AP® Tip

The pH value of a solution may be greater than 14, zero, or even slightly negative at its extremes. If the $[H^+]$ of a solution = 1.0 *M*, then pH = 0.00. For a strong acid, any molarity above 1.0 *M* will result in a negative pH.

On the rectangle shown below, the outside corners are the four values that you may need to find. The mathematical relationships are shown inside the rectangle. You should be able to go both clockwise and counterclockwise around the rectangle.

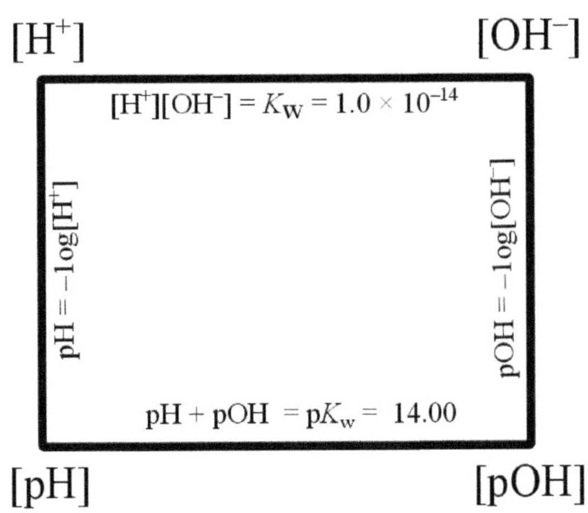

EXAMPLE: If the $[H^+] = 4.19 \times 10^{-4}$, find $[OH^-]$, pOH, and then pH in that order.

SOLUTION:

$$[H^+][OH^-] = K_W \Rightarrow \frac{1.00 \times 10^{14}}{4.19 \times 10^{-4}} = [OH^-]$$

$$[OH^-] = 2.39 \times 10^{-11}$$

$$pOH = -\log[OH^-] \Rightarrow pOH = -\log[2.39 \times 10^{-11}]$$

$$pOH = 10.622$$

$$pH + pOH = 14 \Rightarrow pH = 14 - pOH \Rightarrow pH = 14 - 10.622$$

$$pH = 3.378$$

EXAMPLE: Now it is your turn. If the $[OH^-] = 3.67 \times 10^{-3}$, what are the $[H^+]$, pH, and pOH values?

SOLUTION: $[H^+] = 2.72 \times 10^{-12}$, pH = 11.565, and pOH = 2.435

CALCULATING THE pH OF WEAK ACIDS

The pH of a weak acid cannot be calculated directly from the concentration of the acid since all of the acid does not dissociate to form H_3O^+. The equilibrium reaction of the acid must be considered and a K_a problem worked.

AP® Tip

Always follow these steps when performing calculations involving the dissociation of weak acids or weak bases.

1. Write the reaction of the acid or base with water. (Use Brønsted–Lowry acid–base theory and check charges.)
2. Set up an ICE chart.
3. Write the equilibrium expression in terms of reactant and product concentrations (without numbers).
4. Solve for x, using the method of approximation.
5. Solve for pH. (Be careful with base equilibria: $x = OH^-$. You need to find pOH and then find pH.)
6. Always remember that for weak acids and bases, at equilibrium, $pH = -\log [H_3O^+]$.

EXAMPLE: Calculate the pH of 0.025 M HCN.

SOLUTION: First, write the reaction of the acid with water. Use the Brønsted–Lowry theory to help you write the products. Check that you have the correct charges on the products. Water does not have to be included for weak acids.

$$HCN + H_2O \rightleftarrows H_3O^+ + CN^-$$

Second, set up an ICE chart as you did in the previous chapters for equilibrium problems.

	HCN	+	H_2O	⇄	H_3O^+	+	CN^-
I	0.025		---		0		0
C	$-x$		---		$+x$		$+x$
E	$0.025 - x$		---		x		x

Third, write the equilibrium expression for K_a in the same manner as you did for equilibrium problems in the last two chapters. Plug in the values from the equilibrium line of the ICE chart.

$$K_a = 6.2 \times 10^{-10} = \frac{[H_3O^+][CN^-]}{[HCN]} \qquad \frac{x^2}{(0.025 - x)} \approx \frac{x^2}{0.025}$$

If x is very small compared to the concentration of the acid you are subtracting it from, you can assume [HA − x] is approximately equal to [HA]. One way to determine if this approximation is valid is to compare the magnitude of K to [HA]; if [HA] is greater than K by a factor of 10^3 or more, x can be safely ignored. On the exam, you can always assume that x is small and then check the value of the x that you calculated to see if [HA − x] is within 5% of [HA].

Fourth, solve for x which equals $[H_3O^+]$.

$$x^2 = (6.2 \times 10^{-10})(0.025)$$

$$x = [H_3O^+] = 3.9 \times 10^{-6}\ M$$

Finally, calculate the pH from the value of x, the H_3O^+ concentration.

$$pH = -\log(3.9 \times 10^{-6}) = 5.41$$

As you can see, acid strength and concentration both play a role in the pH of a solution. At the same concentration, the strong acid was pH 1.60, while the weak acid was pH 5.41. The difference is due to the higher $[H_3O^+]$ in the strong acid.

CALCULATING THE PERCENT IONIZATION

The percent ionization (also known as percent dissociation) of an acid (or a base) is the amount of the acid, HA, which has ionized, x, divided by the acid's initial concentration, $[HA]_0$, multiplied by 100.

$$\%\ \text{ionization} = \frac{x}{[HA]_0} \times 100$$

When making assumptions in an equilibrium calculation, it is best to test the assumption by making sure that the percent dissociation is less than or equal to 5%. The test for the assumption is the same as the calculation for the percent dissociation.

EXAMPLE: Calculate the percent ionization of the 0.025 *M* HCN solution from the previous example.

SOLUTION: The [H^+] was found to be 3.9×10^{-6} *M*. This is the $-x$ value in the ICE diagram and is the amount of HA that ionized. To determine the percent ionization: (3.9×10^{-6} *M*/0.025 *M*) × 100 = 0.016% ionized

EXAMPLE: The percent ionization of an acid, HA, which is 0.100 *M* is 2.5%. Calculate the K_a of the acid.

SOLUTION: $0.025 \times 0.100\,M = 2.5 \times 10^{-3}\,M$ (This is the x value that goes in the ICE table.)

	HA	+	H_2O	⇄	H_3O^+	+	A^-
I	0.100		---		0		0
C	-2.5×10^{-3}		---		$+2.5 \times 10^{-3}$		$+2.5 \times 10^{-3}$
E	$0.100 - (2.5 \times 10^{-3})$		---		2.5×10^{-3}		2.5×10^{-3}

$$K_a = \frac{[H_3O^+][A^-]}{[HA]}$$

$$K_a = \frac{(2.5 \times 10^{-3})^2}{0.100 - (2.5 \times 10^{-3})} = 6.4 \times 10^{-5}$$

Note: If you know the value of x, always use it in the $+x$ or $-x$ calculations. Don't assume it is negligible and leave it out.

CALCULATING THE PH OF WEAK BASES

The calculations involving weak base equilibria are similar to the weak acid equilibria problems except that the equation is written for a base reacting with water and the calculation initially involves finding [OH^-]. You will need to find the pOH and then the pH. The smaller the K_b value, the less the weak base will dissociate.

EXAMPLE: Calculate the pH of a 0.50 *M* solution of ethylamine ($C_2H_5NH_2$, $K_b = 5.6 \times 10^{-4}$).

SOLUTION:

	$C_2H_5NH_2$	+	H_2O	⇆	OH^-	+	$C_2H_5NH_3^+$
I	0.50		---		0		0
C	$-x$		---		$+x$		$+x$
E	$0.50 - x$		---		x		x

$$K_b = \frac{[OH^-][C_2H_5NH_3^+]}{[C_2H_5NH_2]} = 5.6 \times 10^{-4} = \frac{x^2}{0.50 - x} \approx \frac{x^2}{0.50}$$

$$x = [OH^-] = 0.0167 \qquad pOH = -\log 0.0167 = 1.78$$

$$pH = 14.00 - 1.78 = 12.22$$

EXAMPLE: The pH of a 0.20 *M* solution of hydrazine (H_2NNH_2) is 11.38. Calculate the K_b for H_2NNH_2.

SOLUTION: Write the reaction with water. Bases accept H^+. Watch charges! One hint to help in writing the reaction is that the pH is 11.38. This basic pH indicates that OH^- must be one of the products.

Fill out the ICE chart under the reaction.

	H_2NNH_2	+	H_2O	⇆	OH^-	+	$H_2NNH_3^+$
I	0.20		---		0		0
C	$-x$		---		$+x$		$+x$
E	$0.20 - x$		---		x		x

You are given the pH, but x equals $[OH^-]$.

Find pOH; pH + pOH = 14.00; pOH = 14.00 – 11.38 = 2.62

$$pOH = -\log [OH^-];\ \text{Find } [OH^-] = 10^{(-pOH)};$$

$$10^{(-2.62)} = 2.4 \times 10^{-3}\ M = x$$

Plug this value of x into the K_b expression:

$$K_b = \frac{[OH^-][H_2NNH_3^+]}{[H_2NNH_2]} = \frac{x^2}{(0.20 - x)}$$

$$K_b = \frac{(2.4 \times 10^{-3})^2}{0.20 - (2.4 \times 10^{-3})} = 2.9 \times 10^{-5}$$

WATER AS AN ACID AND AS A BASE

Water is an amphoteric substance; it can act as an acid or as a base (accept a proton or donate a proton) according to the reaction

$$H_2O + H_2O \rightleftarrows H_3O^+ + OH^-$$

The equilibrium expression for this reaction is

$$K_w = [H_3O^+][OH^-]$$

where K_w is the equilibrium constant for the ionization of water. Water is neutral whenever $[H_3O^+] = [OH^-]$, but that does not mean that the pH is always 7. At 25°C, $[H_3O^+] = [OH^-] = 1.0 \times 10^{-7}\ M$, but since this is an endothermic equilibrium reaction, the concentrations are temperature dependent.

This relationship can be expressed as

$$pK_w = pH + pOH$$

Since pK_w is 14.00 at 25°C, the equation is often written as

$$14.00 = pH + pOH$$

The equation can be generalized for any acid–conjugate base pair in aqueous solution at 25°C as

$$14.00 = pK_a + pK_b \text{ because } K_a \times K_b = K_w$$

POLYPROTIC ACIDS AND MIXTURES OF ACIDS

Polyprotic acids donate more than 1 proton, H^+, and dissociate by losing 1 H^+ at a time. Sulfuric acid, H_2SO_4 is the only strong acid that is polyprotic.

$$H_2SO_4 \rightarrow H^+ + HSO_4^-$$

$$HSO_4^- \rightleftarrows H^+ + SO_4^{2-}$$

If sulfuric acid is 1 *M* or more in concentration, the ionization of the second H^+ is negligible when calculating pH and can be ignored. If the concentration of the sulfuric acid is less than 1 *M*, both hydrogen ions must be considered and the calculation involves a quadratic equation. Because no quadratic equations will be necessary on the AP® exam, exact calculation of the pH of a dilute sulfuric acid solution won't be required. It is certainly possible that an estimate of pH would be required on a multiple-choice question, though.

For other polyprotic acids, it is only necessary to work a K_a problem with the first dissociation to find the pH. The second (and third, if triprotic) dissociation will produce a negligible increase in hydrogen ion concentration.

EXAMPLE: Estimate the pH of a 0.10 *M* solution of H_2SO_4. The K_a for $HSO_4^- = 1.2 \times 10^{-2}$. Circle your best estimate.

0 between 0 and 1.00 1.00 between 1.00 and 2.00 2.00

SOLUTION: H_2SO_4 is a strong acid in its first dissociation.

$$H_2SO_4 \rightarrow H^+ + HSO_4^-$$

If 0.10 *M* sulfuric acid dissociates, $[H^+]$ and $[HSO_4^-]$ will each be 0.10 *M*. From this dissociation alone, pH = –log $[H^+]$ = 1.00. The second dissociation ($HSO_4^- \rightleftarrows H^+ + SO_4^{2-}$) is weak and will only produce a relatively small increase in $[H^+]$, causing the pH to be a little bit lower than 1.00 (circle "between 0 and 1.00").

EXAMPLE: Calculate the $[H_3O^+]$ of a 0.20 *M* solution of vitamin C (ascorbic acid, $H_2C_6H_6O_6$).

For $H_2C_6H_6O_6$, $K_{a_1} = 8.0 \times 10^{-5}$, $K_{a_2} = 1.6 \times 10^{-12}$.

	$H_2C_6H_6O_6$	+	H_2O	⇄	H_3O^+	+	$HC_6H_6O_6^-$
I	0.20		---		0		0
C	$-x$		---		$+x$		$+x$
E	$0.20 - x$		---		x		x

$$H_2C_6H_6O_6 + H_2O \rightleftarrows H_3O^+ + HC_6H_6O_6^-$$

$$K_{a_1} = \frac{[H_3O^+][HC_6H_6O_6^-]}{[H_2C_6H_6O_3]} = 8.0 \times 10^{-5} = \frac{x^2}{0.20 - x} \approx \frac{x^2}{0.20}$$

$$x = 4.0 \times 10^{-3}\ M = [H_3O^+]$$

MIXTURES OF ACIDS

When given a solution consisting of a mixture of a strong acid and a weak acid or a mixture of several weak acids, use the strongest acid in the mixture to determine the $[H^+]$ and pH. The hydrogen ions contributed by the weaker acids would be negligible and can be ignored. If a mixture of strong acids is given, the total moles of hydrogen ions must be calculated and divided by the volume to determine the $[H^+]$. Similar questions could be asked for bases.

ACID–BASE PROPERTIES OF SALTS

PREDICTING THE pH OF SALTS

You may be asked to determine if a salt will form an acidic, basic, or neutral solution by looking at its chemical formula.

"Salts" are ionic compounds that can be made by the reaction of an acid with a base. (Recall that an acid plus a base make a salt and water.) Only a few salts are edible, and only one salt is NaCl (what we commonly think of as "salt"—a better term would be "table salt").

When a salt is dissolved in water, it can form a solution that is acidic, basic, or neutral. To determine which, look at the cation and anion of the salt. The cation of the salt always comes from the base. If the salt starts with a Group 1 (1A) or Group 2 (2A) metal, the base was a strong base [like NaOH or $Sr(OH)_2$]. If it starts with NH_4^+ or a related ion, the base was a weak base (such as NH_3).

To determine which acid could have formed the salt, look at the anion of the salt and add 1 hydrogen ion to it. If the anion is Cl^-, the acid was HCl (strong acid). If the anion is $C_2H_3O_2^-$ (also shown as CH_3COO^-), the acid is acetic acid ($HC_2H_3O_2$ or CH_3COOH), a weak acid. If the anion is SO_4^{2-}, the acid is HSO_4^- (not H_2SO_4), a weak acid.

Use the memory trick below to make a quick determination of whether the salt is acidic, basic, or neutral. This memory trick is not the reason (we'll go into that next), but will serve to quickly identify the type of salt.

Memory Trick

Remember, the strongest one wins! If the salt you are analyzing is made from a strong acid and a weak base, it is acidic. If it is made from a weak acid and a strong base, it is basic. If both the acid and the base are strong, it is neutral. You won't be given a salt made from a weak acid and a weak base, but if you were, you would compare the K_a and K_b values to determine who is "strongest." ***This is only a trick, not a reason, and should never be written down as a justification on a test!***

The memory trick above may lead us to believe that it is the strong acid or strong base that influences the pH of the salt. This is not true; it is actually the ion that came from the weak acid or base that will affect the pH. The real reason that certain salts are acidic or basic is that one (or both) of the ions in the salt hydrolyzes (breaks up) water to produce hydronium or hydroxide ions. The ion that reacts with water will be the one that came from the weak acid or base. The ion that came from the strong acid or base is simply a spectator and will not produce any hydronium or hydroxide ions.

To show exactly why a salt is acidic or basic, write an equation where you take the ion that came from the weak acid or base and add it to water. If the salt is basic, your products are hydroxide ions and the conjugate acid of the ion. If the salt is acidic, your products are a hydronium ion and the conjugate base of the ion.

EXAMPLE: Determine whether an aqueous solution of $KC_2H_3O_2$ is acidic, basic, or neutral. Write an equation to justify your answer if you believe that the salt is acidic or basic.

SOLUTION: A solution of $KC_2H_3O_2$ is basic.

Split the salt into its cation, K^+, and its anion, $C_2H_3O_2^-$. The cation came from the base KOH (a strong base), and the anion came from the acid $HC_2H_3O_2$ (weak acid). Our prediction is that it will be a basic salt. Take the ion that came from the weak acid and add it to water. Because it was predicted to be a basic salt, put a hydroxide ion on the product side. The "leftovers" give us the conjugate acid of our acetate ion, acetic acid. The presence of hydroxide ions shows that the salt produces a basic solution.

$$C_2H_3O_2^- + H_2O \rightleftarrows OH^- + HC_2H_3O_2$$

EXAMPLE: Determine whether an aqueous solution of $LiNO_3$ is acidic, basic, or neutral. Write an equation to justify your answer if you believe that the salt is acidic or basic.

SOLUTION: A solution of $LiNO_3$ is neutral.

Split the salt into its cation, Li^+, and its anion, NO_3^-. The cation came from the base LiOH (a strong base), and the anion came from the acid HNO_3 (strong acid). Our prediction is that it will be a neutral salt. Neither ion will hydrolyze water, and the pH of the water will not be affected.

EXAMPLE: Determine whether an aqueous solution of NH_4I is acidic, basic, or neutral. Write an equation to justify your answer if you believe that the salt is acidic or basic.

SOLUTION: A solution of NH_4I is acidic.

Split the salt into its cation, NH_4^+, and its anion, I^-. The cation came from the base NH_3 (a weak base), and the anion came from the acid HI (strong acid). Our prediction is that it will be an acidic salt. Take the ion that came from the weak base and add it to water. Because it was predicted to be an acidic salt, put a hydronium ion on the product side. The "leftovers" give us the conjugate base of our ammonium ion, ammonia. The presence of hydronium ions shows that the salt produces an acidic solution.

$$NH_4^+ + H_2O \rightleftarrows H_3O^+ + NH_3$$

CALCULATING THE pH OF SALTS

Although salts may be acidic or basic, they will generally only be slightly acidic or slightly basic, with pH values in the 5–9 region. To calculate the pH of a salt, first you must decide whether the salt is acidic, basic, or neutral. If the salt is basic, then its anion is the conjugate base of a weak acid. The anion will undergo hydrolysis. Write an equation for the reaction of that ion with water to form the acid and OH^- ions: $A^- + H_2O \rightleftarrows HA + OH^-$. Then write the equilibrium expression:

$$K_b = \frac{K_w}{K_a} = \frac{[HA][OH^-]}{[A^-]}$$

If the salt is basic, the hydrolysis reaction will always produce a hydroxide ion.

EXAMPLE: Determine the pH of a 0.100 *M* aqueous solution of NaCN. The K_a for HCN is 5.8×10^{-10}.

SOLUTION: In aqueous solutions, NaCN ionizes completely into Na^+ and CN^-. However, the cyanide ions react with water according to the following equation:

$$CN^-(aq) + H_2O(l) \rightleftarrows HCN(aq) + OH^-(aq)$$

Na^+ ions do not react with water so they do not affect the pH. Ions from Group 1 (1A) and 2 (2A) never undergo hydrolysis because they are cations of strong bases. Since we have a basic salt, a K_b problem must be worked. The equilibrium expression for the solution is

$$K_b = \frac{[OH^-][HCN]}{[CN^-]}$$

We need a value for K_b. Since we have the K_a for HCN, we can calculate the value of K_b for CN^-.

$$K_b = \frac{1.0 \times 10^{-14}}{K_a} = \frac{1.0 \times 10^{-14}}{5.8 \times 10^{-10}}$$

$$K_b = 1.7 \times 10^{-5}$$

Constructing an ICE table for filling in the available information, we get

	CN^-	+	H_2O	⇆	HCN	+	OH^-
I	0.100		---		0		0
C	$-x$		---		$+x$		$+x$
E	$0.100 - x$		---		x		x

Since K is much smaller than 0.100, assume that $0.100 - x \approx 0.100$.

$$K_b = \frac{[OH^-][HCN]}{[CN^-]} = 1.7 \times 10^{-5} = \frac{x^2}{0.100 - x} \approx \frac{x^2}{0.100}$$

Solving for x, we have

$$x = [OH^-] = 1.3 \times 10^{-3}$$

$$pOH = 2.89; \quad pH = 14.00 - pOH$$

$$pH = 11.11$$

Math Trick

When using the relationship $K_w = K_a \times K_b$ and solving for either the K_a or the K_b value, the numbers in front of the multiplication symbol (since they will be written in scientific notation) will multiply to give you 10 and the exponents will add to give you a negative 15.

EXAMPLE: $K_a = 2 \times 10^{-4}$ so the K_b will be 5×10^{-11} (since $2 \times 5 = 10$ and $-4 + -11 = -15$).

The same relationship exists when using $[H^+][OH^-] = K_w$, so if the $[H^+] = 3.0 \times 10^{-3}$, then the $[OH^-]$ will be 3.3×10^{-12} since $3.0 \times 3.3 = 10$ and $-3 + -12 = -15$.

The general order of pH must be understood, although there can certainly be many exceptions to this order.

GENERAL ORDER OF PH

Strong Bases (SB)
Weak Bases (WB)
Basic Salts (BS)
Neutral Salts (NS)
Acidic Salts (AS)
Weak Acids (WA)
Strong Acids (SA)

When asked to rank a mixture of compounds by pH, label each, as shown above, and then arrange them.

EXAMPLE: Arrange the following 0.1 *M* solutions in order of decreasing pH: NaOH, NaCN, KCl, CH_3NH_2, H_3PO_4, NH_4NO_3, HBr.

SOLUTION: Look at the structure of each compound and identify and label as to type of acid, base, or salt.

NaOH – SB, NaCN – BS, KCl – NS, CH_3NH_2 – WB, H_3PO_4 – WA, NH_4NO_3 – AS, HBr – SA

Rank from highest pH (most basic) to lowest pH (most acidic), following the general order given above. If you are given a list of compounds with no additional information, such as K_a or K_b values, you may assume that they will follow the general order.

$NaOH < CH_3NH_2 < NaCN < KCl < NH_4NO_3 < H_3PO_4 < HBr$

MULTIPLE-CHOICE QUESTIONS

No calculators are to be used in this section.

1. A student was testing the pH of two different acid solutions and found them both to have a pH of 3.25. Which of the following must be true?
   (A) Both solutions must be weak acids.
   (B) Both solutions must have the same molarity.
   (C) Both solutions must have the same percent dissociation.
   (D) Both solutions must have the same hydronium ion concentration.

2. The autoionization of water is an endothermic process. Which statement is correct concerning the pH of pure water at 100°C?
   (A) The pH would be 7 because the $[H_3O^+] = [OH^-]$.
   (B) The pH would be 7 because pure water has no H_3O^+ ions or OH^- ions.
   (C) The pH would be greater than 7 because the reaction shifts to produce more H_3O^+ and OH^-.
   (D) The pH would be less than 7 because the reaction shifts to produce more H_3O^+ and OH^-.

3. Which of the following solutions would have the lowest percent ionization?
   (A) 0.10 *M* HCl
   (B) 1.0 *M* HCl
   (C) 0.10 *M* HF ($K_a = 7.2 \times 10^{-4}$)
   (D) 1.0 *M* HF ($K_a = 7.2 \times 10^{-4}$)

Questions 4–6 refer to the following information.

$H_3PO_4 \rightleftarrows H^+ + H_2PO_4^-$

$H_2PO_4^- \rightleftarrows H^+ + HPO_4^{2-}$

$HPO_4^{2-} \rightleftarrows H^+ + PO_4^{3-}$

Phosphoric acid, H_3PO_4, dissociates stepwise as shown above.

4. Which of the following is an acid followed by its conjugate base?
   (A) HPO_4^{2-}, $H_2PO_4^-$
   (B) $H_2PO_4^-$, PO_4^{3-}
   (C) H_3PO_4, HPO_4^{2-}
   (D) $H_2PO_4^-$, HPO_4^{2-}

5. In a 0.1 *M* solution of H_3PO_4, which species will be in the greatest concentration?
   (A) H_3PO_4
   (B) $H_2PO_4^-$
   (C) HPO_4^{2-}
   (D) PO_4^{3-}

6. Which of the following changes would increase the phosphate ion concentration in 100 mL of 0.5 *M* phosphoric acid?
   (A) Add 10 mL of 0.5*M* HCl.
   (B) Add 10 mL of 0.5 *M* H_3PO_4.
   (C) Add 10 mL of 0.5 *M* NaH_2PO_4.
   (D) Add a catalyst.

7. A solution is prepared by adding 0.10 mole of potassium fluoride to 1.00 L of water. Which pH is closest to that of the solution?
   (A) 2.00
   (B) 7.00
   (C) 9.00
   (D) 13.00

8. A 0.100 *M* solution of HA, a weak acid, is found to be 4.0% dissociated. Determine the K_a for this acid.
(A) 0.00016
(B) 0.0040
(C) 0.040
(D) 1.6

9. A 100.0-mL sample of 0.23 *M* CH_3NH_2 ($K_b = 4.38 \times 10^{-4}$) has a pH of 12.00. Approximately how much water would need to be added to this sample to drop the pH to 11.00?
(A) 100 mL
(B) 900 mL
(C) 1000 mL
(D) greater than 1000 mL

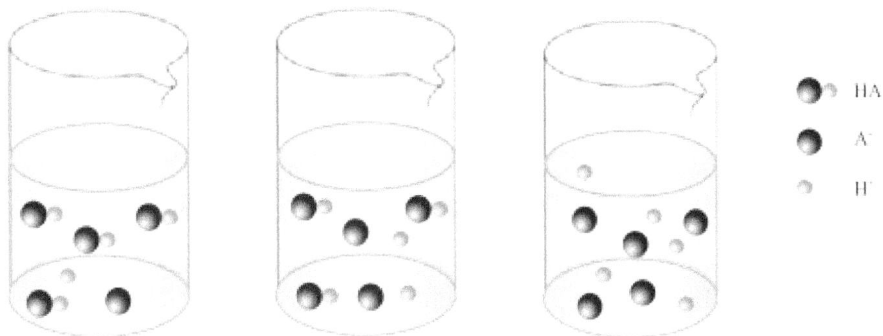

Beaker 1 **Beaker 2** **Beaker 3**

10. The beakers above each contain 100 mL of a solution containing a different acid. All solutions have equal molarities. Which beaker would have the highest pH?
(A) Beaker 1
(B) Beaker 2
(C) Beaker 3
(D) All have the same pH because the molarities are equal.

Questions 11–13 refer to the following information.

Acid	K_a
HF	7.2×10^{-4}
HNO_2	4.0×10^{-4}
HCO_2H	1.8×10^{-4}
$HC_3H_5O_3$	1.4×10^{-4}

11. A 0.010 *M* solution of which of these acids would have the highest pH?
(A) HF
(B) HNO_2
(C) HCO_2H
(D) $HC_3H_5O_3$

12. Given 0.10 M solutions of the four weak acids in the chart above, all with 0.10 M concentration, which will have the greatest percent ionization?
 (A) HF because its K_a is the largest.
 (B) HF because its molar mass is the smallest.
 (C) $HC_3H_5O_3$ because its K_a is the smallest.
 (D) $HC_3H_5O_3$ because its molar mass is the greatest.

13. Which of these is a correct ranking of the strength of the following conjugate bases: F^-, NO_2^-, CO_2H^-, $C_3H_5O_3^-$, H_2O, Cl^-?
 (A) $Cl^- > H_2O > F^- > NO_2^- > CO_2H^- > C_3H_5O_3^-$
 (B) $Cl^- > C_3H_5O_3^- > F^- > NO_2^- > CO_2H^- > H_2O$
 (C) $C_3H_5O_3^- > CO_2H^- > NO_2^- > F^- > H_2O > Cl^-$
 (D) $H_2O > Cl^- > F^- > NO_2^- > CO_2H^- > C_3H_5O_3^-$

Iodine oxidation state	1	3	5	7
Name	Hypochlorous acid	Chlorous acid	Chloric acid	Perchloric acid
Formula	$HClO$	$HClO_2$	$HClO_3$	$HClO_4$
K_a	2.9×10^{-8}	1.1×10^{-2}	5×10^{2}	1×10^{3}

14. As the number of oxygen atoms increases in any series of oxygen acids, such as HXO, HXO_2, HXO_3, … , which of the following is generally true?
 (A) The acid strength increases with an increasing oxidation state of the central atom because it causes each oxygen atom to become more electronegative.
 (B) The acid strength increases with an increasing oxidation state of the central atom because the formal charge of the central atom increases.
 (C) The acid strength increases with greater numbers of oxygen atoms because the hybridization of the central atom increases, thus making it easier to ionize the hydrogen.
 (D) The acid strength increases with greater numbers of oxygen atoms because it causes the strength of the bond between oxygen and hydrogen to increase.

Beaker 1 **Beaker 2**

15. Beaker 1 contains 50.0 mL of 0.10 *M* hydrochloric acid (HCl), while Beaker 2 contains 50.0 mL of 0.10 *M* nitric acid (HNO_3). The two solutions are poured together into a large container and thoroughly mixed. Water is added to bring the final volume up to 1.00 L. What is the pH of the final solution?
(A) 1.00
(B) between 1.00 and 2.00
(C) 2.00
(D) greater than 2.00

FREE-RESPONSE QUESTIONS

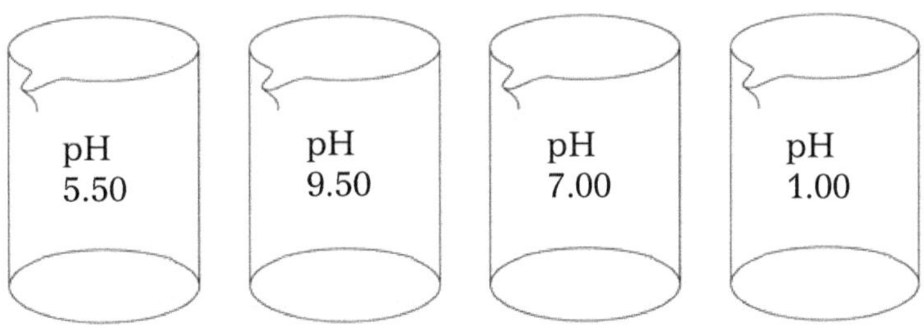

1 **2** **3** **4**

1. The four beakers above each contain 100. mL of different solutions of similar concentrations.
   (a) The K_b for ammonia is 1.8×10^{-5}.
      (i) Which beaker is most likely to contain $NH_3(aq)$? Provide a chemical equation to explain your answer.
      (ii) Calculate the molarity of the solution in the beaker that you chose for (i).
   (b) If the contents of Beakers 3 and 4 are poured together and mixed thoroughly, what will be the resulting pH?
   (c) Explain how it is possible that Beakers 1 and 4 are acids with equal molarities.

(d) Citric acid and malic acid are the most commonly occurring acids found in fruits and vegetables.

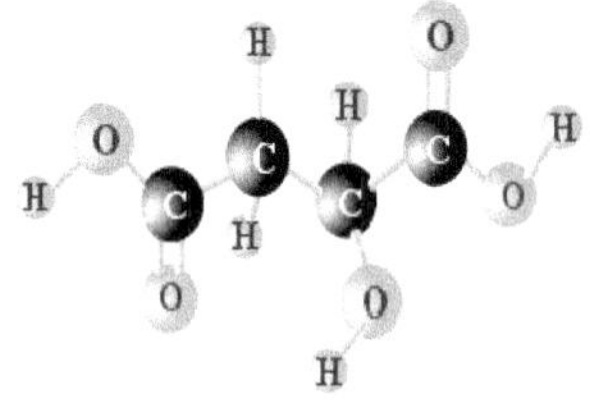

Malic Acid

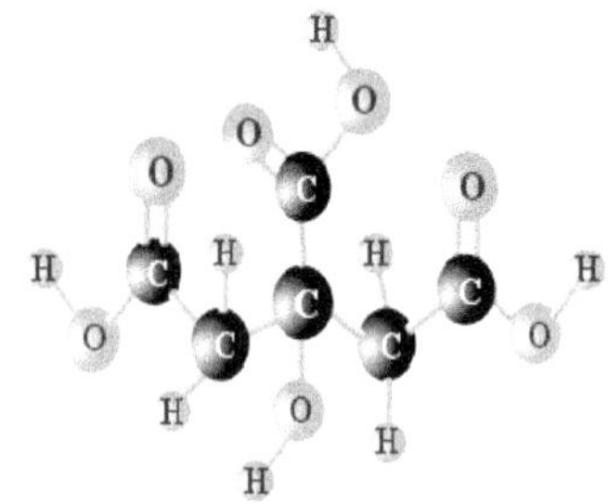

Citric Acid

(i) The juice from a fresh tomato is analyzed and found to have a pH of 4.40. Assuming that the juice is a dilute solution of citric acid, what is the $[H_3O^+]$ concentration of the tomato juice?

(ii) Circle all labile protons in the citric acid formula above.

(e) Malic acid, $H_2C_4H_4O_5$, is the predominant acid found in apples. Assuming that it is the only acid present, a 100.0-mL sample of apple juice was titrated with 0.150 *M* NaOH to determine the malic acid concentration, using an indicator to determine the equivalence point. The table below shows the results of three trials with separate 100.0-mL samples of juice.

	Trial 1	Trial 2	Trial 3
Initial buret reading	0.09 mL	1.31 mL	0.89 mL
Final buret reading	34.02 mL	34.29 mL	33.90 mL
Volume of NaOH used			

(i) Write the balanced net ionic reaction for the neutralization of malic acid by sodium hydroxide.

(ii) Fill in the volumes of NaOH used in the table above.

(iii) Using the data from the table above, determine the average molarity of the malic acid in the apple juice samples.

(iv) The data in Trial 1 differs significantly from that in Trials 2 and 3. What experimental error could have caused this difference?

2. A solution is prepared by adding 10.0 g of potassium acetate ($KC_2H_3O_2$, 98.14 g/mol) to enough water to make 500 mL of solution. The K_a for $HC_2H_3O_2 = 1.8 \times 10^{-5}$.

(a) Write a net ionic equation to show the reaction that occurs when the potassium acetate is dissolved in water.

(b) Set up the appropriate equilibrium expression (K_a or K_b) for the reaction in (a).

(c) Determine the pH of the solution.

(d) If 500. mL of 0.0100 *M* $Ca(NO_3)_2$ is added to the solution above, would a precipitate form? The K_{sp} for $Ca(OH)_2$ is 1.3×10^{-6}.

Answers

MULTIPLE-CHOICE QUESTIONS

1. D Since pH is the negative log of the hydronium ion concentration, all solutions with a given pH will have the same hydronium ion (or hydrogen ion) concentration. A weak acid and a strong acid can have the same pH as long as their concentrations are different. The strong acid will be much more dilute than the weak acid (LO 6.12).

2. D Endothermic reactions shift to products to relieve stress. Higher $[H_3O^+]$ means lower pH, although the $[H_3O^+]$ will still equal the $[OH^-]$ (LO 6.14).

3. D Strong acids, such as HCl, will ionize 100% at either 1.0 *M* or 0.10 *M*. Weak acids ionize to a greater extent when they are diluted (LO 6.12).

4. D An acid and its conjugate base differ by exactly 1 hydrogen ion (LO 3.7).

5. A In a polyprotic acid such as H_3PO_4, the first dissociation will be the strongest and subsequent dissociations will be weaker. Since phosphoric acid is weak in all of its dissociations, the species in the greatest concentration will be the undissociated H_3PO_4 molecule (LO 6.17).

6. C Le Chatelier's principle can be applied to a multistep reaction. If a stress is added to a step of the reaction and that stress drives that step to the products, the later steps are also driven to products through the increase in concentration of their reactants. Adding a strong acid to any of the steps of the phosphoric acid dissociation will drive the overall reaction to the left, decreasing $[PO_4^{3-}]$. Adding more H_3PO_4 of the same concentration will simply increase the volume of the solution and will not cause a change in the $[PO_4^{3-}]$. Adding a catalyst simply speeds up the process and would not cause an increase in equilibrium $[PO_4^{3-}]$. Adding $H_2PO_4^-$ would shift the second step to the right, causing an increase in HPO_4^-. This would shift the third step to the right, increasing the $[PO_4^{3-}]$ (LO 5.16).

7. C Potassium fluoride, KF, is a basic salt. The fluoride ion reacts with water to produce HF and OH^-. A basic salt will have a slightly basic pH but will not have a pH as high as 13 (LO 6.16).

8. A $[HA]_0 = 0.100\ M$ 4.0% of 0.100 *M* is 0.00400 *M*.

   $[H^+] = [A^-] = 0.00400\ M$

   $K_a = ([H^+][A^-])/[HA] = (0.00400)(0.00400)/0.100$
   $= (4 \times 10^{-3})(4 \times 10^{-3})/(10^{-1}) = 1.6 \times 10^{-4}$

Note that although the actual value for [HA] is 0.100 – 0.004 = 0.096 *M*, this approximation is within the 5% rule, and students do not have calculators for this part of the test (LO 6.5).

9. D Because the base is weak, decreasing the concentration by a power of 10 will not decrease the pH by 1, as it would if the base were strong. Since the base does not dissociate 100% and its percent dissociation increases with dilution, the solution would need to be diluted further to allow the pH to drop from 12 to 11. For this particular example, the actual concentration of base required to have a pH of 11 is 0.0023 *M* (LO 6.16).

10. A Beaker 3 contains a strong acid since it is completely ionized. Beakers 1 and 2 contain weak acids, with Beaker 1 being the weakest as it is not showing any ionization in this small sample. Each beaker contains the same number of moles of acid, so the one with the least ionization has the lowest $[H_3O^+]$ and highest pH (LO 6.12).

11. D When all concentrations are equal, the acid with the highest pH (lowest $[H_3O^+]$) will be the weakest acid and will have the smallest K_a value (LO 6.7).

12. A HF has the largest K_a which means its equilibrium lies farthest to products (LO 6.12).

13. C There is an inverse relationship between the strength of an acid and its conjugate base. Since Cl^- is the conjugate base of HCl, a strong acid, Cl^- is the weakest conjugate base in the set. The conjugate base of a strong acid is a weaker base than water. The lower the K_a of a weak acid, the greater its conjugate base strength (LO 6.16).

14. A As the number of oxygen atoms increases in an oxyacid series, the oxidation number of the central atom increases, resulting in the atom becoming more electronegative. This tends to draw electrons away from the oxygen atoms that surround the central atom, thereby making the oxygen atoms more electronegative as well. This makes the O–H bond more polar, increasing the degree of ionization and acid strength (LO 2.2).

15. C 0.0500L × 0.10 mol/L = 5.0×10^{-3} mol of each acid. Both nitric and hydrochloric are strong, monoprotic acids. Adding the moles of H^+ from each acid together results in 1.0×10^{-2} mol of H_3O^+. When diluted to 1 L, the final $[H_3O^+]$ is 1.0×10^{-2} *M* and the pH will be 2.00 (LO 6.15).

FREE-RESPONSE QUESTIONS

1. (a) (i) Beaker 2 is most likely NH_3 because ammonia is a weak base with pH > 7. $NH_3 + H_2O \rightleftarrows NH_4^+ + OH^-$

(LO 6.16)

(ii) pH = 9.50 pOH = 14.00 – 9.50 = 4.50

$[OH^-] = 10^{-4.50} = 3.2 \times 10^{-5}$

	NH_3	+	H_2O	⇄	NH_4^+	+	OH^-
I	?		----		0		0
C	$-x$		----		$+x$		$+x$
E	$? - x$		----		3.2×10^{-5}		3.2×10^{-5}

$x = 3.2 \times 10^{-5}$

$$K_b = \frac{[OH^-][NH_4^+]}{[NH_3]} = \frac{(3.2 \times 10^{-5})^2}{[NH_3]} = 1.8 \times 10^{-5}$$

$$[NH_3] = 0.57 \times 10^{-5}\ M$$

(LO 6.16)

(b) Beaker 4 has a pH of 1.00. $[H^+] = 10^{-1.00} = 1.0 \times 10^{-1}$. Since Beaker 3 has a pH of 7.00, it has no appreciable $[H^+]$ or $[OH^-]$ and will only dilute the solution in Beaker 4 when they are added together. Since they are each 100 mL, the concentration in Beaker 4 will be cut in half and will be 5.0×10^{-2}. pH = $-\log 5.0 \times 10^{-2} = 1.30$ (LO 6.17).

(c) Beaker 1 must be a weak acid, while Beaker 4 is a strong acid. Weak acids do not ionize much and will produce fewer $[H^+]$; therefore, they have higher pH values than strong acids (LO 6.12).

(d) (i) pH = 4.40 $[H_3O^+] = 10^{-4.40} = 4.0 \times 10^{-5}$ (LO 6.16)

(ii)

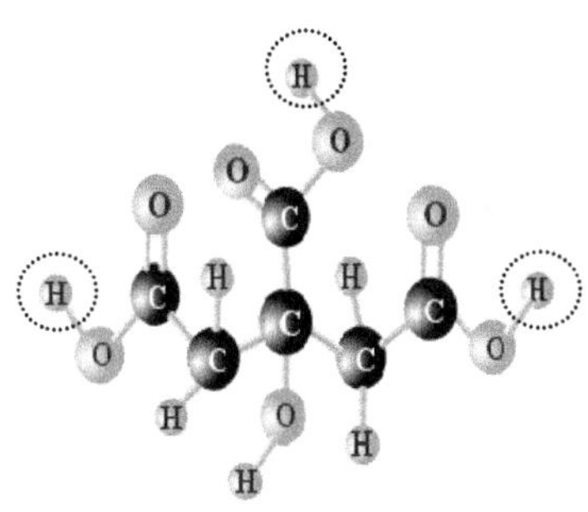

Citric Acid

(LO 2.1)

(e) (i) $H_2C_4H_4O_5 + 2\ OH^- \rightarrow C_4H_4O_5^{2-} + 2\ H_2O$ (LO 3.2)

(ii)

	Trial 1	Trial 2	Trial 3
Initial buret reading	0.09 mL	1.31 mL	0.89 mL
Final buret reading	34.02 mL	34.29 mL	33.90 mL
Volume of NaOH used	**33.93 mL**	**32.98 mL**	**33.01 mL**

(LO 1.20)

(iii)

Trial 1: $0.03393 \text{ L NaOH} \times \dfrac{0.150 \text{ mol NaOH}}{1 \text{ L}} \times \dfrac{1 \text{ mol malic acid}}{2 \text{ mol NaOH}}$
$= 2.54 \times 10^{-3}$ mol malic acid

2.54×10^{-3} mol/0.100 L = 2.54×10^{-2} M malic acid

Trial 2: $0.03298 \text{ L NaOH} \times \dfrac{0.150 \text{ mol NaOH}}{1 \text{ L}} \times \dfrac{1 \text{ mol malic acid}}{2 \text{ mol NaOH}}$
$= 2.47 \times 10^{-3}$ mol malic acid

2.47×10^{-3} mol/0.100 L = 2.47×10^{-2} M malic acid

Trial 3: $0.03301 \text{ L NaOH} \times \dfrac{0.150 \text{ mol NaOH}}{1 \text{ L}} \times \dfrac{1 \text{ mol malic acid}}{2 \text{ mol NaOH}}$
$= 2.48 \times 10^{-3}$ mol malic acid

2.48×10^{-3} mol/0.100 L = 2.48×10^{-2} M malic acid

$[(2.54 \times 10^{-2}) + (2.47 \times 10^{-2}) + (2.48 \times 10^{-2})]/3 =$

Average molarity = 2.50×10^{-2} M

(LO 1.20)

(iv) Trial 1 required more NaOH. The most likely cause of error is that the buret was rinsed with water and not with the NaOH solution before the titration. This dilutes the solution so more NaOH is required (LO 1.20)

2. (a) $C_2H_3O_2^- + H_2O \rightleftarrows OH^- + HC_2H_3O_2$
(LO 3.2)

(b)

$$K_b = \frac{[OH^-][HC_2H_3O_2]}{[C_2H_3O_2^-]}$$

(LO 6.5)

(c)

$$\frac{10.0 \text{ g } KC_2H_3O_2}{0.500 \text{ L}} \times \frac{1 \text{ mol } KC_2H_3O_2}{98.15 \text{ g } KC_2H_3O_2} = 0.204\ M$$

$$K_b = \frac{K_w}{K_a} = \frac{1.0 \times 10^{-14}}{1.8 \times 10^{-5}} = 5.6 \times 10^{-10}$$

	$C_2H_3O_2^-$	+ H_2O	⇄	OH^-	+ $HC_2H_3O_2$
I	0.204	---		0	0
C	$-x$	---		$+x$	$+x$
E	$0.204 - x$	---		x	x

$$K_b = \frac{[x][x]}{[0.204 - x]} \approx \frac{x^2}{0.204}$$

$x = [OH^-] = 1.1 \times 10^{-5}$

$pOH = -\log 1.1 \times 10^{-5} = 4.96$

$pH = 14.00 - 4.96 = 9.04$

(LO 6.16)

(d) $K_{sp} = [Ca^{2+}][OH^-]^2 = 1.3 \times 10^{-6}$

Since the two volumes are the same (both 500 mL), the molarities of the ions are cut in half.

$[Ca^{2+}] = 0.0100/2 = 5.00 \times 10^{-3}$

$[OH^-] = (1.1 \times 10^{-5})/2 = 5.5 \times 10^{-6}$

$Q = (5.00 \times 10^{-3})(5.5 \times 10^{-6})^2 = 1.5 \times 10^{-13}$

Since $1.5 \times 10^{-13} < 1.3 \times 10^{-6}$; $Q < K$, no precipitate forms.

(LO 6.21)

14

BIG IDEA 6: APPLICATIONS OF EQUILIBRIA: BUFFERS AND TITRATIONS

Big Idea 6

Any bond or intermolecular attraction that can be formed can be broken. These two processes are in a dynamic competition, sensitive to initial conditions and external perturbations.

Concepts from the last few chapters will be applied in this section covering the reactions of acids with bases (both strong and weak) and any resulting equilibria. Calculations involved in acid–base titrations and buffer preparation will be demonstrated. Properties of titration curves of various types will be explored. Very little new material is presented, but the applications of equilibria in new situations will be covered. This material is typically the most challenging for students. It is important to remember that buffers and titrations are not new chemistry concepts, just different applications of concepts and calculations already learned.

You should be able to:

- Identify the components of a buffer system and write equations to demonstrate how buffers maintain pH in the presence of strong acids or strong bases.

- Choose components needed to make a buffer of a given pH and perform calculations involving the preparation of a buffer.
- Perform calculations involving strong acid–strong base titrations as well as weak acid–strong base and weak base–strong acid titration calculations.
- Be familiar with types of titration curves and selection of acid–base indicators.

AP® Tip

Be systematic. All solution equilibrium problems have the same fundamental features—a ratio of product concentrations over reactant concentrations raised to their stoichiometric coefficients. Remember that pure substances (solids and liquids) are not involved in equilibrium calculations.

BUFFERS

Buffer systems resist large changes in pH when small amounts of acids or bases are added. Buffer systems in our body and in many natural and manmade products assist with the maintenance of relatively constant pH levels.

Buffers have two components; an acid and its conjugate base. A buffer may either consist of a weak acid and its conjugate base (e.g., HF and F^- or $HC_2H_3O_2$ and $C_2H_3O_2^-$) or a weak base and its conjugate acid (e.g., NH_3 and NH_4^+).

FOUR WAYS TO MAKE A BUFFER:

1. Combine a weak acid with its salt (HCOOH plus NaCOOH).
2. Combine a weak base with its salt (CH_3NH_2 plus CH_3NH_3Cl).
3. React a strong acid with a weak base so that the strong acid is limiting. This produces a weak base and its salt. $H^+ + NH_3 \rightarrow NH_4^+$
4. React a strong base with a weak acid so that the strong base is limiting. This produces a weak acid and its salt. $OH^- + HF \rightarrow F^-$ and H_2O

Buffers cannot consist of strong acids or strong bases and their salts because their conjugate forms are extremely weak and the system is not in equilibrium; it has gone to completion. If any strong acid or strong base remains, the solution is not a buffer.

Buffer		
Substances Needed to Make	Components	Examples
Similar concentrations of weak acid + salt containing the conjugate base	Weak acid and conjugate base	HCN and NaCN
Similar concentrations of weak base + salt containing the conjugate acid	Weak base and conjugate acid	CH_3NH_2 and CH_3NH_3Cl
Excess weak acid + strong base	Weak acid and conjugate base	2 mol of HCN + 1 mol NaOH react to yield 1 mol HCN and 1 mol NaCN
Excess weak base + strong acid	Weak base and conjugate acid	2 mol NH_3 + 1 mol HCl react to yield 1 mol NH_3 and 1 mol NH_4Cl

EXAMPLE: 50.0 mL of 0.20 *M* NH_3 is mixed with 50.0 mL of 0.10 *M* HCl. Is the resulting solution a buffer?

SOLUTION: NH_3 is a weak base and HCl is a strong acid. At first glance, this doesn't look like it could be a buffer since a buffer cannot contain a strong acid or a strong base. However, the HCl will react with the ammonia to produce NH_4Cl. The stoichiometry of the reaction must be considered and the "leftovers" analyzed to see if a buffer is created. Since the volumes are equal, the molarities are cut in half. Stoichiometry should be done in moles. The total volume is 100.0 mL. Multiplying the molarity by 0.1000 L gives the moles of each reactant.

	NH_3	+	HCl	→	NH_4Cl
Before	0.010 mol		0.0050 mol		0
Change	– 0.0050		– 0.0050		+ 0.0050
After	0.0050 mol		0		0.0050 mol

Looking at the products (After), 0.0050 mole of NH_3 and 0.0050 mole of NH_4Cl are present. Since this solution contains a weak base and its conjugate acid, this is a buffer.

EXAMPLE: 50.0 mL of 0.20 *M* $HC_2H_3O_2$ is mixed with 50.0 mL of 0.20 *M* NaOH. Is the resulting solution a buffer?

SOLUTION: $HC_2H_3O_2$ is a weak acid, and NaOH is a strong base. At first glance, this doesn't look like it could be a buffer since a buffer cannot contain a strong acid or a strong base. However, the NaOH will react with the acetic acid to produce sodium acetate and water. The stoichiometry of the reaction must be considered and the "leftovers" analyzed to see if a buffer is created. Since the volumes are equal, the molarities are cut in half. Stoichiometry should be done in moles. The total volume is 100.0 mL. Multiplying the molarity by 0.1000 L gives the moles of each reactant.

	$HC_2H_3O_2$	+	NaOH	→	$NaC_2H_3O_2$	+	H_2O
Before	0.010 mol		0.010 mol		0		----
Change	–0.010		–0.010		+0.010		----
After	0		0		0.010 mol		----

Looking at the products (After), 0.010 mole of $NaC_2H_3O_2$ is present. A salt alone is not a buffer.

EXAMPLE: 50.0 mL of 0.20 *M* $NaHCO_3$ is mixed with 50.0 mL of 0.50 *M* NaOH. Is the resulting solution a buffer?

SOLUTION: $NaHCO_3$ is a basic salt, but it is amphoteric (can both accept and donate a proton) and NaOH is a strong base. At first glance, this doesn't look like it could be a buffer since a buffer cannot contain a strong acid or strong base. However, the NaOH will react with the $NaHCO_3$ to produce sodium carbonate and water. The stoichiometry of the reaction must be considered and the "leftovers" analyzed to see if a buffer is created. Since the volumes are equal, the molarities are cut in half. Stoichiometry should be done in moles. The total volume is 100.0 mL. Multiplying the molarity by 0.1000 L gives the moles of each reactant.

	$NaHCO_3$	+	NaOH	→	Na_2CO_3	+	H_2O
Before	0.010 mol		0.025 mol		0		----
Change	– 0.010		– 0.010		+ 0.010		----
After	0		0.015 mol		0.010 mol		----

Looking at the products (After), 0.010 mol of Na_2CO_3 is present, but no HCO_3^- remains. Excess OH^- is present so this is not a buffer.

Buffer pH is related to the pK_a of the weak acid (pK_a is simply the –log K_a) and the relative amounts of the acid and base forms.

How Does a Buffer Work?

A buffer will react with or "absorb" H^+ or OH^- (from the strong acid or base that is added) and change them into weak acids or weak bases, thus avoiding large pH changes. If a buffer system is composed of NH_3/NH_4^+, the reaction with a strong acid, H^+, would be:

$$H^+ + NH_3 \rightarrow NH_4^+$$
SA WA

The reaction of the same buffer system with a strong base, OH^-, would be:

$$OH^- + NH_4^+ \rightarrow NH_3 + H_2O$$
SB WB

Notice that the H^+ reacted with the more basic part of the buffer and the OH^- reacted with the more acidic part of the buffer. In both reactions, one part of the buffer system was changed into the other part (NH_3 to NH_4^+ and NH_4^+ to NH_3).

EXAMPLE: Write the net ionic equation that explains why pH remains relatively constant when 0.10 mole of HNO_3 is added to 100.0 mL of a buffer consisting of 0.20 *M* $HC_2H_3O_2$ and 0.20 *M* $NaC_2H_3O_2$. Also write the equation for the reaction that occurs to the same buffer system when 0.10 mole of LiOH is added. Assume that volume does not change in either case.

SOLUTION: (*Hint:* When writing equations involving strong acids or strong bases, simply use H^+ and OH^-.)

$H^+ + C_2H_3O_2^- \rightarrow HC_2H_3O_2$
SA WA

Note the one-way arrow whenever a strong acid or base is a reactant. The presence of the SA or SB drives the reaction to completion.

$OH^- + HC_2H_3O_2 \rightarrow H_2O + C_2H_3O_2^-$
SB WB

BUFFER CAPACITY

The concentrations of the two components of a buffer system determine its buffer capacity, the amount of acid or base that the buffer can "absorb" or react with before the pH changes by 1 unit. The higher the concentrations of the buffer components, the more strong acid and/or strong base it can react with.

CALCULATING THE PH OF A BUFFER

The calculation to find the pH of a buffer is similar to all equilibrium calculations *except* that there are now two initial concentrations, one for each part of the buffer, the conjugate acid–base pair. Before starting your calculations, stop and look at what you have. If your buffer system contains a weak acid and its conjugate base, always work a K_a problem. If your buffer system contains a weak base and its conjugate acid, always work a K_b problem. When you read your problem, label what you have (WA, WB, etc).

EXAMPLE: Calculate the pH of a solution that is 0.60 *M* HF and 1.00 *M* KF. K_a for HF is 7.2×10^{-4}.

SOLUTION: First, think about what you have. HF is a weak acid and F^- is its conjugate base so you are going to work a K_a problem. Second, write the equation for the dissociation of the acid just like you would do in any K_a problem. It is up to you whether you include water when working with a weak acid.

$HF + H_2O \rightleftarrows H_3O^+ + F^-$ Or $HF \rightleftarrows H^+ + F^-$

Second, set up an ICE chart. Since KF is a soluble salt, $[F^-] = [KF] = 1.00\ M$.

	HF	+ H_2O	⇄ H_3O^+	+ F^-
I	0.60	----	0	1.00
C	$-x$	----	$+x$	$+x$
E	$0.60 - x$	----	x	$1.00 + x$

Third, write the equilibrium expression for K_a.

Substitute the values from the equilibrium line of the ICE chart into the K_a expression. Check to see if x is small enough to be negligible.

$$K_a = 7.2 \times 10^{-4} = \frac{[H_3O^+][F^-]}{[HF]} = \frac{x\,(1.00 + x)}{(0.60 - x)} \approx \frac{x\,(1.00)}{0.60}$$

Fourth, solve for x which equals $[H_3O^+]$.

$x = [H_3O^+] = 4.3 \times 10^{-4}\ M$

$pH = -\log[H_3O^+]$

$pH = -\log(4.3 \times 10^{-4}) = 3.36$

An alternative method of solving a buffer problem is to use the Henderson–Hasselbalch equation, as shown below. If you use it, you should still set up the ICE box.

$$pH = pK_a + \log\frac{[\text{base}]}{[\text{acid}]}$$

$$pK_a = -\log K_a$$

EXAMPLE: Repeat the example above using the Henderson–Hasselbalch equation.

SOLUTION: Set up the ICE box as you did in the previous method:

$$pH = pK_a + \log\frac{[F^-]}{[HF]}$$

$$pH = -\log(7.2 \times 10^{-4}) + \log\frac{1.00}{0.60}$$

$$pH = 3.14 + 0.22 = 3.36$$

Comparing the pH and pK_a of the weak acid in a buffer system can give a qualitative indication of the predominant species present. When $pH = pK_a$, $[A^-] = [HA]$. When $pH > pK_a$, $[A^-] > [HA]$ and when $pH < pK_a$, $[HA] > [A^-]$. Applications of this relationship include acid–base indicators and protonation of amino acids in polypeptides and proteins.

An amino acid has both an amine group (basic, NH_x) and a carboxyl group (acidic, COOH). Depending upon the pH, the amino acid may lose its labile proton (from the carboxyl group) or accept a proton from water onto its amine group.

EXAMPLE: Identify the forms of the amino acid phenylalanine (structure shown below) in highest concentration at pH 2.0 and 10.0. pK_a = 2.58 (acid), 9.24 (amine).

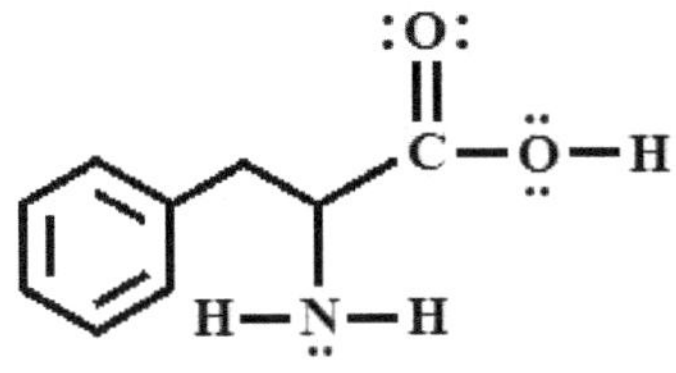

SOLUTION: pH 2.0 is less than the pK_a, so the amino group in phenylalanine will be fully protonated (in its acid form). It will attract the hydrogen from H_3O^+ and will leave H_2O behind.

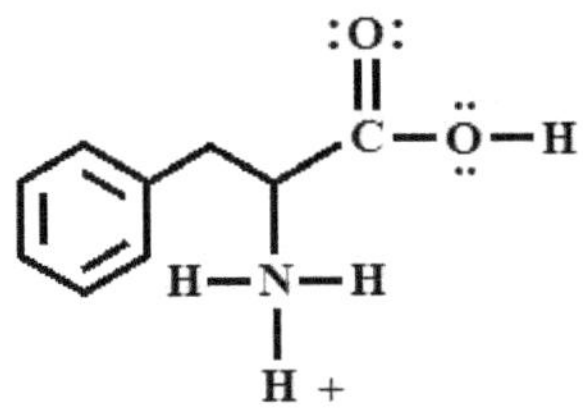

pH 10.0 is greater than the pK_a of the amine, so the molecule will be fully deprotonated (in its basic form)—the hydrogen on the carboxylic acid group will be attracted to OH^- in the solution forming water.

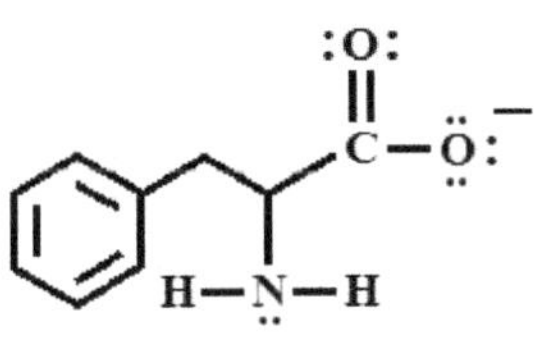

If the pH were between 2.58 and 9.24, the amino acid would be present mainly in its usual form with a neutral amine group and a neutral carboxyl group.

DILUTING A BUFFER

Because pH = pK_a when the base and acid concentrations are equal, a buffer consisting of 1 *M* NH_3 and 1 *M* NH_4^+ would have the same pH as one that consisted of 0.01 *M* NH_3 and 0.01 *M* NH_4^+. When comparing equal volumes of these two buffer systems, the 1 *M* buffer would have 100 times the buffer capacity as the 0.01 *M* buffer, but they would each have the same pH. Since the K_b of NH_3 is 1.8×10^{-5}, the pK_b and the pOH of the buffer would be 4.74. The pH (14 – pOH) would be 9.26.

Diluting a buffer system with water will not change its pH as both components of the buffer are in the same container and would be diluted equally. The ratio of base/conjugate acid stays constant.

PREPARATION OF A BUFFER

A buffer can be prepared from a weak acid and a salt containing its conjugate base or from a weak base and a salt containing its conjugate acid.

EXAMPLE: Calculate the mass of $NaC_2H_3O_2$ required to prepare a buffer of pH 4.55 when added to 0.500 L of 0.67 *M* acetic acid. (Assume no change in volume.) $K_a = 1.8 \times 10^{-5}$ for $HC_2H_3O_2$.

SOLUTION: This problem can be solved using an ICE chart or by using the Henderson–Hasselbalch equation. Once $[C_2H_3O_2^-]$ is calculated, stoichiometry is used to relate concentration to mass.

$$\text{pH} = \text{p}K_a + \log\frac{[C_2H_3O_2^-]}{[HC_2H_3O_2]}$$

$$\text{p}K_a = -\log(1.8 \times 10^{-5}) = 4.74$$

$$4.55 = 4.74 + \log\frac{[C_2H_3O_2^-]}{0.67\ M}$$

$$-0.19 = \log\frac{[C_2H_3O_2^-]}{0.67\ M}$$

$$0.65 = \frac{[C_2H_3O_2^-]}{0.67\ M}$$

$$[C_2H_3O_2^-] = 0.44\ M$$

$$0.500\ \text{L} \times \frac{0.44\ \text{mol}\ C_2H_3O_2^-}{1\ \text{L}} \times \frac{1\ \text{mol}\ NaC_2H_3O_2}{1\ \text{mol}\ C_2H_3O_2^-} \times \frac{82.0\ \text{g}\ NaC_2H_3O_2}{1\ \text{mol}\ NaC_2H_3O_2} = 18\ \text{g}\ NaC_2H_3O_2$$

You may be asked to select components to make a buffer of a given pH from a list of weak acids and/or weak bases, given their K_a or K_b values (or pK_a or pK_b values). The useful buffering range of a buffer is generally considered to be the pK_a ±1. If given K_a or K_b values (and you don't have a calculator), round the value to the nearest power of 10. The exponent will be an estimate of the closest whole number pK_a or pK_b.

Acid	K_a
Acetic acid	1.8×10^{-5}
Salicylic acid	1.1×10^{-3}
Dichloroacetic acid	5.0×10^{-2}
Chlorous acid	1.2×10^{-2}

EXAMPLE: A buffer of pH 3.5 is needed to maintain the pH of a skin treatment medication. Choose an acid from the table above that will be an effective buffer at this pH when combined with its conjugate base in the proper ratio.

SOLUTION: The pK_a of the acid must be ±1 from the pH of the desired buffer. With estimation based on the K_a values, chlorous acid has a pK_a a little less than 2, acetic acid has a pK_a a little less than 5, salicylic acid has a pK_a a little less than 3, and dichloroacetic acid has a pK_a a little more than 1. Based on these values, the salicylic acid and its conjugate base, the salicylate ion, would be the best choice to prepare a buffer. Because the pK_a is lower than the desired pH, the moles of salicylate need to be higher than the moles of salicylic acid.

ADDITION OF STRONG ACID AND STRONG BASE TO A BUFFER

When the $[A^-]/[HA]$ ratio is 1, the pH of the buffer equals the pK_a of the weak acid component of the buffer. Addition of small amounts of strong acid or base will not significantly change this ratio because the large amount of weak acid is available to react with the strong base and the large amount of conjugate base is available to react with the strong acid. It is not until the $[A^-]/[HA]$ ratio changes by a factor of 10 that the pH changes by 1 unit—due to the fact that you are taking the $\log_{10}$ and a change of 10 is needed to change the log by a factor of 1.

TITRATION

A titration is a laboratory procedure, not a new type of chemistry problem! A titration can be used to determine the concentration of an unknown solution. The neutralization reaction will use either a strong acid and/or a strong base so these reactions can be considered to go to completion.

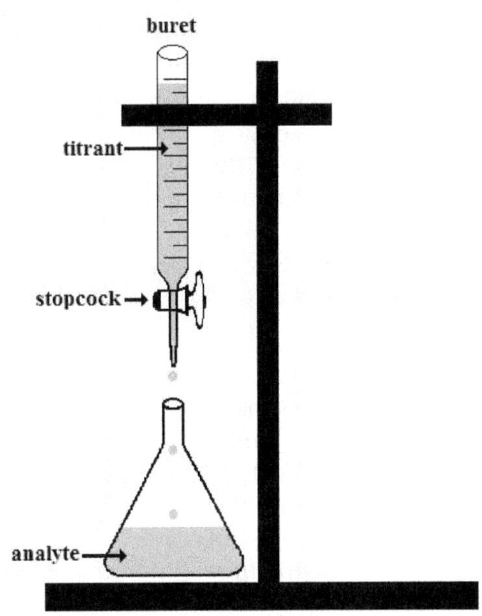

In a traditional titration, the titrant (substance that we know the concentration of) is in the buret. The substance that we are analyzing (unknown concentration) is in a flask or beaker. An indicator or a pH meter is placed in the flask or beaker to signal the equivalence point of the titration.

If a pH meter is used, the volume of titrant used versus the pH can be plotted throughout the progress of the titration. The resulting graph is called a titration curve. An example is shown on the next page. The three important properties of each titration curve will identify what type of substance is being titrated.

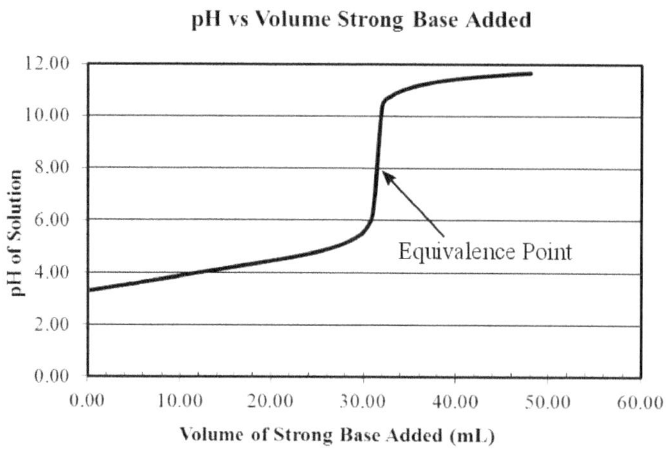

1. **The beginning pH** (is the analyte a weak acid, strong acid, weak base, or strong base?).
2. **The equivalence point.** The equivalence point is the center of the vertical region of the graph and is the point at which pH changes the fastest per volume of titrant added. At the equivalence point, the moles of titrant added = moles of analyte initially present. If the equivalence point has a pH < 7, an acidic salt is present. This indicates that a weak base was titrated with a strong acid. If the pH = 7, a neutral salt is present and the two solutions were a strong acid and a strong base. If the pH > 7, a basic salt is present. This indicates that a weak acid was titrated with a strong base (most common type of titration).
3. **The length of the vertical region.** The stronger the acid, the longer the vertical region. Weak acids with small K_a values have very short vertical regions and are tricky to titrate accurately.

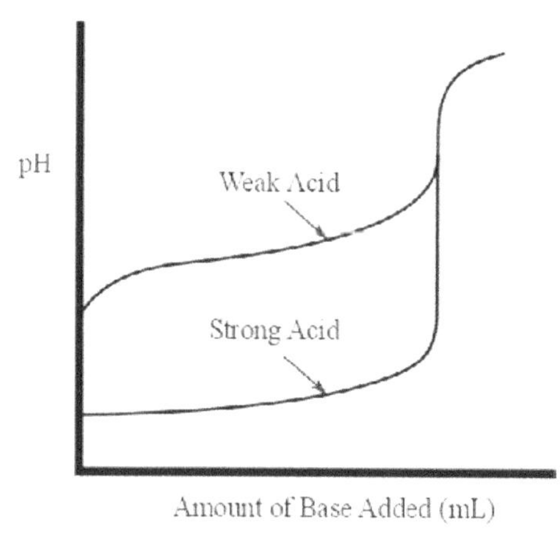

Calculations used to solve these problems will not be new (stoichiometry, K_a, K_b, buffer). At the equivalence point, the moles of titrant added and moles of analyte are present in stoichiometric ratios; neither one is in excess. Near the equivalence point the pH changes rapidly as fewer moles of analyte are present.

The following table summarizes the types of calculations involving acid–base titrations. Examples of each type of titration problem that will appear on the AP® Chemistry exam are shown in the table.

Characteristics of Titrations

Titrations	Before Equivalence Point	At Equivalence Point	After Equivalence Point
Strong acid titrated with a strong base	Stoichiometry only Use excess $[H_3O^+]$ to calculate the pH. pH < 7	Stoichiometry only pH = 7	Stoichiometry only Use excess $[OH^-]$ to calculate the pH. pH > 7
Species in solution which affect the pH (in addition to H_2O)	H_3O^+	H_2O	OH^-
Weak acid titrated with a strong base	Stoichiometry + K_a Weak acid + conjugate base (buffer calculation) pH < 7	Stoichiometry + K_b Salt hydrolysis calculation pH > 7	Stoichiometry only Use excess $[OH^-]$ to calculate pH pH > 7
Species in solution which affect the pH (in addition to H_2O)	HA, A^-	A^-	OH^-
Weak base titrated with a strong acid	Stoichiometry + K_b Weak base + conjugate acid (buffer calculation) pH > 7	Stoichiometry + K_a Salt hydrolysis calculation pH < 7	Stoichiometry only Use excess $[H_3O^+]$ to calculate pH pH < 7
Species in solution which affect the pH (in addition to H_2O)	B, HB^+	HB^+	H_3O^+

Note: Weak acid–weak base titrations are not usually performed as the reaction is not driven to completion, making the chemistry much more complex.

TITRATION OF A STRONG ACID WITH A STRONG BASE

At any point during the titration of a strong acid with a strong base (or a strong base with a strong acid), the pH can be calculated from the molarity of the ion present in excess after the complete reaction. Halfway to the equivalence point, the main species present are H_3O^+, the anion from the acid, and the cation from the base.

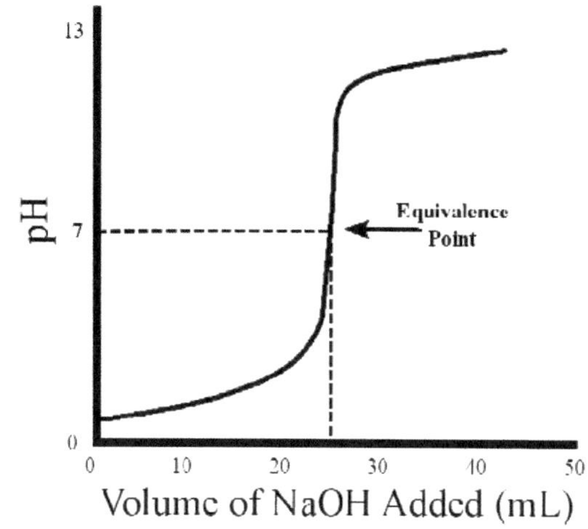

For a strong acid–strong base titration, the pH equals 7 at the equivalence point.

EXAMPLE: Consider the titration of 20.00 mL of 0.200 *M* HBr by 0.100 *M* KOH. Calculate the pH of the resulting solution when the following volumes of KOH have been added.

A. 10.00 mL B. 40.00 mL C. 45.00 mL

SOLUTION: Complete work is only shown for part A. Follow the same processes to find the pH for B and C.

First, using the volume and molarity of the HBr, calculate the moles of H^+ to be titrated. The resulting moles of H^+ appear in the first column in the table at the end of the problem.

$$0.02000\text{ L} \times \frac{0.200\text{ mol HBr}}{1\text{ L}} = 4.00 \times 10^{-3}\text{ mol HBr}$$

Next, for the titration in question, calculate the moles of OH^- added from the volume and concentration of KOH. The resulting moles of OH^- used in each titration appear in the second column of the table.

$$0.01000\text{ L} \times \frac{0.100\text{ mol KOH}}{1\text{ L}} = 1.00 \times 10^{-3}\text{ mol KOH}$$

Calculate the moles of ion in excess, H_3O^+ or OH^-.

	HBr	+	KOH	→	KBr	+	HOH
Before	4.00×10^{-3}		1.00×10^{-3}		0		----
Change	-1.00×10^{-3}		-1.00×10^{-3}		$+1.00 \times 10^{-3}$		----
After	3.00×10^{-3}		0		1.00×10^{-3}		----

SA + NS therefore, –log of [SA] = pH

The results for each titration appear in column 4 in the table on the next page.

Using the total volume in column 5 of the table, and the moles of ion in excess in column 4, determine the molarity of the ion in excess. The molarity of the ion in excess for each of the three titrations is in column 6.

20.00 mL acid + 10.00 mL base = 30.00 mL solution

$$\frac{3.00 \times 10^{-3} \text{ mol}}{0.03000 \text{ L}} = 0.100 \, M \, \text{H}^+ \text{ (from the HBr)}$$

It is the molarity of the ion in excess that determines the pH.

$pH = -\log[H_3O^+]$

$pH = -\log(0.100) = 1.000$

In part B of the example, neither ion is in excess. The pH equals 7 because the titration is at the equivalence point.

Be careful, in part C of the example, the excess ion is the OH^- ion. To calculate the pH, you must determine the pOH and then the pH.

Problem Part	Moles H_3O^+	Moles OH^-	Moles of Ion in Excess	Total Volume (when acid and base react)	Molarity of Ion in Excess	pH
A	4.00×10^{-3}	1.00×10^{-3}	3.00×10^{-3} (H_3O^+)	0.0300 L	0.100 (H_3O^+)	1.000
B	4.00×10^{-3}	4.00×10^{-3}	None Equivalence point	Not needed	None	7.000
C	4.00×10^{-3}	4.50×10^{-3}	5.0×10^{-4} (OH^-)	0.0650 L	0.0077 (OH^-)	11.89

AP® Tip

When you are calculating the pH during a titration, be sure to find the negative log of the concentration of H_3O^+, not just the moles of H_3O^+, which is a common mistake. If the excess ion or the value of x is OH^-, be sure that you calculate the pOH and then the pH.

TITRATION OF A WEAK ACID WITH A STRONG BASE

There are four characteristic types of calculations for the titration of a weak acid with a strong base: before the equivalence point (1), halfway to the equivalence point (2), at the equivalence point (3), and after the equivalence point (4). For points 1–3 of this titration, always do stoichiometric calculations followed by equilibrium calculations.

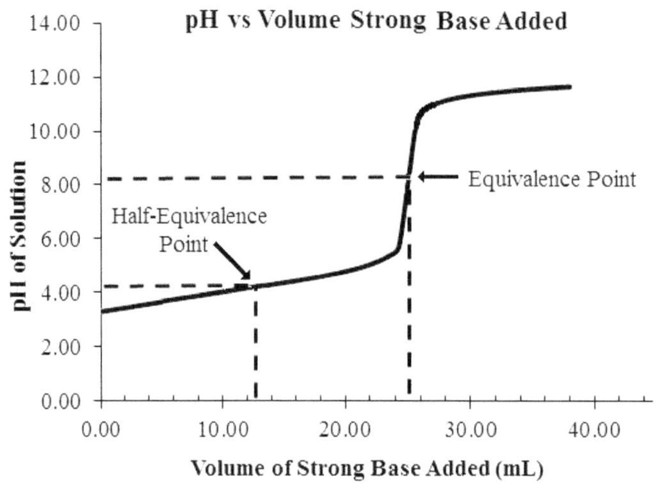

EXAMPLE: A 25.00-mL sample of 0.100 *M* $HC_3H_5O_2$ is titrated with 0.100 *M* NaOH. K_a = 1.3 × 10^{-5} for $HC_3H_5O_2$. Calculate the pH when the following volumes of 0.100 *M* NaOH are added:

A. 8.00 mL B. 12.50 mL C. 25.00 mL D. 30.00 mL

SOLUTION:

<u>Titration Point 1</u>: Calculation of the pH Before the Equivalence Point

Part A of the example involves the titration before the equivalence point. Perform the stoichiometry and record the results in a table under the reaction.

Stoichiometry:

Calculate the initial moles of the weak acid present in the sample.

$0.02500 \text{ L} \times 0.100 \text{ mol/L} = 2.50 \times 10^{-3} \text{ mol } HC_3H_5O_2$

Calculate the moles of strong base, OH^-, added.

$0.00800 \text{ L} \times 0.100 \text{ mol/L} = 8.00 \times 10^{-4} \text{ mol}$

This version of the BCA table adds a row for converting the moles of each remaining substance to molarity (by dividing by the total volume).

The total volume of the solution is now 33.0 mL. Calculate the concentration of each species that remains.

$[HC_3H_5O_2] = 1.70 \times 10^{-3} \text{ mol}/0.0330 \text{ L} = 0.0515\ M$

$[C_3H_5O_2^-] = 8.00 \times 10^{-4} \text{ mol}/0.0330 \text{ L} = 0.0242\ M$

	$HC_3H_5O_2$	+	OH^-	→	$C_3H_5O_2^-$	+	H_2O
Before	2.50×10^{-3} mol		8.00×10^{-4} mol		0		----
Change	-8.00×10^{-4}		-8.00×10^{-4}		$+8.00 \times 10^{-4}$		----
After	1.70×10^{-3} mol		0		8.00×10^{-4} mol		----
Final [ ]	0.0515 *M*		0		0.0242 *M*		----

Equilibrium: WA + CB, therefore, work a K_a

Look at your "leftovers." A weak acid and its conjugate base remain, so a K_a problem is necessary.

This problem is now identical to a "buffer" problem. Make an ICE table to do an equilibrium calculation (K_a).

	$HC_3H_5O_2$	+	H_2O	⇄	H_3O^+	+	$C_3H_5O_2^-$
I	0.0515		----		0		0.0242
C	$-x$		----		$+x$		$+x$
E	$0.0515 - x$		----		x		$0.0242 + x$

You can now perform the equilibrium calculation.

$$K_a = \frac{[H_3O^+][C_3H_5O_2^-]}{[HC_3H_5O_2]}$$

$$1.3 \times 10^{-5} = \frac{[H_3O^+](0.0242 + x)}{0.0515 - x} \approx \frac{[H_3O^+]\,0.0242}{0.0515}$$

$2.77 \times 10^{-5}\ M = [H_3O^+]$

$pH = -\log [H_3O^+] = 4.56$

Alternatively, you can use the Henderson–Hasselbalch equation.

$pH = pK_a + \log[C_3H_5O_2^-]/[HC_3H_5O_2]$

$pH = -\log(1.3 \times 10^{-5}) + [\log(0.0242 + x)/(0.0515 - x)] = 4.56$

Titration Point 2: Halfway to the Equivalence Point

In part B of the example, a 25.00-mL sample of 0.100 *M* $HC_3H_5O_2$ is titrated with 12.50 mL of 0.100 *M* NaOH. K_a is 1.3×10^{-5} for $HC_3H_5O_2$.

Half of the acid being titrated is neutralized. This point is halfway to the equivalence point.

Since half the acid is neutralized and half remains, $[HA] = [A^-]$. The major species present at this point in the titration are H_3O^+, the anion from the acid, the cation from the base, and the weak acid. The total positive charge is equal to the total negative charge; the mixture remains electrically neutral.

$$K_a = \frac{[H^+][A^-]}{[HA]}$$

$pK_a = pH$

$pK_a = -\log K_a = -\log (1.3 \times 10^{-5}) = 4.89$

<u>Titration Point 3</u>: At the Equivalence Point

In part C of the example, a 25.00-mL sample of 0.100 M $HC_3H_5O_2$ is titrated with 25.00 mL of 0.100 M NaOH.

K_a is 1.3×10^{-5} for $HC_3H_5O_2$.

Stoichiometry:

Calculate the initial moles of the weak acid in the sample.

Calculate the moles of strong base, OH^-, added.

$0.02500 \text{ L} \times 0.100 \text{ mol/ L} = 2.50 \times 10^{-3} \text{ mol } OH^- = \text{mol } C_3H_5O_2^-$

The total volume of the solution is now 50.00 mL. Calculate the concentration of each species that remains.

$[C_3H_5O_2^-] = 2.50 \times 10^{-3} \text{ mol}/0.0500 \text{ L} = 0.0500\ M$

	$HC_3H_5O_2$	+	OH^-	→	$C_3H_5O_2^-$	+	H_2O
Before	2.50×10^{-3}		2.50×10^{-3}		0		----
Change	-2.50×10^{-3}		-2.50×10^{-3}		$+2.50 \times 10^{-3}$		----
After	0		0		2.50×10^{-3}		----
Final [ ]	0		0		0.0500		----

CB only, therefore, work a K_b

There is no weak acid present once the strong base added completely reacts. All that is present is A^- (in this case, $C_3H_5O_2^-$), the conjugate base of the weak acid, HA (in this case, $HC_3H_5O_2$). An aqueous solution of A^- will be basic due to its reaction with water.

$A^- + H_2O \rightleftarrows HA + OH^-$

The problem from this point on is a salt hydrolysis problem.

	$C_3H_5O_2^-$	+ HOH	⇄	$HC_3H_5O_2$	+	OH^-
I	0.0500	----		0		0
C	$-x$	----		$+x$		$+x$
E	$0.500 - x$	----		x		x

$$K_b = \frac{K_w}{K_a} = \frac{1.0 \times 10^{-14}}{1.3 \times 10^{-5}} = 7.7 \times 10^{-10}$$

$$K_b = \frac{[OH^-][HC_3H_5O_2]}{[C_3H_5O_2^-]} = \frac{x^2}{0.0500 - x} \approx \frac{x^2}{0.0500}$$

$x = [(7.7 \times 10^{-10})(0.0500)]^{½} = 6.2 \times 10^{-6}\ M\ OH^-$

pOH = 5.21; pH = 14.00 – 5.21 = 8.79

Titration Point 4: After the Equivalence Point

The calculations involved here are identical to those in a strong acid–strong base titration.

In part D of the example, a 25.00-mL sample of 0.100 M $HC_3H_5O_2$ is titrated with 30.00 mL of 0.100 M NaOH. K_a is 1.3×10^{-5} for $HC_3H_5O_2$.

0.03000 L × 0.100 mol/L = 3.00×10^{-3} mol NaOH

0.02500 L × 0.100 mol/L = 2.50×10^{-3} mol $HC_3H_5O_2$

$[OH^-] = 5.0 \times 10^{-4}$ mol/0.0550 L = $9.1 \times 10^{-3}\ M$

Stoichiometry:

	$HC_3H_5O_2$	+	OH^-	→	$C_3H_5O_2^-$	+	H_2O
Before	2.50×10^{-3}		3.00×10^{-3}		0		----
Change	-2.50×10^{-3}		-2.50×10^{-3}		$+2.50 \times 10^{-3}$		----
After	0		5.0×10^{-4}		2.50×10^{-3}		----
Final [ ]	0		9.1×10^{-3}		0.0500		----

SB + CB, therefore, –log of [SB] = pOH

The hydrolysis of $C_3H_5O_2^-$ will not significantly change the $[OH^-]$, so the pH is only dependent on the excess $[OH^-]$.

pOH = 2.04; pH = 11.96

TITRATION OF A WEAK BASE WITH A STRONG ACID

The calculations for a weak base–strong acid titration are very similar to the weak acid–strong base titration.

The differences are these:

- The pH before the equivalence point is greater than 7. (Work a K_b problem.)
- At the half-equivalence point, the pOH is equal to the pK_b.
- At the equivalence point, the pH is less than 7 since the salt formed is acidic. (Work a K_a problem.)
- After the equivalence point, the pH is determined by the molarity of the excess H^+. Use stoichiometry only.

CHARACTERISTICS OF TITRATION CURVES

It is important for you to be able to sketch and identify the key points on titration curves for the AP® exam. The next example will help you to review titration curves for the different types of titrations discussed in this section. Calculations involving titrations of polyprotic acids are beyond the scope of the AP® exam, but qualitative questions may be asked.

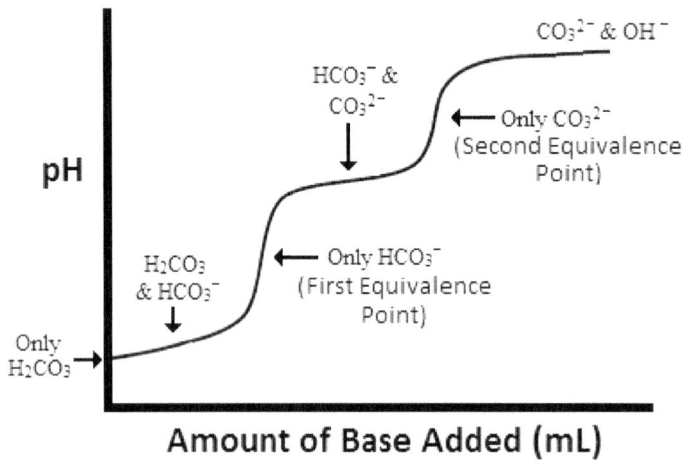

$$H_2CO_3 + OH^- \rightarrow HCO_3^- + H_2O$$

$$HCO_3^- + OH^- \rightarrow CO_3^{2-} + H_2O$$

The graph above shows a titration of carbonic acid (a diprotic acid) with a strong base. Two fairly vertical regions (labeled HCO_3^- and CO_3^{2-}) indicate rapid pH changes resulting from a small change in volume. The centers of these vertical regions are the first and second equivalence points. This indicates 2 *labile protons*. Labile protons are protons prone to separate or be removed from the rest of the molecule; they are the "acidic" protons. The labile protons in H_2CO_3 are circled in the following structure.

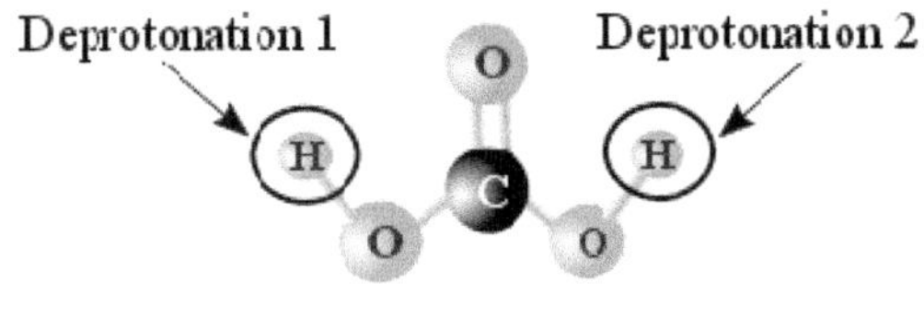

The species present at different points in the titration depend on the pH of the solution. Protonation and deprotonation are an equilibrium situation: a competition between the H_2CO_3, HCO_3^-, CO_3^{2-}, and H_2O. For carbonic acid in aqueous solution at 25°C, $pK_{a_1} = 6.3$ and $pK_{a_2} = 10.3$.

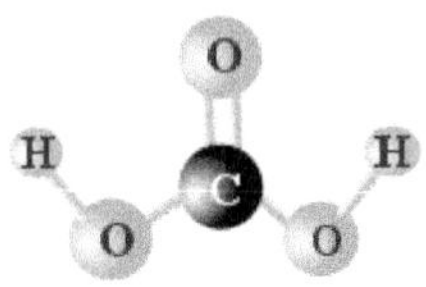
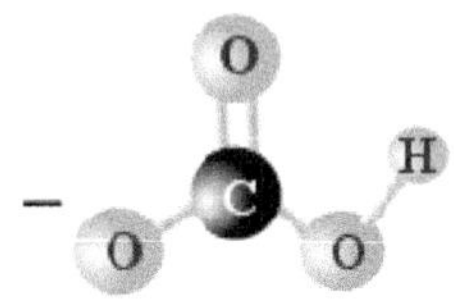

$H_2CO_3 + OH^- \rightarrow HCO_3^- + H_2O \qquad pK_{a_1} = 6.3$

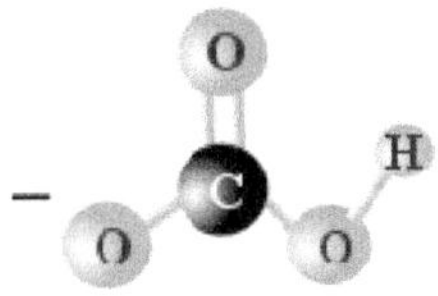
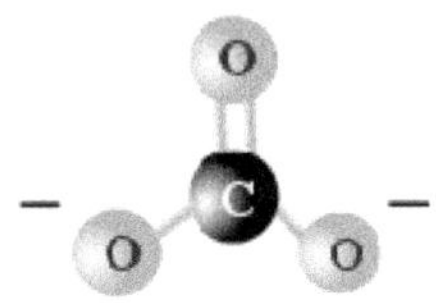

$HCO_3^- + OH^- \rightarrow CO_3^{2-} + H_2O \qquad pK_{a_2} = 10.3$

H_2CO_3 is the predominant species present at pH < 6.3 because the carbonic acid has a greater affinity for protons than water does. Bicarbonate ion (HCO_3^-) is predominant at 6.3 < pH < 10.3 and carbonate ion (CO_3^{2-}) predominates at pH > 10.3.

EXAMPLE: The questions that follow refer to the following three titration curves. All solutions are equimolar.

Figure 1:

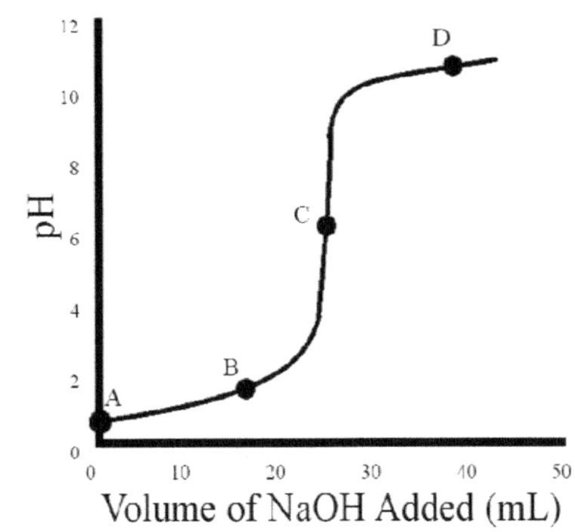

Figure 2:

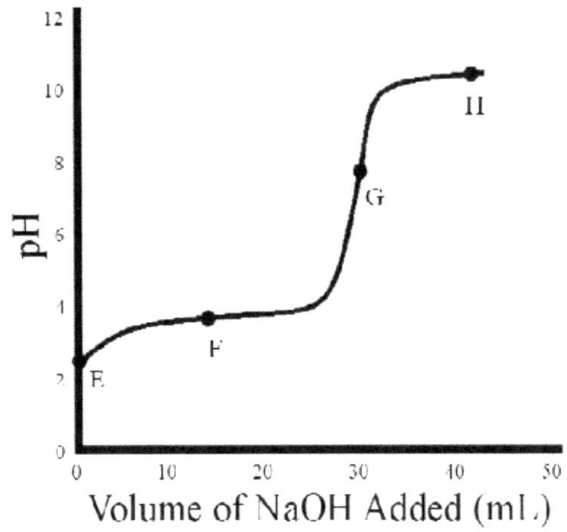

Figure 3:

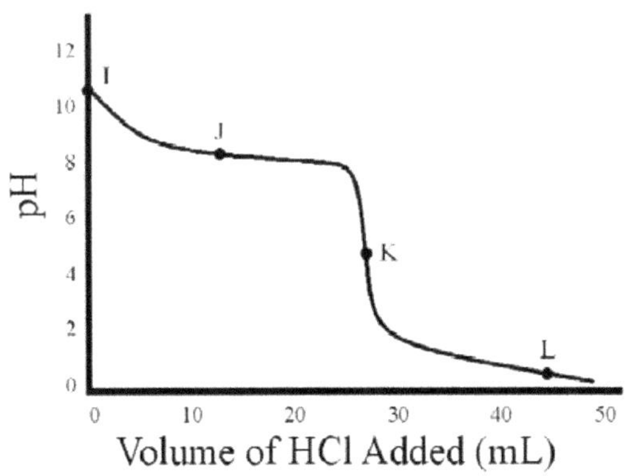

(a) Which titration curve represents a strong acid titrated by a strong base? Explain your answer.
(b) Which titration curve represents a weak acid titrated by a strong base? Explain your answer.
(c) Which titration curve represents a weak base titrated by a strong acid? Explain your answer.
(d) For each lettered part of each curve, A through L, identify the major species that affects the pH.
(e) Which titration curve(s) have buffering regions? Where is that region? Explain.
(f) For Figures 2 and 3, explain the significance of the point on the titration curve exactly halfway between the beginning of the titration and the equivalence point.

SOLUTION:

(a) Figure 1 represents a strong acid titrated by a strong base. The pH is very low when no base has been added. The pH at the equivalence point is 7.00.
(b) Figure 2 represents a weak acid titrated by a strong base. The initial pH is acidic, but slightly higher than the titration curve in Figure 1. The equivalence point occurs at pH 9, signifying the presence of the conjugate base of the weak acid being titrated.

(c) Figure 3 represents a weak base titrated by a strong acid. When no acid has been added, the pH is basic. Notice that at the equivalence point the pH is acidic in this titration.

(d) The points are labeled as follows:

Figure 1:
A strong acid, H^+
B strong acid, H^+
C water (the salt is neutral)
D strong base, OH^-

Figure 2:
E weak acid, HA
F weak acid, HA, and its conjugate base, A^-
G conjugate base, A^-
H strong base, OH^-

Figure 3:
I weak base, B
J weak base, B, and its conjugate acid, HB^+
K conjugate acid, HB^+
L strong acid, H^+

(e) Figure 2 (titration of a weak acid with a strong base) and Figure 3 (titration of a weak base with a strong acid) have buffer regions. These regions are the relatively flat regions occurring between the origin and the equivalence point (near points F and J) and represent points at which the concentration of the weak acid (or weak base) and its conjugate both have substantial concentrations. The pH changes very little during these regions as the conjugate reacts with the titrant to make more of the substance being titrated.

(f) Points F and J, which are each halfway between the beginning of the titration and the equivalence point, represent the points at which the concentration of the substance being titrated (weak acid or weak base) and its conjugate have exactly equal concentrations. At this point, pH = pK_a and pOH = pK_b. This is called the half-equivalence point.

ERROR ANALYSIS WITH TITRATION PROCEDURES

There are several common errors that can occur with titrations, and these are likely to be asked about on the exam.

- PREPARING THE BURET: The buret should be rinsed with distilled water and then several times with the solution it will be filled with (the titrant). Make sure that the tip of the buret is also rinsed with water and with solution in the process (allow to drain through the tip). Rinsing with the titrant prevents the solution from becoming diluted with any water left in the buret. Any water remaining when the titrant is added would result in the titrant being less concentrated and more titrant would be needed. This would make it appear as if the analyte is more concentrated than it really is.
- AIR BUBBLES: Make sure that there are no air bubbles in the buret, especially in the tip. An air bubble may or may not affect

your results. If the air bubble is present during the entire titration, then it would have no effect on the volume of titrant used. If the bubble "disappeared" when the titrant was released, the volume of the titrant used in the calculations would be too high. This would make it appear that the analyte is more concentrated than it really is.

- INDICATOR CHOICE: The indicator chosen for the titration needs to change color at or very near the pH of the equivalence point of the titration. The pK_a of the indicator ±1 will give its useful range. If the indicator changes color too early in the titration, the calculated concentration of the analyte will be too low. If the indicator changes color too late in the titration (past the equivalence point), the calculated concentration of the analyte will be too high.
- ADDITION OF WATER: Addition of water to the analyte to provide sufficient volume will not cause errors in titration because the moles of analyte will not be changed. Addition of water to the titrant will cause the calculated concentration of the analyte to be too high as additional titrant will be needed to react with the analyte. Water should be used to rinse down the sides of the flask during titration (especially near the equivalence point) as titrant may splash onto the sides of the flask instead of into the solution. This will cause no change in results as the moles of titrant do not change.

MULTIPLE-CHOICE QUESTIONS

No calculators are to be used in this section.

Questions 1 and 2 use the following information.

Acid	K_a
Acetic acid	1.8×10^{-5}
Benzoic acid	6.4×10^{-5}
Hydrofluoric acid	7.2×10^{-4}
Lactic acid	1.4×10^{-4}

1. Which of the acids in the table above, along with their conjugate base, would make the most effective buffer to maintain the pH of a solution at 4.95?
(A) Acetic acid
(B) Benzoic acid
(C) Hydrofluoric acid
(D) Lactic acid

2. 100.0 mL of a solution is prepared that is 0.15 *M* in lactic acid ($CH_3CHCOOH$) and 0.10 *M* in sodium lactate ($NaCH_3CHCOO$). Which statement is correct concerning the addition of 100.0 mL of water to this solution?
   (A) The equilibrium shifts towards the lactate ion and the pH decreases.
   (B) The equilibrium shifts towards the lactate ion and the pH increases.
   (C) The equilibrium shifts towards the lactic acid and the pH decreases.
   (D) The equilibrium does not shift and the pH remains constant.

3. Consider the phosphoric acid equilibria.

   $H_3PO_4 + H_2O \rightleftarrows H_3O^+ + H_2PO_4^-$ $\quad K_{a_1} = 7.5 \times 10^{-3}$

   $H_2PO_4^- + H_2O \rightleftarrows H_3O^+ + HPO_4^{2-}$ $\quad K_{a_2} = 6.2 \times 10^{-8}$

   $HPO_4^{2-} + H_2O \rightleftarrows H_3O^+ + PO_4^{3-}$ $\quad K_{a_3} = 4.8 \times 10^{-13}$

   What species is predominant at pH 12?
   (A) H_3PO_4
   (B) $H_2PO_4^-$
   (C) HPO_4^{2-}
   (D) PO_4^{3-}

4. A solution may have its pH maintained in the basic region through the use of a $NaHCO_3/Na_2CO_3$ buffer system. Which equation shows how this buffer system would resist pH changes when a strong base is added?
   (A) $OH^- + HCO_3^- \rightarrow H_2O + CO_3^{2-}$
   (B) $OH^- + H_2CO_3 \rightarrow H_2O + HCO_3^-$
   (C) $2\ OH^- + H_2CO_3 \rightarrow 2\ H_2O + CO_3^{2-}$
   (D) $OH^- + CO_2 \rightarrow H^+ + CO_3^{2-}$

5. 0.0040 mole HCl was added in a titration of 100 mL of a 0.10 *M* NH_3 solution. The resulting solution was thoroughly mixed. Assume that volume does not change. List the species in order of decreasing equilibrium concentration at this point in the titration.
   (A) $[NH_3] > [NH_4^+] > [H^+] > [OH^-]$
   (B) $[NH_3] > [NH_4^+] > [OH^-] > [H^+]$
   (C) $[OH^-] > [NH_3] > [NH_4^+] > [H^+]$
   (D) $[NH_4^+] > [NH_3] > [H^+] > [OH^-]$

6. Two unlabeled Erlenmeyer flasks each contained 25.00 mL of solution at pH 4.00. Each flask was titrated with the same standardized NaOH solution. The volumes needed to reach the endpoint are listed in the table below. What claim can be made about the acids in Flasks A and B?

Flask	Titrant Volume (mL)
A	11.29
B	37.07

(A) The concentration of the acids was the same because both solutions were pH 4.00.
(B) The strength of the acids was the same because both solutions were pH 4.00.
(C) Flask A was a strong acid because less titrant was needed to reach the endpoint.
(D) Flask B was a strong acid because more titrant was needed to reach the endpoint.

7. A buffer with pH of 7.5 is needed. Which combination of concentrations of NaH_2PO_4 ($K_a = 6.2 \times 10^{-8}$) and Na_2HPO_4 ($K_a = 4.8 \times 10^{-13}$) would give a pH closest to 7.5?

	$[NaH_2PO_4]$	$[Na_2HPO_4]$
(A)	0.1 *M*	0.9 *M*
(B)	1.0 *M*	1.0 *M*
(C)	0.10 *M*	0.20 *M*
(D)	0.90 *M*	0.10 *M*

	Trial 1	Trial 2	Trial 3
Initial buret reading	0.35 mL	1.24 mL	24.32 mL
Final buret reading	23.80 mL	24.50 mL	47.69 mL
Volume of 0.100 *M* NaOH added	23.45 mL	23.05 mL	23.37 mL

8. Using phenolphthalein as an indicator, a student performed three trials of a titration of 25.0-mL samples of a weak acid with 0.100 *M* NaOH. He calculated the results from each trial, averaged the three trials, and reported that the weak acid had a concentration of 0.0932 *M*. The teacher told the student that the actual concentration of the weak acid was 0.0475 *M*. What error did the student make?
   (A) The buret was not rinsed with the NaOH solution after cleaning with distilled water and before filling with the NaOH for the titration.
   (B) The acid was actually diprotic and the student did not consider the stoichiometry of the reaction.
   (C) The indicator bromothymol blue (range 6.0–7.6) was used instead of phenolphthalein (range 8.0–9.8).
   (D) After measuring the weak acid for each trial, the student added an additional 25 mL of water to each sample.

9. At what point on this titration curve does pH = pK_{a_2}?

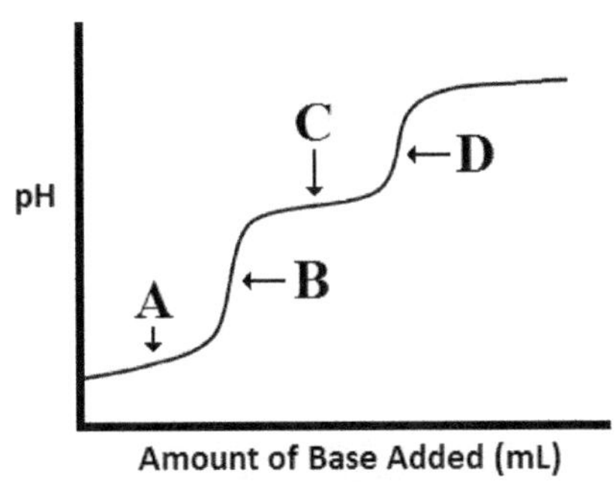

   (A) A
   (B) B
   (C) C
   (D) D

10. A buffer can maintain relatively constant pH upon addition of small amounts of a strong base because
    (A) only large amounts of base will cause an appreciable pH change in an aqueous solution.
    (B) the strong base is balanced by the weak acid and conjugate base in the buffer.
    (C) the weak acid reacts with the strong base to make additional conjugate base.
    (D) only the addition of an acid would change the pH of a buffer system.

11. A solution is prepared that is 0.25 *M* in HX and 0.10 *M* in X^-. The K_a of HX is 4.0×10^{-4}. What is the pH of the solution?
    (A) 2.00
    (B) 3.00
    (C) 4.00
    (D) 5.00

Questions 12 and 13 use the following information.

The titration curve below represents the titration of a weak acid, HX with 0.1 *M* NaOH.

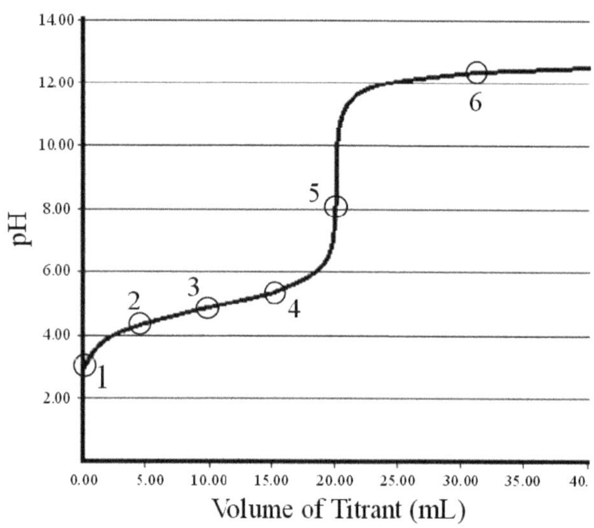

12. At which point in the titration described above would the major species present be as shown in the box to the right? Water molecules and spectator ions are omitted for clarity.
    (A) 2
    (B) 3
    (C) 4
    (D) 5

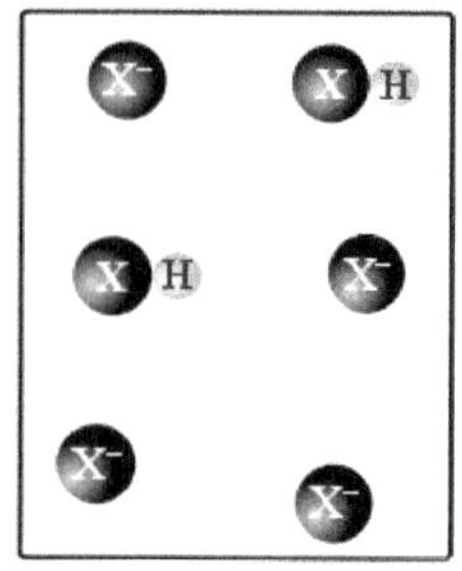

13. At which point in the titration described above would the major species present be as shown in the box to the right? Water molecules and spectator ions are omitted for clarity.
    (A) 3
    (B) 4
    (C) 5
    (D) 6

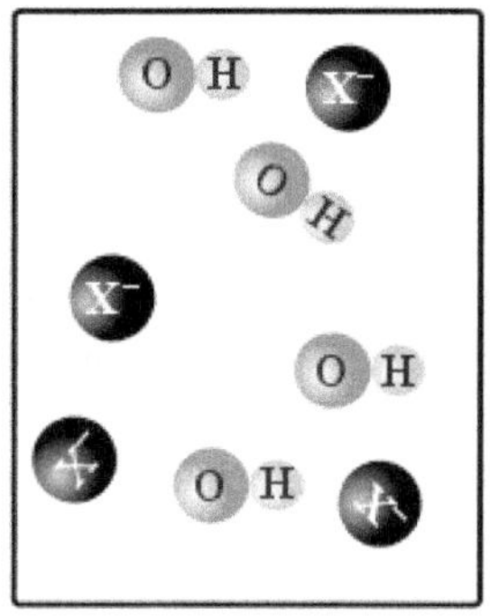

14. A researcher needs to prepare a pH 5 buffer. What solutions would be best?
    (A) 0.20 *M* boric acid ($K_a = 5.8 \times 10^{-10}$) + 0.10 *M* sodium borate
    (B) 0.20 *M* ammonia ($K_b = 1.8 \times 10^{-5}$) + 0.10 *M* ammonium chloride
    (C) 0.20 *M* propanoic acid ($K_a = 1.3 \times 10^{-5}$) + 0.10 *M* sodium hydroxide
    (D) 0.20 *M* lactic acid ($K_a = 1.38 \times 10^{-4}$) + 0.10 *M* hydrochloric acid

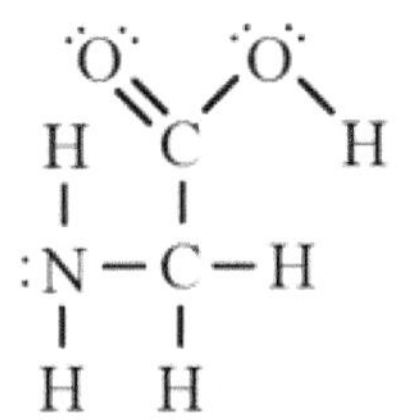

pK_a carboxylic acid = 2.34
pK_a protonated amine = 9.6

15. One form of the amino acid, glycine, is shown above. At which pH would this be the predominant form of glycine present?
    (A) pH = 2.1
    (B) pH = 2.3
    (C) pH = 7.3
    (D) pH = 9.6

FREE-RESPONSE QUESTIONS

1. Methanoic (formic) acid is responsible for the pain of bee stings, some ant bites, and stinging nettles. It was first isolated in the destructive distillation of ants and the Latin name for ants (*formica*, which also provides the family name of ants as *Formicidae*), $K_a = 1.8 \times 10^{-4}$.
    (a) Circle the labile proton in the structure.

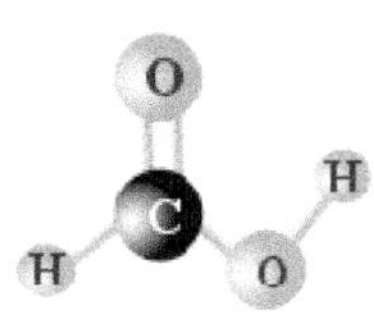

(b) A species of ants that are starting to invade the southern part of the United States is called "crazy ants." They appear to be replacing native "fire ants" because of the ability of the crazy ant to neutralize the effects of fire ant venom when the two species do battle. Crazy ant venom contains formic acid, while fire ant venom is thought to contain approximately 46 different proteins. It is still unknown exactly how the formic acid works against the fire ant venom. One of the proteins first found in fire ant venom is solenopsin, whose structure is shown below. Based on the structures of formic acid and of solenopsin, propose a hypothesis as to how the formic acid may protect the crazy ant from the effects of the solenopsin.

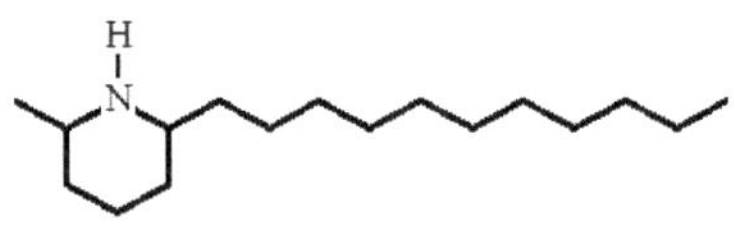

(c) 15.00 mL of formic acid was titrated to its equivalence point with 29.07 mL of 0.0896 *M* potassium hydroxide.
   (i) Write the net ionic equation for the reaction.
   (ii) Calculate the original concentration of the formic acid solution.
   (iii) Is the pH at the equivalence point less than 7, equal to 7, or greater than 7? Explain your answer.

(d) Calculate the pH of a mixture of 1.5 *M* formic acid and 0.500 *M* sodium formate.

(e) A "home remedy" for bee stings is to prepare a paste of baking soda ($NaHCO_3$) and water and apply to the sting. Write an equation to show how this might help reduce the impact of the sting.

Bond	Energy (kJ/mol)
C–H	413
C–O	358
C=O	745
O–H	467

(f) Given the bond energies in the table above, calculate ΔH°_{rxn} for the ionization of formic acid according to the following equation:

$$HCOOH + H_2O \rightleftarrows H_3O^+ + COOH^-$$

(g) Calculate ΔG° for the ionization of formic acid.

(h) Calculate ΔS° for the ionization of formic acid.

2. Because of the poisonous nature of $Cl_2(g)$, swimming pools are often chlorinated by dissolving sodium hypochlorite or calcium hypochlorite in the pool water, where the hypochlorite ion reacts with water to produce hypochlorous acid and hydroxide ions. The hypochlorous acid is an effective disinfectant.

$$OCl^- + H_2O \rightleftarrows HOCl + OH^- \qquad K = 2.9 \times 10^{-7}$$

$$HOCl + H_2O \rightleftarrows H_3O^+ + OCl^- \qquad K = 3.5 \times 10^{-8}$$

(a) Calculate the concentration of OCl^- in a sample if 5.00 g of $Ca(OCl)_2$ is dissolved in 2.00 L of water. Assume no changes in volume.
(b) The optimum pH range for swimming pool water is achieved when approximately equal concentrations of HOCl and OCl^- are present. Calculate the pH at which [HOCl] = [OCl^-].
(c) In order to maintain a proper concentration of HOCl to disinfect the water, other chemicals may be needed to raise or lower the pH of the pool water. If the concentration of the HOCl is too low and the concentration of the OCl^- is too high, which compound should you add to achieve roughly equal concentrations?
(i) Circle your answer.

$HCl(aq)$ $\qquad$ $Na_2CO_3(s)$

(ii) Justify your answer.

Answers

MULTIPLE-CHOICE QUESTIONS

1. A The useful buffering range of a buffer is the pK_a of its acid ±1. Since only K_a values are given for each of the acids, the pK_a must be estimated. The pK_a of an acid with K_a of 1×10^{-5} is 5.0. The acetic acid with K_a of 1.8×10^{-5} is the closest to having a pK_a of 5 (LO 6.18).

2. D Adding water to a buffer system does not change the pH because the concentration of both components is decreased equally. No shift in the reaction occurs, and the ratio of conjugate base to acid remains constant. Because pH = pK_a + log (base/acid), pH does not change unless the conjugate base to acid ratio changes (LO 6.8).

3. D The third dissociation has a K_a value that is between 1×10^{-12} and 1×10^{-13} so the pK_a is between 12 and 13 when the base and acid components of the reaction are equal. Since the pK_a is greater than 12, the concentration of the PO_4^{3-} must be the greatest (LO 6.19).

4. A A strong base (OH^-) will react with the more acidic part of the buffer to make the more basic part of the buffer (LO 6.20).

5. B Do a stoichiometry (BCA) table and then an equilibrium (ICE) table.

	NH_3	+	H^+	→	NH_4^+
Before	0.010 mol		0.0040 mol		0
Change	–0.0040 mol		–0.0040 mol		+0.0040 mol
After	0.0060 mol		0		0.0040 mol

Divide each mole quantity by 0.10 L to get molarity.

	NH_3	+	H_2O	⇄	NH_4^+	+	OH^-
I	0.060 M		----		0.040 M		0
C	$-x$		----		$+x$		$+x$
E	$0.060\ M - x$		----		$0.040\ M + x$		x

$[NH_3] = 0.060\ M - x \quad [NH_4^+] = 0.040\ M + x$

$[OH^-] = x \quad [H^+] \approx 0$

The titration has not yet reached the half-equivalence point. The concentration of ammonia remaining is still higher than the concentration of ammonium formed (LO 6.17).

6. C At the same pH, a strong acid has a lower concentration because all of the acid is dissociated. In order for the weak acid to ionize to have the same $[H_3O^+]$, the concentration of HA must be much higher (LO 6.12).

7. C The K_a of the acid component of the buffer ($H_2PO_4^-$) is 6.2×10^{-8} (between 1×10^{-7} and 1×10^{-8}) so the pK_a is also between 7 and 8 and is closer to the 7 than the 8. This indicates that to get a pH of 7.5, a higher concentration of the more basic component is needed (LO 6.18).

8. B Because the calculated molarity was twice the expected molarity, the student must not have taken into consideration (or known) that the acid was diprotic. Adding extra water after measurement of the acid sample would not affect results because the moles of acid did not change (LO 6.13).

9. C The pH = pK_a at the half-equivalence point, so pH = pK_{a_2} at the point halfway between the first and second equivalence points (LO 6.13).

10. C Buffers have weak acid-conjugate base equilibrium so small additions of strong base will react with the weak acid forming additional conjugate base (LO 6.20).

11. B

	HX +	+	H_2O	⇄	X^-	H_3O^+
I	0.25 *M*		----		0.10 *M*	0
C	$-x$		----		$+x$	$+x$
E	0.25 *M* – *x*		----		0.10 *M* + *x*	*x*

$$K_a = \frac{[H_3O^+][X^-]}{[HX]} = 4.0 \times 10^{-4} = \frac{(0.10 + x)(x)}{0.25 - x} \approx \frac{0.10x}{0.25}$$

$x = 1.0 \times 10^{-3}\ M = [H_3O^+]$ $\quad$ $pH = -\log 1.0 \times 10^{-3} = 3.00$

(LO 6.16)

12. C The diagram shows 4 X^- and 2 HX particles, so this represents a point in between the half-equivalence and the equivalence point (LO 6.11).

13. D The diagram shows 4 X^- and 4 OH^- ions. The excess hydroxide ions indicate that the reaction is past the equivalence point (LO 6.11).

14. C Propanoic acid has the K_a that is closest to 1×10^{-5} so the pH of the buffer can easily be made to 5. The NaOH (limiting reagent) will react with half of the propanoic acid leaving equal concentrations of propanoic acid and propanoate ions (LO 6.11).

15. C At a near neutral pH, the neutral form of the amino acid will predominate (LO 6.19).

FREE-RESPONSE QUESTIONS

1. (a)

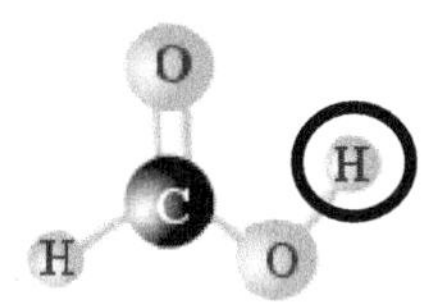

The labile proton is the one that will be ionized. Since oxygen is more electronegative, it pulls electrons away from hydrogen. This allows hydrogen to become more attracted to water (LO 6.19).

(b) The solenopsin has a basic NH group which can be neutralized by the formic acid (LO 3.7).

(c) (i) $HCOOH + OH^- \rightarrow HCOO^- + H_2O$ (LO 3.2)

(ii)

$$0.02907\text{ L} \times \frac{0.0896\text{ mol KOH}}{1\text{ L}} \times \frac{1\text{ mol HCOOH}}{1\text{ mol KOH}} = 2.605 \times 10^{-3}\text{ mol HCOOH}$$

$$\frac{2.605 \times 10^{-3}\text{ mol HCOOH}}{0.01500\text{ L}} = 0.1737\ M\text{ HCOOH}$$

(LO 6.13)

(iii) The pH will be greater than 7. Since it is a weak acid–strong base titration, the conjugate base of the weak acid is one of the products. At the equivalence point, the basic salt will be the predominant species, resulting in a slightly basic pH (LO 6.13).

(d)

	HCOOH	+	H_2O	⇄	$HCOO^-$	+	H_3O^+
I	1.5 M		----		0.500 M		0
C	$-x$		----		$+x$		$+x$
E	1.5 $M - x$		----		0.500 $M + x$		x

$$K_a = \frac{[H_3O^+][HCOO^-]}{[HCOOH]} = 1.8 \times 10^{-4} = \frac{(0.500 + x)(x)}{1.5 - x} \approx \frac{0.500x}{1.5}$$

$x = 5.4 \times 10^{-4}\ M = [H_3O^+]$ pH = $-\log(5.4 \times 10^{-4}) = 3.27$

(LO 6.16)

(e) $NaHCO_3 + HCOOH \rightarrow NaCOOH + H_2O + CO_2$ (LO 3.2)

(f) $HCOOH + H_2O \rightleftarrows H_3O^+ + COOH^-$

Bonds Broken	Energy (kJ/mol)	Bonds Formed	Energy (kJ/mol)
3 O–H	3(467)	4 O–H	4(467)
1 C–H	413		
1 C=O	745	1 C=O	745
1 C–O	358	1 C–O	358
Total	2917 kJ	Total	2971 kJ

$\Delta H = \Sigma$ Bond energies of bonds broken – Σ Bond energies of bonds formed

$\Delta H = 2917 - 2971 = -54$ kJ/mol

(LO 5.8)

(g) $\Delta G = -RT \ln K$

$\Delta G = -8.314(298)\ln (1.8 \times 10^{-4})$

$\Delta G = 2.1 \times 10^{4}$ J/mol

(LO 5.14)

(h) $\Delta G = \Delta H - T\Delta S$

2.1×10^{4} J = –54,000 J – 298 K(ΔS)

ΔS = –250 J/(K·mol)

(LO 5.13)

2. (a)

$$\frac{5.00 \text{ g } Ca(ClO)_2}{2.00 \text{ L}} \times \frac{1 \text{ mol } Ca(ClO)_2}{142.98 \text{ g } Ca(ClO)_2} \times \frac{2 \text{ mol } ClO^-}{1 \text{ mol } Ca(ClO)_2} = 0.0350\ M\ ClO^-$$

(LO 1.4)

(b) The first reaction represents the K_b of OCl^- and the second reaction represents the K_a of HOCl. The concentrations of HOCl and OCl^- will be equal when pH = pK_a of the second reaction: $-\log(3.5 \times 10^{-8}) = 7.46$ (LO 6.17).

(c) (i) Circle the HCl (LO 6.9).

(ii) Adding a strong acid will remove hydroxide ions (product side), decreasing Q and causing the reaction to shift toward the product side to replace HOCl and OH^-, reestablishing equilibrium (LO 6.9).

Part III

Practice Tests

Practice Test 1

These questions are representative of the AP® Chemistry examination, but keep in mind that it is impossible to predict exactly how well you will do on the actual exam. The first section of this test is 50% of your total test grade. Time yourself so that you finish this part in 90 minutes. Remember that on the actual examination, there is no penalty for guessing, so you should answer every question.

AP® CHEMISTRY EXAMINATION
Section I: Multiple-Choice Questions
Time: 90 minutes
Number of Questions: 60

DIRECTIONS: Each of the questions or incomplete statements below is followed by four suggested answers or completions. Select the one that is best in each case. **No calculators can be used in this section.** A periodic table and a formula chart with constants are provided at the back of this book. Assume that the conditions are a temperature of 298 K and pressure of 1.0 atm with all solutions aqueous unless otherwise specified.

1. Which is the correct Lewis structure for I_3^-?

(A) 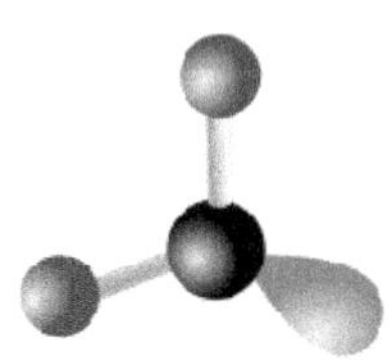

(B) 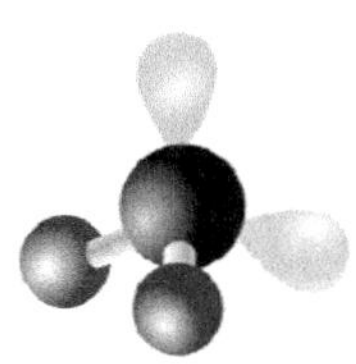

(C) 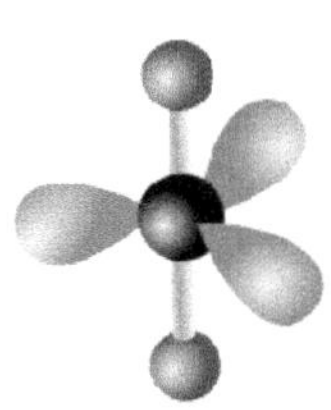

(D) 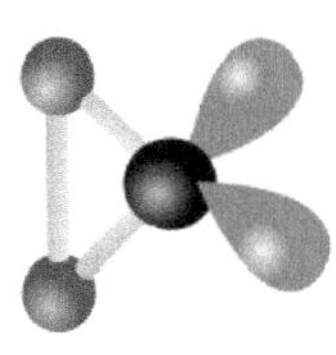

Questions 2–5 refer to the following information:

$2\,Al(s) + 6\,HCl(aq) \rightarrow 3\,H_2(g) + 2\,AlCl_3(aq)$

Aluminum metal pellets are placed in a 1 *M* solution of hydrochloric acid. The barometric pressure is 757 torr and at the room temperature of 27°C, water vapor pressure is 21 torr. The H_2 gas produced in the reaction is collected by water displacement, and its volume is measured.

2. If 0.500 L of H_2 is collected, the partial pressure of the H_2, in torr, is
(A) 368.
(B) 730.
(C) 736.
(D) 757.

GO ON TO NEXT PAGE

3. In a second experiment, an equal mass of powdered aluminum is substituted for the aluminum pellets. How will the results differ?
   (A) Hydrogen gas will be produced at a faster rate because the powder has greater surface area.
   (B) Hydrogen gas will be produced more slowly because the small size of the particles causes fewer collisions.
   (C) Hydrogen gas will be produced at a faster rate because there is more entropy in the powder than in the pellets and this reaction is driven by entropy.
   (D) Hydrogen gas will be produced more slowly because this reaction is driven by enthalpy and an increase in the entropy will decrease the free energy.

4. $H_2(g)$ is considered to behave almost "ideally," while $H_2O(g)$ deviates greatly from "ideal" behavior. What is the most important factor that explains this observation?
   (A) The molar mass of $H_2O(g)$ is greater than that of $H_2(g)$.
   (B) $H_2(g)$ molecules are nonpolar, while $H_2O(g)$ molecules are highly polar.
   (C) The volume of $H_2(g)$ is less than that of an equal number of $H_2O(g)$ molecules.
   (D) Water molecules have a lower velocity than hydrogen molecules at any given temperature.

5. Under which of the following conditions would the hydrogen gas deviate most from ideal behavior?
   (A) 100 K and 0.20 atm
   (B) 100 K and 20 atm
   (C) 800 K and 0.20 atm
   (D) 800 K and 20 atm

Questions 6–8 refer to the following information:

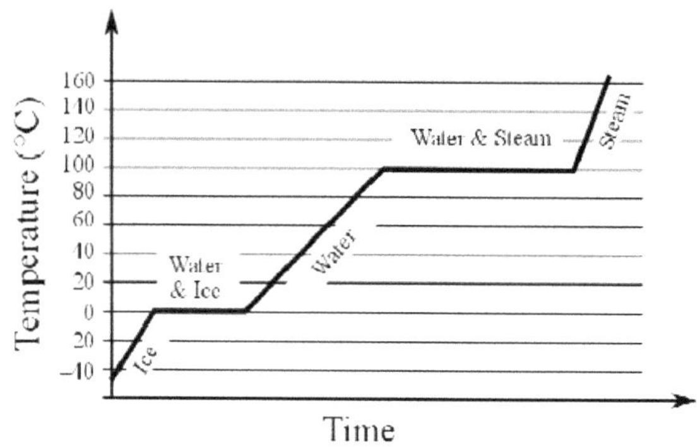

6. In the water heating curve above, why is the slope in the section for the heating of water different from the slope for the heating of steam?
   (A) It takes more energy to vaporize water than to melt ice.
   (B) It takes more energy to melt ice than to vaporize water.
   (C) Liquid water has a lower specific heat capacity than does steam.
   (D) Liquid water has a higher specific heat capacity than does steam.

7. A 50.0 g sample of steam at 120.0°C is condensed to liquid water at 100.0°C. What is the energy change involved? The specific heat capacity of steam is 2.08 J/(g °C), and the heat of vaporization of water is 2260 J/g.
   (A) –115 kJ
   (B) –2.08 kJ
   (C) 2.08 kJ
   (D) 115 kJ

8. Why is the "water and steam" line so much longer than the "ice and water" line?
   (A) There is a greater increase in entropy when going from a liquid to a gas than when going from a solid to a liquid.
   (B) Ice does not sublime at the temperature and pressure that this experiment was carried out at; therefore, it requires more energy to change it into a gas.
   (C) All of the intermolecular forces need to be overcome when going from a liquid to a gas. When going from a solid to a liquid, only part of the intermolecular forces needs to be overcome.
   (D) Because ice is less dense than liquid water, ice will float in water. As a result, it takes less energy to melt the ice since it is not surrounded by liquid water which acts as an insulator.

9. Until the late 1970s, most gasoline was "leaded," which meant that it contained tetraethyl lead, $(CH_3CH_2)_4Pb$, as a means of improving the gasoline performance cheaply. Bromine compounds were also added to the gasoline, resulting in the production of lead bromide that became part of the automobile exhaust. What is the concentration of bromide ions in a saturated solution of lead(II) bromide ($K_{sp} = 4.0 \times 10^{-6}$)?
   (A) 1.0×10^{-3} *M*
   (B) 2.0×10^{-3} *M*
   (C) 1.0×10^{-2} *M*
   (D) 2.0×10^{-2} *M*

$Na^+ + e^- \rightarrow Na(s)$	E° = – 2.71 V
$Zn^{2+} + 2\ e^- \rightarrow Zn(s)$	E° = –0.76 V
$Fe^{2+} + 2\ e^- \rightarrow Fe(s)$	E° = –0.44 V
$Fe^{3+} + 3\ e^- \rightarrow Fe(s)$	E° = –0.036 V
$O_2(g) + 4\ H^+ + 4\ e^- \rightarrow 2\ H_2O(l)$	E° = 1.23 V
$Cl_2(g) + 2\ e^- \rightarrow 2\ Cl^-$	E° = 1.36 V

10. The engines of boats that sail in salt water have an iron shaft that is protected from corrosion by the use of a zinc collar. How does this zinc collar protect the shaft from corrosion?
   (A) The collar does not oxidize as easily as iron.
   (B) The collar acts as a sacrificial anode (is more easily oxidized than iron).
   (C) The collar acts as a sacrificial cathode (is more easily reduced than iron).
   (D) The collar prevents air from reacting with the iron.

GO ON TO NEXT PAGE

11. CO_2 has a boiling point of –78.5 °C, while CS_2 has a boiling point of 46.3 °C. What accounts for the difference in their boiling points?
    (A) CS_2 is a longer molecule than CO_2 due to sulfur being a larger atom. CO_2 has a more compact electron cloud that is more polarizable.
    (B) CS_2 has a molar mass of 76.14 g/mol, and CO_2 has a molar mass of 44.01 g/mol. The higher the molar mass, the lower the molecular velocity, so fewer particles will have the kinetic energy necessary to break the intermolecular force of attraction and will need to be heated to a higher temperature to break them.
    (C) Even though S and O each have 6 valence electrons, S has more core electrons and thus is more polarizable than O, which increases the dispersion forces in CS_2, resulting in a higher boiling point.
    (D) CO_2 has resonance structures that are nonpolar. These structures contribute their zero dipole moment to make the overall molecule less polar, which decreases the inter-molecular forces and lowers the boiling point.

12. How many grams of chromium metal are in 58.5 g of $K_2Cr_2O_7$? (Molar mass = 294 g/mol)
    (A) 10.4 g of Cr
    (B) 15.6 g of Cr
    (C) 20.8 g of Cr
    (D) 208 g of Cr

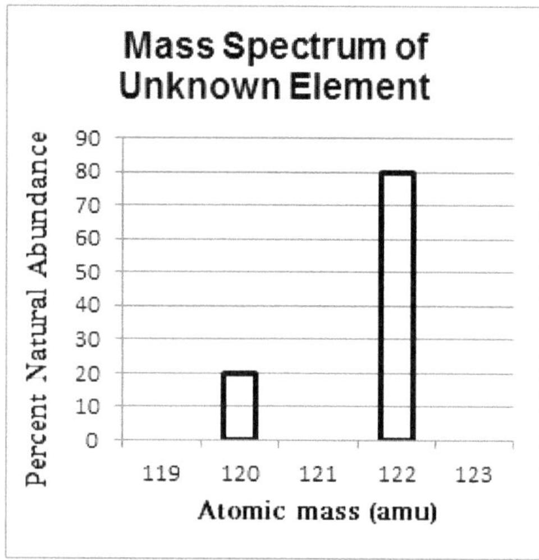

13. An unknown element has two naturally occurring isotopes, as shown in the mass spectrum above. What is the average atomic mass for this unknown element?
    (A) 120.4 amu
    (B) 121.0 amu
    (C) 121.5 amu
    (D) 121.6 amu

14. How many calcium atoms are present in a 40.08-mg sample of solid calcium metal?
    (A) 6.022×10^{20}
    (B) 6.022×10^{21}
    (C) 6.022×10^{23}
    (D) 6.022×10^{26}

Questions 15–21 refer to the following information:

A student places small pieces of magnesium metal into five different uninflated balloons and then places each of the balloons over one of five identical Erlenmeyer flasks without allowing any of the magnesium ribbon to fall into the flasks. The flasks each contain 100.0 mL of 2.0 *M* hydrochloric acid. Each balloon is then lifted upright, allowing the magnesium ribbon that it contains to fall into the HCl(*aq*). Shown below are the student's observations of the reactions and the diagrams that the student made in her laboratory notebook.

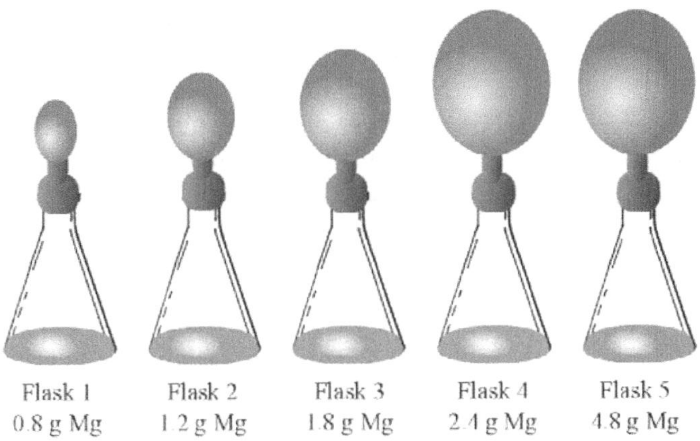

- The hydrochloric acid is a clear, colorless solution with no detectable odor.
- The magnesium ribbon is a soft, shiny metal that can be easily bent and broken with your fingers.
- After the magnesium ribbon is added to the hydrochloric acid, bubbles rapidly form and the magnesium solid disappears in less than 1 minute.
- As the bubbles form, the balloons start to inflate. Not all balloons inflate to the same size.
- The flask has gotten warmer, but not to the point where it is too hot to touch.

15. Which is the correct net ionic equation for this reaction?
    (A) $Mg + 2\ HCl \rightarrow MgCl_2 + H_2$
    (B) $Mg^{2+} + 2\ Cl^- \rightarrow MgCl_2$
    (C) $Mg + 2\ H^+ \rightarrow Mg^{2+} + H_2$
    (D) $Mg + 2\ H^+ + 2\ Cl^- \rightarrow Mg^{2+} + Cl_2 + H_2$

16. In which flask is it most likely that there are no excess reactants?
    (A) Flask 2
    (B) Flask 3
    (C) Flask 4
    (D) Flask 5

17. Which is the correct statement for the reaction?
    (A) ΔG is positive, and the sign for work is positive.
    (B) ΔG is positive, and the sign for work is negative.
    (C) ΔG is negative, and the sign for work is positive.
    (D) ΔG is negative, and the sign for work is negative.

18. If the student hypothesizes that this reaction is driven by both enthalpy and entropy, which of the following matches this hypothesis?
    (A) The ΔH is positive, and ΔS is positive.
    (B) The ΔH is positive, and ΔS is negative.
    (C) The ΔH is negative, and ΔS is positive.
    (D) The ΔH is negative, and ΔS is negative.

19. What is the volume of gas that is produced in Flask 2? Assume that the gas is at STP.
    (A) 0.56 L
    (B) 1.1 L
    (C) 2.2 L
    (D) 4.5 L

GO ON TO NEXT PAGE

20. Which of the following statements about Flask 3 is correct?
    (A) The magnesium ribbon is the excess reactant and is reduced.
    (B) The magnesium ribbon is the excess reactant and is oxidized.
    (C) The magnesium ribbon is the limiting reactant and is reduced.
    (D) The magnesium ribbon is the limiting reactant and is oxidized.

21. The following diagram represents the solution that is in Flask 5 at the conclusion of the trial. Note that only water particles are shown for clarity.

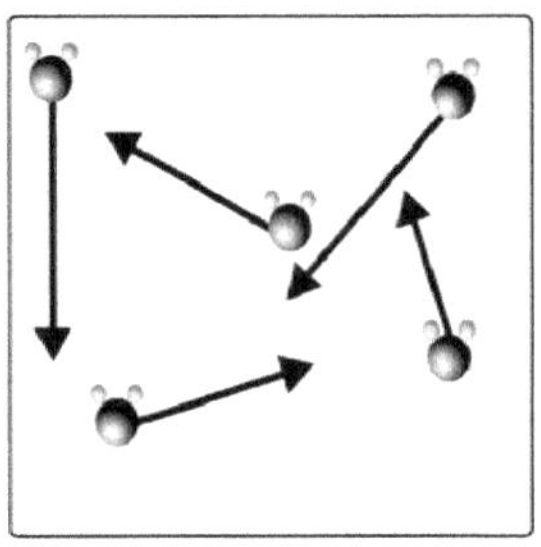

Which diagram below would represent Flask 4?

(A)

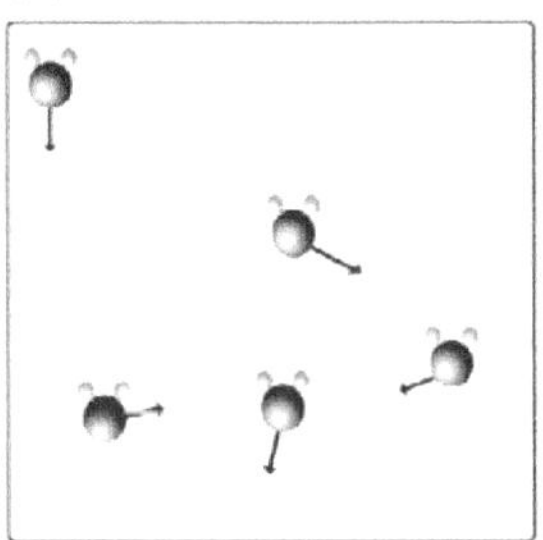

(B)

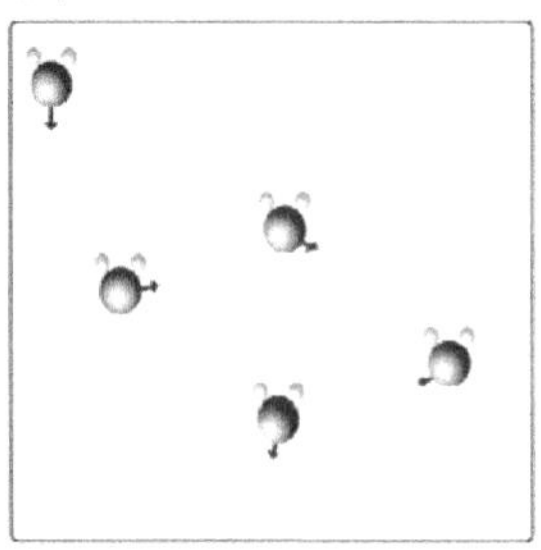

(C)

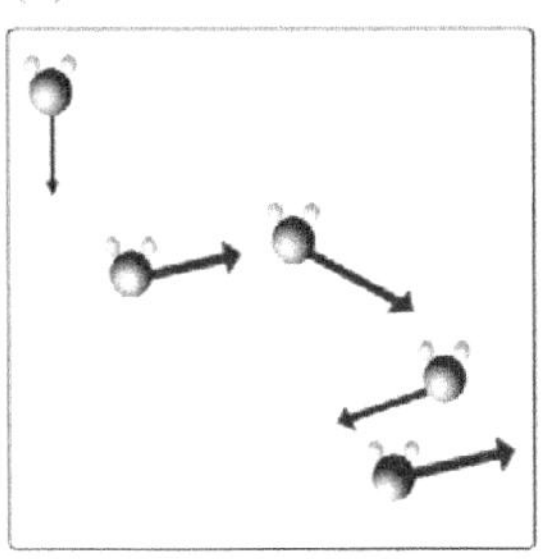

(D)

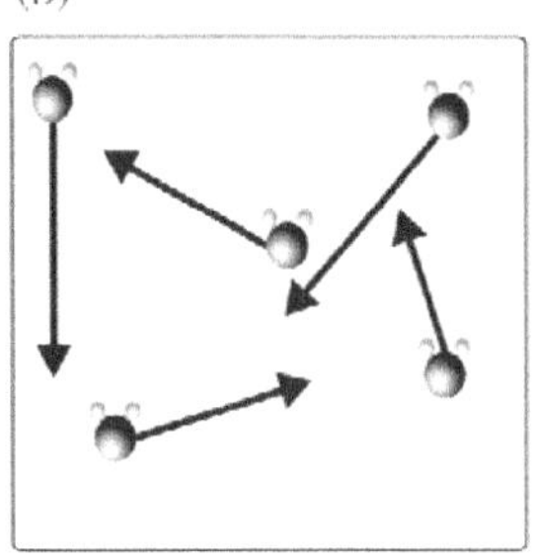

22. Which of the following salts, when dissolved in water containing phenolphthalein indicator ($pK_a \approx 9$, acidic form is colorless and basic form is pink), forms a pink solution?
    (A) KCl
    (B) $CuCl_2$
    (C) Na_2CO_3
    (D) NH_4NO_3

23. When NH_4Cl is dissolved in water, the water temperature decreases. Therefore, the enthalpy of the solution is
    (A) positive, because the temperature of the surroundings has decreased.
    (B) negative, because the temperature of the surroundings has decreased.
    (C) positive, because the temperature of the system has increased.
    (D) negative, because the temperature of the system has increased.

24. In a 0.10 *M* solution of potassium acetate, which of the following is true? The K_a for acetic acid, $HC_2H_3O_2$, is 1.8×10^{-5} at 25°C.
    (A) $[C_2H_3O_2^-] = [OH^-]$
    (B) $[H^+] = [OH^-]$
    (C) $[C_2H_3O_2^-] < 0.10\ M$
    (D) $pK_b = pOH$

Questions 25–26 refer to the reaction $A + 2\,B \rightarrow C + 2\,D$, and these data:

Exp. No.	Initial [A]	Initial [B]	Initial Rate $\Delta[A]/\Delta t$ in mol $L^{-1}s^{-1}$
1	0.100 M	0.050 M	7.00×10^{-6}
2	0.200 M	0.050 M	1.40×10^{-5}
3	0.100 M	0.100 M	7.00×10^{-6}

25. What is the rate law for this reaction?
    (A) Rate = k[A][B]
    (B) Rate = k[A][B]^2
    (C) Rate = k[A]
    (D) Rate = k[B]

26. Determine the rate of appearance of Product D in Experiment 2.
    (A) 7.00×10^{-6} mol L^{-1} s^{-1}
    (B) 3.50×10^{-6} mol L^{-1} s^{-1}
    (C) 1.40×10^{-5} mol L^{-1} s^{-1}
    (D) 2.80×10^{-5} mol L^{-1} s^{-1}

Questions 27–28 refer to the following material:

Second Period Elements	Covalent Radii, nm
Li	0.152
Be	0.111
B	0.088
C	0.077
N	0.070
O	0.066
F	0.064
Ne	0.070

27. Determining the "covalent radii" is one experimental technique used to measure the size of atoms. It is defined as half the distance between 2 covalently bonded atoms. Explain why the atomic radii generally decrease going left to right across Period 2 of the periodic table.
    (A) As you go to the right in Period 2, the elements become more nonmetallic. The more nonmetallic an element is, the more likely it is to share electrons and the radius will decrease.
    (B) As you go to the right in Period 2, the element's valence shell has more electrons. The closer an element is to having a full octet, the more likely it is to share electrons and the radius will decrease.
    (C) As you go to the right in Period 2, the nuclear charge increases. The higher the nuclear charge, the more attraction there is from the nucleus to the valence electrons; hence, the valence electrons will be more tightly held and the radius will decrease.
    (D) As you go to the right in Period 2, the molar mass of the elements increase. The increasing mass causes an increase in density and a decrease in the radius.

GO ON TO NEXT PAGE

28. Explain why the radius of Ne does not follow the general trend of the other elements in Period 2.
(A) Neon has a full octet of valence electrons, with all orbitals being filled. Filled subshells are the most stable, resulting in a larger atom.
(B) Neon is a noble gas and does not take part in bonding, so the covalent radius will be the largest of any Period 2 element.
(C) Neon has the highest number of electrons of any Period 2 element. Therefore, it has the highest electron to electron repulsion, resulting in the atom being larger.
(D) Neon has the highest nuclear charge of any Period 2 element. This results in greater attraction between the nucleus and the valence electrons; hence, the valence electrons will be more tightly held and the radius will increase.

29. Which of the following solutions has the greatest number of ions in solution?
(A) 500. mL of 0.100 M $FeCl_3$
(B) 700. mL of 0.200 M NaOH
(C) 400. mL of 0.100 M $Al(NO_3)_3$
(D) 600. mL of 0.200 M $AlCl_3$

	C_4H_8	C_8H_{18}
Physical state at room temperature	Gas	Liquid
Molar mass (g/mol)	56.11	114.23
Number of double bonds	1	0
Boiling point (°C)	−6.47	125.6
Number of isomers	6	18
Hybridization of carbon atoms	sp^2 and sp^3	sp^3

30. Which statement best explains why C_4H_8 has a lower boiling point than C_8H_{18}?
(A) C_4H_8 has fewer isomers to contribute to the strength of the intermolecular forces; hence, it will have a lower boiling point.
(B) The multiple hybridizations of the carbon atoms in C_4H_8 decrease the polarizability of the electrons, reducing the intermolecular forces and lowering the boiling point.
(C) C_8H_{18} has a higher molar mass than C_4H_8. The higher the molar mass, the stronger the intermolecular forces will be; hence, the boiling point will be higher in C_8H_{18}.
(D) C_8H_{18} is a longer molecule and has more electrons than C_4H_8 so its electron cloud is more polarizable, giving it stronger intermolecular forces and a higher boiling point.

31. What is the percent by mass of nitrogen in sodium cyanide, NaCN?
(A) 14.01%
(B) 24.41%
(C) 28.59%
(D) 49.01%

32. For which of the following is the sign of the enthalpy change different from the others?
(A) $CO_2(s) \rightarrow CO_2(g)$
(B) $H_2O(l) \rightarrow H_2O(s)$
(C) $Br_2(g) \rightarrow 2\ Br(g)$
(D) $H_2O(l) \rightarrow H_2O(g)$

33. The Lincoln penny was introduced in 1909 to commemorate the centennial of Abraham Lincoln's birth. From 1909 to 1982, the penny was 95% copper and 5% zinc and had a density of 8.8 g/mL. In 1982, the government began to make pennies that were 97.5% zinc and 2.5% copper with a density of 7.2 g/mL. Approximately 1200 pennies will fit into a 1.0-L glass jar. A student needs to determine the approximate number of pre-1982 pennies in a jar that is completely filled with a mixture of pre- and post-1982 pennies. His data table is shown below.

Mass of glass jar empty	550 g
Mass of glass jar filled with pennies	8961 g

Approximately how many pre-1982 pennies are in the jar?
(A) 1200
(B) 900
(C) 600
(D) 300

34. Stomach acid is mostly made up of water, potassium chloride, sodium chloride, and hydrochloric acid, HCl. The HCl gives stomach acid its acidic properties and makes up approximately 5% of the stomach acid, which usually has a pH of between 1 and 2. To carry out investigations of new drugs (the rate at which the drug is digested, the level of absorption, etc.), drug companies routinely warm gastric juices to 37°C (normal body temperature for a human) before experimentation. One technique to determine the change in the gastric juices is a "back titration" where a known amount of stomach acid is reacted with a new drug and then titrated with a strong base to determine the amount of acid that remains. A graph of the titration of stomach acid (assume that the only acid present is HCl) with NaOH is shown below. What is a likely reason why the pH at the equivalence point is not 7.0?

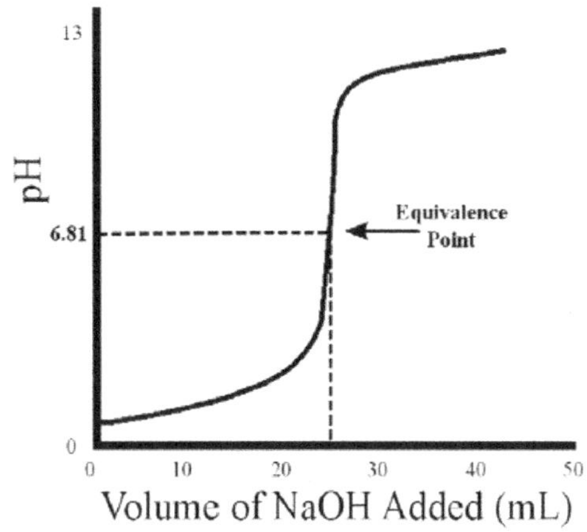

(A) The drug that was being tested added a small amount of weak acid to the gastric juices, thus reducing the pH at the equivalence point.
(B) The drug that was being tested added a small amount of weak base to the gastric juices, thus reducing the pH at the equivalence point.
(C) The body temperature caused the K_w value to increase, thus lowering the point where the $[H^+] = [OH^-]$.
(D) The body temperature caused the K_w value to increase, thus raising the point where the $[H^+] = [OH^-]$.

GO ON TO NEXT PAGE

Questions 35–39 refer to the following information:

Reaction 1: $2\ NOCl(g) \rightleftarrows 2\ NO(g) + Cl_2(g)$
$K = 1.6 \times 10^{-5}$

Reaction 2: $2\ NO(g) \rightleftarrows N_2(g) + O_2(g)$
$K = 1.0 \times 10^{31}$

Both reactions are at the same temperature.

35. Which statement is true concerning the set of reactions shown above?
   (A) Reaction 1 is product favored and its $\Delta G < 0$.
   (B) Reaction 1 is product favored and its $\Delta G > 0$.
   (C) Reaction 1 is reactant favored and its $\Delta G < 0$.
   (D) Reaction 1 is reactant favored and its $\Delta G > 0$.

36. Which statement is correct for the following reaction at the same temperature as the reactions above?
   $2\ NOCl(g) \rightleftarrows N_2(g) + Cl_2(g) + O_2(g)$
   (A) The reaction is non-thermodynamically favored and $K = 1.6 \times 10^{26}$.
   (B) The reaction is thermodynamically favored and $K = 1.6 \times 10^{26}$.
   (C) The reaction is non-thermodynamically favored and $K = 2.6 \times 10^{26}$.
   (D) The reaction is thermodynamically favored and $K = 2.6 \times 10^{31}$.

37. What is the hybridization of nitrogen in the NOCl, and how many sigma and pi bonds does the molecule have?

	Hybridization of N	Number of Sigma Bonds	Number of Pi Bonds
(A)	sp^2	2	1
(B)	sp^2	2	0
(C)	sp^3	3	1
(D)	sp^3	3	0

38. For Reaction 1, what are the correct signs for ΔH_{rxn} and ΔS_{rxn}?

	ΔH_{rxn}	ΔS_{rxn}
(A)	+	+
(B)	+	−
(C)	−	+
(D)	−	−

39. Which statement is correct for Reaction 2?
   (A) The reaction has $\Delta G > 0$ since the K value is very large. The reaction must be endothermic since the value for ΔS has to be positive since two products with pi bonds are produced.
   (B) The reaction has $\Delta G < 0$ since the K value is very large. The reaction must be exothermic since the value for ΔS has to be negative since two products with pi bonds are produced.
   (C) The reaction has $\Delta G > 0$ since the K value is very large. The reaction must be endothermic since the value for ΔS is close to zero and will not affect the value for ΔG.
   (D) The reaction has $\Delta G < 0$ since the K value is very large. The reaction must be exothermic since the value for ΔS is close to zero and will not affect the value for ΔG.

40. Which graph correctly shows both bond energy and bond length between two covalently bonded atoms?

(A)
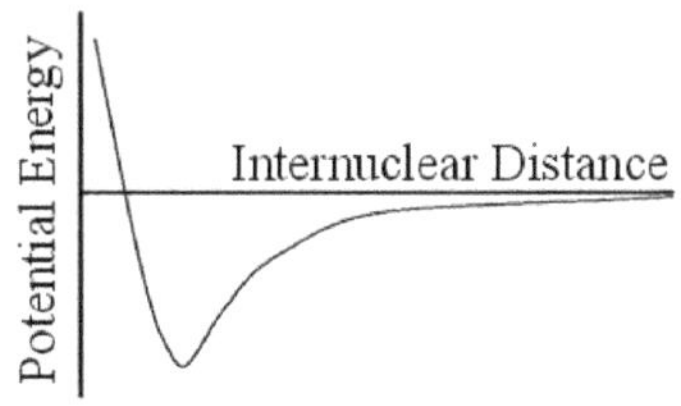

(B)
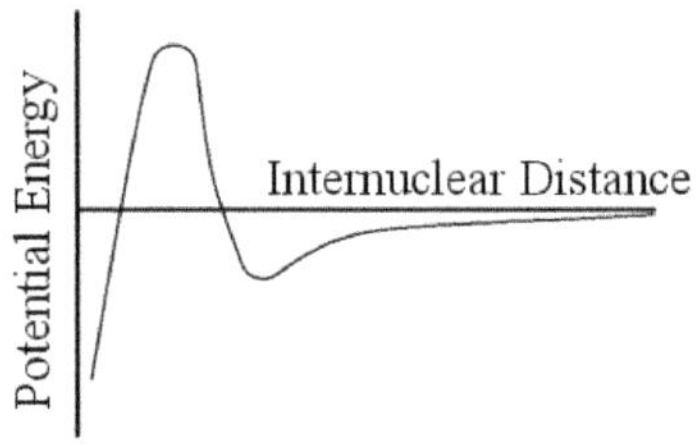

(C)
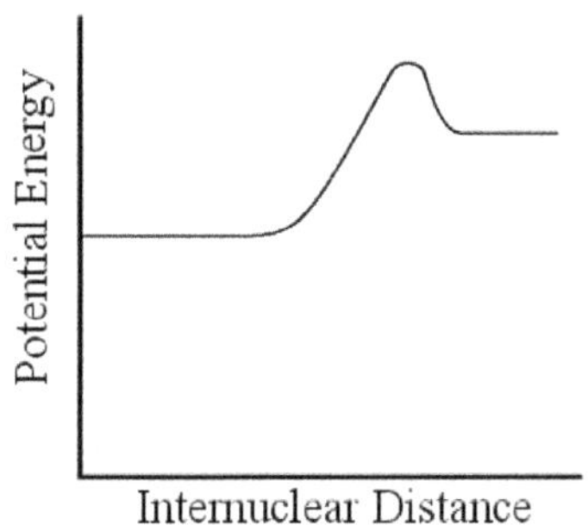

(D)
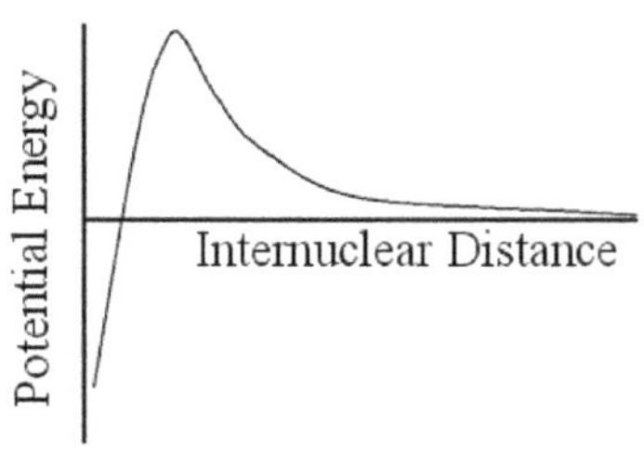

$2\ HI(g) \leftrightarrows H_2(g) + I_2(g)$

41. A student is studying the kinetics of the reaction shown above. In the course of that study, the student made the three graphs shown below.

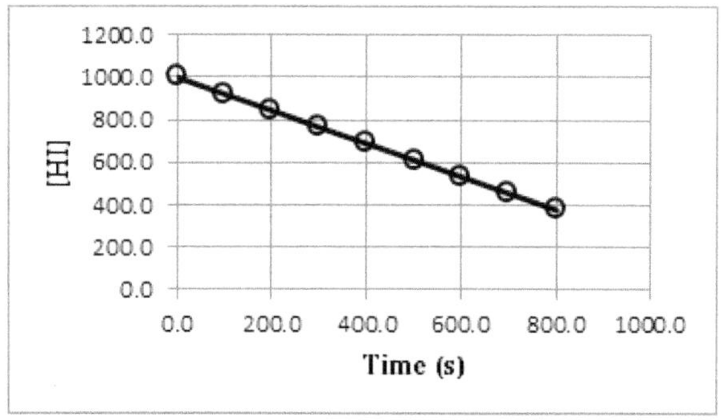

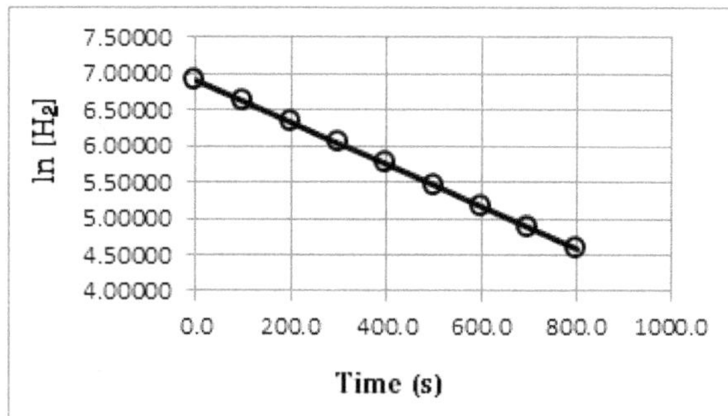

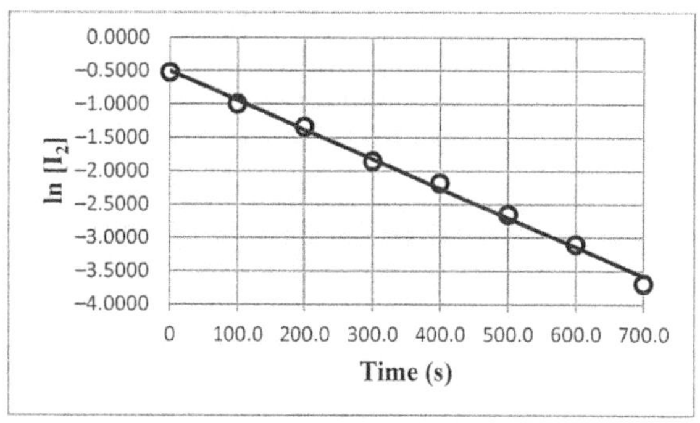

Which statement is correct based on the student's data?

(A) The forward reaction is first order, and the reverse reaction is zero order overall.
(B) The forward reaction is second order, and the reverse reaction is zero order overall.
(C) The forward reaction is zero order, and the reverse reaction is second order overall.
(D) The forward reaction is second order, and the reverse reaction is second order overall.

GO ON TO NEXT PAGE

Questions 42–46 use the following equation:

$PCl_5(g) \leftrightarrows PCl_3(g) + Cl_2(g)$
$\Delta H_{rxn} = 124\ kJ/mol_{rxn}$

42. $PCl_5(g)$ is placed into a rigid 4.0-L container and the pressure is found to be 6.0 atm. As the temperature is increased to 575 K, the pressure increases to 10.0 atm and the system reaches equilibrium. What is the K_p value at 575 K?
(A) 2
(B) 8
(C) 16
(D) 100

43. After 30 minutes at 575 K, the temperature is lowered to 425 K. Which direction is the shift in equilibrium, and what happens to the total pressure?
(A) The reaction shifts to the left, toward the reactants, and the total pressure increases.
(B) The reaction shifts to the right, toward the products, and the total pressure increases.
(C) The reaction shifts to the left, toward the reactants, and the total pressure decreases.
(D) The reaction shifts to the right, toward the products, and the total pressure decreases.

44. Pure samples of the three substances involved in this reaction are condensed to their liquid phase. Which correctly identifies the intermolecular forces involved in each of the substances as they condense?

	PCl_5	PCl_3	Cl_2
(A)	London dispersion forces	London dispersion forces	London dispersion forces
(B)	London dispersion forces	London dispersion forces	Dipole–dipole and London dispersion forces
(C)	London dispersion forces	Dipole–dipole and London dispersion forces	London dispersion forces
(D)	Dipole–dipole and London dispersion forces	London dispersion forces	London dispersion forces

45. At 450 K, it has been found that a sample of $PCl_5(g)$ is 40% decomposed. Which of the following contains a sample of $PCl_5(g)$ at 450 K, and what is the K_p value of that sample? Assume that each particle represents a mole of that substance.

(A)

$K_p = 0.625$

(B)

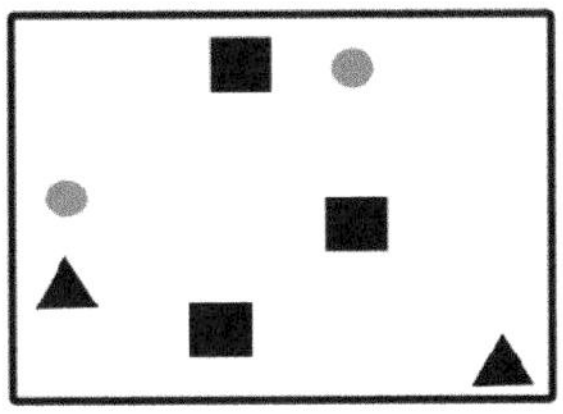

$K_p = 0.750$

(C)

$K_p = 1.60$

(D)

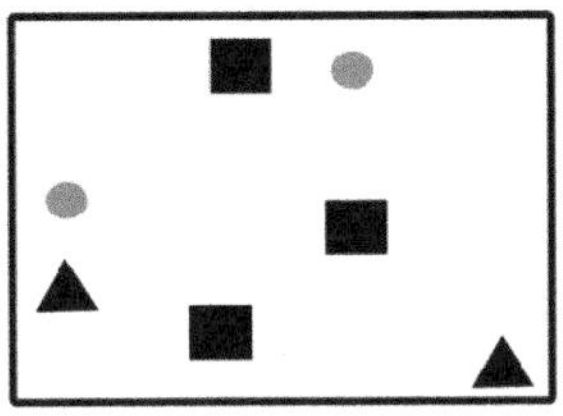

$K_p = 1.33$

PCl_5 = ■ PCl_3 = ▲ Cl_2 = ●

46. Phosphorus pentachloride, PCl_5, has chlorine atoms bonded around phosphorus in both axial and equatorial positions. The bond length for the axial position in phosphorus pentachloride is 214 pm, while the equatorial position has a bond length of 202 pm. Antimony and arsenic form similar compounds to phosphorus pentachloride. What would be the expected bond lengths for $SbCl_5$ and $AsCl_5$?

	$SbCl_5$		$AsCl_5$	
	Equatorial (pm)	**Axial (pm)**	**Equatorial (pm)**	**Axial (pm)**
(A)	227	233.3	211	221
(B)	233.3	227	221	211
(C)	211	221	227	233.3
(D)	221	211	233.3	227

Questions 47–50 use the following information:

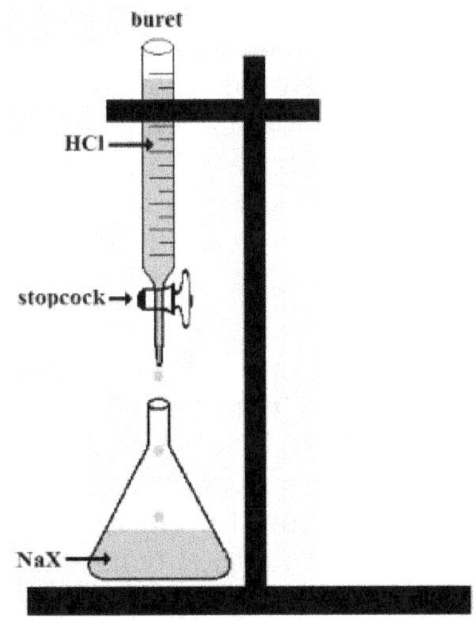

A student dissolves a salt, NaX, that contains a weak base, X^-, in water and titrates it with HCl. The experimental setup is shown above. As the titration proceeds, the solution is monitored with a pH probe, and the resulting pH curve is shown below.

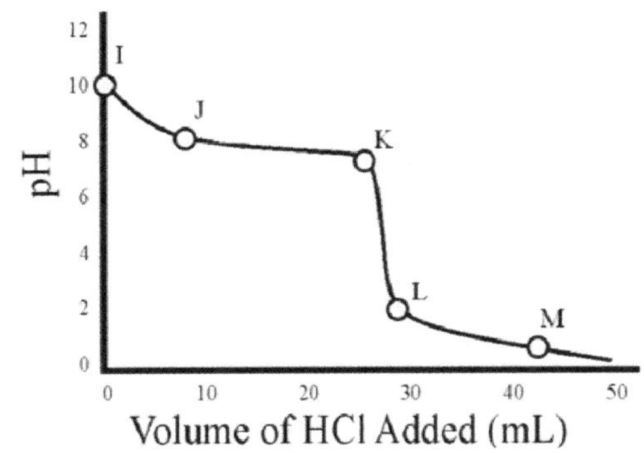

47. Between which two points will information be found that allows the determination of the K_a for the weak conjugate acid, HX?
(A) between I and J
(B) between J and K
(C) between K and L
(D) between L and M

GO ON TO NEXT PAGE

48. During the titration, 2.0 mL of 3 *M* NaOH(*aq*) was accidentally added to the flask below the buret. It was noticed that the pH did not change. Between what two points did the accident occur?
(A) between I and J
(B) between J and K
(C) between K and L
(D) between L and M

49. A particle representation of a point during the titration is shown below. Note that water, sodium ions, and chloride ions are omitted from the picture for clarity. At what point during the titration does this diagram represent?

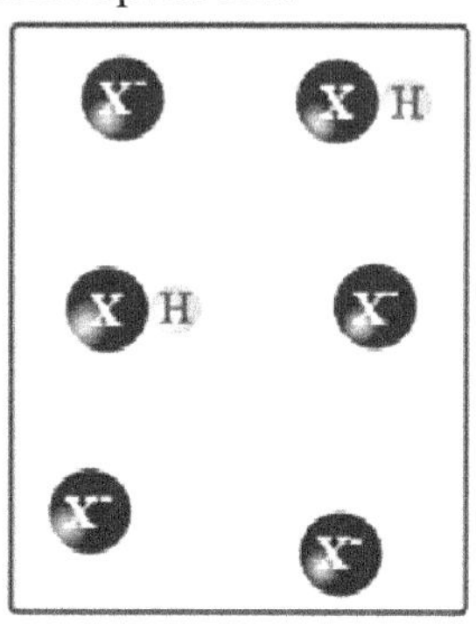

(A) I
(B) J
(C) K
(D) L

50. A particle representation of a point during the titration is shown below. Note that water and sodium ions are omitted from the picture for clarity. At what point during the titration does this diagram represent?

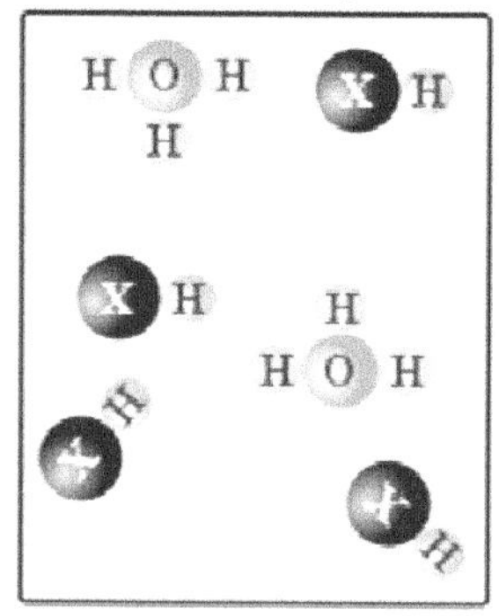

(A) J
(B) K
(C) L
(D) M

51. When excess concentrated ammonia is added to a solution of zinc nitrate, a complex ion forms between the zinc ion and ammonia. It is experimentally determined that the zinc complex ion has a tetrahedral structure. What is the correct balanced equation for this reaction?
(A) $Zn^{2+} + NH_3 \rightarrow [ZnNH_3]^{2+}$
(B) $Zn^{2+} + 2\ NH_3 \rightarrow [Zn(NH_3)_2]^{2+}$
(C) $Zn^{2+} + 3\ NH_3 \rightarrow [Zn(NH_3)_3]^{2+}$
(D) $Zn^{2+} + 4\ NH_3 \rightarrow [Zn(NH_3)_4]^{2+}$

Questions 52–53 refer to the following information:

A student places 50.00 mL of 3.00 *M* HBr(*aq*) into a flask and adds 50.00 mL of 1.00 *M* NaOH(*aq*) solution to it. The student then stirs the solution and measures its pH.

52. What is the resulting pH of the solution?
(A) −1.000
(B) 0.000
(C) 1.000
(D) 2.000

53. Which of the following species is present in the highest concentration?
(A) H_3O^+
(B) Na^+
(C) Br^-
(D) OH^-

$NH_4^+(aq) + HNO_2(aq) \rightarrow N_2(g) + 2\ H_2O(l) + H^+(aq)$

54. Ammonium is a weak acid. However, when placed in solution with a stronger acid, the reaction shown above takes place. The reaction is believed to take place according to the following mechanism:

Step 1: $HNO_2 + H^+ \leftrightarrows H_2O + NO^+$ (fast equilibrium)
Step 2: $NH_4^+ \leftrightarrows NH_3 + H^+$ (fast equilibrium)
Step 3: $NO^+ + NH_3 \rightarrow NH_3NO^+$ (slow)
Step 4: $NH_3NO^+ \rightarrow H_2O + H^+ + N_2$ (fast)

The rate law derived from experimental evidence is:

$$rate = k\frac{[HNO_2][NH_4^+]}{[H_2O]}$$

According to the rate law and proposed mechanism, how does the water affect the rate of the reaction?
(A) It is providing the solvent that acids need to ionize.
(B) It is interfering with the collisions of the reactants, hence slowing down the process.
(C) It is a catalyst, speeding up the process by allowing the formation of hydronium ions.
(D) It is absorbing and dissipating large amounts of heat from the reaction of two acids that are added to the water.

A teacher gives her students a sheet of graph paper with the following diagram on it. Each sphere represents a gas particle moving in a rigid container.

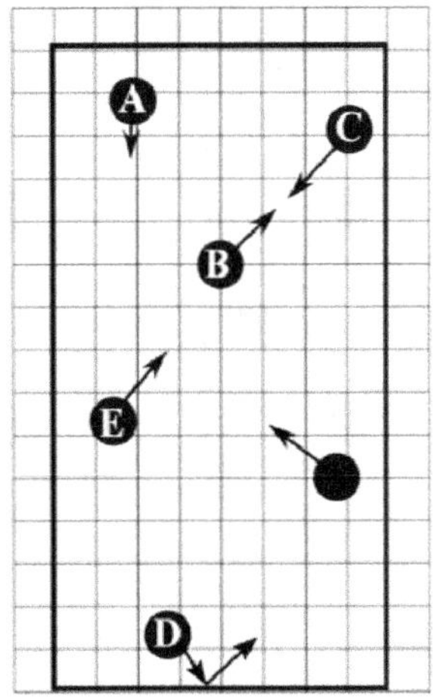

55. Which particle or particles in the diagram are responsible for the pressure of the gases? Base your answer solely on the diagram.
(A) A
(B) D
(C) E
(D) B and C

GO ON TO NEXT PAGE

56. The students were asked to draw, on an identical sheet of graph paper, the effects of reducing the volume of the container by half. Four students' drawings are shown below.

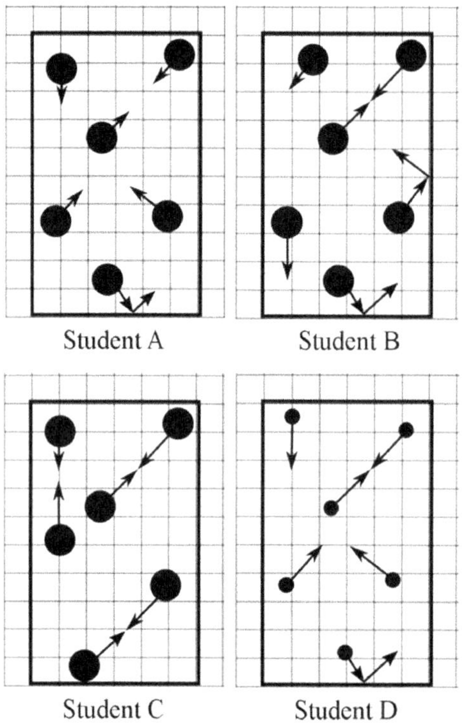

Which student correctly indicated the effects of change of volume on the pressure of the container? Base your answer solely on the diagrams presented by the students.
(A) Student A
(B) Student B
(C) Student C
(D) Student D

	methanol	hexane	water
methanol (CH_3OH)		I	S
hexane (C_6H_{14})	I		I
water	S	I	

57. Which is the **best** explanation as to why methanol is soluble in one solvent, but not the other?
(A) Water and methanol have molar masses that are more similar than hexane and methanol. Like molecules are more soluble in each other.
(B) Strong hydrogen bonds are formed between methanol and water, while only weaker dipole-induced dipole attractions are found between methanol and hexane.
(C) Water and methanol both have a bent shape, rather than the linear shape that hexane has. This bent shape allows the molecules to have greater attraction for each other.
(D) It takes more energy to overcome the attraction between the solvent molecules when water is the solvent. The bonds between hexane molecules are weak dispersion forces, whereas the bonds between water molecules are hydrogen bonds.

58. Consider two solutions: Solution X with a pH of 7 and Solution Y with a pH of 9. What would be the hydronium ion ratio of the two solutions, X/Y?
(A) 7/9
(B) 1/2
(C) 2/1
(D) 100/1

59. The graph below shows the molecular velocity distributions for three gases, Ar, He, and Ne, all at the same temperature.

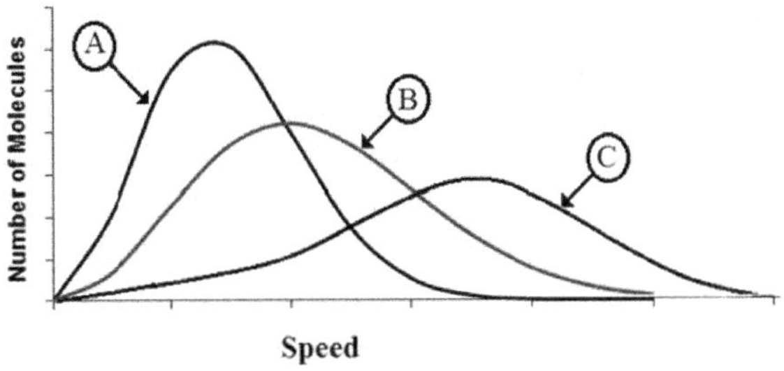

Match the gas with the distribution.

	A	B	C
(A)	Ar	Ne	He
(B)	He	Ne	Ar
(C)	Ne	He	Ar
(D)	Ar	He	Ne

60. Which of the gases exhibits the greatest deviation from ideal behavior?
(A) He, because it is the smallest atom with the least number of electrons.
(B) Ne, because it is the first noble gas with a completed octet.
(C) Ar, because it is the largest atom with the greatest number of electrons.
(D) None, because inert gases are ideal gases.

INTRODUCTION TO SECTION II: FREE-RESPONSE QUESTIONS

Section II of the AP® Chemistry examination counts for 50% of the total test grade and involves several parts. Answering these questions gives you an opportunity to demonstrate your ability to present your material in clear, orderly, and convincing language. Your answers will be graded on the basis of accuracy, the kinds of information you include to support your responses, and the importance of the descriptive material used. Be specific; general, all-encompassing answers will not be graded as well as detailed answers with examples and equations. CLEARLY SHOW THE METHOD USED AND THE STEPS INVOLVED IN ARRIVING AT YOUR ANSWERS. It is to your advantage to do this, since you may obtain partial credit if you do and you will receive little or no credit if you do not. Attention should be paid to significant figures. On the AP® exam, be sure to write all your answers to the questions on the lined pages following each question in the test booklet. Do not write your answers in the white space between questions.

Section II: Free-Response Questions
Time: 105 minutes
Number of Questions: 7

DIRECTIONS: You may use a calculator, the formula chart, and the periodic table located at the back of this book throughout this section.

1. A student is asked to determine the heat of formation of MgO and is given the following materials:
   - coffee cup calorimeter
   - thermometer
   - balance
   - stirring rod
   - magnesium oxide
   - magnesium ribbon
   - 1.00 *M* hydrochloric acid (density 1.00 g/mL)

   The student performs experiments to determine the value of ΔH for the reactions of

   (1) magnesium + hydrochloric acid → $\Delta H_1 = ?$

   (2) magnesium oxide + hydrochloric acid → $\Delta H_2 = ?$

   The data for these two experiments are listed below.

Magnesium + Hydrochloric Acid	
magnesium mass (g)	0.121
hydrochloric acid volume (mL)	50.0
initial temperature (°C)	20.0
final temperature (°C)	30.7

Magnesium Oxide + Hydrochloric Acid	
magnesium oxide mass (g)	0.348
hydrochloric acid volume (mL)	50.0
initial temperature (°C)	14.9
final temperature (°C)	20.0

The student is also given the following information:

(3) $H_2(g) + \frac{1}{2} O_2(g) \rightarrow H_2O(l)$ $\Delta H_3 = -285.8 \text{ kJ mol}^{-1}$

(a) Write the net ionic equations for the following reactions:
  (i) magnesium with hydrochloric acid
  (ii) magnesium oxide with hydrochloric acid

(b) Calculate ΔH_1. Assume the specific heat capacity of the solution is 4.184 J/(g °C) and the density of the solution is 1.0 g/mL.

(c) Calculate ΔH_2. Assume the specific heat capacity of the solution is 4.184 J/(g °C) and the density of the solution is 1.0 g/mL.

(d) Show how the three equations above can be rearranged to calculate $\Delta H_f°$ of magnesium oxide.

(e) Calculate the $\Delta H_f°$ of magnesium oxide.

(f) The accepted value for $\Delta H_f°$ of MgO = –602 kJ/mol. Calculate the percent error.

(g) The calorimeter constant is the amount of heat that the calorimeter must gain to change temperature by 1°. Why would the use of a calorimeter constant improve the accuracy of the results?

2. *Oxalis* is the largest genus in the wood sorrel family of plants whose scientific name is "oxalidaceae." Many of the wild plants in this family have been eaten by humans for more than a thousand years. A characteristic of this family of plants is that they contain large amounts of oxalic acid, $H_2C_2O_4$. Oxalic acid is the simplest of the dicarboxylic acids. The acid dissociation constants are given below.

K_{a1}	K_{a2}
5.9×10^{-2}	6.4×10^{-5}

Answer the following questions about oxalic acid:

O O

H O C C O H

(a) Complete the Lewis dot diagram for oxalic acid. Include any lone pairs of electrons.

(b) Based on the Lewis dot diagram completed in part (a), do all the atoms lay in the same plane? Explain your answer.

(c) Based on your Lewis dot diagram completed in part (a), what is the number of sigma bonds and number of pi bonds in oxalic acid?

(d) The student is asked to determine the pH of a 0.25 *M* solution of oxalic acid. The student's work is shown below.

	$H_2C_2O_4$	+	H_2O	$\rightleftarrows$	$HC_2O_4^{-1}$	+	H_3O^+
I	0.25 *M*		--		0		0
C	$-x$		--		$+x$		$+x$
E	$0.25 - x$ ≈ 0.25		--		x		x

$$K_a = \frac{[HC_2O_4^{-1}][H_3O^+]}{[H_2C_2O_4]} \Rightarrow 5.9 \times 10^{-2} = \frac{x^2}{0.25} \Rightarrow x = 0.12 = [H_3O^+]$$

GO ON TO NEXT PAGE

$$pH = -\log[H_3O^+] \Rightarrow pH = -\log(0.12) \Rightarrow pH = 0.92$$

(i) Is the student's assumption that the concentration of $H_2C_2O_4$ is equal to 0.25*M* at equilibrium a valid assumption? Justify your answer with a calculation.

(ii) The student wrote

Since oxalic acid is a weak acid, the pH should not be less than 1.0.

Is this student's statement correct? Explain your answer.

(e) A recipe for lasagna calls for ½ cup of packed Italian parsley. Assume that this would be 60.0 g of parsley. Parsley is one of the foods with the highest oxalic acid content, 1.70 g $H_2C_2O_4$ per 100. g of parsley. If all of the oxalic acid in the parsley used to make this lasagna was dissolved in 100.00 mL of water, what would the resulting molarity be?

(f) There are five types of kidney stones. Eighty percent of kidney stones are made in whole or in part from calcium oxalate, CaC_2O_4 (molar mass = 128.10 g/mol). While most kidney stones do not need medical attention, a typical kidney stone that does need attention would be one that is 12 mm in diameter, which would have a volume of 1.09 cm^3. Assuming that all of the kidney stone is calcium oxalate, how many ions of oxalate are contained in the kidney stone? The density of calcium oxalate is 2.12 g/cm^3.

(g) Creeping wood sorrel (a plant that is a member of the Oxalis family) is a bioaccumulator of copper, and the plant can be used for phytoremediation of contaminated soils. Phytoremediation is the process by which plants clean contaminated soils by removing the contaminants, such as metals. The equation for the formation of the complex ion $[Cu(C_2O_4)_2]^{2-}$ is shown below.

$$Cu^{2+} + 2\,C_2O_4^{2-} \rightarrow [Cu(C_2O_4)_2]^{2-} \qquad K_{eq} = 1.58 \times 10^{6}$$

Explain how a plant that has large amounts of oxalic acid can remove metal contaminants, such as copper, from the soil. Base your answer on the equation and equilibrium value provided.

3. A standard battery is constructed using the components shown below.

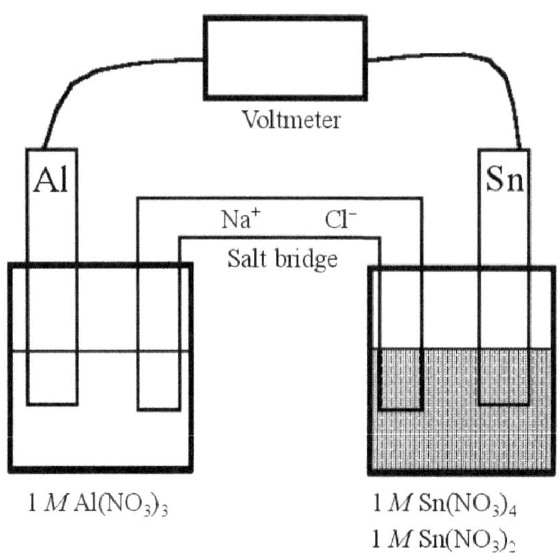

The standard reduction potentials are given below.

$$Sn^{4+} + 2e^- \rightarrow Sn^{2+} \qquad E° = +0.15\ V$$
$$Al^{3+} + 3e^- \rightarrow Al(s) \qquad E° = -1.66\ V$$

(a) Write the balanced chemical equation that occurs as the cell operates.
(b) Determine the standard initial voltage as the cell operates and place that value in the box marked "voltmeter" in the diagram.
(c) Which electrode is the anode?
(d) What happens to the mass of the Sn electrode as the cell operates?
(e) Circle the ion in the salt bridge that migrates toward the Al electrode.
(f) Do the electrons flow from the left electrode to the right electrode or from the right electrode to the left electrode?
(g) Calculate $\Delta G°_{rxn}$ for this cell as it operates.
(h) Given the following thermodynamic values, determine the $\Delta S°_{rxn}$.

Substance	ΔH_f (kJ/mol)
$Sn^{4+}(aq)$	30.5
$Sn^{2+}(aq)$	–8.8
$Al^{3+}(aq)$	–531.37

(i) Which graph shows the correct relationship between ΔG and temperature for this process? Justify your answer.

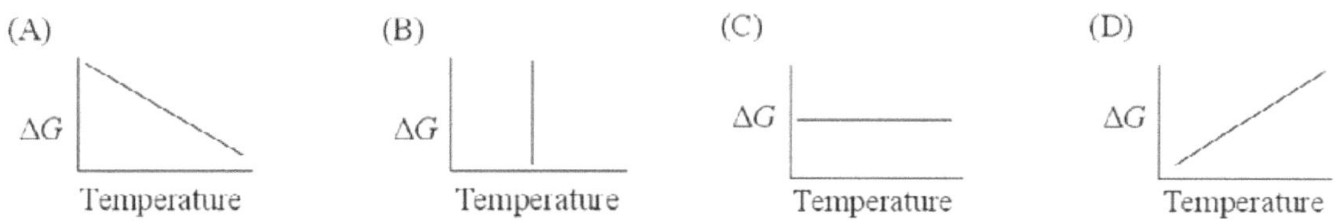

4. Chloroform has an infamous reputation among the public at large due to mystery and crime writers having their criminals soak rags with chloroform and place them over the mouth and nose of a victim. The victim, having struggled for a few seconds, slumps over unconscious. In fact, it would take between 5 and 10 minutes to render someone incapacitated using chloroform. Very few people have actually ever been murdered using it. One exception was William March Rice, the founder of Rice University. In his case, the story is right out of a crime novel—involving forged checks, a fake will, a lawyer who plans the murder, and, in this case, the butler really did do it.
(a) Chloroform is 10.06% carbon and 0.846% hydrogen, and the remainder is chlorine. Determine the empirical formula for chloroform.
(b) Chloroform is a precursor chemical that, after a series of steps, produces polytetrafluorethylene—it is the "nonstick" in older nonstick cookware. Since this is a polymer, part of a Lewis structure is shown below

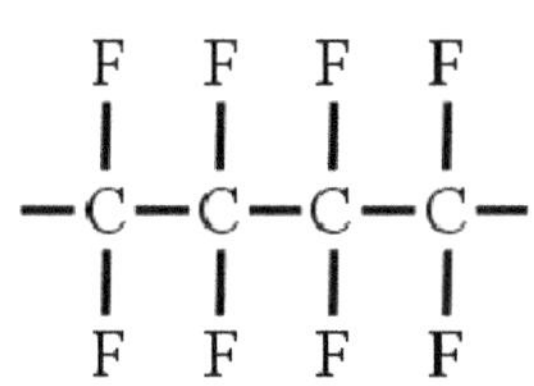

The carbon to fluorine bond can be considered the strongest single bond in organic chemistry. Provide one reason why the C–F bond is so strong.

GO ON TO NEXT PAGE

(c) If a compound reacts chemically with water, it cannot form an aqueous solution and another solvent must be used. Also, if a reaction needs to be cooled to below 0°C, water cannot be used. Chloroform is used as the solvent for the reaction shown below since the two species will react with water and the temperature needs to be very low for this reaction. The useful temperature range of a solvent is between its melting point and its boiling point, which means that chloroform has a range of –63.5°C to 61.15 °C.

$$N_2O_4 \leftrightarrows 2\ NO_2$$

At equilibrium, 250.0 mL of a chloroform solution is found to contain 1.15 moles of N_2O_4 and 4.45×10^{-3} moles of NO_2. Calculate the value for the equilibrium constant, K_c, for this reaction.

5. The first ionization energies, in kJ/mol, for elements in Period 3 of the periodic table are as follows:

Na	Mg	Al	Si	P	S	Cl	Ar
496	738	577	786	1060	999.6	1256	1520

(a) Explain the general trend in first ionization energies as you move across the third period.
(b) Explain why the general trend is not followed at aluminum.
(c) Explain why there is a drop in the first ionization energy of sulfur.
(d) Below are four photoelectron spectroscopy graphs for four different elements from the second period. On the blank graph at the bottom, draw the photoelectron spectroscopy for the element boron. Based on the data presented above and below, label the graphs with the approximate ionization energy values.

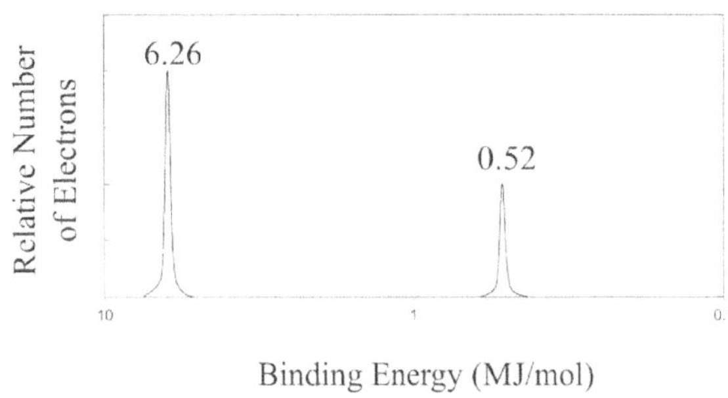

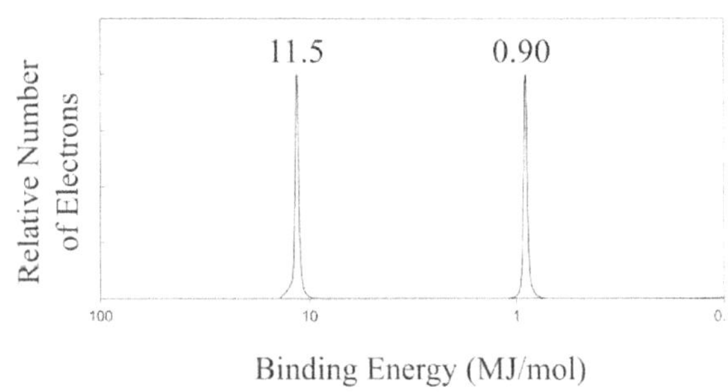

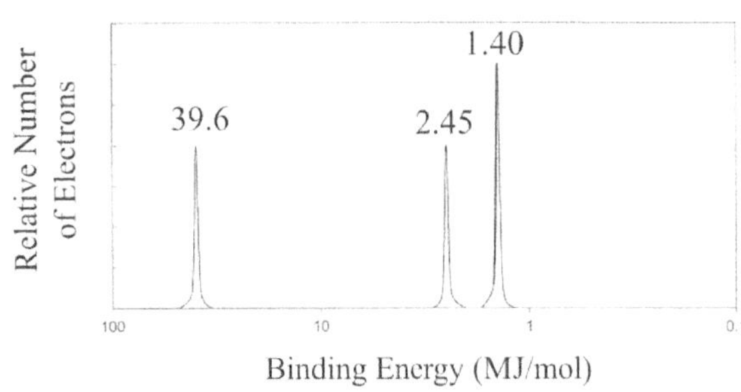

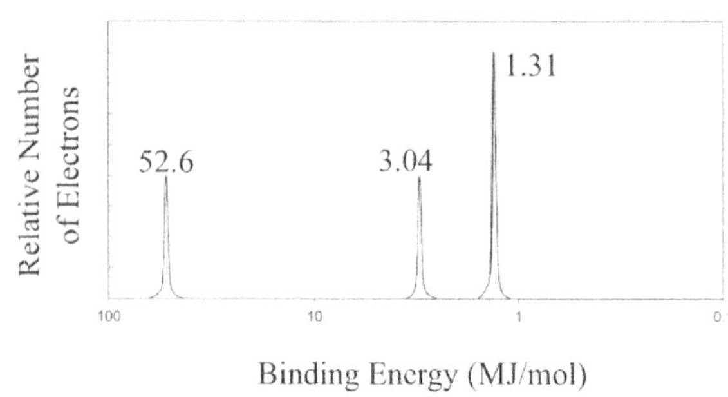

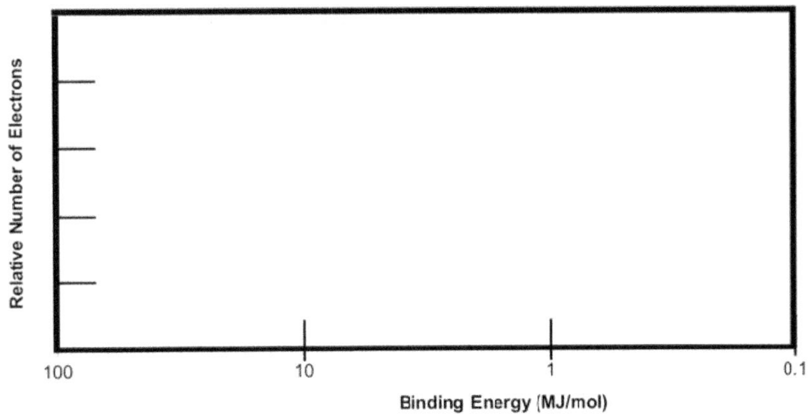

6. A ketone is an organic compound that has a C=O group with two hydrocarbon groups attached to the carbon atom. The hydrocarbon groups are indicated by using an R and an R', as shown in the diagram below.

A student was asked to investigate the boiling point of various ketones. He made the following table and graphed the data shown. He also did a linear regression (best fit line), as shown on the graph below.

Ketone	R Group	R' Group	Molar Mass (g/mol)	Boiling Point (°C)
Dimethyl ketone (acetone)	CH_3	CH_3	58.08	56.05
Ethyl methyl ketone (butanone)	CH_3CH_2	CH_3	72.11	79.64
diethyl ketone (3-pentanone)	CH_3CH_2	CH_3CH_2	86.13	101
Ethyl propyl ketone (3-hexanone)	$CH_3CH_2CH_2$	CH_3CH_2	100.16	123
Dipropyl ketone (4-heptanone)	$CH_3CH_2CH_2$	$CH_3CH_2CH_2$	114.19	143.9
Butyl propyl ketone (4-octanone)	$CH_3CH_2CH_2CH_2$	$CH_3CH_2CH_2$	128.21	

GO ON TO NEXT PAGE

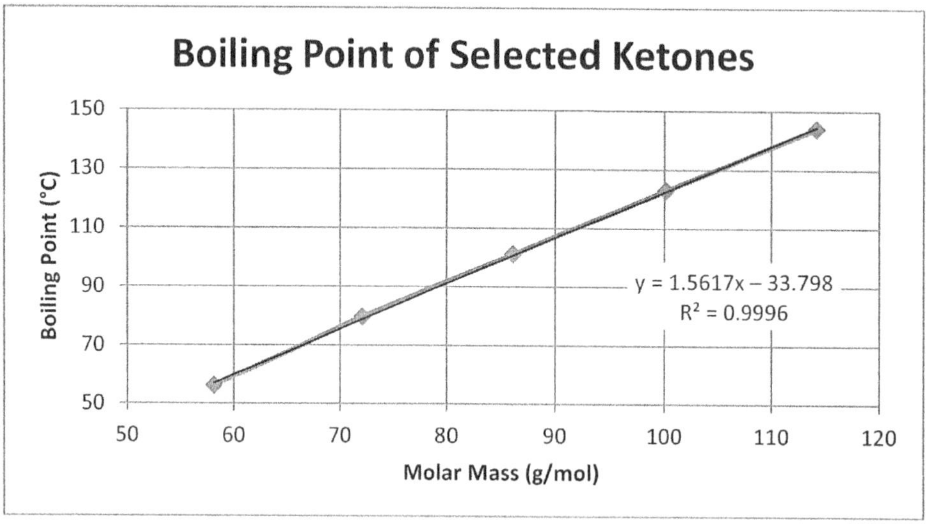

(a) Explain why the boiling point of the ketones increases as the number of carbon atoms in the R and R′ groups increase.

(b) At the R—C(=O)—R′ location, what is the hybridization of the carbon atom? Justify your answer.

(c) Ketones are more soluble in water than other hydrocarbons due to their ability to form hydrogen bonds with water. Draw 1 water molecule in the correct orientation to show acetone's ability to form hydrogen bonds with water. Use a dashed line to show the hydrogen bond.

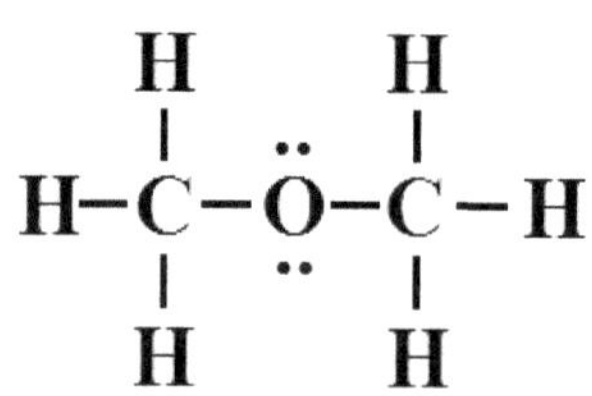

(d) Predict the boiling point of butyl propyl ketone by filling in the missing value in the data table.

7. For the hypothetical reaction

$$A + 3\,B \rightarrow 2\,C + D$$

A student conducts a series of trials to determine the rate law. The experimental data are shown below.

Trials	[A] (mol/L)	[B] (mol/L)	Initial Rate of Appearance of D (mol L^{-1} min^{-1})
1	10.1	2.01	5.00
2	19.8	3.99	80.0
3	40.2	2.00	80.0

(a) Find the order with respect to A. Justify your answer.
(b) Find the order with respect to B. Justify your answer.
(c) Write the rate law.
(d) Determine the value for *k* and specify its units.

END OF EXAMINATION

ANSWERS TO PRACTICE TEST 1

SECTION I: MULTIPLE-CHOICE QUESTIONS

Score your test using the table below.

Determine how many questions you answered correctly. You will find explanations of the answers on the following pages.

1. C	13. D	25. C	37. A	49. B
2. C	14. A	26. D	38. A	50. D
3. A	15. C	27. C	39. D	51. D
4. B	16. C	28. C	40. A	52. B
5. B	17. D	29. D	41. C	53. C
6. D	18. C	30. D	42. B	54. B
7. A	19. B	31. C	43. C	55. B
8. C	20. D	32. B	44. C	56. B
9. D	21. D	33. B	45. D	57. B
10. B	22. C	34. D	46. A	58. D
11. C	23. A	35. D	47. B	59. A
12. C	24. C	36. B	48. B	60. C

CALCULATE YOUR SCORE:

Number answered correctly: ______

WHAT YOUR SCORE MEANS:

Each year, since the test is different, the scoring is a little different. But generally, if you scored 23 or more on the multiple-choice questions, you'll most likely get a 3 or better on the test. If you scored 33 or more, you'll probably score a 4 or better. And if you scored 42 or more, you'll most likely get a 5. Keep in mind that the multiple-choice section is worth 50% of your final grade, and the free-response section is worth 50% of your final grade.

ANSWERS AND EXPLANATIONS

SECTION I: MULTIPLE-CHOICE QUESTIONS

1. ANSWER: C The I_3^- ion must have I as the central atom and two additional I atoms bonded to it. Each iodine atom contributes 7 valence electrons and the charge contributes 1 additional electron for a total of 22. They will form a linear shape, based on the trigonal bipyramidal electronic geometry (LO 2.11).

2. ANSWER: C Because

$$P_{total} = P_{H_2} + P_{H_2O} \Rightarrow 757 = P_{H_2} + 21 \Rightarrow P_{H_2} = 736 \text{ torr}$$

(LO 2.6)

3. ANSWER: A There are four ways to increase the speed of a reaction: (1) add a catalyst, (2) change the temperature, (3) increase concentration, and (4) increase the surface area. The greater the surface area, the greater the number of collisions (LO 4.1).

4. ANSWER: B Ideal gases have particles of zero volume, have no forces between the particles, and are described by the relationship $PV = nRT$. H_2O molecules have hydrogen bonding whereas H_2 molecules only have weak London dispersion forces (LO 2.12).

5. ANSWER: B Ideal gas particles have no attraction for each other, and the particles take up no space. If the temperature is high, the particles move fast enough that there is no attraction between them. If the pressure is very low, the volume of the gas is very high and the gas particles occupy a very small percentage of the volume of the container that the gas occupies (the percentage occupied by the gas particles is negligible compared to the volume of the container that the gas occupies) (LO 2.4).

6. ANSWER: D Specific heat capacity refers to the energy required to change the temperature of 1 mole of the substance (in this case, steam or water) by 1°C (LO 5.6).

7. ANSWER: A $q = mc\Delta T$ and q = (heat of vaporization)(mass) are the two equations that are needed.

$$q = 50.0 \text{ g} \times 20.0°\text{C} \times \frac{2.08 \text{ J}}{\text{g} \times °\text{C}} = 2080 \text{ J released}$$

$$q = 50.0 \text{ g} \times \frac{2260 \text{ J}}{\text{g}} = -113{,}000 \text{ J}$$

$$q_{\text{total}} = -113{,}000 + -2080 = -115{,}080 \text{ J}$$

For the 50.0 × 2260, this may be thought of as 50.0 × (2000 + 200 + 60) = 100,000 + 10,000 + 3000 (LO 5.6).

8. ANSWER: C For a substance to be in the solid or liquid state, the intermolecular forces have to hold the particles close together. When a substance is a gas, the intermolecular forces are not sufficient to hold the particles together (LO 5.9).

9. ANSWER: D $PbBr_2 \rightleftarrows Pb^{2+} + 2\ Br^-$ so the $K_{sp} = [Pb^{2+}][Br^-]^2 \Rightarrow 4.0 \times 10^{-6} = (x)(2x)^2 \Rightarrow 4.0 \times 10^{-6} = 4x^3 \Rightarrow 1.0 \times 10^{-6} = x^3 \Rightarrow (1.0 \times 10^{-6})^{\frac{1}{3}} = x = 1.0 \times 10^{-2}$. Br^- is $2x$ so $[Br^-] = 2.0 \times 10^{-2}$ (LO 6.21 and 3.2).

10. ANSWER: B Because the reaction for the collar is $Zn \rightarrow Zn^{2+} + 2e^-$, it is undergoing oxidation. The oxidation potential for zinc is

+0.76 V while the potentials for oxidation of iron are +0.036 V and 0.44 V. Zinc is preferentially oxidized, preserving the iron. The site of oxidation is known as the anode (LO 3.12).

11. ANSWER: C Both molecules are linear and nonpolar. Each sulfur atom has 16 electrons while each oxygen atom has only 8. This makes CS_2 more polarizable than CO_2, resulting in stronger London dispersion forces and a higher boiling point in CS_2 (LO 2.11).

12. ANSWER: C

$$58.5\text{g } K_2Cr_2O_7 \times \frac{1 \text{ mol } K_2Cr_2O_7}{294 \text{ g } K_2Cr_2O_7} \times \frac{2 \text{ mol Cr}}{1 \text{ mol } K_2Cr_2O_7} \times \frac{52.0 \text{ g Cr}}{1 \text{ mol Cr}} = 20.8 \text{ g Cr}$$

This problem may seem very hard to do without a calculator, but if you estimate and cross cancel numbers, the problem is very easy:

$$60\text{g } K_2Cr_2O_7 \times \frac{1 \text{ mol } K_2Cr_2O_7}{300 \text{ g } K_2Cr_2O_7} \times \frac{2 \text{ mol Cr}}{1 \text{ mol } K_2Cr_2O_7} \times \frac{50 \text{ g Cr}}{1 \text{ mol Cr}} = 20 \text{ g Cr}$$

(LO 1.4)

13. ANSWER: D No calculator is necessary to solve this problem. More than half of the substance is the higher mass isotope, so the weighted average mass should be greater than 121. Weighted average = (0.80 × 122) + (0.20 × 120) = 121.6. To make the problem easier to do, think of 0.80 × 122 as 0.80 × (100 + 20 + 2) = 80 + 16 + 1.6 (LO 1.14).

14. ANSWER: A The 40.08 mg can be written as 40.08×10^{-3} g.

$$40.08 \times 10^{-3} \text{ g Ca} \times \frac{1 \text{ mol Ca}}{40.08 \text{ g Ca}} \times \frac{6.022 \times 10^{23} \text{ atoms Ca}}{1 \text{ mol Ca}} = 6.022 \times 10^{20} \text{ atoms Ca}$$

(LO 1.4)

15. ANSWER: C HCl is a strong acid, so it will ionize 100% and will be written as H^+ and Cl^-, and $MgCl_2$ is soluble since no observation of the production of a solid and will be written as Mg^{2+} and Cl^-. The total ionic equation is $Mg + 2\ H^+ + 2\ Cl^- \rightarrow Mg^{2+} + 2\ Cl^- + H_2$, with the Cl^- as a spectator ion (LO 3.2).

16. ANSWER: C Because the balloons attached to Flasks 1, 2, 3, and 4 inflate more with more Mg and Flasks 4 and 5 remain the same, even though more Mg is added, the Mg in Flasks 1, 2, and 3 is the limiting reagent. In Flask 5, the Mg is the excess reagent. Flask 4 would appear to be the point where the limiting reagent is completely gone, indicating the stoichiometric point. (LO 3.4).

17. ANSWER: D Since the reaction took place, ΔG is negative. Since the balloon expanded, the system did work on the surroundings and the sign of work is negative (LO 5.4).

18. ANSWER: C In order for both enthalpy and entropy to drive a reaction and make ΔG negative, ΔH must be negative and ΔS must

be positive. $\Delta G = \Delta H - T\Delta S$; $\Delta G = (-\ \#) - (+\ \#)$, which will always produce a negative Δ. There was an increase in the entropy since a gas was produced (the balloon filled with gas). The flask got warmer, so the reaction was exothermic, ΔH is negative (LO 5.14).

19. ANSWER: B Since Mg must be the limiting reagent, we need to base our stoichiometry problem on the amount of Mg.

$$1.2 \text{ g Mg} \times \frac{1 \text{ mol Mg}}{24.3 \text{ g Mg}} \times \frac{1 \text{ mol H}_2}{1 \text{ mol Mg}} \times \frac{22.4 \text{ L H}_2}{1 \text{ mol H}_2} =$$

$$0.1 \text{ g Mg} \times \frac{1 \text{ mol Mg}}{2 \text{ g Mg}} \times \frac{1 \text{ mol H}_2}{1 \text{ mol Mg}} \times \frac{22.4 \text{ L H}_2}{1 \text{ mol H}_2} = 0.1 \times 11.2 = 1.12 = 1.1 \text{ L}$$

(LO 3.3)

20. ANSWER: D Since the balloons attached to Flasks 1, 2, 3, and 4 inflate more with more Mg, the Mg in Flasks 1, 2, 3, and 4 is the limiting reagent. In the reaction, the Mg turns into Mg^{2+}, so the Mg must be losing electrons and is oxidized (LO 3.8 and 3.4).

21. ANSWER: D The volume of the balloons over Flasks 4 and 5 remains the same, even though more Mg is added. The Mg in Flasks 4 and 5 must be the excess reagent, so the HCl is the limiting reagent. Flasks 4 and 5 have the exact same amount of limiting reagent, Mg (which determines the amount of product produced), and thus will have the same amount of heat energy produced and the same particle energy diagram (LO 5.2).

22. ANSWER: C The CO_3^{2-} reacts with H_2O as: $CO_3^{2-} + H_2O \rightarrow HCO_3^- + OH^-$, which gives a basic solution. You can determine whether salts are acidic, basic, or neutral by determining the parent acid and parent base. For salt Na_2CO_3, the parent base NaOH is strong and the parent acid is weak H_2CO_3, so the salt is basic. Remember that the strong one always wins, SB & WA, pH will be more than 7. *This is a memory aid and does not explain the chemistry of why this occurs* (LO 6.16).

23. ANSWER: A The temperature of the surroundings decreases because heat energy transfers from the surroundings to the system. The system is the reaction, which requires heat, so it must be endothermic (LO 5.7).

24. ANSWER: C The only ion present that will react will be the acetate ion, which will undergo a hydrolysis reaction with water.

	$C_2H_3O_2^-$	+	H_2O	$\rightleftarrows$	$HC_2H_3O_2$	+	OH^-
I	0.10		---		0		0
C	$-x$		---		$+x$		$+x$
E	$0.10 - x \approx 0.10$		---		x		x

The solution will be basic (therefore, H^+ and OH^- cannot be equal); the $[C_2H_3O_2^-] \approx 0.10$ and $[HC_2H_3O_2] = [OH^-] = x$ and $x < 0.10\ M$ (LO 6.6).

25. ANSWER: C In Experiments 1 and 2, [B] is constant and the [A] changed by a factor of 2. The initial rate went up by a factor of 2. Since $2^1 = 2$, A is first order. In Experiments 1 and 3, [A] was constant, [B] goes up by 2, and the initial rate remains the same. B is zero order. The rate law is Rate = k[A] (LO 4.1).

26. ANSWER: D

$$\frac{1.4 \times 10^{-5} \text{ mol A}}{\text{L}} \times \frac{2 \text{ mol D}}{1 \text{ mol A}} = \frac{2.8 \times 10^{-5} \text{ mol D}}{\text{L}}$$

(LO 3.3 and 4.1)

27. ANSWER: C As the number of protons increases, there is an increase in the attraction from the nucleus, holding the valence electrons more tightly and pulling them closer to the nucleus (LO 1.9).

28. ANSWER: C As you go to the right, the number of protons increases, so the attraction between nucleus and valence electrons increases, which would reduce the radius. However, the number of electrons also increases, which increases the amount of electron to electron repulsion, which increases the radius. Neon has the most electron to electron repulsion of any element in Period 2. The data support that the electron to electron repulsion is more important in neon than the increase of the proton to valence electron attaching (LO 1.9).

29. ANSWER: D The greater the number of moles of ions:
(A) 0.500 L × 0.100 M × 4 ions in $FeCl_3$ = 0.20 mol ions
(B) 0.700 L × 0.200 M × 2 ions in NaOH = 0.28 mol ions
(C) 0.400 L × 0.100 M × 4 ions in $Al(NO_3)_3$ = 0.16 mol ions
(D) 0.600 L × 0.200 M × 4 ions in $AlCl_3$ = 0.48 mol ions
(LO 1.4)

30. ANSWER: D Both C_4H_8 and C_8H_{18} are nonpolar; therefore, they only have London dispersion forces as intermolecular forces. The more electrons and the larger the electron cloud, the more polarizable the molecule is, the stronger the intermolecular forces, and the higher the boiling point (LO 2.11).

31. ANSWER: C While you do not have a calculator, this problem is very easy. Percent of nitrogen is

$$\frac{14.0 \text{ g N}}{49 \text{ g total mass}} \cong \frac{28}{100} = 28\%$$

(LO 1.2)

32. ANSWER: B Three of the choices are endothermic. Going from solid to gas phase or liquid to gas phase requires an input of energy, as does bond breaking. When water freezes (goes to a lower energy state), energy is released (LO 5.8 and 5.9).

33. ANSWER: B First, determine the density of the pennies in the jar. Mass of pennies = 8961 – 550 = 8411 g. Density = 8411 g/1000 mL = 8.411 g/mL. One method to solve this type of problem is to create a chart that puts both densities on the chart and then fills in five key points. The key points are assuming that one of the items constitutes 100% of the sample (and therefore, the other one is 0%), 75–25% split, 50–50% split, 25–75%, and finally 0–100% split.

Post-1982 Density = 7.2	100%	75%	50%	25%	0%
Pre-1982 Density = 8.8	0%	25%	50%	75%	100%
Final average	7.2	7.6	8.0	8.4	8.8

The density that was calculated is 8.4 g/mL, and the chart matches that up with a 25%–75% split (900/1200). Note that this chart technique works for average atomic mass problems also (LO 1.3).

34. ANSWER: D The increase in temperature causes the K_w value to increase as the ionization of water is an endothermic process. $K_w = [H^+][OH^-]$ If the K_w value increases, then the $[H^+]$ and $[OH^-]$ must increase. As the $[H^+]$ increases, the pH will decrease (LO 6.14).

35. ANSWER: D The K value for Reaction 1 is < 1 and since $\Delta G = -RT\ln K$ and the $\ln K$ is negative, ΔG must be positive (LO 6.25).

36. ANSWER: B

Reaction 1: $2\,NOCl(g) \rightleftarrows$ ~~$2\,NO(g)$~~ $+ Cl_2(g)$ $K = 1.6 \times 10^{-5}$

Reaction 2: ~~$2\,NO(g)$~~ $\rightleftarrows N_2(g) + O_2(g)$ $K = 1.0 \times 10^{31}$

Overall: $2\,NOCl(g) \rightleftarrows N_2(g) + Cl_2(g) + O_2(g)$

$$K = K_1 \times K_2 = (1.6 \times 10^{-5}) \times (1.0 \times 10^{31}) = 1.6 \times 10^{(-5\,+\,31)}$$

(LO 5.17)

37. ANSWER: A The Lewis structure for NOCl is

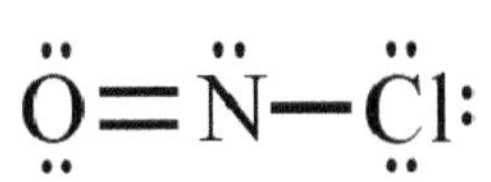

The first bond between any 2 atoms is always a sigma bond, and the second or third bond is a pi bond. There are three areas of electron density (i.e., multiple bonds count as a single bond for geometry and hybridization) that will result in sp^2 hybridization (LO 2.21).

38. ANSWER: A The reaction goes from 2 moles of gas reactants to 3 moles of gas products; hence, an increase in entropy, so ΔS is positive. Since $K < 1$, the reactants are favored and ΔG must be positive. Since $\Delta G = \Delta H - T\Delta S$, (+ #) = ($\Delta H$) – (+ #), so ΔH must always be positive (LO 5.13).

39. ANSWER: D The reaction goes from 2 moles of gas reactants to 2 moles of gas products; hence, the value for ΔS will be close to zero.

Since $K > 1$, the products are favored and ΔG must be negative. Since $\Delta G = \Delta H - T\Delta S$, $(-\ \#) = (\Delta H) -$ (very small number), so ΔH must be negative (LO 6.25 and 5.13).

40. ANSWER: A The formation of a bond always releases energy (LO 5.1).

41. ANSWER: C If the ln[ ] versus time graph is a straight line, then it would indicate first-order kinetics. If the [ ] versus time graph is a straight line, then it would indicate zero order (LO 4.2).

42. ANSWER: B

	PCl_5	$\rightleftarrows$	PCl_3	+	Cl_2
I	6.0		0		0
C	$-x$		$+x$		$+x$
E	$6.0 - x$		x		x

$$P_{total} = P_{PCl_5} + P_{PCl_3} + P_{Cl_2} \Rightarrow 10.0 = 6 - x + x + x \Rightarrow x = 4$$

$P_{PCl_5} = 2 \qquad P_{PCl_3} = 4 \qquad P_{Cl_2} = 4$

$$K_p = \frac{(P_{PCl_3})(P_{Cl_2})}{P_{PCl_5}} \Rightarrow K_p = \frac{(4)(4)}{2} = 8$$

(LO 6.5)

43. ANSWER: C As the temperature is reduced, the reaction will shift to increase temperature. Exothermic reactions increase temperature, so the reaction will shift to the left. Since the reaction is shifting from 2 moles of gas to 1 mole of gas, the total pressure will decrease (LO 6.8).

44. ANSWER: C First, draw the Lewis structures for the 3 molecules. PCl_5 and Cl_2 are nonpolar and will exhibit only London dispersion forces. PCl_3 is polar and will exhibit dipole to dipole interactions (LO 2.13).

45. ANSWER: D If 40% decomposed, then the amount of PCl_3 and Cl_2 present must represent that 40%. In A and C, the PCl_3 and Cl_2 are both 4 and 40% of 10 = 4. There must have been 10 PCl_5 present initially that have decomposed, forming 6 moles of PCl_5, not the 10 that are there. In B and D, the amount of PCl_3 and Cl_2 present is 2; 40% of 5 = 2. There must have been 5 PCl_5 present initially that have decomposed, forming 3 moles of PCl_5.

$$K_p = \frac{(P_{PCl_3})(P_{Cl_2})}{P_{PCl_5}} \Rightarrow K_p = \frac{(2)(2)}{3} = 1.33$$

(LO 6.5)

46. ANSWER: A Since the elements in Group 15 are in the order of P, As, and Sb, Sb should be the largest atom in the set and would have the longest bond lengths. Also, in PCl_5, the axial bond length was longer than the equatorial bond length. This should be true in the other compounds, as well (LO 1.11).

47. ANSWER: B The K_a or the K_b can be found at the half-equivalence point where pH = pK_a or pOH = pK_b (LO 6.13).

48. ANSWER: B The fact that a small amount of base did not change the pH indicates that you are in the buffer region of the curve (LO 6.20).

49. ANSWER: B The diagram shows 4 X^- and 2 HX, which indicates that point I started with 6 X^- so the half-equivalence point has not yet been reached (LO 6.13 and 6.19).

50. ANSWER: D The diagram indicates that excess H_3O^+ is present and no X is present. The titration therefore must be substantially past the equivalence point. Note that point F cannot be the answer as the volume of additional acid added past the equivalence point is very small (LO 6.13 and 3.1).

51. ANSWER: D The tetrahedral structure indicates that there must be four areas of electron density (LO 3.1 and 2.21).

52. ANSWER: B

	HBr	+	NaOH	→	NaBr	+	HOH
B	0.150 mol		0.0500 mol		0		0
C	– 0.0500		– 0.0500		+ 0.0500		+ 0.0500
A	0.100 mol		0		0.0500		0.0500

The total volume is 100.00 mL, so the molarity of the HBr is 1.00 *M*. The NaBr does not affect the pH in this reaction. Since HBr is a strong acid, it will ionize 100% and the $[H^+]$ = 1.00 *M*. pH = $-\log[H^+]$, pH = $-\log[H^+]$ = 0 (LO 3.4 and 6.11).

53. ANSWER: C Using the BCA table from problem 52, the $[H_3O^+]$ = 1.00 *M*, $[Br^-]$ = 1.50 *M*, $[Na^+]$ = 0.500 *M*, and the concentration of $[OH^-]$ will be virtually zero with such a high concentration of H_3O^+ (LO 3.4).

54. ANSWER: B Since the water is located in the denominator, it is acting as an inhibitor in the process. Since it is a product, water would indicate that it is interfering with the collisions between the reactants (LO 4.4 and 4.7).

55. ANSWER: B Pressure is force/area. The pressure of a gas inside a container is determined by the force with which the particles hit the walls of the container (LO 2.4).

56. ANSWER: B The volume of the container was reduced by half, which would result in the pressure doubling since volume and pressure are inversely related. The one diagram that indicates that two particles are colliding with the walls of the container is B (LO 2.6).

57. ANSWER: B Methanol has an –OH group, which can form hydrogen bonds with water. Hexane is nonpolar and has London dispersion forces and can only form weaker dipole-induced dipole interactions with the methanol (LO 2.15 and 2.16).

58. ANSWER: D Keep in mind that the pH scale is a log scale based on 10; pH = –log [H$^+$]. X/Y = $10^{-7}/10^{-9} = 10^2/1$ = 100/1 (LO 6.12).

59. ANSWER: A Average speeds increase as the molecular masses decrease (LO 2.4).

60. ANSWER: C Ar is the largest atom, but the reason it deviates from ideal gas behavior is its larger number of electrons, which makes it more polarizable (LO 2.12).

SECTION II: FREE-RESPONSE QUESTIONS

Question 1: Answers

(a) (i) $Mg(s) + 2\,H^+(aq) \rightarrow Mg^{2+}(aq) + H_2(g)$ (LO 3.2)

(ii) $MgO(s) + 2\,H^+(aq) \rightarrow Mg^{2+}(aq) + H_2O(l)$ (LO 3.2)

(b) Use $q = mc\Delta T$ to calculate the heat gained by the surroundings.

Mass of solution = 50.0 + 0.121 = 50.121 g = 50.1 g

$$q = 50.1\text{ g} \times \frac{4.184\text{ J}}{\text{g°C}} \times 10.7\text{°C} = 2240\text{ J}$$

The energy absorbed by the surroundings must be equal and opposite in sign to the energy released by the reaction. Convert Mg mass to Mg moles and divide the energy released by the moles of Mg.

$$\text{Moles Mg} = 0.121\text{ g Mg} \times \frac{1\text{ mol Mg}}{24.30\text{ g Mg}} = 0.00498\text{ mol Mg}$$

$$\Delta H_1 = \frac{-2240\text{ J}}{0.00498\text{ mol Mg}} \times \frac{1\text{ kJ}}{1000\text{ J}} = -4.50 \times 10^2\text{kJ/mol}_{rxn}$$

(LO 5.7)

(c) Use $q = mc\Delta T$ to calculate the heat gained by the surroundings.

Mass of solution = 50.0 + 0.348 = 50.348 g = 50.3 g

$$q = 50.3\text{ g} \times \frac{4.184\text{ J}}{\text{g°C}} \times 5.1\text{°C} = 1100\text{ J}$$

The energy absorbed by the surroundings must be equal and opposite in sign to the energy released by the reaction. Convert MgO mass to MgO mol and divide the energy released by the moles of MgO.

$$\text{Moles MgO} = 0.348 \text{ g MgO} \times \frac{1 \text{ mol MgO}}{40.30 \text{ g MgO}} = 0.00864 \text{ mol MgO}$$

$$\Delta H_2 = \frac{-1100 \text{ J}}{0.00864 \text{ mol Mg}} \times \frac{1 \text{ kJ}}{1000 \text{ J}} = -1.27 \times 10^2 \text{kJ/mol}_{rxn}$$

(LO 5.7)

(d) Mg + ~~$2\,HCl$~~ $\rightarrow$ ~~$MgCl_2$~~ + ~~H_2~~ $\qquad \Delta H_1 = -4.50 \times 10^2 \text{ kJ/mol}_{rxn}$

~~$MgCl_2$~~ + ~~H_2O~~ $\rightarrow MgO$ + ~~$2\,HCl$~~ $\qquad \Delta H_2 = 1.27 \times 10^2 \text{ kJ/mol}_{rxn}$

~~H_2~~ + $\tfrac{1}{2}\,O_2 \rightarrow$ ~~H_2O~~ $\qquad \Delta H_3 = -2.858 \times 10^2 \text{ kJ/mol}_{rxn}$

$Mg + \tfrac{1}{2}\,O_2 \rightarrow MgO$

Note: The second reaction is reversed so the sign on ΔH_2 is reversed (LO 5.6).

(e) Summing the equations and enthalpies from (d) gives

$$\Delta H_f = (-4.50 \times 10^2) + (1.27 \times 10^2) + (-2.858 \times 10^2) = -6.09 \times 10^2 \text{ kJ/mol}_{rxn}$$

(LO 5.6)

(f) % error = [(lab value – theoretical value) ÷ theoretical value] × 100
% error = $[(-6.09 \times 10^2 \text{ kJ mol}^{-1}) - (-6.02 \times 10^2 \text{ kJ mol}^{-1})/-6.02 \times 10^2 \text{ kJ mol}^{-1}] \times 100$
= –1.2% error (note 2 significant figures from the subtraction) (LO 5.7).

(g) Including the calorimeter constant in the calculations would account for any heat gained or lost by the apparatus itself (the calorimeter includes the thermometer as well as the coffee cup calorimeter) and therefore provides a more accurate value for ΔH (LO 5.7).

2. (a)

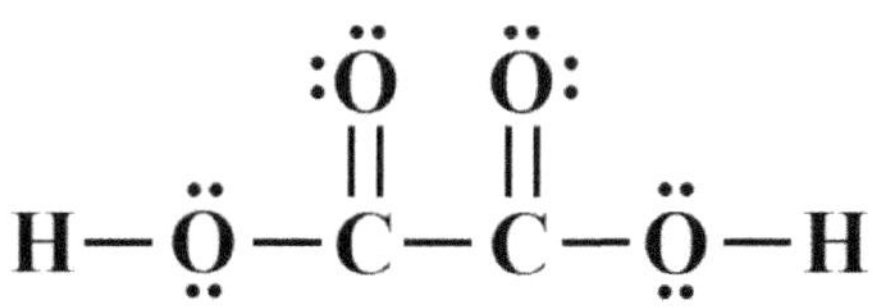

(LO 2.21)

(b) No, they do not lie in the same plane. The C–O–H bonds at the end are sp^3 hybridized (they have four centers of electron density) (LO 2.21).

(c) 7 sigma bonds and 2 pi bonds (LO 2.21)

(d) (i) You need to find the percent ionization and see if it is less than 5%.

$$\frac{0.12}{0.25} \times 100 = 48\% \text{ ionization}$$

Since the ionization is more than 5%, you cannot ignore the "x" on the $0.25 - x$ for the ionization of the oxalic acid and must solve this problem using the quadratic formula. However, on this AP® exam, if you are asked to solve a similar problem, then you will be able to ignore the "x," as it will be negligible (LO 6.6).

(ii) When the K_a is very large for a weak acid and the concentration is not extremely dilute, the pH can be slightly less than 1 (LO 2.2).

(e) $$60.0 \text{ g parsley} \times \frac{1.70 \text{ g } H_2C_2O_4}{100. \text{ g parsley}} \times \frac{1 \text{ mol } H_2C_2O_4}{90.04 \text{ g } H_2C_2O_4} \times \frac{1}{0.10000 \text{ L}} = 0.113M$$

(LO 1.4)

(f) $1.09\ \text{cm}^3 \times \dfrac{2.12\ \text{g}\ H_2C_2O_4}{1\ \text{cm}^3} \times \dfrac{1\ \text{mol}\ H_2C_2O_4}{128.10\ \text{g}\ H_2C_2O_4} = 0.0180\ \text{mol}\ H_2C_2O_4$

$$0.0180\ \text{mol}\ H_2C_2O_4 \times \frac{1\ \text{mol}\ C_2O_4^{2-}}{1\ \text{mol}\ H_2C_2O_4} \times \frac{6.022 \times 10^{23}\ \text{ions}\ C_2O_4^{2-}}{1\ \text{mol}\ C_2O_4^{2-}} = 1.08 \times 10^{22}\ \text{ions}\ C_2O_4^{2-}$$

(LO 1.4)

(g) Since K_{eq} is very large, the reaction favors the products and effectively goes to completion. A large amount of oxalic acid can remove a large amount of copper, in the form of the complex ion; therefore, the copper is stored in the plant tissues and not in the soil (LO 6.7).

3. (a) $3\ Sn^{4+} + 2\ Al(s) \rightarrow 2\ Al^{3+} + 3\ Sn^{2+}$
(LO 3.2)
(b)

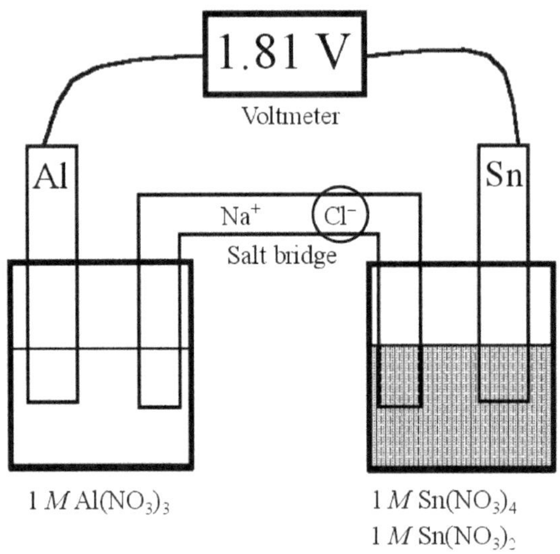

(LO 3.12)
(c) Al (LO 3.13)
(d) The mass remains unchanged. Unlike other batteries, the half-reaction involving tin is $Sn^{4+} + 2\ e^- \rightarrow Sn^{2+}$. No metal is involved in this half-reaction, so the electrode's mass remains unchanged (LO 3.12).
(e) See (b) above. Cl^- should be circled (LO 3.13).
(f) From left to right, electrons flow from the anode to the cathode (remember FAT CAT) (LO 3.13).

(g) $\Delta G° = -nFE° \Rightarrow \Delta G° = -(6\ \text{mol}\ e^-)\left(\dfrac{96485\text{C}}{\text{mol}\ e^-}\right)(1.81\ \text{V}) \Rightarrow$

$\Delta G° = -1{,}047{,}827\ \text{J} = -1.05 \times 10^3\ \text{kJ/mol}_{rxn}$

(LO 5.14)

(h) We must first find $\Delta H°_{rxn}$ using

$$\Delta H°_{rxn} = \Sigma\Delta H°_{f_{products}} - \Sigma\Delta H°_{f_{reactants}}$$

$$\Delta H°_{rxn} = [(3)(-8.8) + (2)(-531.37)] - [(2)(0) + (3)(30.5)]$$

$$\Delta H°_{rxn} = -1180.6 \text{ kJ/mol}_{rxn}$$

Use $\Delta G° = \Delta H° - T\Delta S°$ to solve for $\Delta S°$. Remember that the temperature is 298.15 K since this is a standard cell.

$$\Delta S° = \frac{\Delta H° - \Delta G°}{T} \Rightarrow \Delta S° = \frac{-1{,}180{,}600 \; - \; -1{,}050{,}000 \text{ J/mol}}{298.15 \text{ K}} = \frac{-438 \text{ J}}{\text{mol} \times \text{K}}$$

(LO 5.13)

(i) The correct graph is D. Since the value of ΔH was found to be negative, this is an exothermic reaction and formation of products will be less favorable at a high temperature. The value of ΔG is negative and will increase (become less negative or even positive) at a higher temperature (LO 6.9).

4. (a) If you assume a 100.0 g sample, then you will have 10.06 g C, 0.846 g H, and 89.09 g Cl.

$$10.06 \text{ g C} \times \frac{1 \text{ mol C}}{12.01 \text{ g C}} = 0.8376 \text{ mol C} \qquad \frac{0.8376 \text{ mol C}}{0.8376} = 1 \text{ mol C}$$

$$0.846 \text{ g H} \times \frac{1 \text{ mol H}}{1.008 \text{ g H}} = 0.839 \text{ mol H} \qquad \frac{0.839 \text{ mol H}}{0.8376} = 1 \text{ mol H}$$

$$89.09 \text{ g Cl} \times \frac{1 \text{ mol Cl}}{35.45 \text{ g Cl}} = 2.513 \text{ mol Cl} \qquad \frac{2.513 \text{ mol Cl}}{0.8376} = 3.00 \text{ mol Cl}$$

Empirical formula is $CHCl_3$ (LO 1.2).

(b) There are several reasons why the C–F bond is very strong. F is a very small atom, so the length of the C–F is one of the shortest bonds in organic chemistry. Another reason is that fluorine has the highest electronegativity of any element (F = 4.0 and C = 2.5). This results in a very polar bond, which has a very high ionic character and introduces partial charges to the bond ($^{\delta-}$F–C$^{\delta+}$) and those partial charges attract each other and increase the strength of the bond (LO 2.18).

(c)

$$\frac{4.45 \times 10^{-3} \text{ mol } NO_2}{0.250 \text{ L}} = 0.0178\,M \qquad \frac{1.15 \text{ mol } N_2O_4}{0.250 \text{ L}} = 4.60\,M$$

$$K_c = \frac{[NO_2]^2}{[N_2O_4]} = \frac{(0.0178)^2}{4.60} = 6.89 \times 10^{-5}$$

(LO 6.5)

5. (a) Across the third period from Na to Ar, the number of protons is increasing in the nucleus while additional electrons are being added to the same energy level. Hence, the effective nuclear charge is increasing with a consequently stronger attraction for the valence electrons and an increase in ionization energy (LO 1.6).
   (b) The electron ionized in the case of Mg is a 3*s* electron, whereas in the case of Al, it is a 3*p* electron. 3*s* electrons penetrate closer to the nucleus than do 3*p* electrons (3*p* electrons are higher in energy than 3*s* electrons); hence, 3*p* electrons have less attraction for the nucleus and the first electron added to 3*p* is easier to ionize (LO 1.7).
   (c) The electron ionized in sulfur is paired with another electron in the same orbital, whereas in phosphorus the electron comes from a singly occupied orbital. The ionization energy of the sulfur electron is less because of the repulsion between the 2 electrons in the same orbital. *Note:* Yes, sulfur has a stronger Z_{eff} than phosphorus does, but it is not enough to overcome the increase in the electron to electron repulsion. These questions are data driven, and the data show that it is easier to ionize an electron in S (LO 1.5).
   (d)

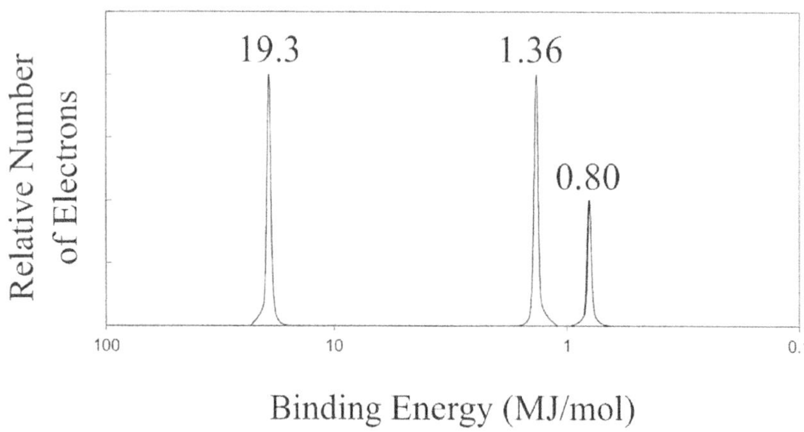

(LO 1.8)

6. (a) As the R groups get larger, there are more polarizable electrons, so the intermolecular forces (London dispersion forces) increase in the R and R′, resulting in a higher attraction between the particles and an increasing boiling point (LO 2.11).
   (b) sp^2 (LO 2.21)
   (c)

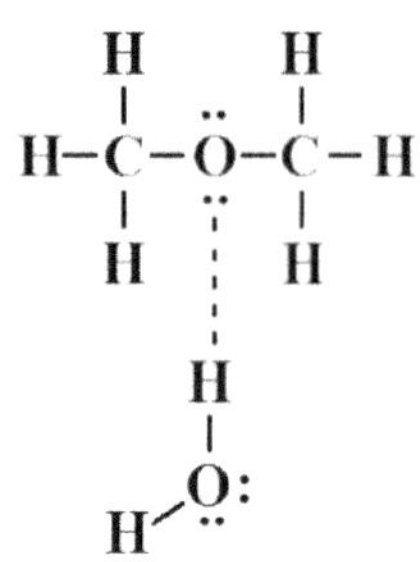

(LO 2.8)

(d) $y = 1.5617x - 33.798$
$y = 1.5617(128.21) - 33.798$
$y = 166.4°C$
(LO 2.11)

7. (a) In Trials 1 and 3, [B] is constant and [A] goes from 10.1 to 40.2, up by a factor of 4. The initial rate goes from 5.00 to 80.0, up by a factor of 16. Since $4^2 = 16$, A is second order (LO 4.1).
(b) In Trials 1 and 2, [A] goes from 10.1 to 19.8, an increase by a factor of 2, which should make the initial rate increase by a factor of 2^2 (since A is second order), becoming $5 \times 2^2 = 20$. However, [B] goes from 2.01 to 3.99, also an increase of 2. The initial rate is 80.0, which is 4 times larger than the rate from changing only A. Since $2^2 = 4$, B must be second order. Another way to look at this: Trials 2 and 3 have the same rate. From trial 2 to trial 3, [A] is roughly doubled while [B] is cut in half. This indicates that the orders of the two reactants must be the same. Since A is second order, B must also be second order.

An alternative approach would be:

$$\frac{Rate_2}{Rate_1} = \frac{k\,[A]^2[B]^x}{k\,[A]^2[B]^x} \Rightarrow \frac{80.0}{5.00} = \frac{\cancel{k}\,[19.8]^2[3.99]^x}{\cancel{k}\,[10.1]^2[2.01]^x}$$

$$16.0 = \frac{(392.04)(3.99)^x}{(102.01)(2.01)^x} \Rightarrow 4.16 = 1.99^x$$

$x = 2.08 \cong 2$

(LO 4.1)
(c) Rate = $k[A]^2[B]^2$
(LO 4.2)
(d) Using Trial 1: rate = $k[A]^2[B]^2$ so

$$k = \frac{rate}{[A]^2[B]^2} \Rightarrow k = \frac{\frac{5.00\text{ mol}}{\text{L} \times \text{min}}}{\left(\frac{10.1\text{ mol}}{\text{L}}\right)^2\left(\frac{2.01\text{ mol}}{\text{L}}\right)^2} = 0.0121\,\frac{\text{L}^3}{\text{mol}^3\text{ min}}$$

Note: The value for k depends on which of the three trials you use to solve for the rate constant, so $k = 0.0121, 0.0124$, or $0.0128\,\frac{\text{L}^3}{\text{mol}^3 \times \text{min}}$ or $M^{-3}\text{ min}^{-1}$
(LO 4.2)

SCORING THE FREE-RESPONSE QUESTIONS

It is difficult to come up with an exact score for this section of the test. However, if you compare your answers to the answers in this book, remembering that each part of the test you answer correctly is worth points even if the other parts of the answer are incorrect, you can get a general idea of the percentage of the questions for which you would get credit. If you believe that you got at least 40 percent of the possible credit, you would probably receive a 3 on this part of the test. If you believe that you would receive close to half or more of the available credit, your score would more likely be a 4 or a 5.

Practice Test 2

These questions are representative of the AP® Chemistry examination, but keep in mind that it is impossible to predict exactly how well you will do on the actual exam. The first section of this test is 50% of your total test grade. Time yourself so that you finish this part in 90 minutes. Remember that on the actual examination, there is no penalty for guessing, so you should answer every question.

AP® CHEMISTRY EXAMINATION
Section I: Multiple-Choice Questions
Time: 90 minutes
Number of Questions: 60

DIRECTIONS: Each of the questions or incomplete statements below is followed by four suggested answers or completions. Select the one that is best in each case. **No calculators can be used in this section.** A periodic table and a formula chart with constants are provided at the back of this book. Assume that the conditions are a temperature of 298 K and pressure of 1.0 atm with all solutions aqueous unless otherwise specified.

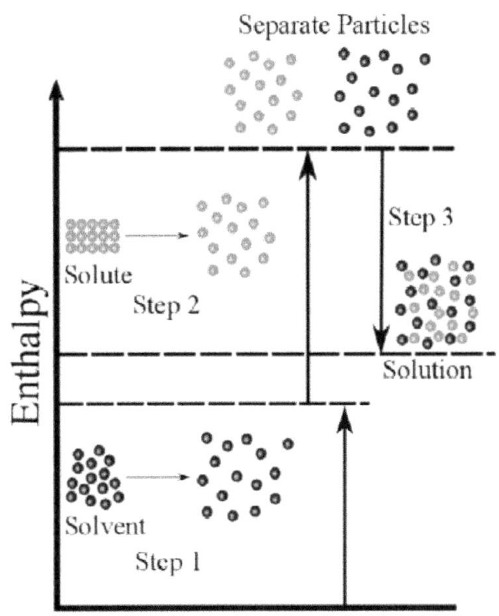

1. Why is the enthalpy change for step 2 greater than for step 1?
   (A) The ΔS is greater for solid solute particles being separated than for liquid solvent particles being separated.
   (B) The strength of the interparticle forces in the solute is greater than the strength of the interparticle forces in the solvent.
   (C) The solvent is closer to the enthalpy of the solution than the solute is.
   (D) The sign for the change in enthalpy of the solute is the same as the sign of the change in enthalpy for the overall reaction.

GO ON TO NEXT PAGE

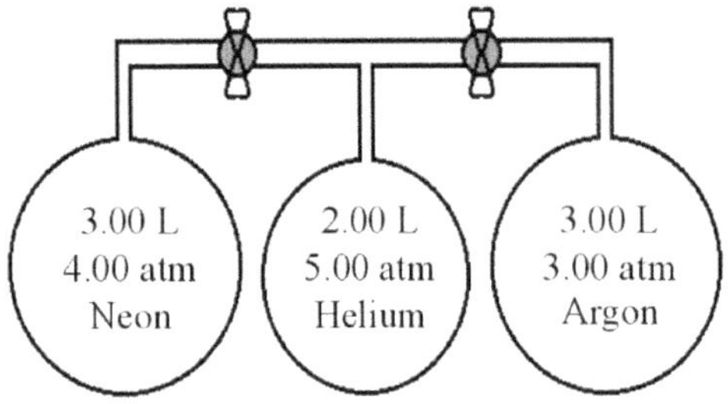

2. With the stopcocks closed, the three bulbs in the apparatus shown above are filled with the quantities indicated of argon, helium, and neon, respectively. Assume the gases behave ideally and that the volumes given include the volumes of the connecting tubes. What is the partial pressure of helium inside the apparatus when the taps are opened?
(A) 1.00 atm
(B) 1.25 atm
(C) 5.00 atm
(D) 20.0 atm

3. Which of the following equations would be the most appropriate equation to use to describe the oxidation–reduction reaction occurring in acidic solution between H_2O_2 and $KMnO_4$?
(A) $MnO_4^-(aq) + 2\,H_2O_2(aq) \rightarrow Mn^{2+}(aq) + 3\,O_2(g) + 2\,H_2O(l)$
(B) $2\,MnO_4^-(aq) + 5\,H_2O_2(aq) + 6\,H^+(aq) \rightarrow 2\,Mn^{2+}(aq) + 5\,O_2(g) + 8\,H_2O(l)$
(C) $2\,KMnO_4(aq) + 5\,H_2O_2(aq) + 3\,H_2SO_4(aq) \rightarrow 2\,MnSO_4(aq) + 5\,O_2(g) + 8\,H_2O(l) + K_2SO_4(aq)$
(D) $2\,MnO_4^-(aq) + 5\,H_2O_2(aq) \rightarrow 2\,Mn^{2+}(aq) + 5\,O_2(g) + 2\,H_2O(l) + 6\,OH^-(aq)$

Questions 4 and 5 refer to the following information:

Ammonia may be reacted with nitrogen monoxide to produce nitrogen gas and water vapor, which is described by the following equation:

$$4\,NH_3(g) + 6\,NO(g) \leftrightarrows 5\,N_2(g) + 6\,H_2O(g)$$

4. In one particular experiment, equal molar amounts of the reactants were mixed and concentrations of the reactants and products were plotted against time. The graph below was obtained. Identify the components A, B, C, and D.

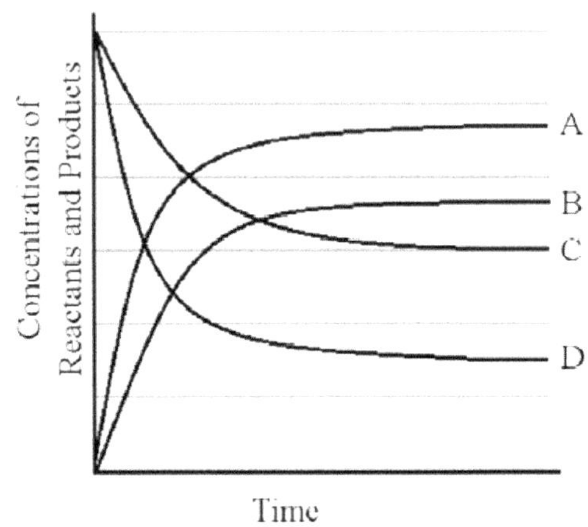

	A	B	C	D
(A)	NH_3	NO	N_2	H_2O
(B)	H_2O	NO	N_2	NH_3
(C)	NH_3	N_2	NO	H_2O
(D)	H_2O	N_2	NH_3	NO

5. At one stage in the reaction, nitrogen is produced at a rate of 30. mol $L^{-1}\,s^{-1}$. At what rate is nitrogen monoxide used up at this stage?
(A) 6.0 mol $L^{-1}\,s^{-1}$
(B) 24 mol $L^{-1}\,s^{-1}$
(C) 30. mol $L^{-1}\,s^{-1}$
(D) 36 mol $L^{-1}\,s^{-1}$

Questions 6 and 7 refer to the following reaction:

$$3\ F_2(g) + Cl_2(g) \rightleftarrows 2\ ClF_3(g)$$

6. When the system is at equilibrium at 298 K, $[F_2] = 2.0\ M$; $[Cl_2] = 2.5\ M$; $[ClF_3] = 3.0\ M$. Calculate the value of K_c, the equilibrium constant.
(A) 0.15
(B) 0.40
(C) 0.45
(D) 0.60

7. The value for $\Delta H_f°$ of $ClF_3(g)$ is -163.2 kJ/mol$_{rxn}$. If the system's volume is decreased at constant temperature, in which direction, if any, will the system shift to restore equilibrium, and what will happen to the value of K?
(A) The reaction will not shift, and K remains constant.
(B) The reaction shifts toward the products, increasing the value of K.
(C) The reaction shifts toward the reactants, decreasing the value of K.
(D) The reaction shifts toward the products, and K remains constant.

8. What is the molar solubility of cadmium sulfide, CdS? (K_{sp} of CdS = 3.6×10^{-29})
(A) $1.8 \times 10^{-16}\ M$
(B) $3.6 \times 10^{-16}\ M$
(C) $6.0 \times 10^{-15}\ M$
(D) $6.0 \times 10^{-14}\ M$

9. Below are representations of energy levels (drawn to scale) in the Bohr model of the hydrogen atom. Which electron making a transition, denoted by the arrows, results in the emission of a photon of shortest wavelength?

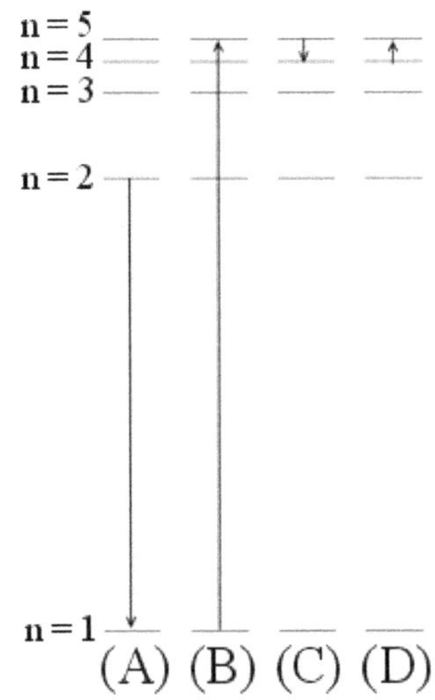

10. Sodium and magnesium are both found in the third period of the periodic table. Which of these elements has the larger atomic radius, and what best explains the difference in their atomic radii?
(A) Na has the larger radius. The nuclear charge of Na is smaller than that of Mg, providing lower attraction for the electrons and allowing them to spread out.
(B) Na has the larger radius. Atomic size decreases going across the periodic table from left to right.
(C) Mg has the larger radius. The nuclear charge of Mg is larger than that of Na, repelling the electrons and making the atom larger.
(D) Mg has the larger radius. Atomic size increases going across the periodic table from left to right.

GO ON TO NEXT PAGE

Element	First Ionization Energy (kJ/mol)	Second Ionization Energy (kJ/mol)	Third Ionization Energy (kJ/mol)
W	495	4560	6910
X	735	1445	7730
Y	1060	1890	2905
Z	1255	2295	3850

11. Which element represented in the table above would most easily form an ion with a 2+ charge?
(A) W
(B) X
(C) Y
(D) Z

12. Compared to phosphine, PH_3, ammonia, NH_3, would be expected to have
(A) a lower vapor pressure.
(B) a lower boiling point.
(C) a lower melting point.
(D) lower solubility in water.

$$Zn(s) + 2\ HCl(aq) \rightarrow ZnCl_2(aq) + H_2(g)$$

13. A student investigates factors affecting the rate of the reaction shown above by collecting and measuring the hydrogen gas produced through water displacement. If 1.0 *M* HCl solution is reacted with 5.0 g Zn for a period of 3.0 minutes, which combination of zinc particle size and temperature would be expected to produce the highest volume of hydrogen? In all cases, some Zn(*s*) and HCl(*aq*) remained after 3.0 minutes.
(A) zinc pieces with 0.5 cm average size at 25°C
(B) zinc pieces with 0.05 cm average size at 25°C
(C) zinc pieces with 0.5 cm average size at 50°C
(D) zinc pieces with 0.05 cm average size at 50°C

14. Various types of spectroscopy can investigate properties of matter such as electron arrangement in atoms, types of bonds and functional groups, intensity of colored solutions, and masses and relative abundance of isotopes. What type of spectroscopy can best be used to determine whether the carbon atoms in an organic molecule are sp, sp^2, or sp^3 hybridized?
(A) visual light spectroscopy
(B) mass spectrometry
(C) infrared (IR) spectroscopy
(D) photoelectron spectroscopy (PES)

Questions 15 and 16 refer to the following information:

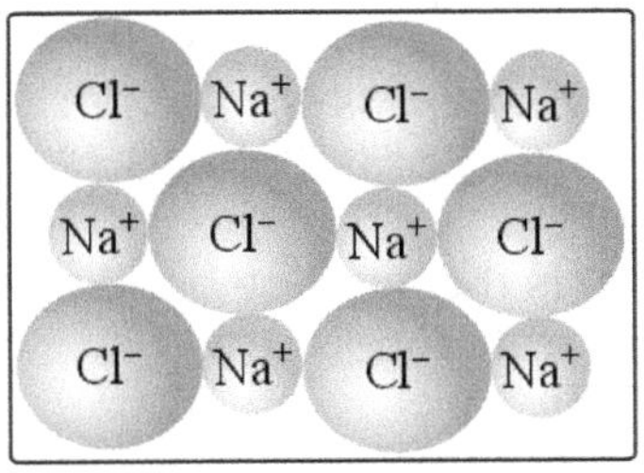

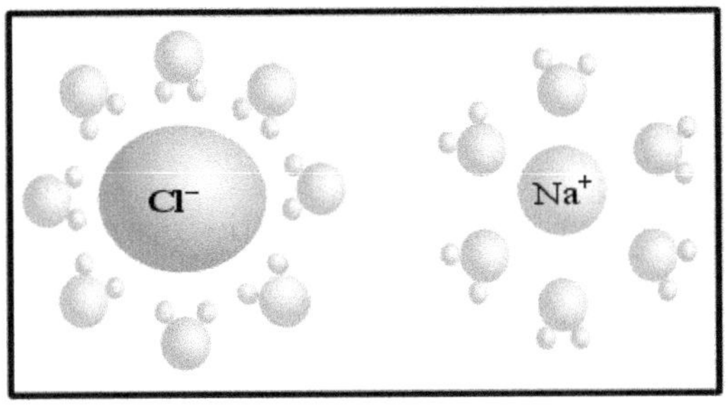

15. Sodium chloride is more commonly referred to as "table salt." Sodium chloride is an ionic solid, as shown on the top. After dissolving in water, it is as shown on the bottom. Which is the correct statement?
    (A) The entropy of the ions increases, and the entropy of the water molecules increases.
    (B) The entropy of the ions increases, and the entropy of the water molecules decreases.
    (C) The entropy of the ions decreases, and the entropy of the water molecules increases.
    (D) The entropy of the ions decreases, and the entropy of the water molecules decreases.

16. Which phrase best describes the interactions occurring between the sodium ions and the water molecules in solution?
    (A) ion–hydrogen bond
    (B) ion–dipole
    (C) dipole–induced dipole
    (D) dipole–hydrogen bond

Half-Reaction	$E°$ (V)
$Cd^{2+}(aq) + 2\ e^- \rightarrow Cd(s)$	–0.40
$Ag^+(aq) + e^- \rightarrow Ag(s)$	0.80

17. A voltaic cell is prepared consisting of solid silver and solid cadmium electrodes in solutions of 1.0 M $AgNO_3(aq)$ and 1.0 M $Cd(NO_3)_2(aq)$, respectively. Which metal will be the cathode, and what will be the standard voltage of the cell?
    (A) Ag; $E°_{cell} = 0.40$ V
    (B) Ag; $E°_{cell} = 1.20$ V
    (C) Cd; $E°_{cell} = 0.40$ V
    (D) Cd; $E°_{cell} = 2.00$ V

Questions 18 and 19 refer to the following reaction:

$2\ CH_3OH(g) + 3\ O_2(g) \rightarrow$
$2\ CO_2(g) + 4\ H_2O(l)$

18. When the reaction above occurs, the temperature of the surroundings increases. Which statement is correct?
    (A) The ΔH_{rxn} is negative because the bond energy of the products is higher than the bond energy of the reactants.
    (B) The ΔH_{rxn} is negative because the bond energy of the reactants is higher than the bond energy of the products.
    (C) The ΔH_{rxn} is positive because the bond energy of the products is higher than the bond energy of the reactants.
    (D) The ΔH_{rxn} is positive because the bond energy of the reactants is higher than the bond energy of the products.

GO ON TO NEXT PAGE

19. Which is true for entropy in the above reaction?
    (A) ΔS_{rxn} is negative because the products have fewer moles of gas than the reactants.
    (B) ΔS_{rxn} is negative because there are more moles of products than of reactants.
    (C) ΔS_{rxn} is positive because the products have fewer moles of gas than the reactants.
    (D) ΔS_{rxn} is positive because there are more moles of products than of reactants.

20. Which correctly shows the electron configuration of the Fe^{3+} ion?
    (A) $1s^2 2s^2 2p^6 3s^2 3p^6 3d^5$
    (B) $1s^2 2s^2 2p^6 3s^2 3p^6 4s^2 3d^3$
    (C) $1s^2 2s^2 2p^6 3s^2 3p^6 4s^2 3d^6$
    (D) $1s^2 2s^2 2p^6 3s^2 3p^6 4s^2 3d^9$

21. What is the average bond order in the nitrate ion, NO_3^-?
    (A) 1.0
    (B) 1.33
    (C) 1.5
    (D) 2.0

22. Which of the following descriptions of the energies involved in the formation of a solution is correct?

	Change with Solute Breakup		Change with Solvent Breakup		Change with Solute–Solvent Interaction	
	ΔH	ΔS	ΔH	ΔS	ΔH	ΔS
(A)	–	+	–	+	–	+
(B)	+	+	+	+	–	–
(C)	+	+	+	+	–	+
(D)	+	–	+	–	+	–

23. Sucrose, $C_{12}H_{22}O_{11}$, has a molar mass of 342 g/mol. How many atoms of carbon are there in 684 g of sucrose?
    (A) 3.01×10^{23} atoms
    (B) 6.02×10^{23} atoms
    (C) 1.20×10^{24} atoms
    (D) 1.45×10^{25} atoms

24. Analysis of a sample of an oxide of chromium is reported as 26 g of chromium and 12 g of oxygen. From these data, determine the empirical formula of this compound.
    (A) CrO
    (B) CrO_3
    (C) Cr_2O_3
    (D) Cr_3O_4

Weak Acid or Acidic Salt	Molar Mass (g/mol)	K_a
$HC_9H_7O_4$	180.2	3.3×10^{-4}
$HC_2H_3O_2$	60.1	1.8×10^{-5}
$KHSO_4$	136.2	1.2×10^{-2} (for HSO_4^-)
KH_2PO_4	136.1	6.2×10^{-8} (for $H_2PO_4^-$)

25. To determine the identity of a solution of an unknown weak acid, a student performed the following steps in the laboratory:
    1. Measured out 100.0 mL of the solution of the unknown weak acid.
    2. Divided the 100.0-mL sample into two 50.0-mL portions, setting one portion aside for later (called the reserved solution).
    3. Added two drops of phenolphthalein indicator (changes from colorless to pink between pH 8 and 10) to one portion.
    4. Added 0.1 *M* NaOH dropwise while swirling, stopping when solution stays light pink.
    5. The pH of the pink solution was tested and found to be 8.54.
    6. The reserved solution of the unknown weak acid was thoroughly mixed with the pink solution to make 100.0 mL of solution.
    7. The pH of the combined solution was determined to be 4.74.

    Using the student's laboratory results and the information in the table above, determine the identity of the unknown weak acid or acidic salt.
    (A) $HC_2H_3O_2$
    (B) $HC_9H_7O_4$
    (C) $KHSO_4$
    (D) KH_2PO_4

26. The values for the change in enthalpy, ΔH, and the activation energy, $E_{a(forward)}$, for an endothermic reaction are known. The value of $E_{a(reverse)}$ for the reverse reaction equals
    (A) $E_{a(forward)}$.
    (B) $-(E_{a(forward)})$.
    (C) $-\Delta H + E_{a(forward)}$.
    (D) $\Delta H + E_{a(forward)}$.

$2\ SO_2(g) + O_2(g) \leftrightarrows 2\ SO_3(g)$

$K_p = 3.0 \times 10^{10}$ at 500°C

27. Which statement is correct for the following reaction at the same temperature as the reaction above?
    $SO_3(g) \rightleftarrows SO_2(g) + \frac{1}{2}\ O_2(g)$
    (A) The reaction is thermodynamically nonfavored and $K = 5.8 \times 10^{-6}$.
    (B) The reaction is thermodynamically favored and $K = 5.8 \times 10^{-6}$.
    (C) The reaction is thermodynamically nonfavored and $K = 3.0 \times 10^{10}$.
    (D) The reaction is thermodynamically favored and $K = 3.0 \times 10^{10}$.

Questions 28–30 refer to the system of reactions below.

$2\ NO(g) \rightleftarrows N_2O_2(g)$	*Fast*
$N_2O_2(g) + H_2(g) \rightarrow N_2O(g) + H_2O(g)$	*Slow*
$N_2O(g) + H_2(g) \rightarrow N_2(g) + H_2O(g)$	*Fast*

28. What is the rate law for the reaction according to this mechanism?
    (A) Rate = $k[N_2O_2][H_2]$
    (B) Rate = $k[NO]^2$
    (C) Rate = $k[N_2O_2]^2[H_2]^2$
    (D) Rate = $k[NO]^2[H_2]$

GO ON TO NEXT PAGE

29. How does the oxidation number of nitrogen change throughout the process?
(A) +2 to +1 to 0
(B) +1 to +2 to +1 to +5
(C) +2 to +4 to +2 to 0
(D) +4 to +2 to +1 to 0

30. The overall equation for the reaction is
(A) $N_2O(g) + H_2(g) \rightarrow N_2(g) + H_2O(g)$.
(B) $2\ NO(g) + 2\ H_2(g) \rightarrow N_2(g) + 2\ H_2O(g)$.
(C) $2\ NO(g) + H_2(g) \rightarrow N_2(g) + H_2O(g)$.
(D) $NO(g) + H_2(g) \rightarrow N(g) + H_2O(g)$.

Questions 31–35 refer to the following information:

Consider the reaction $SO_2Cl_2(g) \leftrightarrows SO_2(g) + Cl_2(g)$, $K_p = 5$ at a temperature of 112°C. In the diagrams below, the following molecules will be represented as indicated.

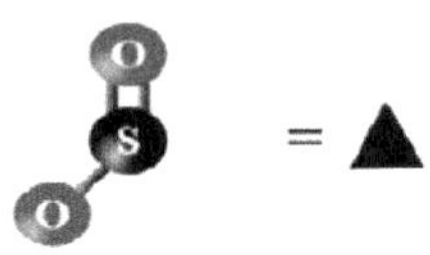

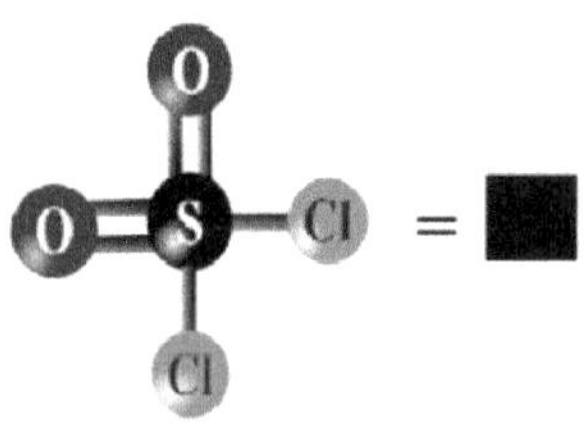

31. Shown below is the same container taken at four different time periods, at 112°C.

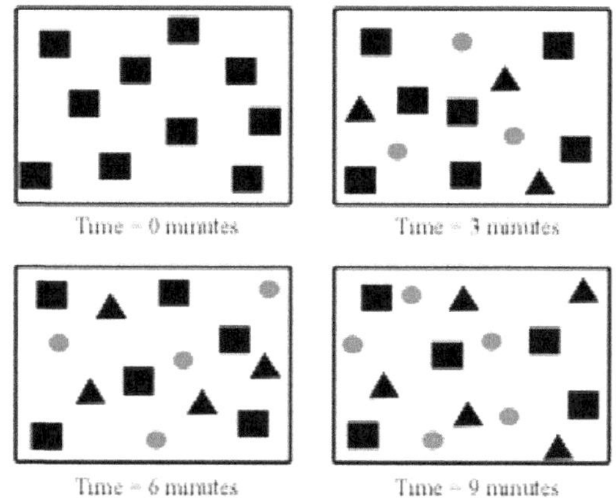

What would the container look like at 15 minutes?

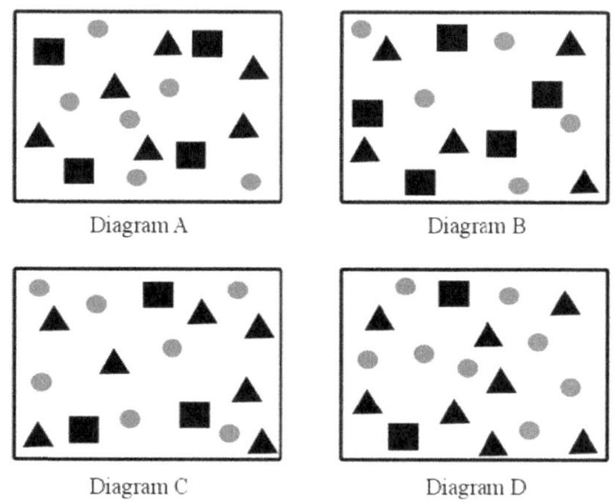

(A) Diagram A
(B) Diagram B
(C) Diagram C
(D) Diagram D

32. A student makes the following statement:

    "Since the SO_2Cl_2 molecule has both SO_2 and Cl_2 in it already, the sign of ΔH_{rxn} will be positive since breaking bonds always requires energy and there are no new bonds being formed."

    Evaluate the student's hypothesis and provide the justification for your answer.
    (A) The student's hypothesis is correct as stated since 1 molecule of SO_2Cl_2 decomposed to form 1 molecule of SO_2 and 1 molecule of Cl_2.
    (B) The student's hypothesis is correct since only 1 pi bond in the reactant is broken and no new pi bonds are formed.
    (C) The student's hypothesis is not correct since this reaction has 3 bonds broken and new bonds formed.
    (D) The student's hypothesis is not correct since the ΔS of this reaction is positive; the ΔH must be negative to account for the ΔG being negative as this reaction is thermodynamically favored.

33. The original pressure (at time = 0) was 6.0 atm. What is the total pressure when the system reaches equilibrium?
    (A) 3.0 atm
    (B) 6.0 atm
    (C) 9.0 atm
    (D) 12 atm

34. When the system reaches equilibrium, the three gases are separated into three different identical rigid 2.0-L containers at 400. K. Which gas will have the highest pressure?
    (A) SO_2Cl_2
    (B) SO_2
    (C) Cl_2
    (D) They will all have the same pressure since an equal number of moles of gas at the same temperature and volume will have the same pressure.

35. After equilibrium has been established, the container has additional $Cl_2(g)$ injected into it and it is allowed to reestablish equilibrium at 112°C. Which is the correct diagram after equilibrium has been reestablished?

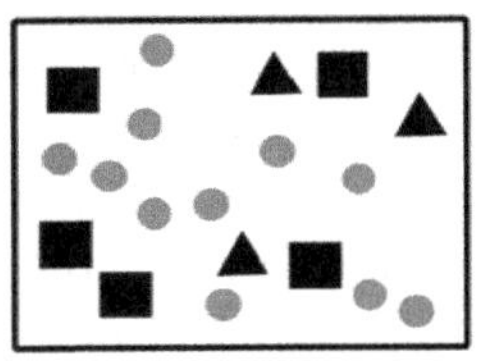

Diagram A

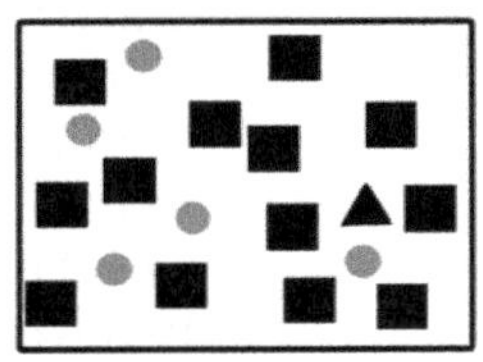

Diagram B

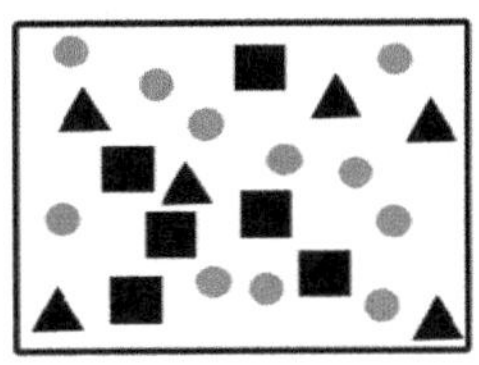

Diagram C

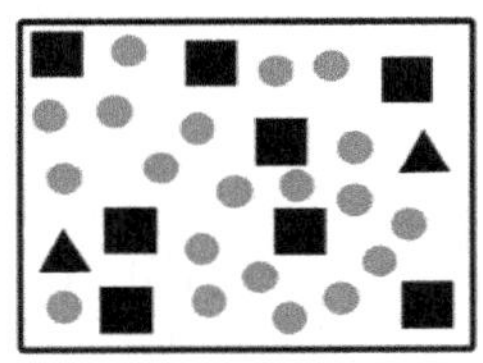

Diagram D

    (A) Diagram A
    (B) Diagram B
    (C) Diagram C
    (D) Diagram D

$H_3PO_4 + H_2O \leftrightarrows H_3O^+ + H_2PO_4^-$
$K_{a1} = 7.5 \times 10^{-3}$
$H_2PO_4^- + H_2O \leftrightarrows H_3O^+ + HPO_4^{2-}$
$K_{a2} = 6.2 \times 10^{-8}$
$HPO_4^{2-} + H_2O \leftrightarrows H_3O^+ + PO_4^{3-}$
$K_{a3} = 4.8 \times 10^{-13}$

36. The phosphate buffer system is very important in human body cells, including the blood. Blood pH is approximately 7.4 in healthy people. What are the predominant forms of phosphate in the blood at this pH?
(A) H_3PO_4 and $H_2PO_4^-$
(B) H_3PO_4 and HPO_4^{2-}
(C) $H_2PO_4^-$ and HPO_4^{2-}
(D) HPO_4^{2-} and PO_4^{3-}

37. A rigid tank having a volume of 10.0 L is filled with an equimolar mixture of H_2, O_2, NH_3, and CO_2 at 25°C. A pin hole is made in the tank, and gases are allowed to escape for 60 minutes. At the end of the 60 minutes, the gas pressure was found to have dropped from 5.0 atm down to 2.0 atm. Assuming that the gases behaved ideally and that no reactions occurred between the gases, which gas would remain in the tank in the greatest proportion by mole at the end of the 60 minutes?
(A) CO_2
(B) H_2
(C) NH_3
(D) O_2

38. In column chromatography, the substance to move through the column at the fastest rate
(A) has the strongest intermolecular attractions for the mobile phase.
(B) has the weakest intermolecular attraction for the mobile phase.
(C) has the same polarity as the stationary phase, resulting in strong intermolecular repulsions for the stationary phase.
(D) has the opposite polarity as the stationary phase, resulting in strong intermolecular attractions for the stationary phase.

Questions 39 and 40 use the following graph, which represents the titration of HX with NaOH:

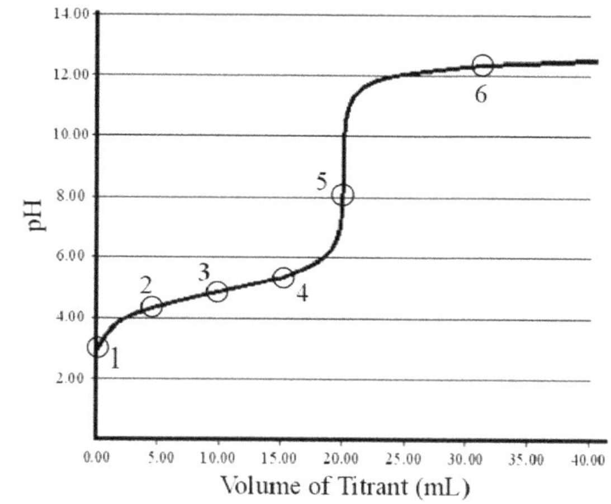

39. At which point in the titration shown above would a basic salt be the main determinant of the pH of the solution?
(A) 3
(B) 4
(C) 5
(D) 6

40. At which point in the titration described above would the major species present be as shown in the box below? Water molecules and spectator ions are omitted for clarity.

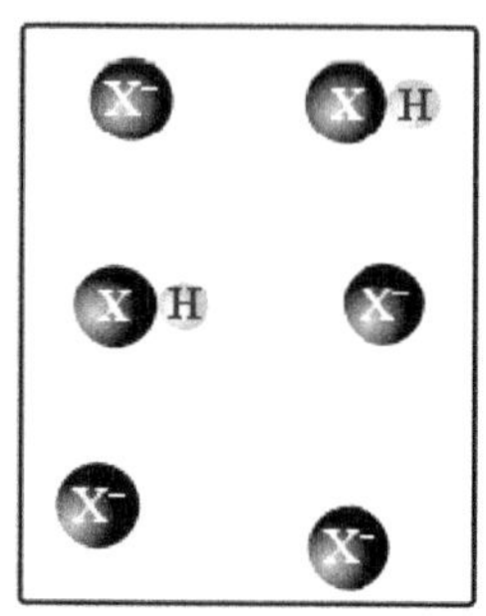

(A) 1
(B) 2
(C) 3
(D) 4

41. A 100.0-mL sample of 0.10 *M* potassium dihydrogen phosphate (KH_2PO_4) is added to 50.0 mL of 0.10 *M* NaOH. The pH of the resulting solution is measured and found to be 7.21. When 1.0 mL of 0.50 *M* HCl is added to the solution, there is no change in pH. Which equation best shows why the pH of the solution did not change when the hydrochloric acid was added?
(A) $OH^- + H_2PO_4^- \rightarrow HPO_4^{2-} + H_2O$
(B) $H_3O^+ + H_2PO_4^- \rightarrow H_3PO_4 + H_2O$
(C) $H_3O^+ + PO_4^{3-} \rightarrow HPO_4^{2-} + H_2O$
(D) $H_3O^+ + HPO_4^{2-} \rightarrow H_2PO_4^- + H_2O$

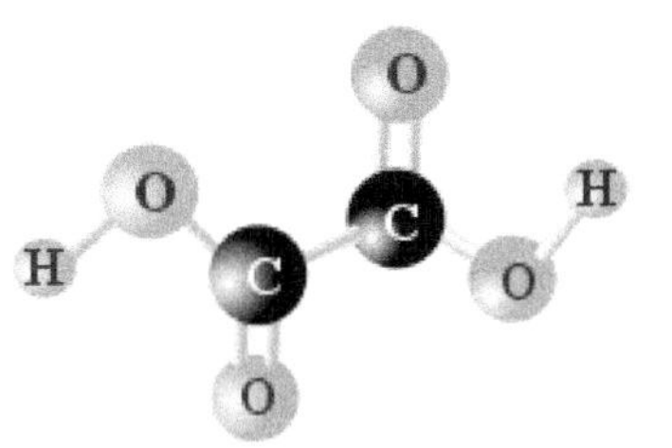

42. Oxalic acid (structure shown above) is found in low concentrations in rhubarb, spinach, and certain other plant materials. To determine the concentration of a solution of oxalic acid, 25.00 mL of the solution is titrated with 0.10 *M* NaOH. The oxalic acid was completely reacted with the sodium hydroxide upon the addition of 37.50 mL of the sodium hydroxide solution. What is the molarity of the oxalic acid solution?
(A) 0.033 *M*
(B) 0.038 *M*
(C) 0.075 *M*
(D) 0.15 *M*

43. As part of a laboratory test, a student had to determine the identity of unknown solids, based on their physical properties. For Sample A, the student recorded the data below. Which of the following substances is likely to be Sample A?
- The solid does not conduct electricity.
- The solid does not dissolve in water or in alcohol.
- The solid has a melting point so high that a Bunsen burner will not melt it.
- The solid is not malleable.

(A) $SiO_2(s)$
(B) $KCl(s)$
(C) $C_{12}H_{22}O_{11}(s)$
(D) $Zn(s)$

Questions 44 and 45 refer to the following information:

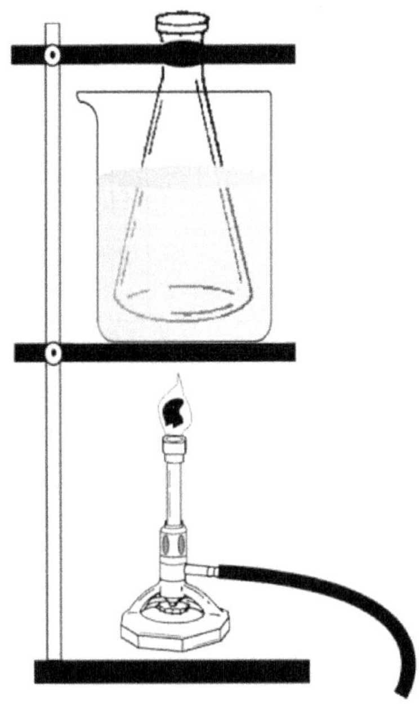

The molar mass of an unknown liquid may be found by vaporizing a small sample of the liquid and determining the mass, volume, temperature, and pressure of the vapor. While water is heating in a large beaker, a dry flask and aluminum foil cap are weighed. A 2.0-mL sample of the volatile liquid is placed in the flask and a pinhole is made in the aluminum cap. The flask is then submerged into the beaker of hot water.

GO ON TO NEXT PAGE

Once the water has reached its boiling point, the flask remains submerged in the boiling water for 10 minutes. During this time, all liquid in the flask should vaporize with the excess vapor escaping out the pinhole. After removal from the water, the flask with cover is immediately dried and cooled to room temperature before weighing to determine the vapor mass. The vapor volume is found by filling the flask completely with water and then pouring the water into a large graduated cylinder. The room temperature and barometric pressure are also read and recorded.

Liquid	Sample X	Sample Y
Mass of condensed vapor remaining	0.750 g	1.089 g
Volume of flask	300. mL	300. mL
Room pressure	760. mm Hg	760. mm Hg
Room temperature	23.0°C	23.0°C

44. Which statement is a correct comparison between Sample X and Sample Y?
    (A) Sample Y contains more molecules than does Sample X.
    (B) The molar mass of Sample Y is higher than the molar mass of Sample X.
    (C) The average velocity of Sample Y molecules is higher than the average velocity of Sample X molecules.
    (D) The average kinetic energy of Sample X molecules is higher than the average kinetic energy of Sample Y molecules.

45. A student attempted to determine the molar mass of a pure liquid sample and collected the data shown in the previous table. What is the most likely identity of liquid X?
    (A) C_3H_6O
    (B) H_2O
    (C) CH_3OH
    (D) C_6H_{14}

Element	Atomic Radii (pm)	Atomic mass (amu)
Carbon	77	12.01
Hydrogen	37	1.008
Nitrogen	70	14.01
Oxygen	66	16.00
Titanium	176	47.87

46. Commercially Pure Titanium (CPTi) is used in dental implants. It is a mixture consisting of very pure titanium, along with hydrogen, carbon, nitrogen, and oxygen. Which diagram best represents this type of alloy?

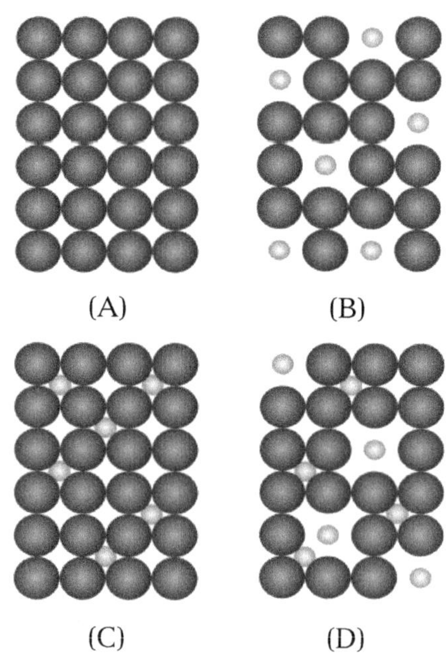

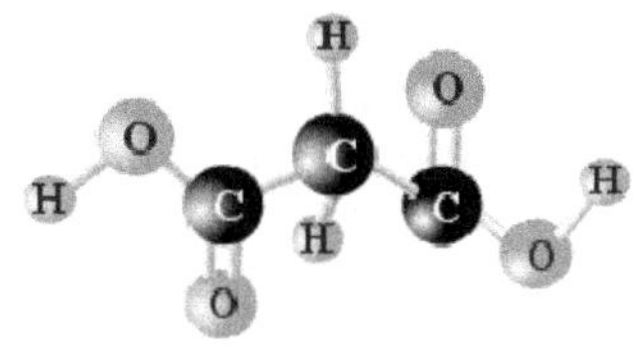

Malonic Acid

47. Suppose 50.0 mL of 0.10 *M* NaOH is added to 30.0 mL of a 0.10 *M* solution, malonic acid, $CH_3(COOH)_2$, whose structure is shown above. After mixing, which species will be found in the highest concentration?
(A) OH^-
(B) $CH_3(COOH)_2$
(C) $CH_3(COOH)(COO^-)$
(D) $CH_3(COO^-)_2$

48. The equilibrium constant for the ionization of hypochlorous acid is 3.0×10^{-8}. If a 1.0 *M* solution is at pH 2.0, what is the equilibrium condition of the system?

$HClO + H_2O \rightleftarrows H_3O^+ + ClO^-$

(A) The system will shift left because $Q < K$.
(B) The system will shift left because $Q > K$.
(C) The system will shift right because $Q < K$.
(D) The system will shift right because $Q > K$.

$H^+(aq) + OH^-(aq) \rightarrow H_2O(l)$

$\Delta H_{neut} = ?$

49. The enthalpy of neutralization of a strong acid with a strong base can be determined by reacting the two in a calorimeter and monitoring the temperature change. A 50.0-mL sample of 1.0 *M* NaOH is added to 50.0 mL of 1.0 *M* H_2SO_4, with both solutions having an initial temperature of 23.3°C. After mixing, the solution temperature increased to a maximum temperature of 29.8°C. What is the heat of neutralization for the reaction above? Assume the density of the final solution is 1.0 g/mL and that the specific heat of the solution is 4.2 J/g°C.
(A) 55 kJ/mol
(B) 28 kJ/mol
(C) –28 kJ/mol
(D) –55 kJ/mol

GO ON TO NEXT PAGE

The following reaction applies to Questions 50–54:

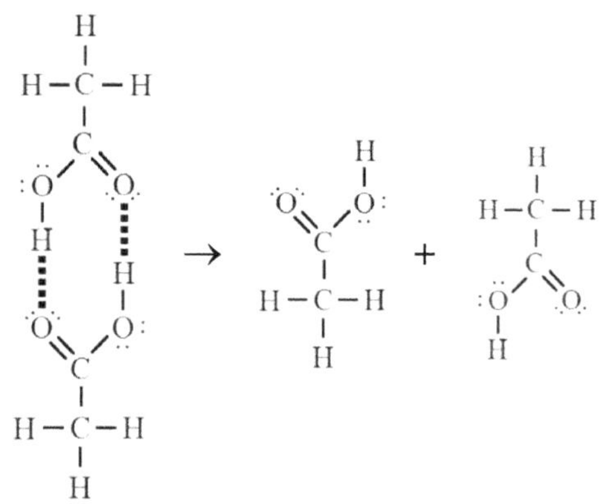

50. A dimer is a small molecule that consists of two structurally similar, or identical, monomers (the smaller units that join together to form a dimer) joined by bonds that can be either strong or weak. Carboxylic acids (formic acid, acetic acid, etc.) can form dimers when not in water. For example, acetic acid forms a dimer in the gas phase, as shown in the diagram above. When the dimer separates into its monomers, the process is called dissociation. What holds the dimer together, and what is the sign of ΔH_{rxn} for the reaction shown above?
    (A) London forces hold the dimer together, and the sign of ΔH_{rxn} is positive.
    (B) London forces hold the dimer together, and the sign of ΔH_{rxn} is negative.
    (C) Hydrogen bonding holds the dimer together, and the sign of ΔH_{rxn} is positive.
    (D) Hydrogen bonding holds the dimer together, and the sign of ΔH_{rxn} is negative.

51. What is the sign of ΔS for the reaction represented to the left, and what will drive the reaction to be thermodynamically favored?
    (A) $\Delta S_{rxn} = +$, and the reaction will be driven by enthalpy.
    (B) $\Delta S_{rxn} = +$, and the reaction will be driven by entropy.
    (C) $\Delta S_{rxn} = -$, and the reaction will be driven by enthalpy.
    (D) $\Delta S_{rxn} = -$, and the reaction will be driven by entropy.

52. What is the hybridization of each carbon atom in acetic acid?
    (A) sp and sp^2
    (B) sp and sp^3
    (C) sp^2 and sp^3
    (D) sp^3 and sp^3

53. The K_a for acetic acid is 2.00×10^{-4} at the temperature that the dimer exists. If 0.500 mole of the dimer is bubbled into 125 mL of water and all of the dimer dissolves and forms acetic acid monomers, determine the percent of ionization of the acetic acid in the resulting solution. Assume that the total volume of solution is 125 mL.
    (A) 0.125%
    (B) 0.250%
    (C) 0.375%
    (D) 0.500%

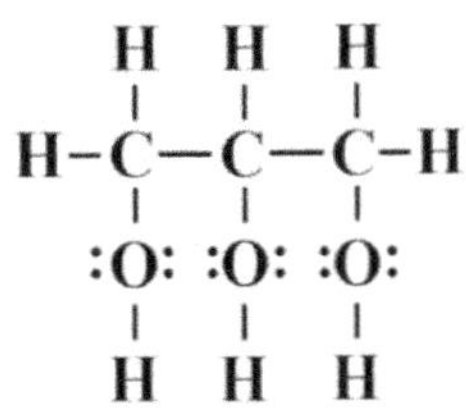

54. Glycerin is a very syrupy liquid and is often found in lotions, as well as in "bubble solution" that children use to blow bubbles. The structure of glycerin is shown above. Which is the most important reason that glycerin has such high viscosity?
    (A) Glycerin molecules have many electrons, resulting in great polarizability and strong intermolecular attractions.
    (B) Glycerin molecules hydrogen bond within a single glycerin molecule, forming a twisted shape that causes the liquid to tangle up on itself.
    (C) Glycerin molecules form many hydrogen bonds between adjacent molecules, causing the liquid to pour very slowly.
    (D) Glycerin molecules are very polar. This polarity causes strong attraction between adjacent molecules and causes the liquid to be resistant to flow.

Questions 55 and 56 refer to the following information:

Molar masses are shown below each compound.

$$2\ NaHCO_3(s) \rightarrow Na_2CO_3(s) + H_2O(g) + CO_2(g)$$

84.0 g	106.0 g	18.0 g	44.0 g

55. A 5.00-g sample of baking soda ($NaHCO_3$) is decomposed by heating according to the equation above. What is the total mass of products, assuming complete decomposition of the baking soda?
    (A) The products have a mass greater than 5.00 g because the molar masses of the products are greater than the molar masses of the baking soda.
    (B) The products have a mass greater than 5.00 g because there are 3 moles of products and only 2 moles of reactants.
    (C) The products have a mass equal to 5.00 g, assuming careful experimental procedures.
    (D) The product has a mass less than 5.00 g because the water and carbon dioxide are gases and have no mass.

GO ON TO NEXT PAGE

Mass of evaporating dish	25.55 g
Mass of evaporating dish + Mixture (before heating)	27.55 g
Mass of evaporating dish + Mixture (after first heating)	26.98 g
Mass of evaporating dish + Mixture (after second heating)	26.93 g
Mass of evaporating dish + Mixture (after third heating)	26.93 g

56. In a second experiment, a sample of a mixture of baking soda and sodium chloride is heated to constant mass. The results are shown in the table above. What can be concluded from the results?
    (A) The mixture was 84% $NaHCO_3$ and 16% NaCl.
    (B) The mixture was 42% $NaHCO_3$ and 58% NaCl.
    (C) The mixture was 31% $NaHCO_3$ and 69% NaCl.
    (D) The mixture was 74% $NaHCO_3$ and 26% NaCl.

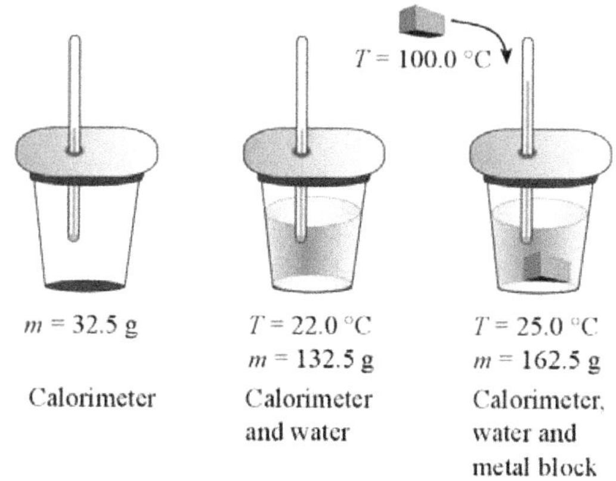

Substance	Specific Heat Capacity (J/g°C)
H_2O	4.2
Al	0.89
Sn	0.54
Cu	0.39
Pb	0.13

57. A student attempts to determine the identity of an unknown metal by finding its specific heat capacity. The specific heat capacity of water is 4.2 J/g°C. The data collected by the student are shown above. What is the correct identity of the unknown metal?
    (A) Al
    (B) Cu
    (C) Pb
    (D) Sn

Ion	Ionic Radius (pm)
K^+	133
Na^+	95
Ca^{2+}	99
Mg^{2+}	65
Cl^-	181

58. Which of these salts has the most exothermic lattice energy?
(A) KCl
(B) NaCl
(C) $CaCl_2$
(D) $MgCl_2$

$H_2(g) + \frac{1}{2} O_2(g) \rightarrow H_2O(l)$

$\Delta G^\circ = -237 \text{ kJ/mol}_{rxn}$

59. Why is the reaction shown above not observed to occur to any great extent at room temperature?
(A) Reactions with a negative free energy change are not thermodynamically favorable.
(B) This reaction is only thermodynamically favorable at high temperatures.
(C) The reaction is under thermodynamic control, which prevents the reaction from occurring.
(D) The reaction is under kinetic control and requires a large activation energy to start the reaction.

$CH_3OH + CO \rightarrow CH_3COOH$

60. The majority of acetic acid produced in the world is made through a process (reaction shown above) that uses a rhodium catalyst, along with an iodide cocatalyst. The element rhodium can act as either a homogeneous or a heterogeneous catalyst, depending on the reaction. Which best explains how the catalyst works with this process?

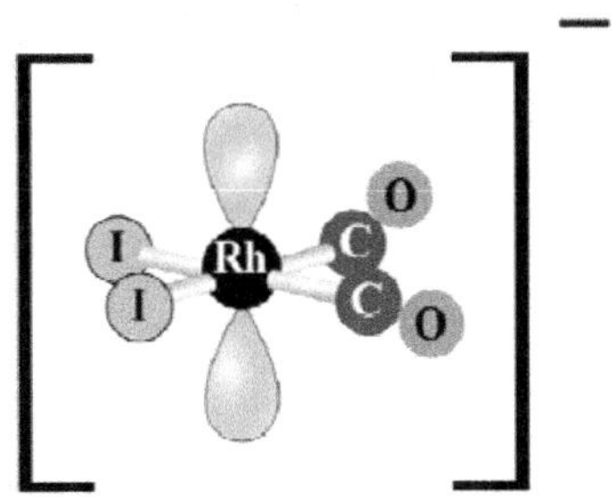

(A) The catalyst lowers the activation energy of the reaction by participating in the reaction. It is a reactant in an early step of the reaction and is regenerated during a later step. This is an example of homogeneous catalysis.
(B) The catalyst lowers the ΔH of the reaction by participating in the reaction. It is a reactant in an early step of the reaction and is regenerated during a later step. This is an example of homogeneous catalysis.
(C) The catalyst lowers the activation energy of the reaction by acting as a site for reactants to become adsorbed on to. This creates more favorable molecular orientation. This is an example of heterogeneous catalysis.
(D) The catalyst lowers the ΔH of the reaction by acting as a site for reactants to become adsorbed on to. This creates more favorable molecular orientation. This is an example of heterogeneous catalysis.

GO ON TO NEXT PAGE

INTRODUCTION TO SECTION II: FREE-RESPONSE QUESTIONS

Section II of the AP® Chemistry examination counts for 50% of the total test grade and involves several parts. Answering these questions gives you an opportunity to demonstrate your ability to present your material in clear, orderly, and convincing language. Your answers will be graded on the basis of accuracy, the kinds of information you include to support your responses, and the importance of the descriptive material used. Be specific; general, all-encompassing answers will not be graded as well as detailed answers with examples and equations. CLEARLY SHOW THE METHOD USED AND THE STEPS INVOLVED IN ARRIVING AT YOUR ANSWERS. It is to your advantage to do this, since you may obtain partial credit if you do and you will receive little or no credit if you do not. Attention should be paid to significant figures. On the AP® exam, be sure to write all your answers to the questions on the lined pages following each question in the test booklet. Do not write your answers in the white space between questions.

Section II: Free-Response Questions
Time: 105 minutes
Number of Questions: 7

DIRECTIONS: You may use a calculator, the formula chart, and the periodic table located at the back of this book throughout this section.

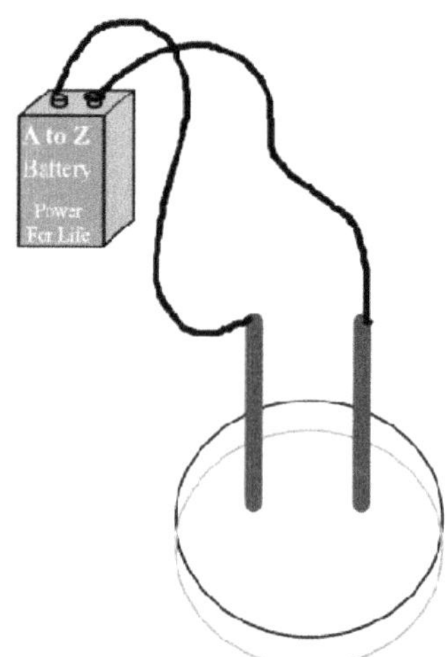

Half-Reaction	$E°$ (V)
$I_2 + 2\,e^- \rightarrow 2\,I^-$	0.54
$K^+ + e^- \rightarrow K$	–2.92
$O_2 + 4\,H^+ + 4\,e^- \rightarrow 2\,H_2O$	1.23
$2\,H_2O + 2\,e^- \rightarrow H_2 + 2\,OH^-$	–0.83

1. A 20.0-mL sample of a 1.0 *M* KI solution containing two drops of phenolphthalein indicator is placed into a shallow glass plate (petri dish). Two inert electrodes are immersed in the solution and connected to a 9V battery.

   (a) After several seconds, evidence of a reaction is observed.
   - (i) Write the balanced net ionic equation for the half-reaction that occurs at the anode.
   - (ii) Write the balanced net ionic equation for the half-reaction that occurs at the cathode.
   - (iii) Calculate $E°$ for the reaction.
   - (iv) Calculate the $\Delta G°$ for the reaction. Is the reaction thermodynamically favored at 25°C? Justify your answer.

A current of 0.200 amps is passed through the 20.0 mL solution for 15.0 minutes.

(b) Calculate the number of moles of electrons passed through the solution.

(c) The iodide ion is very soluble in water; however, I_2 has very limited solubility. Explain.

(d) The solubility of I_2 in water can be increased by the addition of KI or NaI. The following reaction is observed to occur: $I_2 + I^- \rightarrow I_3^-$.
Draw the complete Lewis structure for I_3^-.

(e) Organic compounds containing iodine are often used as contrast agents for medical procedures as they block X-rays. Using principles of atomic structure, discuss why iodine is able to stop X-rays from penetrating through it and the other halogens do not have this property.

(f) Br_2 is a liquid at room temperature, while I_2 is a solid. Explain.

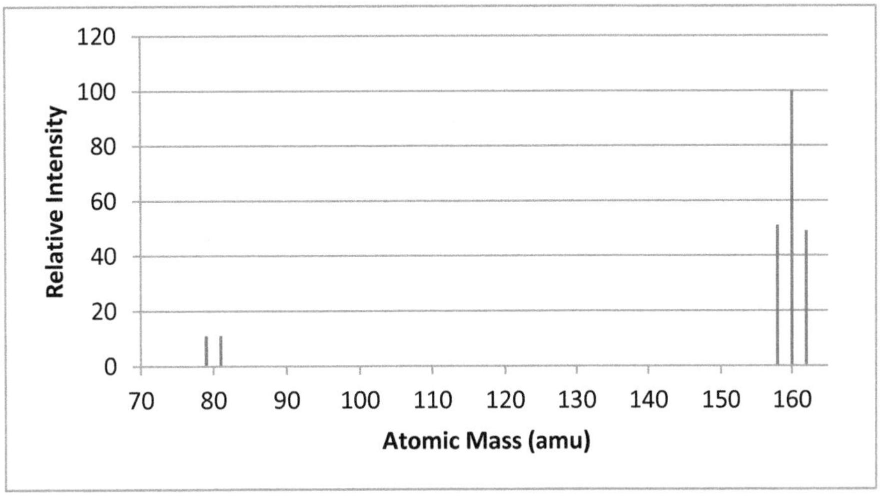

(g) The mass spectrometry diagram for bromine is shown above. How many stable isotopes does the element have?

$$PCl_3(g) + Cl_2(g) \rightleftarrows PCl_5(g)$$

2. An evacuated flask is filled with 0.30 atm PCl_3, 0.60 atm Cl_2, and 0.10 atm PCl_5 at 285°C.
   (a) Write the K_p expression for the reaction above.
   (b) At equilibrium, it is found that the total pressure in the flask is 1.08 atm. Calculate the equilibrium partial pressures of each gas.
   (c) Determine the equilibrium constant (K_p) for the reaction at these conditions.
   (d) Predict the effect that increasing the volume of the system will have on the mole fraction of PCl_5 in the mixture. Justify your answer.
   (e) The reaction as written is exothermic.
      (i) Predict the effect that decreasing the temperature of the system will have on the mole fraction of PCl_5 in the mixture. Justify your answer.
      (ii) How will decreasing the temperature affect the value of K_p?
   (f) Draw Lewis diagrams for the following molecules and predict their shapes.
      (i) PCl_3
      (ii) PCl_5
   (g) When $PCl_3(g)$ is added to water, it reacts very exothermically to produce two acids, as shown below.

$$PCl_3(g) + 3\ H_2O(l) \rightarrow 3\ HCl(aq) + H_3PO_3(aq)$$

   If 0.0158 mole of PCl_3 is added to 100.0 mL of water, what is the pH of the solution formed?

GO ON TO NEXT PAGE

3. Cyanamide, CN_2H_2, is an interesting compound used extensively in agriculture (as a plant growth regulator) and in the synthesis of a variety of pharmaceutical drugs.
   (a) Two resonance structures for cyanamide are shown below.

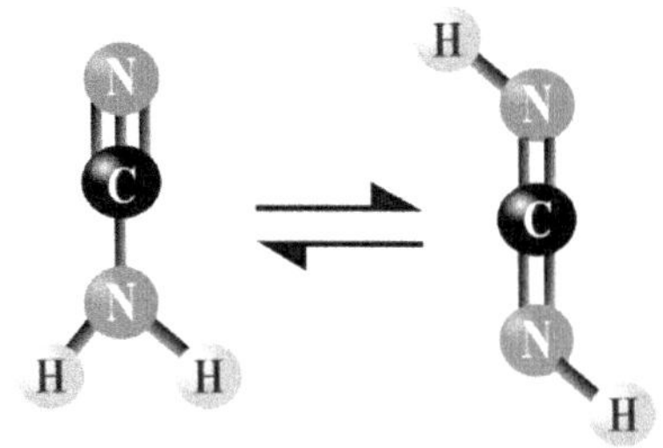

Structure 1 **Structure 2**

   (i) Which structure, if either, is favored, according to formal charge? Justify your answer.
   (ii) Select one of the two structures, clearly indicate which structure you selected, and state the H–N–C bond angle.
   (iii) Which structure has all of the atoms lying in the same plane (i.e., lays flat)? Circle your answer.

   Structure 1 Structure 2

   (b) The calcium salt of cyanamide, $CaCN_2$, is mainly used as a fertilizer. When placed in water, $CaCN_2$ hydrolyzes water to produce NH_3 and calcium carbonate ($CaCO_3$). Write the balanced equation for this reaction.
   (c) If 5.00 g of calcium cyanamide is added to water to make a total volume of 150.0 mL of solution and allowed to react completely, what is the concentration of the ammonia that results? Assume all of the ammonia produced is dissolved in the solution.
   (d) What is the resulting pH of the solution? The K_b for ammonia is 1.8×10^{-5}. Assume that the presence of calcium carbonate does not affect the pH of the solution.
   (e) Ammonia is more economically produced using the Haber process, which involves the exothermic reaction of hydrogen gas with nitrogen gas to produce ammonia. Which set of conditions favors both kinetics and product yield when producing ammonia? Circle the best choice.

   High temperature and high pressure Low temperature and high pressure

   High temperature and low pressure

4. $NO_2 \rightarrow NO + \frac{1}{2} O_2$
   The decomposition of nitrogen dioxide is studied, and the following data are collected:

Time(s)	$[NO_2]$ (M)
0.0	0.200
5.0	0.034
10.0	0.018
15.0	0.014
20.0	0.0094

(a) From the data in the table above, what does the half-life of the reaction tell you about the order of the reaction?

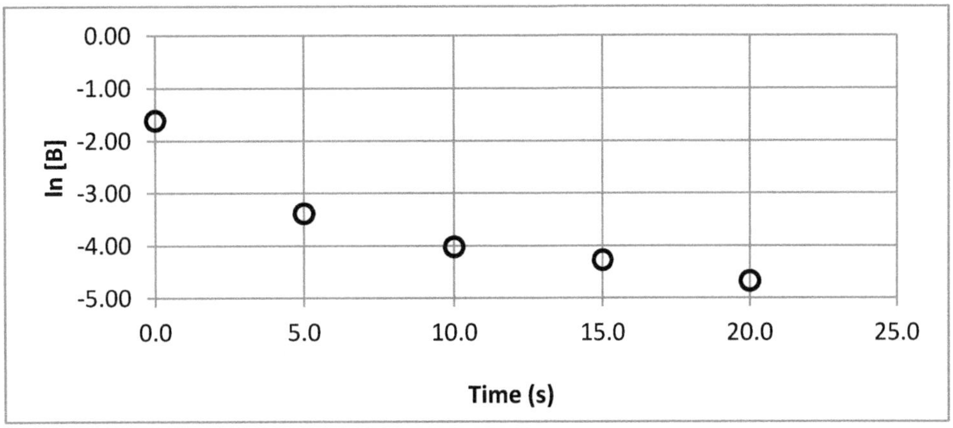

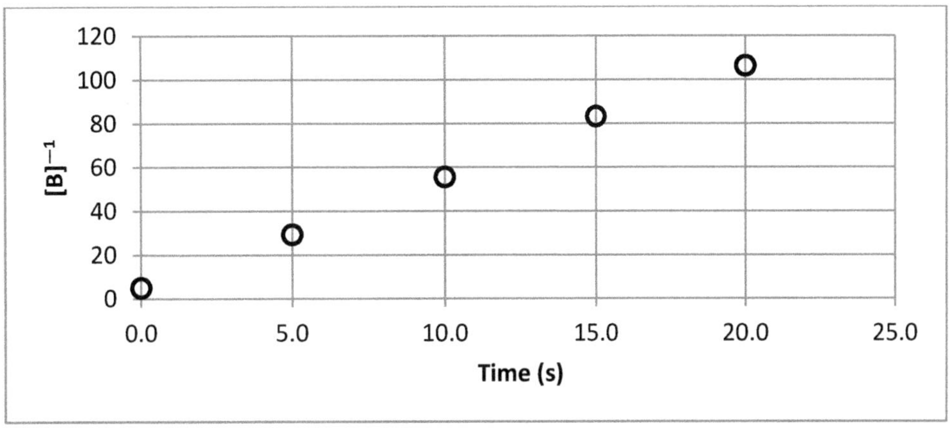

(b) The natural log of concentration versus time data and reciprocal of concentration versus time data were graphed, as shown above. What is the order of the reaction? Justify your answer.

(c) What is the value of the specific rate constant, k? Include units with your answer.

(d) NO_2 is a very reactive gas. Why does it react so readily with free chlorine atoms? Consider molecular structure when answering.

5. The reaction of calcium nitrate with sodium carbonate produces a white precipitate. A student investigated the stoichiometry of this reaction by adding carefully measured volumes of each reactant into five separate, identical test tubes, stirring each tube well and allowing the tubes to sit undisturbed for 30 minutes to allow any precipitate to settle. The student then measured the height of the precipitate in each tube to the nearest 0.1 cm and recorded the data in a table like the one below.

Tube	Volume of 2.0 *M* $Ca(NO_3)_2$ (mL)	Volume of 1.0 *M* Na_2CO_3 (mL)	Height of Precipitate (cm)
1	5	1	
2	4	2	
3	3	3	
4	2	4	
5	1	5	

GO ON TO NEXT PAGE

(a) Write the net ionic equation for the reaction occurring between calcium nitrate and sodium carbonate.
(b) Which tube(s) had the highest quantity of precipitate? Justify your answer.
(c) The K_{sp} for calcium carbonate is 8.7×10^{-9}.
   (i) Determine the solubility of calcium carbonate in moles/L in pure water.
   (ii) If an excess of sodium carbonate remains in the solution, will the solubility of calcium carbonate be affected? Justify your answer.

6. A 0.495 M solution of nitrous acid, HNO_2, has a pH of 1.83. Assume that the temperature remains constant at 25°C.
   (a) What is the percent ionization of nitrous acid in this solution?
   (b) Calculate the value of K_a for nitrous acid.
   (c) Calculate the pH of the solution formed by adding 1.00 g of $NaNO_2$ (molar mass = 69.0 g/mol) to 750. mL of 0.0125 M solution of nitrous acid. Assume no volume change.
   (d) 10.0 mL of 0.50 M NaOH is added to the solution in part (c). The pH changes only very slightly. Explain.

7. Calcium oxide has a much higher melting point (2580°C) than potassium fluoride (858°C).
   (a) Use principles of interparticle attractions to explain the difference in melting points between CaO and KF.
   (b) Solid calcium oxide is often used around construction sites after a heavy rain, as it reacts with water, forming calcium hydroxide and acts to dry the soil. This reaction also releases a large quantity of heat.
      (i) Write a balanced equation for the reaction of calcium oxide with water.
      (ii) Which is higher for this reaction—the bond energies of the reactants or the bond energies of the product? Explain.
      (iii) What is the sign of ΔS for this reaction? Explain.

END OF EXAMINATION

ANSWERS TO PRACTICE TEST 2

SECTION I: MULTIPLE-CHOICE QUESTIONS

Score your test using the table below.

Determine how many questions you answered correctly. You will find explanations of the answers on the following pages.

1. B	13. D	25. A	37. A	49. D
2. B	14. C	26. C	38. A	50. C
3. B	15. B	27. A	39. C	51. B
4. D	16. B	28. D	40. C	52. C
5. D	17. B	29. A	41. D	53. D
6. C	18. A	30. B	42. C	54. C
7. D	19. A	31. B	43. A	55. C
8. C	20. A	32. C	44. B	56. A
9. A	21. B	33. C	45. A	57. D
10. A	22. B	34. C	46. C	58. D
11. B	23. D	35. D	47. D	59. D
12. A	24. C	36. C	48. B	60. A

CALCULATE YOUR SCORE:

Number answered correctly: ______

WHAT YOUR SCORE MEANS:

Each year, since the test is different, the scoring is a little different. But generally, if you scored 23 or more on the multiple-choice questions, you'll most likely get a 3 or better on the test. If you scored 33 or more, you'll probably score a 4 or better. And if you scored 42 or more, you'll most likely get a 5. Keep in mind that the multiple-choice section is worth 50% of your final grade, and the free-response section is worth 50% of your final grade.

ANSWERS AND EXPLANATIONS

SECTION I: MULTIPLE-CHOICE QUESTIONS

1. ANSWER: B Both steps 1 and 2 involve overcoming of interparticle forces. Step 1 is breaking apart the solvent (in water, that means overcoming hydrogen bonding). Step 2 is breaking apart the solute (in this diagram, that means breaking ionic bonds). If it takes more energy to break apart the solute, then the energy for step 2 is higher than that for step 1 (LO 5.8).

2. ANSWER: B $P_1V_1 = P_2V_2$, 5.0 atm(2.0 L) = P_2(8.0 L), P_2 = 1.25 atm (LO 2.6)

3. ANSWER: B A redox reaction generally is given as a balanced net ionic equation. Both charge and mass must be balanced. Because this reaction occurs in acidic solution, H^+ must be used to balance it (LO 3.2)

4. ANSWER: D The reactants diminish in concentration (C and D). D [NO] goes down 6/4 faster than C [NH_3]. The products are formed (A and B). A [H_2O] is formed 6/5 times faster than B [N_2] (LO 3.4).

5. ANSWER: D Rate at which NO is used = Rate of production of N_2 × (6/5) = 36 mol/L/s. You can also do a full stoichiometry problem for this reaction.

$$\frac{30 \text{ mol } N_2}{L \times s} \times \frac{6 \text{ mol NO}}{5 \text{ mol } N_2} = \frac{36 \text{ mol NO}}{L \times s}$$

(LO 3.4)

6. ANSWER: C

$$K = \frac{[ClF_3]^2}{[F_2]^3[Cl_2]} = \frac{(3)^2}{(2)^3(2.5)} = 0.45$$

(LO 6.5)

7. ANSWER: D Decreasing volume will force the reaction to the side with fewer moles of gas, the right. Changing pressure with temperature constant will not change the value of K (LO 6.8).

8. ANSWER: C

$K_{sp} = [Cd^{2+}][S^{2-}] = 3.6 \times 10^{-29}$

$$[Cd^{2+}] = \sqrt{3.6 \times 10^{-29}} = \sqrt{36 \times 10^{-30}} = 6.0 \times 10^{-15}$$

(LO 6.21)

9. ANSWER: A The arrow must point down to show emission and the largest difference corresponds to the greatest difference in energy, which means highest frequency and shortest wavelength (LO 1.6).

10. ANSWER: A The radius decreases across a period because the Z (number of positive charges in the nucleus) increases. *Note:* B is not correct—you can never use a trend to explain a trend on the exam (LO 1.9).

11. ANSWER: B Group 2 elements have an ns^2 valence configuration and will form 2+ cations. These elements will have relatively low first and second ionization energies but much higher third ionization energies (LO 1.8).

12. ANSWER: A NH_3, with hydrogen bonding, has stronger intermolecular forces than PH_3, with dipole–dipole attractions. Stronger IMFs produce lower vapor pressure (LO 2.1).

13. ANSWER: D Reactions occur more rapidly with small particle size (more collisions) and high temperature (more collisions with greater force) (LO 4.1).

14. ANSWER: C IR spectroscopy differentiates types of covalent bonds and functional groups. This would allow determination of hybridization (LO 1.15).

15. ANSWER: B As the strong attraction between ions is overcome and the ions are free to move about, their entropy increases greatly. Some of the hydrogen bonds are broken between water molecules, but they are strongly attracted to the ions so entropy of water decreases slightly (LO 5.12).

16. ANSWER: B The attraction between water and the sodium ions is ion–dipole (LO 2.8).

17. ANSWER: B The reaction with the less positive reduction potential will be reversed to become the oxidation reaction. Reduction occurs at the cathode so the silver ion will be reduced. 0.80 V + 0.40 V = 1.20 V (LO 3.12).

18. ANSWER: A If the temperature of the surroundings increases, the reaction is releasing heat and has a negative ΔH. An exothermic reaction has higher bond energy (more stable) in the products than in the reactants (LO 5.8).

19. ANSWER: A ΔS is negative as entropy decreases. Five moles of gas produce 2 moles of gas plus 4 moles of liquid. Although the number of particles increases slightly, the phase change contributes greater disorder (LO 5.12).

20. ANSWER: A The Fe^{3+} ion loses 3 electrons, the first two from the 4*s* sublevel and the third from the 3*d* sublevel (LO 1.5).

21. ANSWER: B

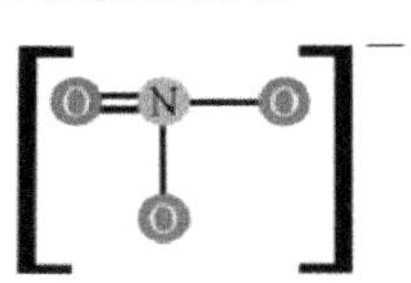

There are 2 N–O single bonds and 1 N–O double bond. The ion would have 3 resonance structures so the average bond order is 1.33 (LO 2.21).

22. ANSWER: B Breaking apart the solute and breaking apart the solvent both require energy and increase disorder; ΔH is positive and ΔS is positive. The interaction between solute and solvent releases energy and increases order so ΔH is negative and ΔS is negative (LO 6.24).

23. ANSWER: D

$$684\text{ g sucrose} \times \frac{1\text{ mol sucrose}}{342\text{ g sucrose}} \times \frac{12\text{ mol carbon}}{1\text{ mol sucrose}} \times \frac{6.02 \times 10^{23}\text{ atoms}}{1\text{ mol carbon}}$$

$= 1.4 \times 10^{25}$ carbon atoms

Math note: Once you start to cross cancel, you get the following problem:

$$2\text{ g sucrose} \times \frac{1\text{ mol sucrose}}{1\text{ g sucrose}} \times \frac{12\text{ mol carbon}}{1\text{ mol sucrose}} \times \frac{6.02 \times 10^{23}\text{ atoms}}{1\text{ mol carbon}}$$

If you will look at it as $2 \times 6 \times 12 = 12 \times 12 = 144$, you don't need to worry about the exponents, as there is only one answer choice that begins with 1.4 (LO 1.4).

24. ANSWER: C 26 g Cr/52.00 = 0.5 mol Cr, 12 g O/16.00 = 0.75 mol O, Cr:O = 0.5:0.75 = 2:3 = Cr_2O_3 (LO 1.2).

25. ANSWER: A The titrated sample is at the equivalence point (all A^-), while the reserved sample has not been reacted (all HA). After pouring them together, the mixture is at the half-equivalence point of a titration with [HA] = [A^-]. At this point, pH = pK_a. $10^{-4.74} = 1.8 \times 10^{-5}$ (LO 6.13).

26. ANSWER: C It helps to draw a reaction diagram to answer this. $E_a + (-\Delta H) = E_a$ for reverse rxn

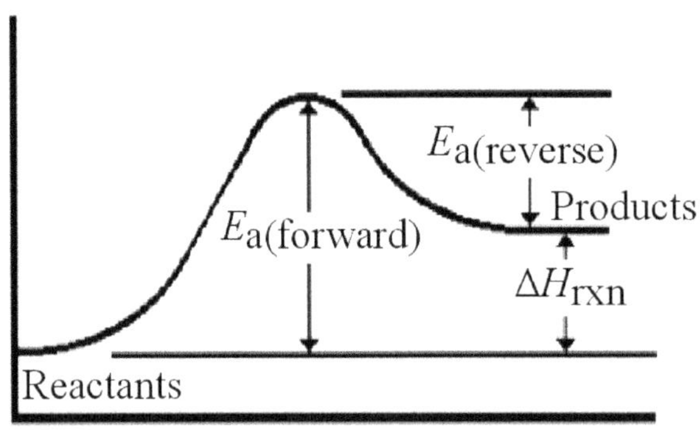

(LO 4.6)

27. ANSWER: A The reaction is the reverse of the given reaction with all coefficients divided by two. When reversing a reaction, the reciprocal of the original K value is used. When the coefficients are divided by 2, the K value must be raised to the ½ power (square root must be taken).

$\sqrt{\frac{1}{3.0 \times 10^{10}}} = 5.8 \times 10^{-6}$.

Since K is less than 1, the reaction is nonfavorable.

Math note:

$$\frac{1}{3.0 \times 10^{10}} = \frac{10 \times 10^{-1}}{3.0 \times 10^{10}} = \frac{10}{3} \times 10^{(-1-10)} = 3.33 \times 10^{-11}$$

so:

$$\sqrt{3.33 \times 10^{-11}} \approx \sqrt{3.6 \times 10^{-11}} = \sqrt{36 \times 10^{-12}} = 6 \times 10^{-6}$$

(LO 5.17 and 6.2)

28. ANSWER: D Slowest step: Rate = $k[N_2O_2][H_2]$, but N_2O_2 is a reaction intermediate, so can't be measured. First step is equilibrium: $K_{eq} = [N_2O_2]/[NO]^2$, so $[N_2O_2] = K_{eq}[NO]^2$ and Rate = $k\,[NO]^2[H_2]$ (LO 4.7).

29. ANSWER: A Oxidation numbers of nitrogen (NO = +2, N_2O_2 = +2, N_2O = +1, N_2 = 0) (LO 3.8)

30. ANSWER: B Sum the three equations, eliminating terms that are the same on both sides (intermediates).

$$2\,NO(g) \rightarrow \cancel{N_2O_2}(g)$$
$$\cancel{N_2O_2}(g) + H_2(g) \rightarrow \cancel{N_2O}(g) + H_2O(g)$$
$$\cancel{N_2O}(g) + H_2(g) \rightarrow N_2(g) + H_2O(g)$$
$$2\,NO(g) + 2\,H_2(g) \rightarrow N_2(g) + 2\,H_2O(g)$$

(LO 4.8)

31. ANSWER: B Assume that pressure is proportional to number of particles.

$$K_p = \frac{(P_{SO_2})(P_{Cl_2})}{(P_{SO_2Cl_2})} = 5$$

At 9 minutes, $Q = (5 \times 5)/5 = 5$, so $Q = K$. Since it is at equilibrium at 9 minutes, it will look the same at 15 minutes (LO 6.6).

32. ANSWER: C One pi bond between S and O must be broken, as well as 2 S–Cl bonds. A Cl–Cl bond must be formed (LO 5.8).

33. ANSWER: C At time zero, there were 10 particles of gas. At equilibrium, there are 15. The pressure increases by a factor of 1.5 (LO 6.6).

34. ANSWER: C Since there are the same number of particles of each gas at equilibrium, the ideality of the gas must be the only factor causing pressure to differ. The gas with the most ideal behavior (Cl_2, which is nonpolar) will have the highest pressure (LO 2.14 and 2.16).

35. ANSWER: D The K_p value must again be 5. $(2 \times 20)/8 = 5$ (LO 6.8)

36. ANSWER: C The pK_{a2} (7.2) for phosphoric acid is closest to the desired pH (7.4). If the two components of the buffer were in equal concentrations, the pH would equal the pK_a. For this buffer, more of the basic component would be used (LO 6.17).

37. ANSWER: A CO_2 is the heaviest gas and will diffuse at the lowest rate (LO 5.2).

38. ANSWER: A The substance with the greatest attraction (most similar IMFs) to the solvent (mobile phase) will travel the fastest. The substance with the greatest attraction to the stationery phase will travel the slowest (LO 2.7).

39. ANSWER: C At the equivalence point of the titration of a weak acid, a basic salt is the main component and will determine pH (LO 6.13).

40. ANSWER: C At the half-equivalence point of the titration of a weak acid with a strong base, the moles of HX will equal the moles of X^- (LO 6.11).

41. ANSWER: D NaOH reacts with $H_2PO_4^-$ to produce HPO_4^{2-}; HPO_4^{2-} reacts with H^+ to make additional $H_2PO_4^-$ (LO 6.20).

42. ANSWER: C

$$0.03750\text{ L} \times \frac{0.10\text{ mol NaOH}}{1\text{ L}} \times \frac{1\text{ mol H}_2\text{C}_2\text{O}_4}{2\text{ mol NaOH}} = 0.001875\text{ mol H}_2\text{C}_2\text{O}_4$$

0.001875 mol/0.02500 L = 0.075 M oxalic acid

(LO 1.20)

43. ANSWER: A The properties given are those of a covalent network solid, such as SiO_2 (LO 2.1).

44. ANSWER: B Since the pressure, temperature, and volume of the flask are the same for both vapors, the number of moles of gas are the same; hence, the number of particles would be the same. Since the mass of Y is higher than X, Y must have a higher molar mass than X (LO 2.5).

45. ANSWER: A Molar mass = (dRT)/P, density = 0.750 g/0.300 L = 2.5, Molar mass = (2.5 g/L × 0.08206(L × atm/mol × K) × 296 K)/1. Round to (2.5 × 0.08 × 300)/1 = 60. The closest compound is C_3H_6O with molar mass of 58 (LO 2.6).

Note: An alternative approach would be to use the molar volume of the gas. Since the temperature and pressure are very close to STP, you can use 1.0 mol = 22.4 L.

$$0.300\text{ L} \times \frac{1\text{ mol}}{22.4\text{ L}} \approx \frac{1}{70}\text{ mol}$$

and molar mass is

$$\frac{0.750\ \text{g}}{\frac{1}{70}\ \text{mol}} = \frac{\frac{3}{4}}{\frac{1}{70}} = \frac{3}{4} \times \frac{70}{1} \approx \frac{53\ \text{g}}{\text{mol}}$$

and then you select the closest compound with that molar mass.

46. ANSWER: C The elements H, C, N, and O are all substantially smaller than titanium so they would form an interstitial alloy where they fit into holes between the titanium atoms (LO 2.25).

47. ANSWER: D 0.005 mole of OH^- and 0.003 mole of malonic acid (which the diagram indicates is a diprotic acid) are initially present. There are not enough moles of hydroxide ions to remove both hydrogen ions from all of the malonic acid. After removal of the first hydrogen ion from the malonic acid, 0.002 mole of OH^- is left and 0.003 mole of $CH_3(COOH)(COO^-)$ is formed. The remaining OH^- reacts with 0.002 mol of $CH_3(COOH)(COO^-)$, leaving 0.001 mole of $CH_3(COOH)(COO^-)$ and producing 0.002 mole of $CH_3(COO^-)_2$. The species in the highest concentration will be $CH_3(COO^-)_2$ (LO 6.17).

48. ANSWER: B If pH = 2.0, $[H_3O^+]$ and $[ClO^-] = 1.0 \times 10^{-2}$. $Q = (1 \times 10^{-2})^2/1.0 = 1.0 \times 10^{-4}$, $Q > K$, so the reaction shifts left to reestablish equilibrium (LO 6.16).

49. ANSWER: D 0.05 L (1.0 M) = 0.050 mol OH^-, 0.05 L (1.0 M) × 2 = 0.10 mol H^+, OH^- is limiting. $q = mc\Delta T$ = 100.0 g (4.2 J/g°C)(6.5°C) = 2730 J, 2730 J/0.050 mol = 55,000 J = 55 kJ heat released, ΔH_{neut} = –55 kJ/mol (LO 6.5).

50. ANSWER: C Hydrogen bonds form between the hydrogen atom on 1 molecule and a lone pair of electrons on an oxygen atom on an adjacent molecule. Energy is required to break the hydrogen bond, so the process is endothermic and has a positive ΔH value (LO 2.13).

51. ANSWER: B There are 2 moles of products and only 1 mole of reactants so entropy increases (ΔS is positive), which is a favorable factor. Because the reaction requires energy to break the bonds between the monomers, ΔH is positive, which is an unfavorable factor. The reaction must be driven forward by entropy. A more mathematical approach would be using $\Delta G = \Delta H - T\Delta S$. Since ΔH is positive and ΔS is positive, then ΔG = (– #) – (+ #). In order to have $\Delta G < 0$, the reaction must be driven by $T\Delta S$ term, that is, $T\Delta S > \Delta H$ (LO 5.13).

52. ANSWER: C The CH_3 carbon has 4 regions of electron density so it is sp^3 hybridized. The carbonyl carbon has only 3 regions of electron density, so it is sp^2 hybridized (LO 2.21).

53. ANSWER: D 0.500 mol/0.125 L = 4.00 M × 2 = 8.00 M

	CH_3COOH	⇆	H^+	+ CH_3COO^-
I	8.00		0	0
C	−x		+x	+x
E	8.00 − x		x	x

$$K_a = \frac{[H^+][CH_3COO^-]}{[CH_3COOH]} = 2.0 \times 10^{-4} = \frac{x^2}{8.00 - x} = 16 \times 10^{-4} = x^2$$

$$x = \sqrt{16 \times 10^{-4}} = 4.0 \times 10^{-2}$$

$$\% = \frac{4.0 \times 10^{-2}}{8} \times 10^2 = \frac{4}{8} = 0.50$$

(LO 6.6)

54. ANSWER: C Glycerin has 3 hydroxyl groups that allow extensive hydrogen bonding between adjacent glycerin molecules, as well as with water molecules. The hydrogen bonding between glycerin molecules is its most important IMF, although it also has dispersion forces (LO 2.13).

55. ANSWER: C The mass of the products and reactants must be equal (law of conservation of mass) (LO 1.18).

56. ANSWER: A 27.55 − 26.93 = 0.62 g lost (H_2O + CO_2). Since mole ratio of H_2O to CO_2 is 1:1, 0.62 g/(18.0 g + 44.0 g) = 0.01 mol of H_2O and 0.01 mol of CO_2. There are 2 moles of $NaHCO_3$ for every mole of water so $NaHCO_3$ = 0.02 mole. 0.02 mol $NaHCO_3$ × 84.0 g/mol = 1.68 g $NaHCO_3$, 1.68 g $NaHCO_3$/2.00 g mixture × 100 = 84% $NaHCO_3$ (LO 1.3).

57. ANSWER: D ΔT_{water} = 25.0 − 22.0 = 3.0°C, ΔT_{metal} = 100.0 − 25.0 = 75.0°C, $mass_{water}$: 132.5 − 32.5 = 100.0 g, $mass_{metal}$ 162.5 − 132.5 = 30.0 g metal, $q = mc\Delta T$, (100.0 g)(4.2 J/g°C)(3.0°C) = (30.0 g)(c_{metal})(75.0°C) c_{metal} = 0.56 J/g°C

The metal is Sn. The specific heat value for Sn in the chart is 0.54 J/g°C, but the value of 0.56 is close enough, allowing for experimental error. Estimation and rounding can be used to get this answer, exact calculations are not needed.

Note: If you use the equation $\Delta T = T_f - T_i$, then your ΔT_{metal} will be negative and you should make the statement that $q_{water} = -q_{metal}$ (LO 5.5).

58. ANSWER: D Lattice energy is determined by Coulomb's law. The most important factor is charge. The higher the charge of ions, the greater the attraction between oppositely charged ions will be. Secondly, small ions can get closer to each other and have greater attraction. Magnesium has a 2+ charge and is smaller than calcium

so $MgCl_2$ will release the most energy when the ions come together to form the solid crystal (LO 2.14).

59. ANSWER: D Since ΔG is quite negative, it is thermodynamically favored. Since it is not observed, it must be under kinetic control. This reaction must require a large activation energy (such as a spark) to start the reaction (LO 5.18).

60. ANSWER: A Catalysts work by lowering the activation energy of a reaction. They may be heterogeneous (different phase than the reactants) or homogeneous (same phase as the reactants). The diagram here shows the metal atom as the center of a complex ion which would be dissolved in solution rather than as a solid site for the reaction to occur on, so it is a homogeneous catalyst in this situation (LO 3.12).

SECTION II: FREE-RESPONSE QUESTIONS

Question 1: Answers

(a) (i) $2\ I^- \rightarrow I_2 + 2\ e^-$ (LO 3.13)

(ii) $2\ H_2O + 2\ e^- \rightarrow H_2 + 2\ OH^-$ (LO 3.13)

(iii) $-0.54\ V + (-0.83\ V) = -1.37\ V$ (LO 3.12)

(iv) $\Delta G° = -nFE = -2 \text{ mol } e^-\ (96{,}485 \text{C/mol } e^-)(-1.37\ V) = 264{,}000 \text{ J/mol}_{rxn}$

The reaction is not thermodynamically favored because ΔG is positive. Electrolysis is needed because of the sign of ΔG (LO 5.14).

(b)

$$15.0 \text{ min} \times \frac{60 \text{ s}}{1 \text{ min}} \times \frac{0.200 \text{ C}}{1 \text{ s}} \times \frac{1 \text{ mol } e^-}{96{,}485 \text{ C}} = 1.87 \times 10^{-3} \text{ mol } e^-$$

(LO 3.12)

(c) The iodide ion is charged and will dissolve very easily in water by forming ion–dipole attractions. I_2 is nonpolar and has only weaker dipole–induced dipole attractions with water (LO 2.15).

(d)

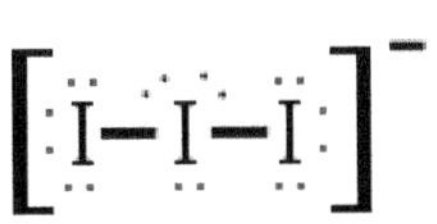

(LO 2.21)

(e) I_2 is a small, very dense molecule. It has many electrons packed in a small area that absorb X-rays without emission. The density of the lighter halogens is much less (LO 1.9).

(f) Bromine and iodine are both nonpolar molecules with only London dispersion forces. Iodine has many more electrons than bromine so is more polarizable and has stronger dispersion forces, causing it to be a solid at room temperature (LO 2.11).

(g) The mass spectrogram is for bromine in its natural state, a molecule rather than an atom. There are two smaller peaks at 79 and 81 amu, which represent the two isotopes. There are three higher peaks that represent the three possible combinations of isotopes found in Br_2 (79 + 79, 79 + 81, 81 + 81) (LO 1.14).

Question 2: Answers

(a)

$$K_p = \frac{(P_{PCl_5})}{(P_{PCl_3})(P_{Cl_2})}$$

(LO 6.5)

(b)

	$PCl_3(g)$	+	$Cl_2(g)$	⇄	$PCl_5(g)$	+	P_{total}
Initial	0.30 atm		0.60 atm		0.10 atm		1.00 atm
Change	$+x$		$+x$		$-x$		$+x$
Equil	$0.30 + x$		$0.60 + x$		$0.10 - x$		1.08 atm
	0.30 + 0.08 = 0.38		0.60 + 0.08 = 0.68		0.10 – 0.08 = 0.02		$x = 0.08$

Therefore, at equilibrium, P_{PCl_5} (0.10 – 0.08) = 0.02 atm

P_{PCl_3} (0.30 + 0.08) = 0.38 atm

P_{PCl_2} (0.60 + 0.08) = 0.68 atm

(LO 6.6)

(c)

$$K_p = \frac{0.02}{(0.38)(0.68)} = 0.077 = 0.08$$

(LO 6.5)

(d) Increasing the volume of the container decreases the pressure and favors the process where more moles of gas are produced, so the reverse reaction is favored. The equilibrium shifts to the left; the mole fraction of PCl_5 decreases (LO 6.8).

(e) (i) For an exothermic reaction, decreasing the temperature shifts the reaction to the right, favoring the products. The mole fraction of PCl_5 increases (LO 6.8).

(ii) Since the reaction shifts toward products because of a temperature change, K_p will increase (LO 6.10).

(f)

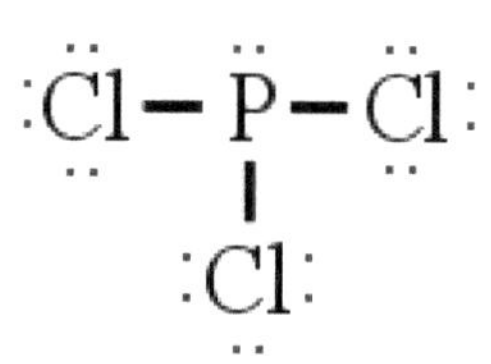

(i)

Trigonal pyramidal

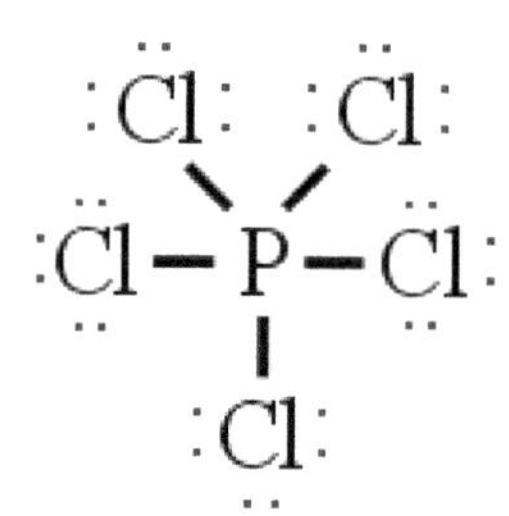

(ii)

Trigonal bipyramidal

(LO 2.21)

(g) 0.0158 mol PCl_3 × 3 mol HCl/1 mol PCl_3 = 0.0474 mol HCl/0.1000 L = 0.474 M HCl
Since H_3PO_3 is a weak acid, its $[H^+]$ is negligible in comparison to that of HCl.
pH = –log 0.474 = 0.324 (LO 6.12)

Question 3: Answers

(a) (i) The formal charge on each carbon and nitrogen in both structures is zero. Neither structure is favored on the basis of formal charge (LO 2.21).

(ii) Structure 1: Bond angle < 109.5°
Structure 2: Bond angle < 120°
In either case, don't forget that there is a lone pair of electrons on the nitrogen atom (LO 2.21).

(iii) Structure 2 should be circled (LO 2.21).

(b) $CaCN_2 + 3\ H_2O \rightarrow 2\ NH_3 + CaCO_3$
(LO 3.2)

(c)

$$5.0\text{ g } CaCN_2 \times \frac{1\text{ mol } CaCN_2}{80.07\text{ g } CaCN_2} \times \frac{2\text{ mol } NH_3}{1\text{ mol } CaCN_2} = 0.125\text{ mol } NH_3$$

$$\frac{0.125\text{ mol } NH_3}{0.150\text{ L}} = 0.833\ M\ NH_3$$

(LO 3.3)

(d)

$$K_b = \frac{[NH_4^+][OH^-]}{[NH_3]} = 1.8 \times 10^{-5} = \frac{x^2}{0.833\text{-}x} \approx \frac{x^2}{0.833}$$

$x = [OH^-] = 3.9 \times 10^{-3}$

pOH = 2.41, pH = 14.00 – 2.41 = 11.59

(LO 6.16)

(e) High temperature and high pressure should be circled. Although high temperature causes the reaction to shift to the reactant side, it is necessary to have sufficient particle collisions with sufficient energy. The high pressure will shift the reaction to produce more products (LO 6.9).

Question 4: Answers

(a) The half-life is not constant so the reaction cannot be first order (LO 4.3).

(b) The reaction is second order, because the graph of the reciprocal of concentration versus time is a straight line (LO 4.2).

(c) The value of k will be the slope of the best-fit line of 1/concentration versus time. 10.164 L/(mol·s) (LO 4.2).

(d) Both NO_2 and Cl have an odd number of electrons and are electron deficient. They achieve greater stability when they can share electrons with each other to go to a lower energy state (LO 2.1).

Question 5: Answers

(a) $Ca^{2+} + CO_3^{2-} \rightarrow CaCO_3$ (LO 3.2)

(b) The greatest amount of precipitate is formed in Tube 4. Calcium and carbonate ions react in a 1:1 ratio. Since the calcium nitrate is twice as concentrated (2.0 *M*) as the sodium carbonate (1.0 *M*), excess calcium ion will be present in Tubes 1–3. Tube 4 has a stoichiometric ratio (same number of moles of Ca^{2+} as CO_3^{2-}), and Tube 5 has an excess of carbonate ions with the low calcium ion concentration limiting the amount of precipitate.

Tube	Volume of 2.0 *M* $Ca(NO_3)_2$ (mL)	Millimoles of Ca^{2+}	Volume of 1.0 *M* Na_2CO_3 (mL)	Millimoles of CO_3^{2-}	Millimoles of $CaCO_3(s)$
1	5	10	1	1	1
2	4	8	2	2	2
3	3	6	3	3	3
4	2	4	4	4	4
5	1	2	5	5	2

(LO 3.3)

(c) (i) $CaCO_3 \rightleftarrows Ca^{2+} + CO_3^{2-}$

$K_{sp} = [Ca^{2+}][CO_3^{2-}] = 8.7 \times 10^{-9}$

$8.7 \times 10^{-9} = x^2$

$x = 9.3 \times 10^{-5}\ M$

(LO 6.21)

(ii) The solubility of calcium carbonate will be decreased because the presence of a common ion (CO_3^{2-}) decreases the solubility of a salt. The increased concentration of carbonate shifts the reaction to the undissolved $CaCO_3$ (LO 6.23).

Question 6: Answers

(a) pH = 1.83, $[H^+] = 10^{-1.83} = 1.5 \times 10^{-2}\ M$

$$\%\ ionization = \frac{1.5 \times 10^{-2}}{0.495} \times 100 = 3.0\%$$

(LO 6.12 and 6.15)

(b)

$$K_a = \frac{[H^+][NO_2^-]}{[HNO_2]} = \frac{(1.5 \times 10^{-2})^2}{0.495 - 0.015} \approx \frac{(1.5 \times 10^{-2})^2}{0.480} = 4.7 \times 10^{-4}$$

In this type of problem where you are given the amount of ionization for the weak acid (or base), you should do the subtraction and not just assume that the ionization is negligible. However, if you did not include the subtraction, the problem is worked out below (LO 6.5).

(c)

	HNO_2	⇄	H^+	+	NO_2^-
I	0.0125 *M*		0		0.0193 *M*
C	$-x$		$+x$		$+x$
E	$0.0125 - x$		x		$0.0193 + x$

$$K_a = \frac{(x)(0.0193 - x)}{0.0125 - x} \approx \frac{(x)(0.0193)}{0.0125} = 4.7 \times 10^{-4}$$

$x = [H^+] = 3.0 \times 10^{-4}$

$pH = -\log 3.0 \times 10^{-4} = 3.52$

(LO 6.17)

(d) The addition of 0.005 mole of hydroxide ions is not enough to exceed the buffer capacity of the solution. $OH^- + HNO_2 \rightarrow NO_2^- + H_2O$
(LO 6.20)

Question 7: Answers

(a) CaO has a much higher lattice energy than KF. Coulomb's law states that the force of attraction between two oppositely charged particles is directly proportional to the product of the absolute value of the charges and inversely proportional to distance between the ion centers.

$$E = (2.31 \times 10^{-19}\ \text{J·nm})\left(\frac{Q_1 Q_2}{r}\right)$$

Because of their charges, Ca^{2+} and O^{2-} ions are attracted about four times more strongly than K^+ and F^- ions. Ionic size is a lesser factor with the calcium ion being smaller than the potassium ion. The fluoride ion is slightly smaller than the oxide ion (LO 2.14).

(b) (i) $CaO(s) + H_2O(l) \rightarrow Ca(OH)_2(s)$ (LO 3.2)

(ii) Since the reaction releases heat, it is exothermic. The bond energy of the products is higher than that of the reactants (LO 5.8).

(iii) The sign of ΔS is negative as 1 mole of solid is produced from 1 mole of solid plus 1 mole of liquid. The product is more organized than the reactants (LO 5.12).

SCORING THE FREE-RESPONSE QUESTIONS

It is difficult to come up with an exact score for this section of the test. However, if you compare your answers to the answers in this book, remembering that each part of the test you answer correctly is worth points even if the other parts of the answer are incorrect, you can get a general idea of the percentage of the questions for which you would get credit. If you believe that you got at least 40 percent of the possible credit, you would probably receive a 3 on this part of the test. If you believe that you would receive close to half or more of the available credit, your score would more likely be a 4 or a 5.

PERIODIC TABLE OF THE ELEMENTS

1	2	3	4	5	6	7	8	9	10	11	12	13	14	15	16	17	18
1 **H** 1.008																	2 **He** 4.00
3 **Li** 6.94	4 **Be** 9.01											5 **B** 10.81	6 **C** 12.01	7 **N** 14.01	8 **O** 16.00	9 **F** 19.00	10 **Ne** 20.18
11 **Na** 22.99	12 **Mg** 24.30											13 **Al** 26.98	14 **Si** 28.09	15 **P** 30.97	16 **S** 32.06	17 **Cl** 35.45	18 **Ar** 39.95
19 **K** 39.10	20 **Ca** 40.08	21 **Sc** 44.96	22 **Ti** 47.87	23 **V** 50.94	24 **Cr** 52.00	25 **Mn** 54.94	26 **Fe** 55.85	27 **Co** 58.93	28 **Ni** 58.69	29 **Cu** 63.55	30 **Zn** 65.38	31 **Ga** 69.72	32 **Ge** 72.63	33 **As** 74.92	34 **Se** 78.97	35 **Br** 79.90	36 **Kr** 83.80
37 **Rb** 85.47	38 **Sr** 87.62	39 **Y** 88.91	40 **Zr** 91.22	41 **Nb** 92.91	42 **Mo** 95.95	43 **Tc** (97)	44 **Ru** 101.1	45 **Rh** 102.91	46 **Pd** 106.42	47 **Ag** 107.87	48 **Cd** 112.41	49 **In** 114.82	50 **Sn** 118.71	51 **Sb** 121.76	52 **Te** 127.60	53 **I** 126.90	54 **Xe** 131.29
55 **Cs** 132.91	56 **Ba** 137.33	57 ***La** 138.91	72 **Hf** 178.49	73 **Ta** 180.95	74 **W** 183.84	75 **Re** 186.21	76 **Os** 190.2	77 **Ir** 192.2	78 **Pt** 195.08	79 **Au** 196.97	80 **Hg** 200.59	81 **Tl** 204.38	82 **Pb** 207.2	83 **Bi** 208.98	84 **Po** (209)	85 **At** (210)	86 **Rn** (222)
87 **Fr** (223)	88 **Ra** (226)	89 **†Ac** (227)	104 **Rf** (267)	105 **Db** (270)	106 **Sg** (271)	107 **Bh** (270)	108 **Hs** (277)	109 **Mt** (276)	110 **Ds** (281)	111 **Rg** (282)	112 **Cn** (285)	113 **Nh** (285)	114 **Fl** (289)	115 **Mc** (288)	116 **Lv** (293)	117 **Ts** (294)	118 **Og** (294)

*Lanthanoid Series	58 **Ce** 140.12	59 **Pr** 140.91	60 **Nd** 144.24	61 **Pm** (145)	62 **Sm** 150.4	63 **Eu** 151.97	64 **Gd** 157.25	65 **Tb** 158.93	66 **Dy** 162.50	67 **Ho** 164.93	68 **Er** 167.26	69 **Tm** 168.93	70 **Yb** 173.05	71 **Lu** 174.97
†Actinoid Series	90 **Th** 232.04	91 **Pa** 231.04	92 **U** 238.03	93 **Np** (237)	94 **Pu** (244)	95 **Am** (243)	96 **Cm** (247)	97 **Bk** (247)	98 **Cf** (251)	99 **Es** (252)	100 **Fm** (257)	101 **Md** (258)	102 **No** (259)	103 **Lr** (262)

AP® CHEMISTRY EQUATIONS AND CONSTANTS

Throughout the exam the following symbols have the definitions specified unless otherwise noted.

L, mL	= liter(s), milliliter(s)	mm Hg	= millimeters of mercury
g	= gram(s)	J, kJ	= joule(s), kilojoule(s)
nm	= nanometer(s)	V	= volt(s)
atm	= atmosphere(s)	mol	= mole(s)

ATOMIC STRUCTURE

$E = h\nu$

$c = \lambda\nu$

E = energy

ν = frequency

λ = wavelength

Planck's constant, $h = 6.626 \times 10^{-34}$ J s

Speed of light, $c = 2.998 \times 10^{8}$ m s^{-1}

Avogadro's number $= 6.022 \times 10^{23}$ mol^{-1}

Electron charge, $e = -1.602 \times 10^{-19}$ coulomb

EQUILIBRIUM

$$K_c = \frac{[C]^c[D]^d}{[A]^a[B]^b}, \text{ where } a\,A + b\,B \rightleftarrows c\,C + d\,D$$

$$K_p = \frac{(P_C)^c(P_D)^d}{(P_A)^a(P_B)^b}$$

$$K_a = \frac{[H^+][A^-]}{[HA]}$$

$$K_b = \frac{[OH^-][HB^+]}{[B]}$$

$$K_w = [H^+][OH^-] = 1.0 \times 10^{-14} \text{ at } 25°C$$

$$= K_a \times K_b$$

$$pH = -\log[H^+],\ pOH = -\log[OH^-]$$

$$14 = pH + pOH$$

$$pH = pK_a + \log\frac{[A^-]}{[HA]}$$

$$pK_a = -\log K_a,\ pK_b = -\log K_b$$

Equilibrium Constants

K_c (molar concentrations)

K_p (gas pressures)

K_a (weak acid)

K_b (weak base)

K_w (water)

KINETICS

$$\ln[A]_t - \ln[A]_0 = -kt$$

$$\frac{1}{[A]_t} - \frac{1}{[A]_0} = kt$$

$$t_{1/2} = \frac{0.693}{k}$$

k = rate constant

t = time

$t_{1/2}$ = half-life

GASES, LIQUIDS, AND SOLUTIONS

$PV = nRT$

$P_A = P_{total} \times X_A$, where $X_A = \dfrac{\text{moles A}}{\text{total moles}}$

$P_{total} = P_A + P_B + P_C + \ldots$

$n = \dfrac{m}{M}$

$K = °C + 273$

$D = \dfrac{m}{V}$

KE per molecule $= \dfrac{1}{2}mv^2$

Molarity, M = moles of solute per liter of solution

$A = abc$

P = pressure
V = volume
T = temperature
n = number of moles
m = mass
M = molar mass
D = density
KE = kinetic energy
v = velocity
A = absorbance
a = molar absorptivity
b = path length
c = concentration

Gas constant, R = 8.314 J mol^{-1} K^{-1}
= 0.08206 L atm mol^{-1} K^{-1}
= 62.36 L torr mol^{-1} K^{-1}

1 atm = 760 mm Hg = 760 torr

STP = 273.15 K and 1.0 atm

Ideal gas at STP = 22.4 L mol^{-1}

THERMODYNAMICS/ELECTROCHEMISTRY

$q = mc\Delta T$

$\Delta S° = \sum S°\ \text{products} - \sum S°\ \text{reactants}$

$\Delta H° = \sum \Delta H°_f\ \text{products} - \sum \Delta H°_f\ \text{reactants}$

$\Delta G° = \sum \Delta G°_f\ \text{products} - \sum \Delta G°_f\ \text{reactants}$

$\Delta G° = \Delta H° - T\Delta S°$
$= -RT \ln K$
$= -nFE°$

$I = \dfrac{q}{t}$

q = heat
m = mass
c = specific heat capacity
T = temperature
$S°$ = standard entropy
$H°$ = standard enthalpy
$G°$ = standard Gibbs free energy
n = number of moles
$E°$ = standard reduction potential
I = current (amperes)
q = charge (coulombs)
t = time (seconds)

Faraday's constant,
F = 96,485 coulombs per mole of electrons

1 volt $= \dfrac{1\ \text{joule}}{1\ \text{coulomb}}$